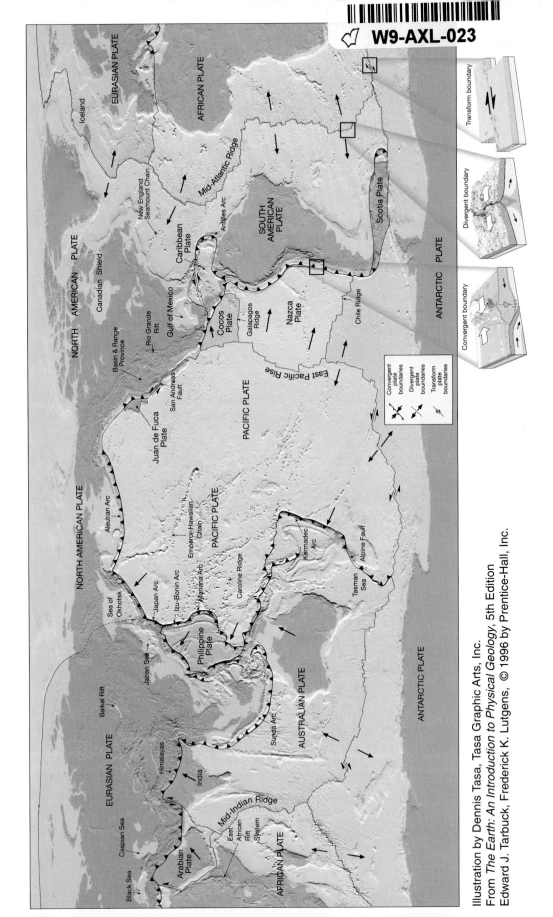

EURASIAN PLATE

Iceland

AFRICAN PLATE

NORTH AMERICAN PLATE

Canadian Shield

New England Seamount Chain

Mid-Atlantic Ridge

Antilles Arc

Caribbean Plate

SOUTH AMERICAN PLATE

Scotia Plate

Basin & Range Province

Rio Grande Rift

Gulf of Mexico

Cocos Plate

Galapagos Ridge

Nazca Plate

Chile Ridge

ANTARCTIC PLATE

San Andreas Fault

Juan de Fuca Plate

East Pacific Rise

PACIFIC PLATE

Aleutian Arc

Emperor-Hawaiian Chain

PACIFIC PLATE

Kermadec Arc

Alpine Fault

Japan Arc

Izu-Bonin Arc

Mariana Arc

Caroline Ridge

Tasman Sea

Sea of Okhotsk

Convergent plate boundaries
Divergent plate boundaries
Transform plate boundaries

Baikal Rift

Japan Sea

Philippine Plate

Sunda Arc

AUSTRALIAN PLATE

EURASIAN PLATE

Caspian Sea

Himalayas

India

Mid-Indian Ridge

ANTARCTIC PLATE

Black Sea

Arabian Plate

East African Rift System

AFRICAN PLATE

Transform boundary

Divergent boundary

Convergent boundary

Illustration by Dennis Tasa, Tasa Graphic Arts, Inc.

From *The Earth: An Introduction to Physical Geology*, 5th Edition

Edward J. Tarbuck, Frederick K. Lutgens, © 1996 by Prentice-Hall, Inc.

Plate Tectonics and Crustal Evolution

Plate Tectonics and Crustal Evolution

Fourth edition

Kent C. Condie
New Mexico Institute of Mining and Technology
Socorro, New Mexico

BUTTERWORTH
HEINEMANN

Butterworth-Heinemann
Linacre House, Jordan Hill, Oxford OX2 8DP
A division of Reed Educational and Professional Publishing Ltd

 A member of the Reed Elsevier plc group

OXFORD   BOSTON   JOHANNESBURG
MELBOURNE   NEW DELHI   SINGAPORE

First published 1976
Second edition 1982
Third edition 1989
Reprinted (with corrections and additions) 1993
Fourth edition 1997

© Kent C. Condie 1997

British Library Cataloguing in Publication Data
A catalogue record for this book is available from the British Library

ISBN 0 7506 3386 7

Library of Congress Cataloguing in Publication Data
A catalogue record for this book is available from the Library of Congress

Printed and bound in Great Britain by The Bath Press, Bath

Contents

Preface   ix

Chapter 1   Plate tectonics   1
A perspective   1
Structure of the Earth   4
Seafloor spreading   5
Plate boundaries   6
   Introduction   6
   Divergent boundaries (ocean ridges)   7
   Transform faults and fracture zones   7
   Triple junctions   8
   Convergent boundaries (subduction zones)   8
   Collisional boundaries   11
   Trench–ridge interactions   12
The Wilson Cycle   12
Stress distribution within plates   13
Plate motions   13
   Cycloid plate motion   13
   Plate velocities in the last 150 My   14
   Plate velocities from paleomagnetism   15
   Space geodetic measurements of plate
     velocities   16
Plate driving forces   17
Geomagnetism   17
   Rock magnetization   17
   Reversals in the Earth's magnetic field   18
   Paleomagnetism   20
Hotpots and plumes   23
   General characteristics   23
   Hotpot tracks   24
Plate tectonics and organic evolution   27
Supercontinents   28
   Introduction   28
   Gondwana and Pangea   30
   Rodinia   32
Interactions of Earth systems   33
Summary statements   34
Suggestions for further reading   35

Chapter 2   The Earth's crust   36
Introduction   36

Crustal types   36
   Oceanic crust   36
   Transitional crust   38
   Continental crust   39
Continent size   39
Seismic crustal structure   40
   The Moho   40
   Crustal layers   40
   Crustal types   40
Crustal gravity   44
Crustal magnetism   45
Crustal electrical conductivity   45
Heat flow   47
   Heat flow distribution   47
   Heat production and heat flow in the
     continents   48
   Heat flow and crustal thickness   49
   Age dependence of heat flow   50
Exhumation and cratonization   50
   Introduction   50
   Unravelling P–T–t histories   51
   Some typical P–T–t paths   52
   Cratonization   52
Processes in the continental crust   53
   Rheology   53
   The role of fluids and crustal melts   54
Crustal composition   54
   Approaches   54
   Seismic-wave velocities   55
   Seismic reflections in the lower continental
     crust   57
   Sampling of Precambrian shields   58
   Use of fine-grained terrigenous sediments   58
   Exhumed crustal blocks   58
   Crustal xenoliths   59
An estimate of crustal composition   59
   Continental crust   59
Crustal provinces and terranes   62
   Introduction   62
   Crustal province and terrane boundaries   64

The United Plates of America   67
Summary statements   67
Suggestions for further reading   68

Chapter 3   Tectonic settings   69
Introduction   69
Ocean ridges   69
  Ocean-ridge basalts   69
Ophiolites   70
  General features   70
Tectonic settings related to mantle plumes   74
  Submarine plateaux and aseismic ridges   74
  Continental flood basalts   76
  Volcanic islands   76
  Giant mafic dyke swarms   77
Cratons and passive margins   77
Continental rifts   79
  General features   79
  Rock assemblages   80
  Rift development and evolution   80
  Mechanisms of rifting   82
Arc systems   83
  Subduction-related rock assemblages   83
  High- and low-stress subduction zones   87
  Arc processes   87
  High-pressure metamorphism   89
  Igneous rocks   89
  Compositional variation of arc magmas   90
Orogens   92
  Two types of orogens   92
  Orogenic rock assemblages   93
  Tectonic elements of a collisional orogen   94
  Sutures   95
  Foreland and hinterland basins   96
  The Himalayas   96
  Idealized orogenic scenarios   97
Uncertain tectonic settings   98
  Anorogenic granites   98
  Archean greenstones   99
Mineral and energy deposits   104
  Introduction   104
  Mineral deposits   105
  Energy deposits   107
Summary statements   108
Suggestions for further reading   109

Chapter 4   The Earth's mantle and core 110
Introduction   110
Seismic structure of the mantle   110
  Upper mantle   110
  Lower mantle   111
Geoid anomalies   113
Electrical conductivity of the mantle   113
Temperature distribution in the mantle   114
Composition of the mantle   116
  Introduction   116
  Seismic velocity constraints   116
  Upper mantle mineral assemblages   117
  Chemical composition of the mantle   118

The lithosphere   119
  Introduction   119
  Thickness of continental lithosphere   120
  Seismic anisotropy   120
  Thermal structure of Precambrian
    continental lithosphere   121
  Age of the subcontinental lithosphere   122
  Origin of the mantle lithosphere   123
The low-velocity zone (LVZ)   123
The transition zone   123
  The 410-km discontinuity   123
  The 660-km discontinuity   125
The lower mantle   126
  General features   126
  The D″ layer   126
Mantle plumes   127
Mantle upwellings, plumes, and
  supercontinents   129
Mantle geochemical components   130
  Introduction   130
Identifying mantle components   130
  Summary   130
  The Dupal anomaly   131
  Mixing regimes in the mantle   132
  Scale of mantle heterogeneities   133
  An overview   133
Convection in the mantle   134
  The nature of convection   134
  Passive ocean ridges   134
  The layered convection model   134
  Towards a convection model for the Earth   136
The core   137
  Introduction   137
  Core temperature   137
  The inner core   138
  Composition of the core   138
  Age of the core   138
  Generation of the Earth's magnetic field   139
  Origin of the core   140
Summary statements   141
Suggestions for further reading   143

Chapter 5   Crustal and mantle evolution 144
Introduction   144
The Earth's thermal history   144
The Earth's primitive crust   144
  Origin of the first crust   144
  Composition of the primitive crust   146
The Earth's oldest rocks   147
Crustal origin   150
How continents grow   151
Plate tectonics with time   153
  Petrotectonic assemblages   153
  Seismic reflection profiles   154
  Archean transcurrent faults   154
Episodic age distributions   154
  Introduction   154
  Granitoid age distributions   155
  Greenstones age distributions   156
  Episodic ages and supercontinents   157

Continental growth rates	158		Paleoclimates	201
Introduction	158		Introduction	201
The role of recycling	159		Glaciation	202
Freeboard	161		Precambrian climates	203
Continental growth in the last 200 My	161		Phanerozoic climates	206
Towards a continental growth rate model	162		Paleoclimates and supercontinents	207
Secular changes in the crust–mantle system	163		Paleoclimates and mantle plumes	207
Changes in composition of the upper			Living systems	209
continental crust	163		General features	209
Alkaline igneous rocks	168		Origin of life	209
Anorthosites	169		The first fossils	212
Ophiolites	169		Stromatolites	213
Greenstones	169		Appearance of eukaryotes	215
Mafic igneous rocks	170		The origin of metazoans	215
Komatiites	172		Late Proterozoic metazoans	216
Blueschists	173		The Cambrian explosion and the	
Summary of chemical changes at the			appearance of skeletons	217
Archean–Proterozoic boundary	173		Evolution of Phanerozoic life forms	217
The komatiite effect	174		Biologic benchmarks	218
The TTG effect	174		Mass extinctions	218
The subduction effect	175		Introduction	218
Evolution of the crust–mantle system	176		Impact extinction mechanisms	219
Archean plate tectonics	176		Episodicity of mass extinctions	220
Towards an Archean model	176		Extinctions at the K/T boundary	220
Changes at the Archean–Proterozoic			The Permian/Triassic extinction	224
boundary	177		Summary statements	225
Episodic ages and mantle dynamics:			Suggestions for further reading	226
a possible connection?	177			
Summary statements	179		**Chapter 7 Comparative planetary evolution**	**228**
Suggestions for further reading	180		Introduction	228
			Impact chronology in the inner Solar System	228
			Members of the Solar System	229
Chapter 6 The atmosphere, oceans, climates,			The planets	229
and life	**181**		Satellites and planetary rings	235
Introduction	181		Comets	237
General features of the atmosphere	181		Asteroids	238
The primitive atmosphere	182		Meteorites	239
The degassed atmosphere	183		Chemical composition of the Earth and Moon	242
Excess volatiles	183		Age of the Earth and Solar System	244
Composition of the early atmosphere	183		Comparative evolution of the atmospheres of	
Growth rate of the atmosphere	184		Earth, Venus and Mars	245
The faint young Sun paradox	185		The continuously habitable zone	246
The carbonate–silicate cycle	186		Condensation and accretion of the planets	247
The history of atmopheric CO_2	187		The Solar Nebula model	247
The origin of oxygen	187		Heterogeneous accretion models	249
Oxygen controls in the atmosphere	187		Homogeneous accretion models	249
Geologic indicators of ancient atmospheric			Accretion of the Earth	250
oxygen levels	188		Origin of the Moon	250
The growth of atmospheric oxygen	191		Fission models	251
Phanerozoic atmospheric history	191		Double planet models	252
The carbon isotope record	192		Capture models	252
The sulphur isotope record	194		Giant impactor model	253
The oceans	194		The Earth's rotational history	253
Introduction	194		Comparative planetary evolution	254
Growth rate of the oceans	195		Summary statements	256
Sea level	195		Suggestions for further reading	258
Changes in the composition of seawater				
with time	197		*References*	259
The dolomite–limestone problem	200			
Archean carbonates	200		*Index*	279
Sedimentary phospates	201			

Preface

This book has grown and evolved from a course I teach at New Mexico Tech, which is taken chiefly by seniors and graduate students in the Earth Sciences. The rapid accumulation of data on plate tectonics, mantle evolution, and the origin of continents in the last decade has necessitated continued updating of the course. The book is written for an advanced undergraduate or graduate student, and it assumes a basic knowledge of geology, biology, chemistry, and physics that most students in the Earth Sciences acquire during their undergraduate education. It also may serve as a reference book for various specialists in the geological sciences who want to keep abreast of scientific advances in this field. I have attempted to synthesize and digest data from the fields of oceanography, geophysics, geology, planetology, and geochemistry, and to present this information in a systematic manner, addressing problems related to the evolution of the Earth's crust and mantle over the last 4 Gy. The role of plate tectonics in the geological past is examined in the light of geologic evidence, and examples of plate reconstructions are discussed.

Since the third edition was published a wealth of information on plate tectonics, mantle structure, and continental origin and evolution has appeared in scientific journals. As with the third edition, to accommodate this new information, it was necessary to rewrite more than seventy-five per cent of the text and add a large number of new figures. Major additions and/or revisions in the fourth edition occur for such topics as mantle plumes, seismic discontinuities in the mantle, supercontinents, interactions between Earth systems, atmosphere and ocean evolution, episodisity of orogeny and continental growth, and planetary origin.

In order to better to accommodate new results and ideas, some topics have been eliminated from the third edition (for instance, most of the material on methodology in the third edition Chapter 1), and many new topics have been added. Also, the book has been completely reorganized and now all planetary topics occur in the last chapter. In Chapter 1 it is assumed that the reader is familiar with the basic tenets of plate tectonics and hence, rather than gloss over all topics related to plate tectonics in a cursory manner, selected topics are discussed in detail. Major topics include plate boundaries, plate motions, hotspots, geomagnetism, Earth systems, and supercontinents. In Chapter 2, geologic and geophysical properties of crustal types are reviewed, heat flow, exhumation, cratonization, and crustal composition are discussed in detail, and the concept of crustal provinces and terranes is introduced. Chapter 3 on Tectonic settings follows the same format as the tectonic setting chapter in the third edition. New tectonic settings are added (giant dyke swarms, submarine plateaux) and collisional orogens are discussed in greater depth. Because the Precambrian chapter is eliminated in the fourth edition, uncertain tectonic settings found in the Precambrian (Archean greenstones, anorogenic granites) are included in Chapter 3. Chapter 4, The Earth's mantle and core, has been significantly revised and enlarged from our rapidly increasing data base from seismic tomography, high pressure experimental data, and computer modelling. New topics are added on the lithosphere, mantle plumes, and geochemical domains in the mantle, and the discussions of mantle convection, seismic discontinuities, and mantle composition are significantly revised and updated. The discussion of the core is also completely revised.

In Chapter 5 both crustal and mantle evolution are discussed, with emphasis on the interaction between the two systems. New sections include the Earth's oldest rocks, plate tectonics with time, and Archean plate tectonics. In addition, the sections on episodic age distribution, continental growth rate, and secular changes in composition of the crust–mantle system have been significantly revised in the light of new data and models. Because of the wealth of information on the origin and interaction of the atmosphere, ocean, and living systems, Chapter 6 is almost twice the size of the other chapters. All topics are revised in this chapter and new topics are added on Phanerozoic atmospheric history, carbon isotopes, changes in the composition of seawater with time, and the relation of paleoclimates to supercontinents and mantle plumes. The discussion of the origin and early evolution of life has been significantly revised and enlarged, as is the discussion of mass extinctions and asteroid impacts. All planetary topics are now included in Chapter 7, in which all major bodies in the Solar System are discussed. Sections on asteroids, meteorites, and the age of the Earth are significantly revised, and Venus is compared with the Earth in the light of the new Magellan data base.

Perhaps the greatest challenge with the fourth edition has been to keep the book to a reasonable length for a textbook without eliminating important topics. To do this, some topics have been discussed in greater detail than others, depending on their relevance to crustal and mantle evolution. Some topics readily available in other advanced texts have not been included, for instance, mathematical treatments of geophysical and geochemical data, and methods by which geological, geochemical, and geophysical data are gathered. Although the reference section may seem large, I have attempted to cite only the major papers and some of the minor ones that have strongly influenced me in regard to interpretations set forth in the text. More extensive bibliographies can be found in these papers and in the titles suggested in the further reading sections at the end of each chapter.

Kent C. Condie
Department of Earth and Environmental Science
New Mexico Institute of Mining and Technology
Socorro, New Mexico

Chapter 1

Plate tectonics

A perspective

Plate tectonics, which has so profoundly influenced geologic thinking since the early 1970s, provides valuable insight into the mechanisms by which the Earth's crust and mantle have evolved. **Plate tectonics** is a unifying model that attempts to explain the origin of patterns of deformation in the crust, earthquake distribution, continental drift, and mid-ocean ridges, as well as providing a mechanism for the Earth to cool. Two major premises of plate tectonics are:

1 the outermost layer of the Earth, known as the lithosphere, behaves as a strong, rigid substance resting on a weaker region in the mantle known as the asthenosphere
2 the lithosphere is broken into numerous segments or plates that are in motion with respect to one another and are continually changing in shape and size (Figure 1.1/Plate 1).

The parental theory of plate tectonics, **seafloor spreading**, states that new lithosphere is formed at ocean ridges and moves away from ridge axes with a motion like that of a conveyor belt as new lithosphere fills in the resulting crack or rift. The mosaic of plates, which range from 50 to over 200 km thick, are bounded by ocean ridges, subduction zones (in part collisional boundaries), and transform faults (boundaries along which plates slide past each other) (Figure 1.1/Plate 1, cross-sections). To accommodate the newly-created lithosphere, oceanic plates return to the mantle at subduction zones such that the surface area of the Earth remains constant. Harry Hess is credited with proposing the theory of seafloor spreading in a now classic paper finally published in 1962, although the name was earlier suggested by Robert Dietz in 1961. The basic idea of plate tectonics was proposed by Jason Morgan in 1968.

Many scientists consider the widespread acceptance of the plate tectonic model as a 'revolution' in the Earth Sciences. As pointed out by J. Tuzo Wilson in 1968, scientific disciplines tend to evolve from a stage primarily of data gathering, characterized by transient hypotheses, to a stage where a new unifying theory or theories are proposed that explain a great deal of the accumulated data. Physics and chemistry underwent such revolutions around the beginning of the twentieth century, whereas the Earth Sciences entered such a revolution in the late 1960s. As with scientific revolutions in other fields, new ideas and interpretations do not invalidate earlier observations. On the contrary, the theories of seafloor spreading and plate tectonics offer for the first time unified explanations for what, before, had seemed unrelated observations in the fields of geology, paleontology, geochemistry, and geophysics.

The origin and evolution of the Earth's crust is a tantalizing question that has stimulated much speculation and debate dating from the early part of the nineteenth century. Some of the first problems recognized, such as how and when did the oceanic and continental crust form, remain a matter of considerable controversy even today. Results from the Moon and other planets indicate that the Earth's crust may be a unique feature in the Solar System. The rapid accumulation of data in the fields of geophysics, geochemistry, and geology since 1950 has added much to our understanding of the physical and chemical nature of the Earth's crust and of the processes by which it evolved. Evidence favours a source for the materials composing the crust from within the Earth. Partial melting of the Earth's mantle produced magmas that moved to the surface and formed the crust. The continental crust, being less dense than the underlying mantle, has risen isostatically above sea level and hence is subjected to weathering and erosion. Eroded materials are partly deposited on continental margins, and partly returned to the mantle by subduction to be recycled and perhaps again become part of the crust at a later time. Specific processes by which the crust formed and evolved are not well-known, but boundary conditions for crustal processes are constrained by an ever-increasing data base. In this book, physical and chemical properties of the

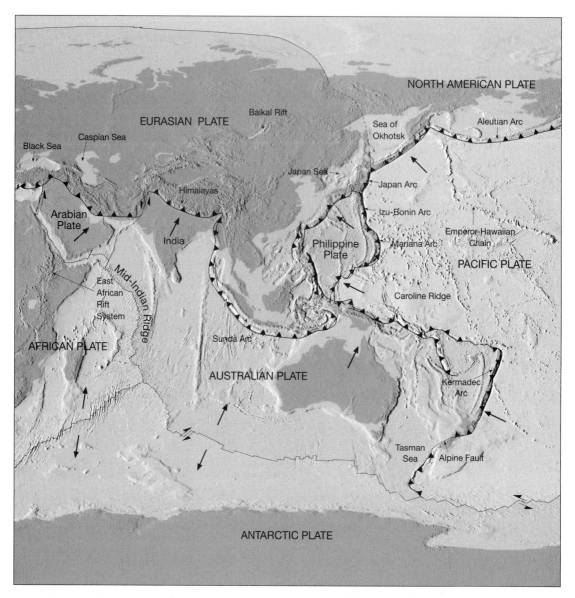

Figure 1.1 Map of the major lithospheric plates on Earth. Arrows are directions of plate motion. Filled barbs, convergent plate boundaries (subduction zones and collisional orogens); single lines, divergent plate boundaries (ocean ridges) and transform faults. Cross-sections show details of typical plate boundaries. Artwork by Dennis Tasa, courtesy of Tasa Graphic Arts, Inc.

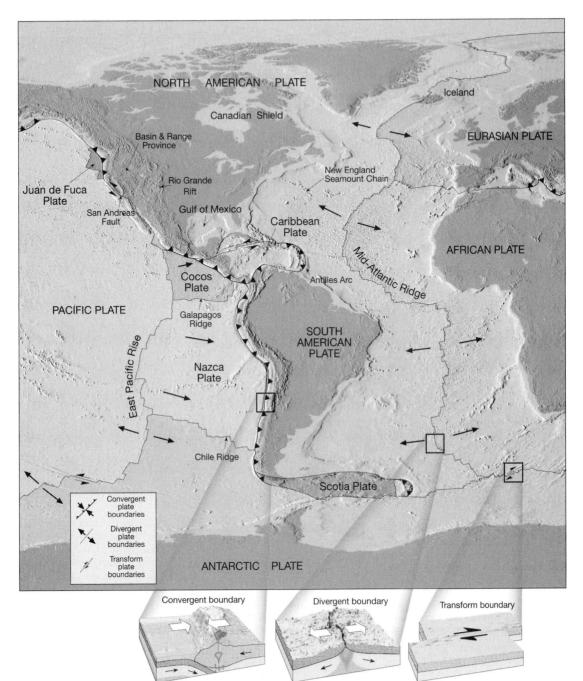

NORTH   AMERICAN   PLATE

Iceland

EURASIAN PLATE

Canadian Shield

Basin & Range
Province

Rio Grande
Rift

New England
Seamount Chain

Juan de Fuca
Plate

AFRICAN PLATE

San Andreas
Fault

Gulf of Mexico

Caribbean
Plate

Mid-Atlantic Ridge

Cocos
Plate

Antilles Arc

PACIFIC PLATE

Galapagos
Ridge

SOUTH
AMERICAN
PLATE

East Pacific Rise

Nazca
Plate

Chile Ridge

Scotia Plate

	Convergent plate boundaries
	Divergent plate boundaries
	Transform plate boundaries

ANTARCTIC   PLATE

Convergent boundary   Divergent boundary   Transform boundary

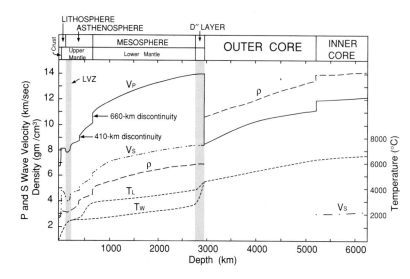

Figure 1.2 Distribution of average compressional (Vp) and shear wave (Vs) velocities and average calculated density (ρ) in the Earth. Also shown are temperature distributions for whole-mantle convection (TW) and layered-mantle convection (TL).

Earth are described, and crustal origin and evolution are discussed in the light of mantle dynamics and plate tectonics. Included also is a discussion of the origin of the atmosphere, oceans, and life, which are all important facets of Earth history. Finally, the uniqueness of the Earth is contrasted with the other planets.

Structure of the Earth

First of all we need to review what is known about the structure of planet Earth. The internal structural of the Earth is revealed primarily by compressional (P-wave) and shear (S-wave) waves that pass through the Earth in response to earthquakes. Seismic-wave velocities vary with pressure (depth), temperature, mineralogy, chemical composition, and degree of partial melting. Although the overall features of seismic-wave velocity distributions have been known for some time, refinement of data has been possible in the last ten years. Seismic-wave velocities and density increase rapidly in the region between 200 and 700 km deep. Three first-order seismic discontinuities divide the Earth into crust, mantle and core (Figure 1.2): the **Mohorovicic discontinuity**, or **Moho**, defining the base of the crust; the **core–mantle interface** at 2900 km; and, at about 5200 km, the **inner-core/outer-core interface**. The core comprises about sixteen per cent of the Earth by volume and thirty-two per cent by mass. These discontinuities reflect changes in composition or phase, or both. Smaller, but very important velocity changes at 50–200 km, 410 km, and 660 km provide a basis for further subdivision of the mantle, as discussed in Chapter 4.

The major regions of the Earth can be summarized as follows with reference to Figure 1.2:

1 The **crust** consists of the region above the Moho, and ranges in thickness from about 3 km at some oceanic ridges to about 70 km in collisional orogens.

2 The **lithosphere** (50–300 km thick) is the strong outer layer of the Earth, including the crust, that reacts to many stresses as a brittle solid. The **asthenosphere**, extending from the base of the lithosphere to the 660-km discontinuity, is by comparison a weak layer that readily deforms by creep. A region of low seismic-wave velocities and high attenuation of seismic-wave energy, the **low-velocity zone (LVZ)**, occurs at the top of the asthenosphere and is from 50–100 km thick. Significant lateral variations in density and in seismic-wave velocities are common at depths of less than 400 km.

3 The **upper mantle** extends from the Moho to the 660-km discontinuity, and includes the lower part of the lithosphere and the upper part of the asthenosphere. The region from the 410-km to the 660-km discontinuity is known as the **transition zone**. These two discontinuities, as further discussed in Chapter 4, are caused by two important solid-state transformations: from olivine to wadsleyite at 410 km and from spinel to perovskite + magnesiowustite at 660 km.

4 The **lower mantle** extends from the 660-km discontinuity to the 2900-km discontinuity at the core-mantle boundary. For the most part, it is characterized by rather constant increases in velocity and density in response to increasing hydrostatic compression. Between 220–250 km above the core–mantle interface a flattening of velocity and density gradients occurs, in a region known as the **D″ layer**, named after the seismic wave used to define the layer. The lower mantle is also referred to as the **mesosphere**, a region that is strong, but relatively passive in terms of deformational processes.

5 The **outer core** will not transmit S-waves and is interpreted to be liquid. It extends from the 2900-km to the 5200-km discontinuity.

6 The **inner core**, which extends from 5200-km discontinuity to the centre of the Earth, transmits S-waves, although at very low velocities, suggesting that it is near the melting point.

There are only two layers in the Earth with anomalously low seismic velocity gradients: the LVZ at the base of the lithosphere and the D″ layer just above the core (Figure 1.2). These layers coincide with very steep temperature gradients, and hence are thermal boundary layers within the Earth. The LVZ is important in that plates are decoupled from the mantle at this layer: plate tectonics could not exist without an LVZ. The D″ layer is important in that it may be the site at which mantle plumes are generated.

Considerable uncertainty exists regarding the temperature distribution in the Earth. It is dependent upon such features of the Earth's history as:

1 the initial temperature distribution
2 the amount of heat generated as a function of both depth and time
3 the nature of mantle convection
4 the process of core formation.

Most estimates of the temperature distribution in the Earth are based on one of two approaches, or a combination of both: models of the Earth's thermal history involving various mechanisms for core formation, and models involving redistribution of radioactive heat sources in the Earth by melting and convection processes.

Estimates using various models seem to converge on a temperature at the core–mantle interface of about 4500 ± 500 °C and the centre of the core 6700 to 7000 °C. Two examples of calculated temperature distributions in the Earth are shown in Figure 1.2. Both show significant gradients in temperature in the LVZ and the D″ layer. The layered convection model also shows a large temperature change near the 660-km discontinuity, since this is the boundary between shallow and deep convection systems in this model. The temperature distribution for whole-mantle convection, which is preferred by most scientists, shows a rather smooth decrease from the top of the D″ layer to the LVZ.

Seafloor spreading

Seafloor spreading was proposed to explain linear magnetic anomalies on the sea floor by Vine and Matthews in 1963. These magnetic anomalies (Figure 1.3), which had been recognized since the 1950s but for which no satisfactory origin had been proposed, have steep flanking gradients and are remarkably linear and continuous, except where broken by fracture systems (Harrison, 1987). Vine and Matthews (1963) proposed that these anomalies result from a combination of seafloor spreading and reversals in the Earth's magnetic field, the record of reversals being preserved in the magnetization in the upper oceanic crust. The model predicts that lines of alternate normally and reversely magnetized crust should parallel ocean ridge crests, with the pronounced magnetic contrasts between them causing the observed steep linear gradients. With the Geomagnetic Time Scale determined from paleontologically-dated deep sea sediments, Vine (1966) showed that the linear oceanic magnetic

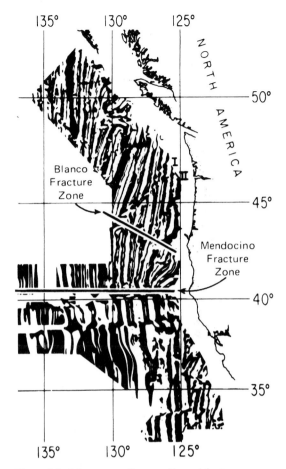

Figure 1.3 Linear magnetic anomalies and fracture zones in the NE Pacific basin. Positive anomalies in black. After Raff and Mason (1961).

anomalies could be explained by seafloor spreading. No other single observation in the last 50 years has had such a profound effect on geology. With this observation, we entered a new scientific era centred around a dynamic Earth.

Ocean ridges are accretionary plate boundaries where new lithosphere is formed from upwelling mantle as the plates on both sides of ridges grow in area and move away from the axis of the ridge (Figure 1.1/Plate 1, cross sections). In some instances, such as the South Atlantic, new ocean ridges formed beneath supercontinents, and thus as new oceanic lithosphere is produced at a ridge the supercontinent splits and moves apart on each of the ridge flanks. The average rate of oceanic lithosphere production over the past few million years is about 3.5 km²/y and, if this rate is extrapolated into the geologic past, the area covered by the present ocean basins (sixty-five per cent of the Earth's surface) would be generated in less than 100 My. In fact, the oldest ocean floor dates only to about 160 Ma, because older oceanic plates have been subducted into the mantle.

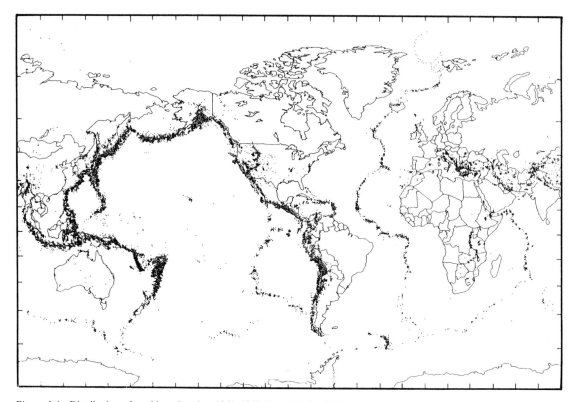

Figure 1.4   Distribution of world earthquakes 1961–1969. From National Earthquake Information Center Map NEIC-3005.

Plate boundaries

Introduction

Earthquakes occur along rather narrow belts (Figure 1.4), and these belts mark boundaries between lithospheric plates. There are four types of seismic boundaries, distinguished by their epicentre distributions and geologic characteristics: ocean ridges, subduction zones, transform faults, and collisional zones. Provided a sufficient number of seismic recording stations with proper azimuthal locations are available, it is possible to determine the directions of first motion at sites of earthquake generation, which in turn provides major constraints on plate motions.

Modern plates range in size from $< 10^4$ km^2 to over 10^8 km^2 and plate margins do not usually coincide with continental margins (Figure 1.1/Plate 1). Seven major plates are recognized: the Eurasian, Antarctic, North American, South American, Pacific, African and Australian plates. Intermediate-size plates (10^6–10^7 km^2) include the Philippine, Arabian, Nasca, Cocos, Caribbean and Scotia plates. In addition, there are more than twenty plates with areas of 10^5–10^6 km^2. Both plate theory and first-motion studies at plate boundaries indicate that plates are produced at ocean ridges, consumed at subduction zones, and slide past each other along transform faults

(Figure 1.1/Plate 1, cross-sections). At collisional zones, plates carrying continents may become sutured together. Plates diminish or grow in area depending on the distribution of convergent and divergent boundaries. The African and Antarctic plates, for instance, are almost entirely surrounded by active spreading centres and hence are growing in area. If the surface area of the Earth is to be conserved, other plates must be diminishing in area as these plates grow, and this is the case for plates in the Pacific area. Plate boundaries are dynamic features, not only migrating about the Earth's surface, but changing from one type of boundary to another. In addition, new plate boundaries can be created in response to changes in stress regimes in the lithosphere. Also, plate boundaries disappear as two plates become part of the same plate, for instance after a continent–continent collision. Small plates ($< 10^6$ km^2) occur most frequently near continent–continent or arc–continent collisional boundaries and are characterized by rapid, complex motions. Examples are the Turkish–Aegean, Adriatic, Arabian, and Iran plates located along the Eurasian–African continent–continent collision boundary, and several small plates along the continent–arc collision border of the Australian–Pacific plates. The motions of small plates are controlled largely by the compressive forces of larger plates.

Continental margins are of two types: active and passive. An **active continental margin** is found where

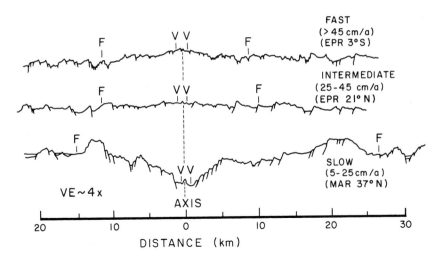

either a subduction zone or a transform fault coincides with continent–ocean interface. Examples are the Andean and Japan continental–margin arc systems and the San Andreas transform fault in California. **Passive continental margins** occur along the edges of opening ocean basins like the Atlantic basin. These margins are characterized by minimal tectonic and igneous activity.

Divergent boundaries (ocean ridges)

The interconnected ocean-ridge system is the longest topographic feature on the Earth's surface, exceeding 70 000 km in length. Typical ocean ridges are 3000–4000 km wide, with up to several kilometres of relief in the axial rift zone. Ocean ridges are characterized by shallow earthquakes limited to axial rift zones. These earthquakes are generally small in magnitude, commonly occur in swarms and appear to be associated with intrusion and extrusion of basaltic magmas. First-motion studies indicate that rift earthquakes are produced dominantly by vertical faulting as is expected if new lithosphere is being injected upwards. Most faulting occurs in the depth range of 2–8 km and some ruptures extend to the sea floor.

The median valley of ocean ridges varies in geological character due to the changing importance of tectonic extension and volcanism. In the northern part of the Mid-Atlantic ridge, stretching and thinning of the crust dominate in one section, while volcanism dominates in another. Where tectonic thinning is important, faulting has exposed gabbros and serpentinites from deeper crustal levels. Volcanic features range from large ridges (> 50 km long) in sections of the median valley where volcanism has dominated, to small volcanic cones in sections dominated by extension. The axial topography of fast- and slow-spreading ridges varies considerably. A deep axial valley with flanking mountains characterizes slow-spreading ridges, while relatively low relief, and in some instances a topographic high, characterize fast-spreading ridges (Figure 1.5). Model studies sug-

gest that differences in horizontal stresses in the oceanic lithosphere may account for the relationship between ridge topography and spreading rate (Morgan et al., 1987). As oceanic lithosphere thickens with distance from a ridge axis, horizontal extensional stresses can produce the axial topography found on slow-spreading ridges. In fast-spreading ridges, however, the calculated stresses are too small to result in appreciable relief. The axis of ocean ridges is not continuous, but may be offset by several tens to hundreds of kilometres by transform faults (Figure 1.1/Plate 1). Evidence suggests that ocean ridges grow and die out by lateral propagation. Offset magnetic anomalies and bathymetry consistent with propagating rifts, with and without transform faults, have been described along the Galapagos ridge and in the Juan de Fuca plate (Hey et al., 1980).

Transform faults and fracture zones

Transform faults are plate boundaries along which plates slide past each other and plate surface is conserved. They uniquely define the direction of motion between two bounding plates. Ocean-floor transform faults differ from transcurrent faults in that the sense of motion relative to offset along an ocean ridge axis is opposite to that predicted by transcurrent motion (Wilson, 1965) (Figure 1.6). These offsets may have developed at the time spreading began and reflect inhomogeneous fracturing of the lithosphere. Transform faults, like ocean ridges, are characterized by shallow earthquakes (< 50 km deep). Both geophysical and petrological data from ophiolites cut by transforms suggest that most oceanic transforms are 'leaky', in that magma is injected along fault surfaces producing strips of new lithosphere (Garfunkel, 1986). Transforms cross continental or oceanic crust and may show apparent lateral displacements of many hundreds of kilometres. First-motion studies of oceanic transform faults indicate lateral motion in a direction away from ocean ridges (Figure 1.6). Also, as predicted by seafloor spreading, earthquakes are restricted to areas

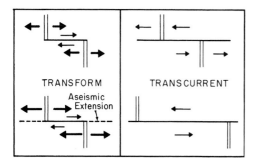

Figure 1.6 Motion on transform and transcurrent faults relative to an ocean ridge axis (double vertical lines). Note that the amount of offset increases with transcurrent motion, while it remains constant with transform motion. Bold arrows refer to spreading directions, small arrows to plate motions.

between offset ridge axes. Transform faults may produce large structural discontinuities on the sea floor, and in some cases structural and topographic breaks known as **fracture zones** mark the locations of former ridge–ridge transforms on the sea floor. There are three types of transform faults: ridge–ridge, ridge–trench, and trench–trench faults. Ridge–ridge transform faults are most common, and these retain a constant length as a function of time, whereas ridge–trench and trench–trench transforms decrease or increase in length as they evolve.

Studies of oceanic transform-fault topography and structure indicate that zones of maximum displacement are very localized (< 1 km wide) and are characterized by an anatomizing network of faults. Steep transform valley walls are composed of inward-facing scarps associated with normal faulting. Large continental transform faults form where pieces of continental lithosphere are squeezed within intracontinental convergence zones such as the Anatolian fault in Turkey. Large earthquakes (M > 8) separated by long periods of quiescence occur along 'locked' segments of continental transforms, whereas intermediate-magnitude earthquakes characterize fault segments in which episodic slippage releases stresses. Large earthquakes along continental transforms appear to have a period of about 150 years, as indicated by records from the San Andreas fault in California.

Ridge segments between oceanic transforms behave independently of each other. This may be caused by instability in the convective upcurrents that feed ocean ridges, causing these upcurrents to segment into regularly-spaced rising diapirs, with each diapir feeding a different ridge segment. Transforms may arise at the junctions of ridge segments because magma supply between diapirs is inadequate for normal oceanic crustal accretion. The persistence of transforms over millions of years indicates that asthenospheric diapirs retain their integrity for long periods of time. It appears from the use of fixed hotspot models of absolute plate motion that both ridge axes and transforms migrate together at a rate of a few centimetres per year. This, in turn, requires that mantle

diapirs migrate, and suggests that ocean-ridge segments and diapirs are decoupled from underlying mantle flow.

Triple junctions

Triple junctions are points where three plates meet. Such junctions are a necessary consequence of rigid plates on a sphere, since this is the common way a plate boundary can end. There are sixteen possible combinations of ridge, trench, and transform-fault triple junctions (McKenzie and Morgan, 1969), of which only six are common. Triple junctions are classified as stable or unstable, depending on whether they preserve their geometry as they evolve. The geometric conditions for stability are described with vector velocity triangles in Figure 1.7 and only RRR triple junctions are stable for all orientations of plate boundaries. It is important to understand evolutionary changes in triple junctions, because changes in their configuration can produce changes that superficially resemble changes in plate motions. Triple junction evolution is controlled by the lengths of transform faults, spreading velocities, and the availability of magma.

Convergent boundaries (subduction zones)

Convergent plate boundaries are defined by earthquake hypocentres that lie in an approximate plane and dip beneath arc systems. This plane, known as the **seismic zone** or **Benioff zone**, dips at moderate to steep angles and extends in some instances to the 660-km seismic discontinuity. The seismic zone is interpreted as a brittle region in the upper 10–20 km of descending lithospheric slabs. Modern seismic zones vary significantly in hypocentre distribution and in dip (Figure 1.8). Some, such as the seismic zone beneath the Aleutian arc, extend to depths < 300 km while others extend to the 660-km discontinuity (Figure 1.8, a and b respectively). In general, seismic zones are curved surfaces with radii of curvature of several hundred kilometres and with irregularities on scales of < 100 km. Approximately planar seismic zones are exceptional. Seismic gaps in some zones (e.g., c and e) suggest, although do not prove, fragmentation of the descending slab. Dips range from 30° to 90°, averaging about 45°. Considerable variation may occur along strike in a given subduction zone, as exemplified by the Izu-Bonin arc system in the Western Pacific. Hypocentres are linear and rather continuous on the northern end of this arc system, (d), becoming progressively more discontinuous toward the south. Near the southern end of the arc, the seismic zone exhibits a pronounced gap between 150 and 400 km depth, (c). A large gap in hypocentres in descending slabs, such as that observed in the New Hebrides arc, (e), may indicate that the tip of the slab broke off and settled into the mantle. Because some slabs appear to penetrate the 660-km discontinuity (Chapter 4), the lack of earthquakes below 700 km probably reflects the depth of the brittle–ductile transition in descending slabs. An excellent cor-

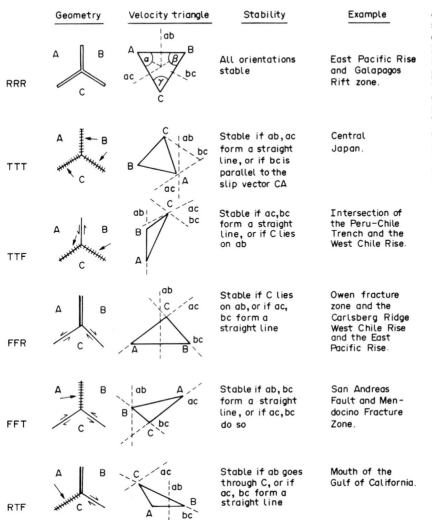

Geometry	Velocity triangle	Stability	Example

RRR — All orientations stable — East Pacific Rise and Galapagos Rift zone.

TTT — Stable if ab, ac form a straight line, or if bc is parallel to the slip vector CA — Central Japan.

TTF — Stable if ac, bc form a straight line, or if C lies on ab — Intersection of the Peru–Chile Trench and the West Chile Rise.

FFR — Stable if C lies on ab, or if ac, bc form a straight line — Owen fracture zone and the Carlsberg Ridge West Chile Rise and the East Pacific Rise.

FFT — Stable if ab, bc form a straight line, or if ac, bc do so — San Andreas Fault and Mendocino Fracture Zone.

RTF — Stable if ab goes through C, or if ac, bc form a straight line — Mouth of the Gulf of California.

Figure 1.7 Geometry and stability requirements of six common triple junctions. Dashed lines ab, bc, and ac in the velocity triangles join points, the vector sum of which leave the geometry of AB, BC, and AC, respectively, unchanged. The junctions are stable only if ab, bc, and ac meet at a point. Key: track symbol, trench; double line, ocean ridge; single line, transform fault.

relation exists between the length of seismic zones and the product of plate convergence rate and age of the downgoing slab.

First-motion studies of earthquakes in subduction zones indicate variation in movement both with lateral distance along descending slabs and with slab depth. Seaward from the trench in the upper part of the lithosphere where the plate begins to bend, shallow extensional mechanisms predominate. Because of their low strength, sediments in oceanic trenches cannot transmit stresses, and hence are usually flat-lying and undeformed. Seismic reflection profiles indicate, however, that rocks on the landward side of trenches are intensely folded and faulted. Thrusting mechanisms dominate at shallow depths in subduction zones (20–100 km). At depths < 25 km, descending slabs are characterized by low seismicity. Large-magnitude earthquakes are generally thrust-types and occur at depths > 30 km (Shimamoto, 1985). During large earthquakes, ruptures branch off the slab and

extend upwards forming thrusts that dip away from the trench axis. Calculations of stress distributions at < 300 km depth show that compressional stresses generally dominate in the upper parts of descending slabs, whereas tensional stresses are more important in the central and lower parts (Figure 1.9). At 300–350 km depth in many slabs compressional stresses are very small, whereas at depths > 400 km a region of compressional stress may be bounded both below and above by tensional stress regions.

The seismicity in descending slabs is strongly correlated with the degree of coupling between the slab and the overriding plate (Shimamoto, 1985). The low seismicity in descending slabs at depths < 25 km may reflect relatively high water contents and the low strength of subducted hydrous minerals, both of which lead to decoupling of the plates and largely ductile deformation. At greater depths, diminishing water and hydrous mineral contents (due to slab devolatilization) result in greater

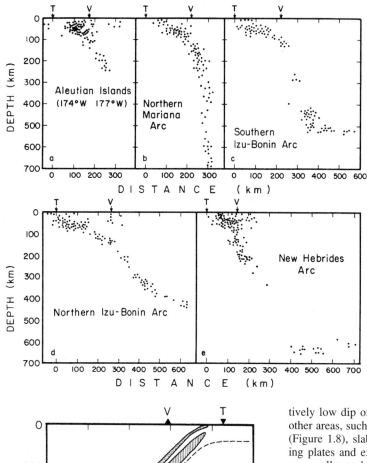

Figure 1.8 Vertical cross-sections of hypocentre distributions beneath modern arc–trench systems. Each diagram shows earthquakes for 7–10 year periods between 1954 and 1969. T = trench axis; V = recently active volcanic chain. Distance is measured horizontally from each trench axis in kilometres. Hypocentre data from many sources, principally from National Earthquake Information Center, US Coast and Geodetic Survey.

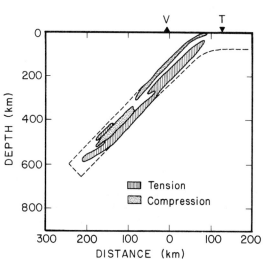

Figure 1.9 Calculated distribution of stresses in a descending slab. Subduction rate, 8 cm/y and dip = 45 °. After Goto et al. (1985). V, volcanic front; T, trench.

coupling of overriding and descending plates, and thus to the onset of major earthquakes. Coupling also varies between descending slabs. In some continental-margin subduction zones (e.g., Peru–Chile, Alaska), coupling is very strong, resulting in large earthquakes and a rela-

tively low dip of the descending slab (Figure 1.10). In other areas, such as the Kurile and Mariana arc systems (Figure 1.8), slabs are largely decoupled from overriding plates and extend to great depths, and earthquakes are smaller and less frequent. In some instances, subducting slabs are forced beneath the lithosphere in the overriding plate, a situation known as **buoyant subduction** (Figure 1.10). Buoyant subduction occurs when the lithosphere is forced to sink before it becomes negatively buoyant (i.e., in < 50 My today), and thus it tends to resist subduction into the asthenosphere. Underplated buoyant slabs eventually sink into the mantle when they cool sufficiently and their density increases.

Seismic tomographic studies of subduction zones reveal a detailed three-dimensional structure of descending slabs. For instance, P-wave tomographic images of the Japan subduction zone correlate well with major surface geological features in Japan (Figure 1.11). Seismic velocities are several percentage points higher in the descending slab than in the surrounding mantle, and results indicate that the slab boundary is a sharp seismic discontinuity (Zhao et al., 1992). Moreover, low-velocity anomalies occur in the crust and mantle wedge over the descending slab, a feature also common in other subduction zones. The two low-velocity zones in the crust correlate with active volcanism in the Japan arc, and probably reflect magma plumbing systems. Deeper low-velocity zones (> 30 km) may represent partly-melted ultramafic rocks, formed in response to the upward trans-

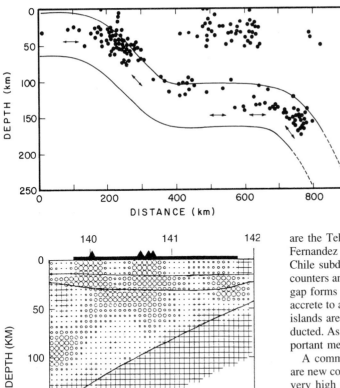

Figure 1.10 Buoyant subduction beneath the Peru–Chile arc in central Peru. Earthquake first-motions shown by arrows are tensional. Black dots are earthquake hypocentres. Modified after Sacks (1983).

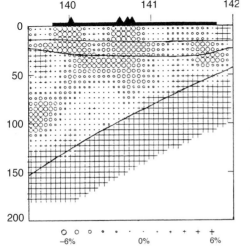

Figure 1.11 Cross-section of the Japan subduction zone showing perturbations of P-wave velocities from normal mantle (in percentages). Crosses and circles show fast and slow velocities respectively. Solid triangles are active volcanoes. Base of crust shown at 25–30 km. Units on horizontal axis are degrees of E longitude. After Zhao et al. (1992).

fer of volatiles from the descending plate, which lowers the melting points of mantle-wedge silicates.

Another interesting question about subduction is, what happens when a submarine plateau or aseismic ridge encounters a subduction zone? Because they resist subduction they may produce a cusp in the arc system, as illustrated for instance by the intersection of the Caroline ridge with the Mariana arc south of Japan (Figure 1.1/Plate 1). Paleomagnetic and structural geologic data from the Mariana arc support this interpretation, indicating that the arc was rotated at its ends by collision of these ridges between 30 and 10 Ma (McCabe, 1984). Also, volcanic and seismic gaps in arc systems commonly occur at points of collision between submarine plateaux and ridges with arcs (McGeary et al., 1985). Examples

are the Tehuantepec, Cocos, Carnegie, Nazca and Juan Fernandez ridges along the Middle American and Peru–Chile subduction systems. When a plateau or ridge encounters an arc, subduction stops and a volcanic/seismic gap forms in the arc. In most instances, plateaux/ridges accrete to arcs, and only small ridges and some volcanic islands are negatively buoyant and can actually be subducted. As we shall see in Chapter 5, this may be an important mechanism by which continents grow laterally.

A commonly asked question is, just where and how are new convergent boundaries initiated? Because of the very high stress levels necessary for the oceanic lithosphere to rupture, it is likely that pre-existing zones of weakness in the lithosphere provide sites for new subduction zones. Of the three proposed sites for initiation of new subduction zones, i.e., passive continental margins, transform faults/fracture zones, and extinct ocean ridges, none can simply convert to subduction zones by the affect of gravitational forces alone (Mueller and Phillips, 1991). Hence, additional forces are needed to convert these sites into subduction zones. One possible source is the attempted subduction of buoyant material (such as a submarine plateau) at a trench, which can result in large compressional forces in both subducting and overriding plates. This is the only recognized tectonic force sufficient to trigger nucleation of a new subduction zone. Transform faults and fracture zones are likely sites for subduction initiation in that they are common in the vicinity of modern subduction zones and are weaker than normal oceanic lithosphere.

Collisional boundaries

Deformation fronts associated with collisional boundaries are widespread, as exemplified by the India–Asia boundary which extends for at least 3000 km northeast of the Himalayas. Earthquakes are chiefly < 100 km deep and first-motion studies indicate a variety of fault types. Thrust fault mechanisms generally dominate near sutures, such as the Indus suture in the Himalayas. Transcurrent faulting is common in the overriding plate as illustrated by the large strike-slip faults produced in China and

Tibet during the India collision. In addition, extensional faulting may extend great distances beyond the suture in the overriding plate. For instance, the Baikal rift in southern Siberia appears to have formed in response to the India collision 55 Ma.

A plate boundary in the early stages of an arc–continent collision is illustrated by the Sunda arc system in eastern Indonesia (Figure 1.1/Plate 1). Australia is beginning to collide with this arc as the Australian plate is subducted beneath the arc. In fact, a large bend in the descending slab beneath the island of Timor may be produced by subduction of continental crust. Numerous hypocentres at 50–100 km depth are interpreted to reflect the beginning of detachment of the descending slab as continental crust resists further subduction. Farther to the east, Australia collided with the arc system and is accreted and sutured to the arc on the northern side of New Guinea. Earthquakes and active volcanism in northern New Guinea are interpreted to reflect initiation of a new subduction zone dipping to the south, in the opposite direction of subduction prior to collision with the Australian plate.

Trench–ridge interactions

It is interesting to consider what happens when an ocean ridge approaches and finally collides with a subduction zone, as the Chile and Juan de Fuca ridges are today (Figure 1.1/Plate 1). If a ridge is subducted, the arc should move 'uphill' and become emergent as the ridge crest approaches, and it should move 'downhill' and become submerged as the ridge passes down the subduction zone (DeLong and Fox, 1977). Corresponding changes in sedimentation should accompany this emergence-submergence sequence of the arc. Ridge subduction may also lead to cessation of subduction-related magmatism as the hot ridge is subducted. This could be caused by reduced frictional heating in the subduction zone or by progressive loss of volatiles from a descending slab as a ridge approaches. Also, the outer arc should undergo regional metamorphism as the hot ridge crest is subducted. All three of these phenomena are recorded in the Aleutian arc and support the subducted ridge model. Ridge subduction may also result in a change in stress regime in the overriding plate from dominantly compressional to extensional, and in the opening of a back-arc basin (Uyeda and Miyashiro, 1974). The subduction of active ridges may lead to formation of new ridges in the descending plate at great distances from the convergent margin. For instance, the rifting of Antarctica from Australia, which began almost 50 Ma, coincided with the subduction of a ridge system along the northern edge of the Australian–Antarctic plate.

In the case of the Chile ridge, which is being subducted today, there are few effects until the ridge arrives at the trench axis. As the ridge approaches the Chile trench, the rift valley becomes filled with sediments and finally disappears beneath the toe of the trench (Cande et al.,

1987). Also, the landward slope of the trench steepens and narrows in the collision zone. At the trench–ridge collision site, the accretionary prism is reduced in size by more than seventy-five per cent and part of the basement beneath the Andean arc appears to have been eroded away and subducted. Two factors seem important in decreasing the volume of the accretionary prism:

1   the topographic relief on the ridge may increase the rate of subduction erosion carrying material away from the bottom of the accretionary prism
2   subduction of oceanic ridges caused by transform faults may mechanically weaken the base of the accretionary prism, making it more susceptible to removal by subduction erosion.

Heat flow increases dramatically in the collision zone and then decays after collision to typical arc values. Also, an ophiolite (fragment of oceanic crust) was tectonically emplaced in the Andean fore-arc region during an earlier stage of the collision 3 Ma, supporting the idea, more fully discussed in Chapter 3, that ridge–trench collisions may be an important way in which ophiolites are emplaced. The intriguing question of what happens to the convective upcurrent beneath the Chile ridge as it is subducted remains poorly understood.

The Wilson Cycle

The opening and closing of an oceanic basin is known as a **Wilson Cycle** (Burke et al., 1976), named after J. Tuzo Wilson who first described it in 1966. He proposed that the opening and closing of a proto-Atlantic basin in the Paleozoic accounted for unexplained changes in rock types, fossils, orogenies, and paleoclimates in the Appalachian orogenic belt. A Wilson Cycle begins with the rupture of a continent along a rift system, such as the East African rift today, followed by the opening of an ocean basin with passive continental margins on both sides (Figure 1.12, a and b). The oldest rocks on passive continental margins are continental rift assemblages. As the rift basin opens into a small ocean basin, as the Red Sea is today, cratonic sediments are deposited along both of the retreating passive margins (b), and abyssal sediments accumulate on the sea floor adjacent to these margins. Eventually, a large ocean basin such as the Atlantic may develop from continued opening. When the new oceanic lithosphere becomes negatively buoyant, subduction begins on one or both margins and the ocean basin begins to close (c and d). Complete closure of the basin results in a continent–continent collision, (e), such as occurred during the Permian when Baltica collided with Siberia forming the Ural Mountains. During the collision, arc rocks and oceanic crust are thrust over passive-margin assemblages. The geologic record indicates that the Wilson Cycle has occurred many times during the Phanerozoic. Because lithosphere is weakened along collisional zones, rifting may open new ocean basins near older sutures, as evidenced by

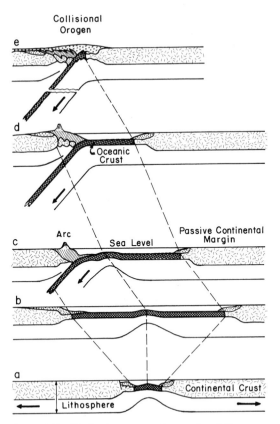

Figure 1.12 Idealized sequence of events in a Wilson Cycle (beginning at the bottom).

the opening to the modern Atlantic basin approximately along the Ordovician Iapetus suture.

Stress distribution within plates

Stress distributions in the lithosphere can be estimated from geological observations, first-motion studies of earthquakes, and direct measurement of in-situ stress (Zoback and Zoback, 1980). Stress provinces in the crust show similar stress orientations and earthquake magnitudes and have linear dimensions ranging from hundreds to thousands of kilometres. Maximum compressive stresses for much of North America trend E–W to NE–SW (Figure 1.13). In cratonic regions in South America most stresses trend E–W to NW–SE. Western Europe is characterized by dominantly NW–SE compressive stresses, while much of Asia is more nearly N–S. Within the Australian plate, compressive stresses range from N–S in India to nearly E–W in Australia. Horizontal stresses are variable in Africa but suggest a NW–SE trend for maximum compressive stresses in West Africa and an E–W trend for the minimum compressive stresses in East Africa. Except at plate boundaries, oceanic lithosphere is characterized by variable compressive stresses.

The overall pattern of stresses in both the continental and oceanic lithosphere is consistent with the present distribution of plate motions as deduced from magnetic anomaly distributions on the sea floor.

Examples of stress provinces in the United States are shown in Figure 1.13. Most of the central and eastern parts of the United States are characterized by compressive stresses, ranging from NW–SE along the Atlantic Coast to dominantly NE–SW in the mid-continent area. In contrast, much of the western United States is characterized by extensional and transcurrent stress patterns, although the Colorado Plateau, Pacific Northwest and the area around the San Andreas fault are dominated by compressive deformation. The abrupt transitions between stress provinces in the western United States imply lower crustal or uppermost mantle sources for the stresses. The correlation of stress distribution and heat-flow distribution in this area indicates that widespread rifting in the Basin and Range province is linked to thermal processes in the mantle. On the other hand, the broad transitions between stress provinces in the central and eastern United States reflect deeper stresses at the base of the lithosphere, related perhaps to drag resistance of the North American plate and to compressive forces transmitted from the Mid-Atlantic ridge.

Plate motions

Cycloid plate motion

The motion of a plate on a sphere can be described in terms of a pole of rotation passing through the centre of the sphere. Because all plates on the Earth are moving relative to each other, the angular difference between a given point on a plate to the pole of rotation of that plate generally changes with time. For this reason, the trajectory of a point on one plate as observed from another plate cannot be described by a small circle around a fixed pole of rotation. Instead, the shape of a relative motion path is that of a spherical cycloid (Cronin, 1991). The trajectory of a point on a moving plate relative to a point on another plate can be described if three variables are known during the period of displacement being considered:

1 the position of the pole of rotation
2 the direction of relative motion
3 the magnitude of the angular velocity.

An example of cycloid motion for a hypothetical plate is given Figure 1.14. Plate 2 rotates around pole P2 with a given angular velocity. P2 appears to move with time relative to an observer on plate 1, tracing the path of a small circle. When the motion of plate 2 is combined with the relative motion of P2, a reference point M on plate 2 traces a figure of rotation around two axes known as a spherical cycloid.

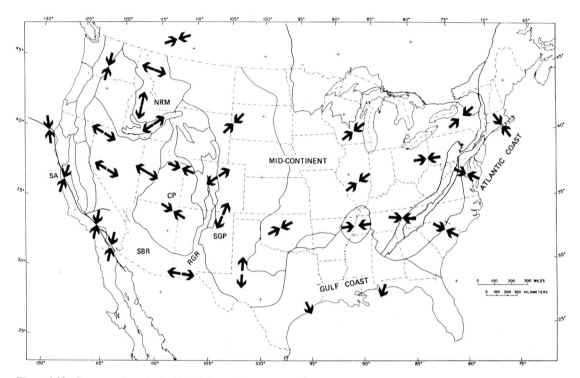

Figure 1.13 Stress provinces in the United States. Arrows are directions of least (outward-directed) or greatest (inward-directed) principal horizontal compression. Province abbreviations: SA, San Andreas; SBR, southern Basin and Range province; RGR, Rio Grande rift; CP, Colorado Plateau; NRM, northern Rocky Mountains; SGP, southern Great Plains. From Zoback and Zoback (1980).

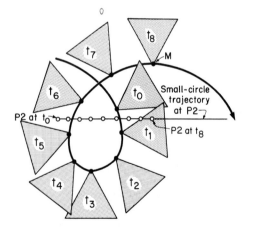

Figure 1.14 Spherical cycloid motion of plate 2 as its pole of rotation P processes along a line that represents a small circle around P1, the pole of rotation of plate 1. M is a point on plate 2. Time progresses from t_0 to t_8. Modified after Cronin (1987).

Plate Velocities in the last 150 My

Rates of plate motion can be quantified for the last 150 My by correlating seafloor magnetic anomalies with the Geomagnetic Time Scale (Figure 1.15). For convenience,

magnetic anomalies are numbered beginning with 1 at ridge axes. From the Geomagnetic Time Scale, each anomaly is assigned an age (for instance anomaly 30 corresponds to an age of 72 Ma, and anomaly 7 to 28 Ma). If the spreading rate has been constant in one ocean basin, it is possible to extend the Geomagnetic Time Scale to more than 4.5 Ma using the magnetic anomaly patterns. Although data indicate that a constant spreading rate is unlikely in any ocean basin, the South Atlantic most closely approaches constancy (~ 1.9 cm/y) and is commonly chosen as a reference to extrapolate the time scale. Paleontologic dates from sediment cores retrieved by the Deep Sea Drilling Project and isotopic ages of basalts dredged or drilled from the ocean floor substantiate an approximately constant spreading rate in the South Atlantic and allow the extension of the magnetic time scale to about 80 Ma. Correlations of magnetic anomalies with distance from ridge axes indicate that spreading rates in the South Indian and North Pacific basins have been more variable and, on the average, faster than the spreading rate of the South Atlantic (Figure 1.15).

Plate velocities also can be estimated from dislocation theory, using data derived from first-motion studies of earthquakes and from observed dip-lengths of subduction zones, if these lengths are assumed to be a measure of the amount of underthrusting during the last 10

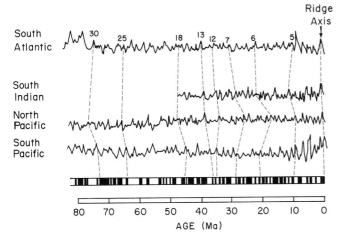

Figure 1.15 Magnetic profiles from the Atlantic, Indian, and Pacific Ocean basins. Geomagnetic Time Scale given beneath the profiles with normal (black) and reversed (white) magnetized bands. Proposed correlations of anomalies are shown with dashed lines. Numbers refer to specific anomalies. Modified after Heirtzler et al. (1968).

My. Still another method used to estimate plate velocities is by using transform faults. Rates and directions of motion can be estimated from the azimuths and amount of offset along transform faults, provided the azimuth and timing of motion can be estimated accurately.

Estimates of plate velocities commonly range within a factor of two of one another using the above methods averaged over several to many millions of years. Rates range from about 1 to 20 cm/y, averaging a few centimetres per year for most plates. Typical velocities of major plates (in cm/y) are as follows: North American, 1.5–2; Eurasian, 2.4–2.7; African, 2.9–3.2; Australian, 6–7.5; and Pacific, 5–7 (DeMets et al., 1990).

From spreading rates estimated from seafloor magnetic anomalies, it is possible to contour the age of the sea floor and several such maps have been published. From these maps, we see that the rate of spreading has varied between crustal segments bounded by transform faults, and has even varied on opposite flanks of the same ridge. The oldest oceanic crust (Jurassic) occurs immediately adjacent to the Izu-Bonin subduction zone south of Japan. Since the rate at which oceanic crust has been produced at ridges during the past several hundred million years is of the order of a few centimetres per year, it is unlikely that crust much older than Jurassic will be found on the ocean floors today. The average age of oceanic crust is about 60 My and the average age it begins to subduct is about 120 My. Fragments of oceanic crust older than Jurassic (ophiolites, Chapter 3) are found in continental orogenic belts where they were tectonically emplaced during orogeny.

From calculated seafloor spreading directions and rates, it is possible to reconstruct plate positions and to estimate rates of plate separation for the last 200 My. One way of illustrating such reconstructions is by the use of flow or drift lines, as shown in Figure 1.16 for the opening of the North Atlantic. The arrows indicate the relative directions of movement. Earlier positions of Africa and Europe relative to North America are also shown, with the corresponding ages in millions of years. The

reconstruction agrees well with geometric fits across the North Atlantic. It is clear from the lines of motion that Europe and Africa have been on two different plates for the last 160 My. Changes in spreading rates and directions, however, occur on both the African and Eurasian plates at the same time, at about 60 and 80 Ma. The separation of Africa and North America occurred primarily between 80 and 180 Ma, whereas separation of Eurasia from North America occurred chiefly in the last 80 My.

Plate velocities from paleomagnetism

If true polar wander has been small compared with the rate of plate motions in the geologic past, it is possible to estimate minimum plate velocities even before 200 Ma from plate motion rates (Bryan and Gordon, 1986; Jurdy et al., 1995). Also, if at least some hotspots have remained relatively fixed, it is possible to estimate plate motions relative to these hotspots (Hartnady and leRoex, 1985). Compared with typical modern continental plate velocities of 2–3 cm/y, during the last 350 My most continental plates have excursions to much faster rates (Figure 1.17). Episodes of rapid plate motion are recorded during the Triassic–Early Jurassic (250–200 Ma) on most continents, and Australia and India show peak velocities at about 150 and 50 Ma, respectively, after they fragmented from Gondwana. Results indicate that in the past continental plates have moved as fast as modern oceanic plates for intervals of 30–70 My. It is noteworthy that maxima in some continental plate velocities occur just after fragmentation from a supercontinent. For instance, the velocity maxima in the early Mesozoic follow the beginning of rifting in Pangea, and the peak velocities for Australia and India follow separation of these continents from Gondwana (Figure 1.17).

Paleomagnetic data from Archean rocks in southern Africa suggest that plates were moving at comparatively slow rates of about 2 cm/y between 3.5 and 2.4 Ga in this region, near the low end of the range of speeds of

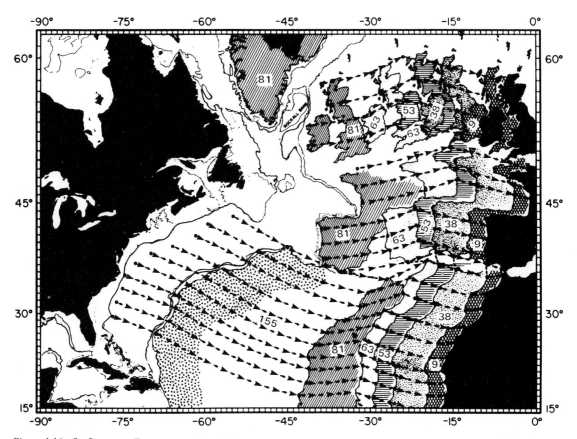

Figure 1.16   Seafloor spreading reconstruction of the opening of the North Atlantic. Black continents are present positions; dates for earlier positions are indicated in millions of years ago. Arrows are flow lines. After Pitman and Talwani (1972).

Phanerozoic plates (Kroner and Layer, 1992). How representative this rate is of the Archean is not known, but it is surprising in that a hotter Archean mantle would seem to be more consistent with faster plate motions.

Space geodetic measurements of plate velocities

Space geodesy is measuring the precise position of sites on the Earth's surface from sources in space, such as radio-wave sources and satellite tracking. Three methods are currently used: very long baseline radio interferometry (VLBI), satellite laser ranging (SRL), and the global positioning system (GPS). VLBI depends on the precise timing of radio-wave energy from extragalactic sources (chiefly quasars) observed by radio telescopes. The radio waves recorded at different sites on the Earth are correlated and used to determine site locations, orientation of the Earth, and azimuths of the radio sources. Arrival times of radio waves are measured with extremely precise hydrogen laser atomic clocks (Robertson, 1991). SRL is based on the round-trip time of laser pulses reflected from satellites that orbit the Earth. Successive observations permit the position of the tracking station

to be determined as a function of time. GPS geodesy uses several high-altitude satellites with orbital periods of twelve hours, and each satellite broadcasts its position and time. When multiple satellites are tracked, the location of the receiver can be estimated to within a few metres on the Earth's surface. Accuracy of the results depends on many factors (Gordon and Stein, 1992), including the length of time over which measurements have been accumulated. To reach accuracies of 1–2 mm for sites that are thousands of kilometres apart requires many years of data accumulation.

Space geodetic measurement are especially important in a better understanding of modern plate tectonics. Results have been used to verify that plate motions are steady on time scales of a few years, to estimate rates and directions of plate motions, to estimate motions of small regions within plate boundary zones, to better understand deformation around plate boundaries, and to estimate rotations about a vertical axis of small crustal blocks. Results are encouraging and indicate that plate velocities averaged over a few years are similar to velocities averaged over millions of years by the methods previously mentioned (Gordon and Stein, 1992; Smith et al., 1994). For instance, SRL velocities for the North

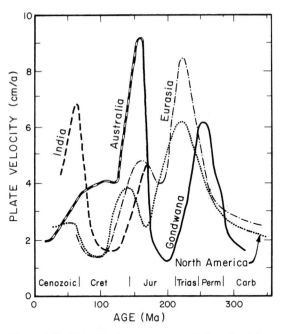

Figure 1.17 Velocities (root mean square) of continental plates calculated from apparent polar wander paths for the last 300 My. Modified after Piper (1987).

American plate of 1.5–2 cm/y compare favourably with magnetic anomaly results averaged over 3 My. Both VLBI and GPS data suggest that the motion of the Pacific plate relative to the Eurasian and North American plates is about ten per cent faster than that estimated from magnetic anomaly data, suggesting that the Pacific plate has speeded up over the past few millions of years (Argus and Heflin, 1995).

Plate driving forces

Although the question of what drives the Earth's plates has stirred a lot of controversy in the past, we now seem to be converging on an answer. Most investigators agree that plate motions must be related to thermal convection in the mantle, although a generally accepted model relating the two processes remains elusive. The shapes and sizes of plates and their velocities exhibit large variations and do not show simple geometric relationships to convective flow patterns. Most computer models, however, indicate that plates move in response chiefly to slab-pull forces as plates descend into the mantle at subduction zones, and that ocean-ridge push forces or stresses transmitted from the asthenosphere to the lithosphere are very small (Vigny et al., 1991; Lithgow-Bertelloni and Richards, 1995). In effect, stress distributions are consistent with the idea that at least oceanic plates are decoupled from underlying asthenosphere (Wiens and Stein, 1985). Ridge-push forces are caused by two factors (Spence, 1987):

1   horizontal density contrasts resulting from cooling and thickening of the oceanic lithosphere as it moves away from ridges
2   the elevation of the ocean ridge above the surrounding sea floor.

The slab-pull forces in subduction zones reflect the cooling and negative buoyancy of the oceanic lithosphere as it ages. The gabbro–eclogite and other high-pressure phase transitions that occur in descending slabs also contribute to slab-pull by increasing the density of the slab.

Using an analytical torque balance method, which accounts for interactions between plates by viscous coupling to a convecting mantle, Lithgow-Bertelloni and Richards (1995) show that the slab-pull forces amount to about ninety-five per cent of the net driving forces of plates. Ridge push and drag forces at the base of the plates are no more than five per cent of the total. Computer models using other approaches and assumptions also seem to agree that slab-pull forces dominate (Vigny et al., 1991; Carlson, 1995). Although slab-pull cannot initiate subduction, once a slab begins to sink the slab-pull force rapidly becomes the dominant force for continued subduction.

Geomagnetism

Rock magnetization

To understand the magnetic evidence for seafloor spreading, it is necessary to understand how rocks become magnetized in the Earth's magnetic field. When a rock forms, it may acquire a magnetization parallel to the ambient magnetic field referred to as **primary magnetization**. Information about both the direction and intensity of the magnetic field in which a rock formed can be obtained by studying its primary magnetization. The most important minerals controlling rock magnetization are magnetite and hematite. However, it is not always easy to identify primary magnetization in that rocks often acquire later magnetization known as **secondary magnetization**, which must be removed by demagnetization techniques prior to measuring primary magnetization.

Magnetization measured in the laboratory is called **natural remanent magnetization** or **NRM**. Rocks may acquire NRM in several ways, of which only three are important in paleomagnetic studies (Bogue and Merrill, 1992; Dunlop, 1995):

1   **Thermal remanent magnetization (TRM)**. TRM is acquired by igneous rocks as they cool through a blocking temperature for magnetization of the constituent magnetic mineral(s). This temperature, known as the **Curie temperature**, ranges between 500 °C and 600 °C for iron oxides, and is the temperature at which magnetization is locked into the rock. The direction of TRM is almost always parallel and proportional in intensity to the applied magnetic field.

2 **Detrital remanent magnetization (DRM)**. Clastic sediments generally contain small magnetic grains, which became aligned in the ambient magnetic field during deposition or during compaction and diagenesis of clastic sediments. Such magnetization is known as DRM.

3 **Chemical remanent magnetization (CRM)**. CRM is acquired by rocks during secondary processes if new magnetic minerals grow. It may be produced during weathering, alteration, or metamorphism.

NRM is described by directional and intensity parameters. Directional parameters include declination, or the angle with respect to true north, and the inclination, or dip from the horizontal. A paleomagnetic pole can be calculated from the declination and inclination determined from a given rock.

Reversals in the Earth's magnetic field

Some rocks have acquired NRM in a direction opposite to that of the Earth's present magnetic field. Such magnetization is known as **reverse magnetization** in contrast to **normal magnetization** which parallels the Earth's present field. Experimental studies show that simultaneous crystallization of some Fe–Ti oxides with different Curie temperatures can cause these minerals to become magnetized with an opposite polarity to the ambient field. This self-reversal magnetization is related to ordering and disording of Fe and Ti atoms in the crystal lattice. Although self-reversal has occurred in some young lava flows, it does not appear to be a major cause of reverse magnetization in rocks. The strongest evidence for this comes from correlation of reverse magnetization between different rock types from widely separated localities. For instance, reversed terrestrial lava flows correlate with reversed deep-sea sediments of the same age. It is clear that most reverse magnetization is acquired during periods of reverse polarity in the Earth's magnetic field.

One of the major discoveries in paleomagnetism is that stratigraphic successions of volcanic rocks and deep-sea sediment cores can be divided into sections that show dominantly reverse and normal magnetizations. **Polarity intervals** are defined as segments of time in which the magnetic field is dominantly reversed or dominantly normal. Using magnetic data from volcanic rocks and deep-sea sediments, the Geomagnetic Time Scale was formulated (Cox, 1969), extending to about 5 Ma (Figure 1.18). Although polarity intervals of short duration (< 50 000 years) cannot be resolved with K–Ar dating of volcanic rocks, they can be dated by other methods in deep-sea sediments, which contain a continuous (or nearly continuous) record of the Earth's magnetic history for the last 100–200 My. The last reversal in the magnetic field occurred about 20 ka (the Laschamp).

Two types of polarity intervals are defined on the basis of their average duration: a polarity event or **subchron** (10^5–10^6 y) and a polarity epoch or **chron** (10^6–10^7 y). A polarity chron may contain several-to-

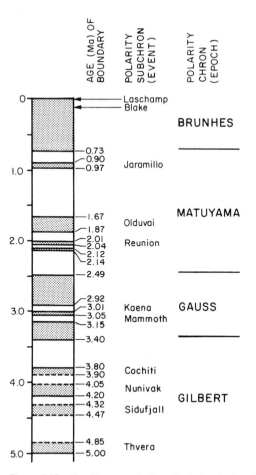

Figure 1.18 The Geomagnetic Time Scale for the last 5 My. Grey pattern, normal polarity; white, reversed polarity.

many polarity subchrons and can be dominantly normal (e.g., the Brunhes), dominantly reversed (e.g., the Matuyama), or mixed (Figure 1.18). Larger intervals (10^7–10^8 y) with few if any reversals are known as **superchrons**. Based on the distribution of oceanic magnetic anomalies, it is possible to extrapolate the Geomagnetic Time Scale to more than 100 Ma. Independent testing of this extrapolation from dated basalts indicates the predicted time scale is correct to within a few percentage points to at least 10 Ma. Results suggest that over the last 80 My the average length of polarity subchrons has decreased with time. Reversals in the Earth's field are documented throughout the Phanerozoic, although the Geomagnetic Time Scale cannot be continuously extrapolated beyond about 200 Ma, the age of the oldest oceanic crust. Reversals, however, have been indentified in rocks as old as 3.5 Ga.

The percentage of normal and reverse magnetization for any increment of time has also varied with time. The Mesozoic is characterized by dominantly normal polarities while the Paleozoic is chiefly reversed (Figure

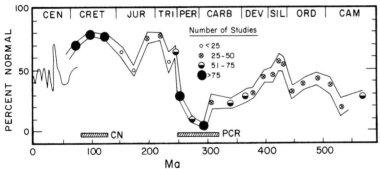

Figure 1.19 Distribution of
magnetic reversals during the
Phanerozoic averaged over 50 My
intervals. Also shown are the
Cretaceous (CN) and Permian–
Carboniferous (PCR) superchrons.
Modified after Piper (1987).

1.19). Periodic variations are suggested by the data at
about 300, 110 and 60 Ma (Irving and Pullaiah, 1976).
Statistical analysis of reversals in the magnetic field
indicate a strong periodicity at about 30 My. Two major
superchrons are identified in the last 350 My. These are
the Cretaceous normal (CN) and Permian–Carboniferous
reversed (PCR) superchrons (Figure 1.19). Statistical
analysis of the youngest and best-defined part of the
Geomagnetic Time Scale (< 185 Ma) shows an almost
linear decrease in the frequency of reversals to the Cretaceous, reaching zero in the CN superchron. The inversion frequency appears to have reached a maximun about
10 Ma and has been declining to the present. Causes of
changes in reversal frequency are generally attributed to
changes in the relief and/or electrical conductivity along
the core–mantle boundary. Both of these parameters are
temperature-dependent and require long-term cyclical
changes in the temperature at the base of the mantle.
This, in turn, implies that heat transfer from the lower
mantle is episodic. A possible source of episodic heat
loss from the core is latent heat released as the inner
core grows by episodic crystallization of iron.

Vine and Matthews (1963) were the first to show that
linear magnetic anomaly patterns on the ocean floors
correlate with reversed and normal polarity intervals in
the Geomagnetic Time Scale. This correlation is shown
for a segment of the East Pacific rise in Figure 1.20. The
correlations with polarity intervals are indicated at the
bottom of the figure. A model profile for a half spreading rate of 4.4 cm/y is also shown and is very similar to
the observed profile. Both the Jaramillo and Olduvai
subchrons produce sizeable magnetic anomalies in the
Matuyama chron. The Kaena and Mammoth subchrons
in the Gauss chron are not resolved, however. The lower
limit of resolution of magnetic subchrons in anomaly
profiles with current methods is about 30 000 years. Although the distribution of magnetic anomalies seems to
correlate well with polarity intervals, the amplitudes of
anomalies can vary significantly between individual profiles. Such variation reflects, in part, inhomogeneous distribution of magnetite in oceanic basalts. Results suggest
that most of the magnetization resides in the upper 0.5
km of basalts in the oceanic crust (Harrison, 1987). Less
than twenty per cent of the magnetization probably occurs
below 0.5 km depth.

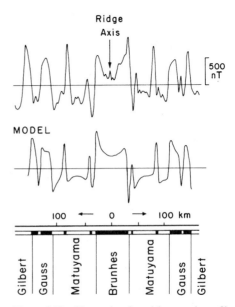

Figure 1.20 Observed and model magnetic profiles across
the East Pacific rise at 51 ° S latitude and corresponding
correlation with magnetic polarity intervals for a spreading
rate of 4.4 cm/y. After Vine (1966).

Several detailed studies across magnetic reversals in
stratigraphic successions provide information on the timing and details of the magnetic field behaviour during
reversals (Bogue and Merrill, 1992). In those few sections where magnetic reversals are well dated, they occur over time intervals of as short as 1000 years and as
long as 10 000 years. The best estimates seem to be near
4000 years for the duration of a reversal. One record for
a section across Tertiary basalt flows at Steens Mountain in Oregon (~ 15 Ma) is shown in Figure 1.21. During
this reversal, the intensity drops significantly and rapid
and irregular changes in inclination and declination occur. In general, the field intensity drops 10–20 per cent
during a reversal and suggests that a decrease in the
dipole field precedes a reversal (Bogue and Merrill,
1992). The actual intensity drop during a transition is
latitude-dependent. Perhaps the most striking observation from the paleointensity record of the magnetic field

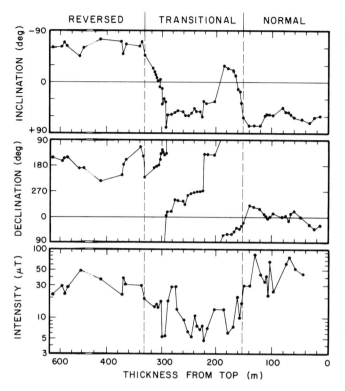

Figure 1.21 Record of a mid-Tertiary magnetic reversal in successive basalt flows from Steens Mountain, Oregon. After Prevot et al. (1985).

is the existence of asymmetric sawtooth patterns associated with reversals (Valet and Meynadier, 1993). The magnetic field decreases slowly during a typical polarity chron (likes the Matuyama), but recovers rapidly after the reversal, producing a sawtooth pattern with time, the origin of which is not well-understood.

In sediment successions, where the record of magnetic reversals is often more complete, two scales of changes in declination are resolved (Valet et al., 1986). Short-term changes (< 100 y), which typically show < 10° change in declination, increase in amplitude to 40–60° during a reversal. This is generally interpreted to reflect an increase in turbulence in the outer core during a magnetic transition. Longer-term changes with a periodicity of 2000–4500 years are observed in both declination and inclination and appear to be caused by variations in the non-dipole field. These are not affected by magnetic reversals. Data suggest that the transitional field during a reversal is almost certainly multipolar.

The origin of magnetic reversals, although poorly understood, will be discussed more fully in Chapter 4.

Paleomagnetism

Paleomagnetism is the study of NRM in rocks and has an important objective: the reconstruction of the direction and strength of the geomagnetic field over geologic time. To accomplish this goal, orientated samples are collected in the field, the NRM is measured in the labor-atory with a magnetometer, and calculated paleopole positions are plotted on equal-area projections. For the paleopole to have significance in terms of ancient plate motions, however, numerous problems must be addressed as follows.

(1) *Secular variation.* Variations in the Earth's magnetic field on time scales > 5 years are known as **secular variation**, and reliable observations on secular variation are available for over 300 years (Lund, 1996). Such variations include a westward drift of the magnetic field of about 0.2 °/y, a decay in the dipole component of about seven per cent since 1845, and drift of the magnetic poles. Using dated archeological remains, it is possible to estimate the trajectory of the North Magnetic Pole for several thousand years. Results show a path around and within a few degrees of the rotational pole, such that the average position of the magnetic pole is equal to the rotational pole. This line of evidence suggests that the magnetic and rotational poles have been coincident in the geologic past.

(2) *Dipole field.* The Earth's present field can be approximated as a geocentric dipole inclined at 11.5° to the rotational axis. The non-dipole field, which remains after subtraction of the dipole field from the measured field, is about five per cent of the total field. In reconstructing ancient pole positions, a dipole field is assumed to have existed in the geologic past. The consistency of pole positions from different parts of a continent at the same geologic time, and the world-wide agreement of

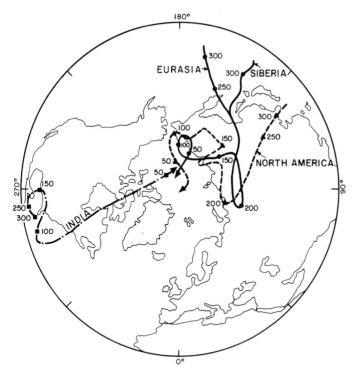

Figure 1.22 Apparent polar wander paths of North America, Eurasia, Siberia, and India for the last 300 My. Ages given in millions of years ago.

paleoclimatic regimes with inferred paleolatitude calculated from paleomagnetic studies, supports this assumption. Because of the axial symmetry of the dipole field, however, paleolongitudes cannot be determined from paleomagnetic data. Thus, plate motions parallel to lines of latitude cannot be detected by the paleomagnetic method.

(3) *True polar wander.* Ancient pole positions are generally illustrated by plotting an apparent polar wander path for a continent (or segment thereof) relative to its present position (Figure 1.22). It is referred to as an **apparent polar wander path** (or **APW path**) because it may result from either plate motions or true polar wander, or both. True polar wander results from changes in the magnetic pole position caused by processes in the liquid core. Movement of the entire lithosphere with respect to the mantle or of the mantle relative to the core can produce the same effect. To separate the effects of plate motion and true polar wander requires one of three approaches: (1) evaluation of the net motion of the entire lithosphere relative to the Earth's spin axis (the vector-sum method), (2) The mean-lithosphere method based on separating random plate motions from the mean motion of the lithosphere, and (3) the hotspot method, based on comparing plate motions to hotspots that are assumed to be fixed relative to the paleomagnetic axis (O'Connor and Duncan, 1990; Duncan and Richards, 1991). Comparison of plate motions assuming a fixed hotspot framework indicates that true polar wander has amounted to about 12° since the Early Tertiary. Because true polar wander affects equally all calculated poles of

a given age in the same way, however, it does not usually influence relative plate motions deduced from APW paths. However, for a single supercontinent, or for two continents with similar APW paths, the affects of true polar wander and plate motion cannot be distinguished.

(4) *Normal or reversed pole.* For an isolated paleopole, it is not possible to tell if the pole is reversed or normal. This decision must be based on other information such as paleoclimatic data or an APW path constructed from other geographic localities, but including the age of the measured sample.

(5) *Deformation.* The most reliable pole positions come from nearly flat-lying rock units. In rocks that acquired NRM prior to deformation, it may be possible to estimate pole positions by 'unfolding' the rocks. Strain associated with folding also must be removed as it affects NRM. It is only in strata which have undergone rather simple folding that these structural corrections can be made. However, in complexly deformed metamorphic terranes, $^{40}Ar/^{39}Ar$ dating indicates that major NRM is usually acquired after deformation, possibly during regional uplift. Thus, in many instances it is possible to obtain meaningful post-deformational paleopoles from strongly deformed and metamorphosed rocks in cratons.

(6) *Dating of NRM.* The Curie temperatures of major magnetic minerals (Fe–Ti oxides) lie between 500 °C and 600 °C, and it is possible to date the magnetization using an isotopic system with a blocking temperature similar to the Curie temperatures. The $^{40}Ar/^{39}Ar$ method has proved useful in this respect, generally using hornblende which has an argon-blocking temperature of about

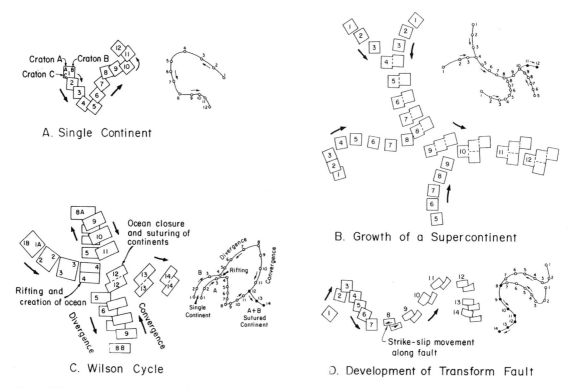

A. Single Continent

B. Growth of a Supercontinent

C. Wilson Cycle

D. Development of Transform Fault

Figure 1.23 Apparent polar wander paths for various tectonic scenarios. Numbers are relative time, 1 being oldest. Modified after Piper (1987).

500 °C. It is very difficult, however, to assign meaningful $^{40}Ar/^{39}Ar$ ages to rocks that have multiple magnetizations. In this light, an exciting new field of research in rock magnetization is related to measurement of magnetic domains on a microscopic scale for individual minerals. From such studies, minerals that carry different fractions of the NRM can be identified and multiple magnetizations can be potentially resolved. Further, using laser $^{40}Ar/^{39}Ar$ dating, it may be possible to date individual NRM components in a rock, and thus to track magnetic poles for individual crystals.

(7) *Separation of remanence components.* One of the most difficult problems in paleomagnetic studies is that of separating, identifying and establishing the sequence of acquisition of multiple magnetizations. This is particularly a problem in metamorphic rocks where TRM or DRM may be overprinted with one or more CRMs of metamorphic origin. Standard techniques for separating magnetic components include the stepwise treatment of samples either by alternating magnetic fields or by heating. Depending on the history of a rock, it may be difficult or impossible to separate and date the various magnetic components, although the laser $^{40}Ar/^{39}Ar$ technique offers promise for overcoming this problem.

In using the paleomagnetic method, APW paths of continents (or segments thereof) are plotted on equal-area projections of the Earth, generally relative to the present positions of continents. If continents show the same APW paths, it is clear that they did not move relative to each other. The fact that paths diverge from each other going back in time, however, indicates that plates have moved separately for the last 100–200 My (Figure 1.22). Europe and Siberia moved independently until the Triassic, when their APW paths converged, reflecting collision and suturing of these plates to the growing Eurasian plate. To illustrate the use of APW paths in reconstructing plate motions, four tectonic scenarios with corresponding APW paths are shown in Figure 1.23. The simplest case is for a single continent in which paleopole positions from various cratons lie on the same APW path (case A). In this case, the cratons have been part of the same continent over the period of time represented by the APW path. Such paths characterize most Precambrian cratons within a given continent for the last 600 My, central and eastern Asia being the major exception. Case B illustrates the aggregation of a supercontinent by successive collisions of continents. Two continents collide at time 4 and are joined by two more continents at times 8 and 10 to form a supercontinent. A Wilson Cycle (the opening and closing of an ocean basin) is illustrated in case C. At time 4, a continent is rifted apart and an ocean basin opens, reaching a maximum width at time 8, when the two continental fragments begin to close again. The original continent is reformed at time 12 by collision of these fragments. The offset in APW paths from times 1–4 reflects the fact that the fragments did not return to their original positions. The opening and closing of the Iapetus Ocean during the Paleozoic is an example of a Wilson Cycle that can be tracked with APW paths. Case D illustrates offset along a major continental transform fault where motion begins at time 8.

One thing that is clear from paleomagnetic studies is that continents grow by terrane collisions and that supercontinents, such as Pangea, have formed and dispersed several times in the geologic past. The paleomagnetic method has the unique property of being able to see through the break up of Pangea 200 Ma. In principle, it is possible to reconstruct continental positions prior to 200 Ma provided the problems listed above can be overcome. Applications to the Precambrian, however, have not met with much success, due principally to the complex thermal histories of most Precambrian terranes and to problems with separating and dating the various magnetizations. Published APW paths for Precambrian continents, often using the same data base, can vary considerably. Despite inherent errors in interpretation, paleomagnetic data constrain the amount of independent motion between adjacent Precambrian crustal provinces. For instance, results from North America and Baltica (Scandinavia) suggest that these continents were part of the same supercontinent during the Early Proterozoic and again, with somewhat different configurations, during the Late Proterozoic. Also, when Baltica is returned to its Early Proterozoic paleomagnetic position, crustal provinces can be matched from northeastern Canada across southern Greenland into the Baltic shield. In this same configuration, the 1.8–1.6 Ga juvenile crustal provinces in southern North America appear to continue eastward into the Baltic shield. Paleomagnetic results also indicate that Africa and the San Francisco craton in Brazil were joined during the Early Proterozoic, and that Australia, India and Antarctica were part of the same supercontinent during the Late Proterozoic.

Hotspots and plumes

General characteristics

The hotspot model (Wilson, 1963), which suggests that linear volcanic chains and ridges on the sea floor form as oceanic crust moves over relatively stationary magma sources, has been widely accepted in the geological community. **Hotspots** are generally thought to form in response to mantle plumes, which rise like salt domes in sediments, through the mantle to the base of the lithosphere (Duncan and Richards, 1991). Partial melting of plumes in the upper mantle leads to large volumes of magma, which are partly erupted (or intruded) at the Earth's surface. Hotspots may also be important in the break up of supercontinents.

Hotspots are characterized by the following features:

1 In ocean basins, hotspots form topographic highs of 500–1200 m with typical widths of 1000–1500 km. These highs are probably indirect manifestations of ascending mantle plumes.
2 Many hotspots are capped by active or recently active volcanoes. Examples are Hawaii and Yellowstone Park in the western United States.
3 Most oceanic hotspots are characterized by gravity highs reflecting the rise of more dense mantle material from the mantle. Some, however, have gravity lows.
4 One or two aseismic ridges of mostly extinct volcanic chains lead away from many oceanic hotspots. Similarly, in continental areas, the age of magmatism and deformation may increase with distance from a hotspot. These features are known as **hotspot tracks**.
5 Most hotspots have high heat flow, probably reflecting a mantle plume at depth.

Somewhere between forty and 150 active hotspots have been described on the Earth. The best documented hotspots have a rather irregular distribution occurring in both oceanic and continental areas (Figure 1.24). Some occur on or near the ocean ridges, such as Iceland, St. Helena and Tristan in the Atlantic basin, while others occur near the centres of plates, such as Hawaii. How long do hotspots last? One of the oldest hotspots is the Kerguelen hotspot in the southern Indian Ocean which began to produce basalts about 117 Ma. Most modern hotspots, however, date from < 100 Ma. The number of hotspots seems to correlate with geoid height as discussed further in Chapter 4. The large number of hotspots in and around Africa and in the Pacific basin correspond to the two major geoid highs (Figure 1.24) (Stefanick and Jurdy, 1984). The fact that the geoid highs appear to reflect processes in the deep mantle or core supports the idea that hotspots are caused by mantle plumes rising from the deep mantle.

A major problem which has stimulated a great deal of controversy is whether hotspots remain fixed relative to plates (Duncan and Richards, 1991). If they have remained fixed, they provide a means of determining absolute plate velocities. The magnitude of interplume motion can be assessed by comparing the geometry and age distribution of volcanism along hotspot tracks with reconstructions of past plate movments based on relative-motion data discussed previously. If hotspots are stationary beneath two plates, then the calculated motions of both plates should follow hotspot tracks observed on them. The close correspondence between observed and modelled tracks on the Australian and African plates (Figure 1.25) supports the idea that hotspots are fixed on these plates and are maintained by deeply-rooted mantle plumes. Also, the almost perfect fit of volcanic chains in the Pacific plate with a pole of rotation at 70 °N, 101 °W and a rate of rotation about this pole of about 1 °/My for the last 10 My suggests that Pacific hotspots have remained fixed relative to each other over this period of time.

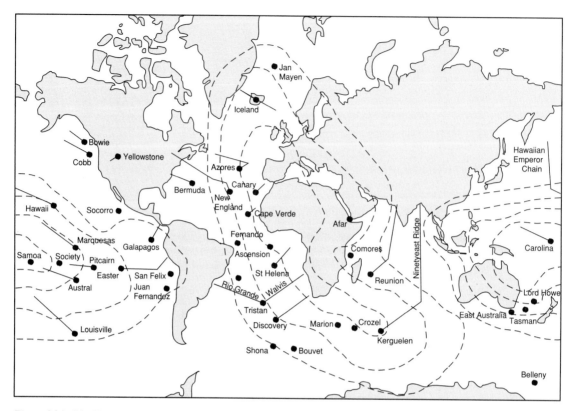

Figure 1.24 Distribution of hotspots (dots) and hotspot tracks (lines). Dashed contours show geoid anomalies. Modified after Crough (1983).

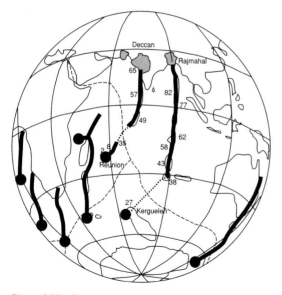

Figure 1.25 Computer-generated hotspot tracks for the Indian Ocean and South Atlantic basins, showing calculated (heavy lines) and observed hotspot trajectories. Current hotspots at black dots and past ages of hotspot basalts given in millions of years ago. After Duncan (1991).

If all hotspots have remained fixed with respect to each other, it should be possible to superimpose the same hotspots in their present positions on their predicted positions at other times in the last 150–200 My. However, except for hotspots in the near proximity of each other or on adjacent plates (as illustrated above), it is not possible to do this, suggesting that hotspots move in the upper mantle (Duncan and Richards, 1991; Van Fossen and Kent, 1992). In comparing Atlantic with Pacific hotspots, there are significant differences between calculated and observed hotspot tracks (Molnar and Stock, 1987). Rates of interplate hotspot motion, however, are more than an order of magnitude less than plate velocities. For instance, using paleolatitudes deduced from seamounts, Tarduno and Gee (1995) show that Pacific hotspots have moved relative to Atlantic hotspots at a rate of only 30 mm/y.

Hotspot tracks

Chains of seamounts and volcanic islands are common in the Pacific basin, and include such well-known island chains as the Hawaiian–Emperor, Line, Society, and Austral islands, all of which are subparallel to either the Emperor or Hawaiian chains and approximately perpendicular to the axis of the East Pacific rise (Figure 1.1/

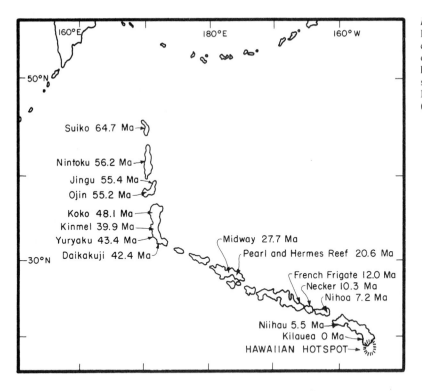

160°E   180°E   160°W

50°N

Suiko 64.7 Ma

Nintoku 56.2 Ma
Jingu 55.4 Ma
Ojin 55.2 Ma

Koko 48.1 Ma
Kinmei 39.9 Ma
Yuryaku 43.4 Ma
Daikakuji 42.4 Ma

30°N

Midway 27.7 Ma
Pearl and Hermes Reef 20.6 Ma

French Frigate 12.0 Ma
Necker 10.3 Ma
Nihoa 7.2 Ma

Niihau 5.5 Ma
Kilauea 0 Ma
HAWAIIAN HOTSPOT→

Figure 1.26 Map of the Hawaiian–Emperor volcanic chain showing locations and ages of last volcanism. The Hawaiian hotspot today is centred just south of the island of Hawaii. Modified after Molnar and Stock (1987).

Plate 1 and Figure 1.24). Closely-spaced volcanoes form aseismic ridges such as the Ninetyeast ridge in the Indian Ocean and the Walvis and Rio Grande ridges in the South Atlantic (Figure 1.24). Isotopic dates demonstrate that the focus of volcanism in the Hawaiian chain has migrated to the southeast at a linear rate of about 10 cm/y for the last 30 My (Figure 1.26). The bend in the Hawaiian–Emperor chain at about 43 Ma may be caused either by a change in spreading direction of the Pacific plate or, possibly, by the Emperor segment of the chain being produced by a different hotspot (Norton, 1995). Similar linear decreases in the age of volcanism occur towards the south-east in the Marquesas, Society, and Austral islands in the South Pacific, with rates of migration of the order of 11 cm/y, and in the Pratt-Welker seamount chain in the Gulf of Alaska at a rate of about 4 cm/y. The life spans of hotspots vary and depend on such parameters as plume size and the tectonic environment into which a plume is emplaced. On the Pacific plate, three volcanic chains were generated by hotspots between 70 and 25 Ma, whereas twelve chains have been generated in the last 25 My.

Hotspots may interact with lithospheric plates in a variety of ways, some of which are illustrated in Figure 1.27. If ocean plate motions relative to hotspots are small, large amounts of magma are erupted, forming a large island with thick crust as exemplified by Iceland and submarine plateaux such as the Shatsky rise and the Ontong-Java plateau in the Pacific basin (Figure 1.27, a). If plate motion is erratic in direction and rate, irregular clusters of volcanoes may form at hotspots (b), as

(1 – Lithosphere   2 – Asthenosphere   3 – Deep mantle)

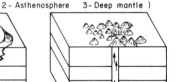

a. No plate/hotspot motion   b. Erratic plate/hotspot motion

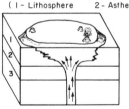

c. Transform intersection   d. Return flow

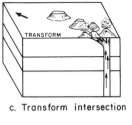

e. Magma storage   f. Bent conduit

Figure 1.27 Possible interactions of lithospheric plates and hotspots. Modified after Epp (1984).

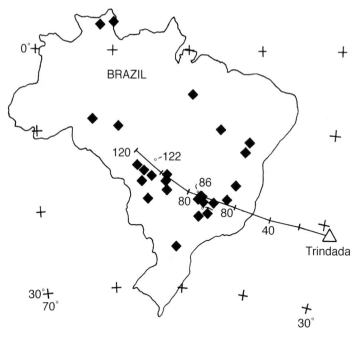

Figure 1.28 Locations of dated kimberlites (small open circles) and major alluvial diamond deposits (black diamonds) in Brazil compared with calculated path of the Trindade hotspot. After Crough et al. (1980).

characterized perhaps by the Mid-Pacific Mountains. Volcanic chains which cross transform faults often show cross-trends of volcanoes parallel to the faults. Such is the case along the Hawaiian ridge where it crosses the Molokai and Murray fracture zones in the Northeast Pacific. Since transform faults are zones of weakness, magma may be injected into them as they pass over hotspots (c). Another interaction is possible if a plume breaks out at the base of the lithosphere and flows from the hotspot back along the weakened hotspot trace (d). This could result in irregularities in the age of volcanism along the hotspot trace as observed in some island chains. Because the lower lithosphere is near its melting temperature, magmas may be stored here for considerable lengths of time. This can lead to eruption of magma (generally small amounts) along the volcanic chain after passing over the hotspot (e). Also hotspot conduit systems may be bent by the plate motion. When conduits are bent to some critical angle, the plume rises and forms a new conduit (f). If an individual conduit is not disconnected from the plume, eruption may continue in volcanoes that have passed over the hotspot. When hotspots form near ocean ridges, they may interact with the ridges, and in some instances, such as the Shona and Discovery hotspots near the Mid-Atlantic ridge (Figure 1.24), they may 'capture' a ridge segment (Small, 1995). Many hotspot–ridge interactions show evidence of distinct ridge jumps in the direction of the hotspot, as the ridge attempts to relocate on the hotspot.

Hotspot tracks also occur in the continents, although less well-defined than in ocean basins due the thicker lithosphere. For example, North America moved northwest over the Great Meteor hotspot in the Atlantic basin

between 125 and 80 Ma (Van Fossen and Kent, 1992). The trajectory of the hotspot is defined by the New England seamount chain in the North Atlantic (Figure 1.1/Plate 1) and by Cretaceous kimberlites and alkalic complexes in New England and Quebec. Dated igneous rocks fall near the calculated position of the hotspot track at the time they formed. In addition, the hotspot appears to have moved south by 11° between 125 and 80 Ma. Geologic data and paleotemperatures indicate that this region was elevated at least 4 km as it passed over the hotspot. When post-Triassic kimberlites from North and South America and Africa are rotated to their position of origin relative to present Atlantic hotspots, the majority appear to have formed within 5° of a mantle hotspot. As an example, the calculated trajectory of the Trindade hotspot east of Brazil (Figure 1.28) matches the locations of three dated kimberlites from Brazil and also roughly coincides with the distribution of alluvial diamond deposits (which are derived from nearby kimberlites). High heat flow, low seismic-wave velocities and densities at shallow depth, and high electrical conductivity at shallow depth beneath Yellowstone National Park in Wyoming reflect a mantle hotspot at this locality (Smith and Braile, 1994). A 600-m high topographic bulge is centred on the Yellowstone caldera and extends across an area 600 km in diameter. Direct evidence of a mantle plume at depth is manifested by anomalously low P-wave velocities that extend to depths of 200 km. The movement of the North American plate over this hotspot during the past 16 My has been accompanied by the development of the Snake River volcanic plain, with the oldest hotspot-related volcanics found in south-eastern Oregon. The movement of volcanism to

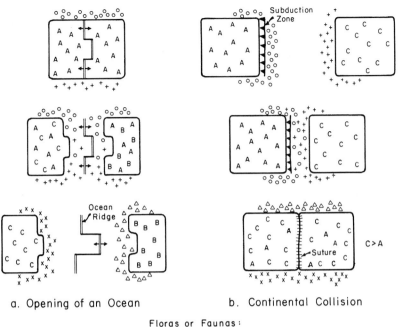

Figure 1.29 Possible evolutionary changes of organisms accompanying the break up (opening of an ocean) and assembly (continental collision) of a supercontinent.

a. Opening of an Ocean

b. Continental Collision

Floras or Faunas:
A, B, C – Terrestrial
o, +, x, △ – Shallow Marine

the northeast at 4.5 cm/y reflects both the southwest movement of the North American plate at 2.5 cm/y and crustal extension of 1–2 cm/y. Other examples of hotspot traces, in part in continental areas, are the Mesozoic granites in Nigeria; the Thulian volcanic chain extending from Iceland to Ireland and recording the opening of the North Atlantic; and the central European volcanic province extending from western Germany to Poland.

Plate tectonics and organic evolution

Plate tectonics has played a major role in the evolution and extinction of some groups of organisms. Modern ocean basins are effective barriers not only to terrestrial organisms, but also to most marine organisms. Larval forms of marine invertebrates, for instance, can only survive several weeks, which means they can travel only 2000–3000 km with modern ocean-current velocities. Hence, the geographic distribution of fossil organisms provides an important constraint on the sizes of oceans between continents in the geologic past. As illustrated in Figure 1.29, when a supercontinent fragments (a), organisms that cannot cross the growing ocean basin become isolated and this can lead to the evolution of diverse groups on each of the separating continents (Hallam, 1974). This affects both marine and terrestrial organisms. Examples are the diversification and specialization of mammals in South America and in Africa following the opening of the South Atlantic in the Late Cretaceous. Similar changes occurred in bivalves of East

Africa and India after the separation of these continents in the Cretaceous and Tertiary. Continental collision (b), on the other hand, removes an oceanic barrier and formerly separate faunal and floral groups mix and compete for the same ecological niches. This leads to extinction of groups that do not successfully compete. Collision also creates a new land barrier between coastal marine populations. If these marine organisms cannot migrate around the continent (due perhaps to climatic barriers), diversification occurs and new populations evolve along opposite coastlines. As an example, during the Ordovician collisions in the North Atlantic many groups of trilobites, graptolites, corals, and brachiopods became extinct. The formation of new arc systems linking continents has the same effect as a collision. For instance, when the Panama arc was completed in the Late Tertiary, mammals migrated between North and South America, which led to the extinction of large endemic mammals in South America. Panama also separated marine populations, which led to divergence of Pacific and Caribbean marine organisms.

The biogeographic distribution of Cambrian trilobites indicates the existence of several continents separated by major ocean basins during the Early Paleozoic. Major faunal province boundaries commonly correlate with suture zones between continental blocks brought together by later collisions. Wide oceans are implied in the Cambrian between Laurentia (mostly North America) and Baltica (northwest Europe), Siberia and Baltica, and China and Siberia, and paleomagnetic data support this interpretation. Minor faunal provincialism within individual blocks probably reflects climatic differences. Studies of Early Ordovician brachiopods indicate that they belonged to at least five distinct geographic provinces, which were reduced to three provinces during the Baltica–Laurentia collision in the Late Ordovician. The opening of the North Atlantic in the Cretaceous resulted in the development of American and Eurasian invertebrate groups from an originally homogeneous Tethyan group. Also similar ammonite populations from East Africa, Madagascar, and India indicate that only shallow seas existed between these areas during the Jurassic.

The similarity of mammals and reptiles in the Northern and Southern Hemispheres prior to 200 Ma demands land connections between the two hemispheres (Hallam, 1973). The break up of Pangea during the early to middle Mesozoic led to diversification of birds and mammals and the evolution of unique groups of mammals (e.g., the marsupials) in the Southern Hemisphere (Hedges et al., 1996). On the other hand, the fact that North America and Eurasia were not completely separated until Early Tertiary accounts for the overall similarity of Northern Hemisphere mammals today. When Africa, India, and Australia collided with Eurasia in the mid-Tertiary, mammalian and reptilian orders spread both ways, and competition for the same ecological niches was keen. This competition led to the extinction of thirteen orders of mammals.

Plant distributions are also sensitive to plate tectonics. The most famous are the *Glossoptera* and *Gangemoptera* flora in the Southern Hemisphere. These groups range in age from Carboniferous to Triassic and occur on all continents in the Southern Hemisphere as well as in northeast China, confirming that these continents were connected at this time. The complex speciation of these groups could not have evolved independently on separate landmasses. The general coincidence of Late Carboniferous and Early Permian ice sheets and the *Glossoptera* flora appears to reflect adaptation of *Glossoptera* to relatively temperate climates and its rapid spread over high latitudes during the Permian. The break up of Africa and South America is reflected by the present distributions of the rain forest tree *Symphonia globulifera* and a semi-arid leguminous herb *Teramnus uncinatus*, both of which occur at approximately the same latitudes on both sides of the South Atlantic (Melville, 1973). Conifer distribution in the Southern Hemisphere reflects continental break up, as is evidenced by the evolution of specialized groups on dispersing continents after supercontinent break up in the early Mesozoic.

Although a correlation between the rate of taxonomic change of organisms and plate tectonics seems to be well established, the actual causes of such changes are less well agreed-upon. At least three different factors have been suggested to explain rapid increases in the diversity of organisms during the Phanerozoic (Hallam, 1973; Hedges et al., 1996):

1 An increase in the areal distribution of particular environments results in an increase in the number of

ecological niches in which organisms can become established. For instance, an increasing diversification of marine invertebrates in the Late Cretaceous may reflect increasing transgression of continents, resulting in an areal increase of the shallow-sea environment in which most marine invertebrates live.

2 Continental fragmentation leads to morphological divergence because of genetic isolation.

3 Some environments are more stable than others in terms of such factors as temperature, rainfall, and salinity. Studies of modern organisms indicate that stable environments lead to intensive partitioning of organisms into well-established niches and to corresponding high degrees of diversity. Environmental instability, decreases in environmental area, and competition of various groups of organisms for the same ecological niche can result in extinctions.

Supercontinents

Introduction

Although continental drift was first suggested in the seventeenth century, it did not receive serious scientific investigation until the beginning of the twentieth century. Wegener (1912) is usually considered the first to have formulated the theory precisely and to have suggested the existence of a supercontinent 200 Ma. A **Supercontinent** is a large continent that includes several or all of the existing continents. Wegener pointed out the close match of opposite coastlines of continents and the regional extent of the Permo–Carboniferous glaciation in the Southern Hemisphere. DuToit (1937) was the first to propose an accurate fit for the continents based on geological evidence. Matching of continental borders, stratigraphic sections, and fossil assemblages are some of the earliest methods used to reconstruct continental positions. Today, in addition to these methods, we have polar wandering paths, seafloor spreading directions, hotspot tracks, paleoclimatic data, and correlation of crustal provinces. The geometric matching of continental shorelines was one of the first methods used in reconstructing continental positions. Later studies have shown that matching of the continents at the edge of the continental shelf or continental slope results, as it should, in better fits than the matching of shorelines, which reflect chiefly the flooded geometry of continental margins. The use of computers in matching continental borders has resulted in more accurate and objective fits. Isotopic dating of Precambrian crustal provinces also provides strong evidence for continental drift and of the existence of a supercontinent in the early Mesozoic. One of the most striking examples is the continuation of the Precambrian provinces from western Africa into eastern South America. As we shall see in Chapter 6, stands of low sea level correlate with supercontinents, and high sea level accompanies break up and dispersal of supercontinents (Kominz and Bond, 1991). One of the most

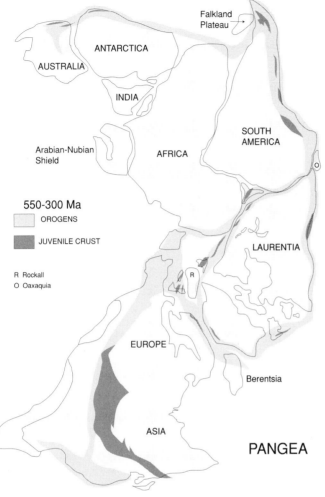

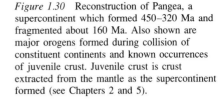

Figure 1.30 Reconstruction of Pangea, a supercontinent which formed 450–320 Ma and fragmented about 160 Ma. Also shown are major orogens formed during collision of constituent continents and known occurrences of juvenile crust. Juvenile crust is crust extracted from the mantle as the supercontinent formed (see Chapters 2 and 5).

definitive matching tools in reconstructing plate positions in a former supercontinent is a piercing point. A **piercing point** is a distinct geologic feature such as a fault or terrane that strikes at a steep angle to a rifted continental margin, the continuation of which should be found on the continental fragment rifted away.

The youngest supercontinent is **Pangea**, which formed between 450 and 320 Ma and includes most of the existing continents (Figure 1.30). Pangea began to fragment about 160 Ma and is still dispersing today, although a new supercontinent may be beginning to form as some major continental collisions have occurred in the last 100 My (such as the India–Tibet collision). **Gondwana** is a Southern Hemisphere supercontinent comprised principally of South America, Africa, Arabia, Madagascar, India, Antarctica, and Australia (Figure 1.31). It formed in the latest Proterozoic and was largely completed by the Early Cambrian (750–550 Ma) (Unrug, 1993). Later it became incorporated in Pangea. **Laurentia**, which is also part of Pangea, includes most of North America, Scotland and Ireland north of the

Caledonian suture, Greenland, Spitzbergen, and the Chukotsk Peninsula of eastern Siberia. The earliest well-documented supercontinent is **Rodinia**, which formed at 1.3–1.0 Ga, fragmented at 750–600 Ma, and appears to have included almost all of the continents in a configuration quite different from Pangea (Figure 1.32). Although the existence of older supercontinents is likely, their configurations are not known. Geologic data strongly suggest the existence of supercontinents in the Early Proterozoic and in the Late Archean. Current thinking is that supercontinents have been episodic, giving rise to the idea of a supercontinent cycle (Nance et al., 1986). A **supercontinent cycle** consists of the rifting and break up of one supercontinent, followed by a stage of reassembly in which dispersed cratons collide to form a new supercontinent, with most or all fragments in different configurations from the older supercontinent (Hartnady, 1991). The assembly process generally takes much longer than fragmentation, and often overlaps in time with the initial phases of rifting that mark the beginning of a new supercontinent dispersal phase. During

GONDWANA

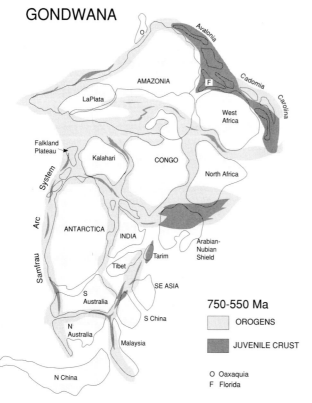

Figure 1.31 Reconstruction of Gondwana, a supercontinent which formed 750–550 Ma and became part of Pangea in the late Paleozoic. For other information, see caption to Figure 1.30.

the Paleozoic, for example, terranes collided along the Pacific margin of Gondwana, while rifting of terranes now found in Laurentia and Eurasia continued along the opposite margin of Gondwana.

The supercontinent cycle provides a record of the processes that control the formation and redistribution of continental crust throughout Earth history. Through magmatism and orogeny associated with the supercontinent cycle, this cycle influences both elemental and isotopic geochemical cycles, climatic distributions, and changing environments which affect the evolution of organisms. We shall return to the supercontinent cycle in later chapters when we discuss mantle processes and history, paleoclimates, and the evolution of the atmosphere–ocean–life system. It is useful at this point to review briefly what we know about the three most recent supercontinents: Gondwana, Pangea and Rodinia.

Gondwana and Pangea

Between 750 and 600 Ma, the large supercontinent Rodinia began to break up, followed soon after by the collision of dispersed cratons to form Gondwana. It is not clear yet if East Gondwana (Australia, India, Antarctica) separated from Rodinia as a coherent block or underwent at least some assembly at the same time as West Gondwana (Africa, South America). The assembly

of most of Gondwana occurred between about 750 and 550 Ma (Figure 1.31), although the earliest collision between the San Francisco craton in Brazil and the Congo craton in Africa occurred in the Early Proterozoic. As Rodinia fragmented, new passive margins formed along the edges of North America, Siberia, and Baltica. In addition, an unknown number of microcontinents may have been rifted from Gondwana at this time. Paleomagnetic and paleoclimatic data indicate that during the early Paleozoic most continents remained at low, equatorial latitudes (Scotese, 1984; Jurdy et al., 1995). Gondwana, however, also extended into southern polar latitudes and was glaciated during the Ordovician. Extensive carbonate platforms in North America, Siberia, and North and South China support paleomagnetic results that indicate near-equatorial positions in the early Paleozoic.

During the Cambrian and Ordovician, subduction zones existed along the south coast of Laurentia, the west coast of Baltica and along the coast of several microcontinents (Avalonia, Cadomia) west of Baltica, separated by a major ocean, the Iapetus Ocean. Florida was part of Gondwana at this time and the United Kingdom was not yet united: i.e., Cadomia (England and Wales) was on one side of the Iapetus Ocean and Scotland on the other. Newfoundland and parts of New England and northeast Canada were also separated by

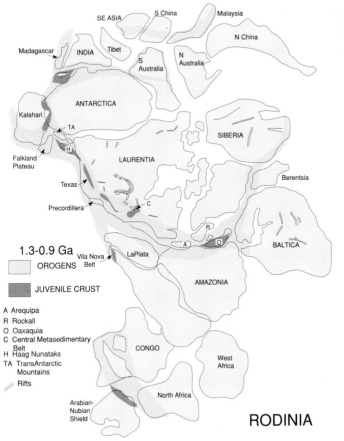

Figure 1.32 Reconstruction of Rodinia, a supercontinent which formed 1.3–1.0 Ga and fragmented at 750–600 Ma. Modified after Hoffman (1991). For other information, see caption to Figure 1.30.

this ocean. The Chukotsk Peninsula in extreme north-east Siberia was connected to northern Alaska (Brooks–Chukotsk) and this landmass was part of Laurentia. Mongolia, which had been rifted from Siberia in the late Proterozoic, had a marginal arc and was closing on Siberia in the earliest Paleozoic, and an extensive arc system ran between Baltica and Siberia (Sengor et al., 1993). Other arcs existed along both coasts of southern Europe, along western Tarim (today in West China), and a very long arc system, the Samfrau arc extended from eastern Australia to southern South America (Figure 1.31). Paleomagnetic data indicate that North and South China were separate plates at this time. As evidenced by the number of Paleozoic accretionary terranes in central and southern Asia, other arcs and microcontinents also probably existed during the early Paleozoic in the region between Australia and Arabia.

During the Ordovician and Silurian, the Iapetus Ocean began to close and Avalonia, Cadomia and Baltica collided with Laurentia, producing a succession of collisional orogens in the Appalachians in North America and the Caledonides in Scandinavia. Also during this time, Siberia moved toward Laurentia, Gondwana was centred approximately over the South Pole and shallow seas covered large portions of the continents. In the Silurian, Mongolia collided with Siberia and several microcontinents collided south of South China. Collision continued along the east coast of Laurentia from the Devonian to the Permian, when Africa collided with southeast Laurentia. Beginning in the Early Carboniferous and ending in the Late Carboniferous, the Hercynian orogen developed by collision along both margins of southern Europe. In the Late Carboniferous and Permian, South America collided with North America causing the Ouachita orogeny. Siberia and Tarim collided with the complex Kazakhstan arc system and North China collided with a small microcontinent in the Late Carboniferous. In addition, Tibet may have been rifted from Gondwana. During the Permian, Kazakhstan collided with Baltica producing the Uralian orogeny, and North China and several other small plates collided with Siberia. By the end of the Permian, Pangea was nearly complete with the Tethys Ocean occupying a large re-entrant between Gondwana and Laurentia (Figure 1.30). Throughout the Paleozoic, the Samfrau arc system was active, propagating with time along the southwest coast of South America and joining with a subduction system in western North America. Thus, a major arc system bounded Pangea along its western and southern margins in the late Paleozoic (directions relative to modern positions). During the early to middle Paleozoic, Australia grew eastward forming the Lachland and New England orogens largely by the addition of accretionary prisms and closure of back-arc basins. Antarctica and the southern tip of South America also grew by similar processes along the Samfrau are system.

Plate motions during the Paleozoic were characterized by continual continental convergence and collision, the net result of which was the aggregation of Pangea, which extended from pole to pole by the Permian (Figure 1.30). About 160 Ma, Pangea, which now included Gondwana, began to fragment (Scotese, 1991; Storey, 1995). By the Triassic, the amalgamation of much of China was complete, and the Golconda allochthon was accreted to the western United States. In the Late Triassic, the Gulf of Mexico began to open as Yucatan and other microplates were rifted away from Texas (Ross and Scotese, 1988). An unknown number of microcontinents and arcs existed in the Tethys Ocean, many of which were rifted from Gondwana. These continued to collide with western North America and southern and eastern Asia during most of the Mesozoic. During the Jurassic and Cretaceous, several terranes were accreted to China and several arc systems were accreted in the northwestern United States. Kolyma and Okhotsk collided with northeast Siberia in the Early Jurassic and the Cretaceous, respectively, thus completing most of Siberia. Continued collisions in the western Tethys led to formation of several terranes that collided with northwest North America in the Late Cretaceous and Early Tertiary.

In the Middle and Late Jurassic, North America and Africa were rifted apart and the North Atlantic began to open. Major opening of the North Atlantic, however, did not occur until the Cretaceous. Also, during the Middle to Late Jurassic, Africa began to rift away from India–Antarctica. In the mid-Cretaceous, the South Atlantic began to open as Africa was rifted away from South America, the western Arctic basin began to open and the Bay of Biscay opened as Iberia (Spain/Portugal) rotated counterclockwise. During the Late Cretaceous, the Labrador Sea began to open as Greenland moved away from North America, and Africa continued its counterclockwise motion closing the Tethys Ocean. Also during the Late Cretaceous, microplates rifted away from North Africa and began to collide with Eurasia, the South Atlantic became a wide ocean, Madagascar reached its present position relative to East Africa, and India began rapid motion northwards towards Asia. Deformation began in the Alps and Carpathians in response to microplate collisions and Iran was rifted from Arabia. During the Jurassic and Cretaceous, the Gulf of Mexico continued to open, the western North America subduction zone propagated southward and a new subduction zone formed along the margin of Cuba. Major deformation occurred in response to collisions and subduction along the Cordilleran orogen in western North and South America and Antarctica. Between 80 and 60 Ma, New

Zealand was rifted from Australia as the Tasman Sea opened.

During the Early Tertiary, India collided with Tibet initiating the Himalayan orogeny, and continued convergence of microplates in the Mediterranean area caused the Alpine orogeny. In addition, the North Atlantic between Greenland and Norway began to open and the Sea of Japan opened. During the mid-Tertiary, Australia was rifted from Antarctica and moved northward, continued convergence in southern Europe led to the formation of large nappes in the Alps and Carpathians, the Rhine graben formed in Germany, and at about 30 Ma the Red Sea began to open as Arabia was rifted from Africa. Collision of the East Pacific rise with western North America led to development of the San Andreas transform fault, and Iceland formed on the mid-Atlantic ridge.

In the Late Tertiary and continuing to the present, oceanic arc systems formed in the South Pacific, subduction migrated into the eastern Caribbean and the Lesser Antilles arc formed, and southward propagation of the Middle America subduction zone resulted in the growth of Panama connecting North and South America (Ross and Scotese, 1988). Arabia collided with Iran in the Late Tertiary and was sutured along the Zagros crush zone. The Basin and Range province and the Rio Grande rift formed in North America, the East African rift system developed in Africa and the Gulf of Aden completely opened. Finally, beginning about 4 Ma, Baja California was rifted from the North American plate along the San Andreas transform fault and the Gulf of California opened.

Rodinia

Correlation of geologic features indicates that a major supercontinent, Rodinia, existed in the Late Proterozoic (Figure 1.32). Matching of Early Proterozoic orogenic belts suggests that Siberia was connected to Laurentia in the Canadian Arctic at this time (Condie and Rosen, 1994), and that Australia–Antarctica was part of western Laurentia (Moores, 1991). The configuration of Australia–Antarctica shown in Figure 1.32 is supported by a possible continuation of the 1-Ga Grenville Front from West Texas into Antarctica and by paleomagnetic data (Powell et al., 1993). The Grenville belt may loop around the East Antarctica craton and continue as the Fraser-Albany orogen in southwest Australia. Also supporting this configuration are similarities in geology and geochronology between the Wopmay orogen in northwest Canada and the Mt Isa and related orogens of Early Proterozoic age in northern Australia (Hoffman, 1991). Similar ages and geology in the Grenville belt in eastern Canada and 1-Ga belts in Scandinavia and western Brazil (the Sunsas belt), further suggest that Amazonia and Baltica lay adjacent to eastern Canada in the Late Proterozoic (Dalziel et al., 1994). Similar ages and geology between the Kalahari in southern Africa and Dronning Maud Land in Antarctica also support a Late Proterozoic connection

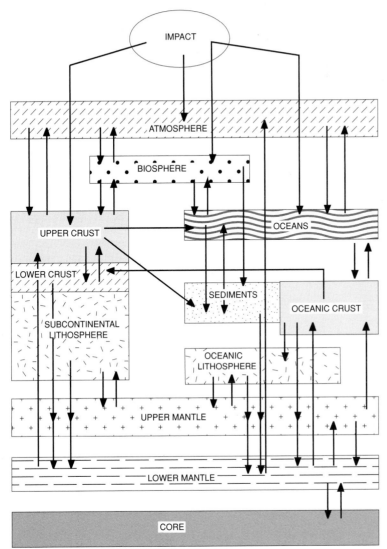

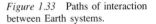

Figure 1.33 Paths of interaction between Earth systems.

here (Moyes et al., 1993). Although paleomagnetic data suggest that the Congo craton was also part of Rodinia, the precise configuration is not yet clear.

The fragmentation of Rodinia appears to have occurred by fan-like collapse of its constituent cratons, involving the counterclockwise rotation of Antarctica–Australia–India and possible clockwise rotation of Amazonia. Such rifting began about 750 Ma in western Laurentia and 600 Ma in eastern Laurentia. Baltica and Siberia were rifted away separately. In the very latest Proterozoic (570–540 Ma), East Gondwana collided with West Gondwana producing an extensive Pan-African orogenic belt in East Africa (Figure 1.31). This 'reassembly' of Rodinia as proposed by Hoffman (1991) turned the constituent cratons inside out as Gondwana formed: the external Paleozoic margins (such as the Samfrau arc system) originated as rift zones in the interior of the

supercontinent. Conversely, the external margins of Rodinia, some of which had been convergent plate margins before 750 Ma, became landlocked within the interior of Gondwana. These margins border the Mozambique and related orogenic belts in West Africa and Brazil.

Interactions of Earth systems

The Earth is a dynamic planet which is the product of a balance between interactions of various terrestrial systems (Figure 1.33). Earth systems are not static but have evolved with time, leading to the habitable planet we reside on. Current and future interactions of these systems will have a direct impact on life, and for this reason it is important to understand how perturbation of one system can affect other systems and how rapidly systems change with time. Short-term climatic cycles are superimposed on and partly controlled by long-term processes in the atmosphere–ocean system, which in turn are affected by even longer-term processes in the mantle and core. Also affecting Earth systems are asteroid and comet impacts, both of which appear to have been more frequent in the geologic past.

Some of the major pathways of interaction between Earth systems and between Earth and extraterrestrial systems are summarized in Figure 1.33. Although many of these interactions will be considered in detail in later chapters, it may be appropriate to preview some of them now. As an example, crystallization of metal in the core may liberate enough heat to generate mantle plumes just above the core–mantle interface. As these plumes rise into the uppermost mantle and begin to melt, large volumes of basalt may be underplated beneath the crust and also erupted at the Earth's surface. Such eruptions may pump significant quantities of CO_2 into the atmosphere which, because of its greenhouse character, may significantly warm the atmosphere leading to climatic changes. This, in turn, may affect the continents (by increasing weathering and erosion rates), the oceans (by increasing the rate of limestone deposition), and life (by leading to extinction of those forms not able to adapt to the changing climates). Thus, through a linked sequence of events, processes occurring in the Earth's core could lead to extinction of life forms at the Earth's surface. Before such changes can affect life, however, **negative feedback** processes may return the Earth systems to some level of equilibrium, or even reverse these changes. For instance, increased weathering rates caused by increased CO_2 levels in the atmosphere may drain the atmosphere of its CO_2, which is then transported by streams to the oceans where it is deposited as limestone. If cooling is sufficient, this could lead to widespread glaciation, which in turn could cause extinction of some life forms. Asteroid or comet impact on the Earth's surface also can have catastrophic effects on the atmosphere, oceans, and life as discussed in Chapter 6.

To prepare for the survival of living systems on planet Earth, it is important to understand the nature and causes of interactions between Earth systems and between Earth and extraterrestrial systems. How fast and how frequently do these interactions occur, and what are the relative rates of forward and reverse reactions? These are important questions that need to be addressed by the present and future generations of scientists.

Summary statements

1   Plate tectonics is a unifying theory for the cooling of the Earth in which lithospheric plates form at ocean ridges, move about on a convecting mantle as they may grow or diminish in size, and return to the mantle at subduction zones.

2   Three major first-order seismic discontinuities occur in the Earth: the Moho (3–70 km), the core–mantle interface (2900 km), and the 5200-km discontinuity. These discontinuities define the crust, mantle, outer core and inner core, in order of increasing depth.

3   In terms of strength and mode of deformation, the Earth is divided into the lithosphere (surface to 200 km deep), a strong outer layer that behaves as a brittle solid; the asthenosphere, extending from the base of the lithosphere to the 660-km discontinuity, a region of low strength with a seismic low-velocity zone in the upper part; and the mesosphere, a strong and homogeneous region extending from the 660-km discontinuity to the base of the mantle.

4   A transition zone in the mantle is bounded by two major seismic discontinuities at 410 km and 660 km depth. At each discontinuity new, high-pressure phases are formed.

5   The Earth has two boundary layers with steep temperature gradients: the LVZ beneath the lithosphere, and the D″ layer at the base of the mantle. Plates move about on the LVZ and mantle plumes may be generated in the D″ layer.

6   Plate boundaries are of four types: ocean ridges where new lithosphere is produced; subduction zones where lithosphere descends into the mantle; transform faults where plates slide past each other; and collisional zones, where continents or arcs have collided.

7   Divergent plate boundaries (ocean ridges) are characterized by small-magnitude, shallow earthquakes with vertical motions reflecting formation of new lithosphere. Topography in axial rifts varies from high relief to little if any relief in going from slow- to fast-spreading ridges. Ocean ridges grow by lateral propagation.

8   Transform faults are characterized by shallow, variable-magnitude earthquakes exhibiting lateral motion. Ridge–ridge transforms retain a constant length with time, whereas other transforms lengthen or shorten with time.

9   Convergent boundaries (subduction zones) are characterized by a dipping seismic zone with variable-magnitude earthquakes and, in some instances, seismic gaps suggestive of plate fragmentation. Fault motions vary with depth in the seismic zone and seismicity is strongly correlated with the degree of coupling of the descending slab and mantle wedge. Plates < 50 My in age may be buoyantly subducted and slide beneath the overriding plate.

10   Buoyant subduction is the only recognized tectonic force sufficient to trigger nucleation of a new subduction zone. New subduction zones commonly form at zones of weakness such as transform faults or fracture zones.

11   Collisional boundaries are characterized by wide (up to 3000 km) zones of lateral deformation with compressive fault motions dominating near sutures and lateral or vertical motions in areas overlying partially subducted plates.

12 A Wilson Cycle, the opening and closing of an ocean basin, has occurred many times during geologic history.

13 The motion of one plate relative to another is that of a spherical cycloid and is a function of the position of the pole of rotation, the direction of relative motion, and the angular velocity of the plate.

14 Plate velocities can be estimated from magnetic anomalies on the sea floor, the azimuths and motions on transform faults, first-motion studies of earthquakes, apparent polar wander paths, and by space geodetic measurements. Average relative velocities of plates range from 1–20 cm/y and the oldest surviving sea floor is about 160 Ma.

15 Computer models indicate that plates move in response chiefly to slab-pull forces, and that ocean-ridge push forces transmitted from the asthenosphere to the lithosphere are very small.

16 Rocks acquire remanent magnetization in the Earth's magnetic field by cooling through the Curie point of magnetic minerals (TRM), during deposition or diagenesis of a clastic sediments (DRM), and during secondary processes if new magnetic minerals are formed (CRM).

17 The Earth's magnetic field has reversed its polarity many times in the geologic past. Normal and reverse polarity intervals in the stratigraphic record allow construction of the Geomagnetic Time Scale. Magnetic reversals show periodicity on several scales and evidence of reversals exists in rocks as old as 3.5 Ga.

18 Polarity intervals correlate with magnetic anomaly distributions on the sea floor allowing seafloor spreading rates to be estimated. The magnetic anomalies are caused by magnetized basalt injected into axial zones of ocean ridges during normal and reversed polarity intervals.

19 During a reversal, which occurs over about 4000 years, the Earth's dipole field decreases in intensity and rapid changes occur in declination and inclination.

20 The two most important problems in using paleomagnetism to reconstruct ancient plate motions are, (1) separation of multiple magnetizations in the same rock, and (2) isotopic dating of the magnetization(s).

21 Apparent polar wander paths show distinct characteristics for various plate tectonic scenarios.

22 Chains of volcanic islands and aseismic ridges on the sea floor appear to have formed as oceanic plates more over hotspots, which are the shallow manifestations of mantle plumes. Similar, although not as well-defined, trajectories are formed when continents move over hotspots. Lifespans of hotspots are < 100 My.

23 Although hotspots appear to have remained fixed beneath a given plate or beneath adjacent plates, distant hotspots have not remained fixed, but move at velocities approximately an order of magnitude less than plate velocities.

24 A supercontinent is a large continent composed of several or all of the existing continents. A supercontinent cycle consists of rifting and break up of one supercontinent, followed by reassembly, in which dispersed cratons collide to form a new supercontinent, with most or all fragments in different configurations from the older supercontinent.

25 The youngest supercontinent is Pangea, which formed between 450 and 320 Ma and includes most of the existing continents. Pangea began to fragment about 160 Ma and is still dispersing today. Gondwana, comprising Southern Hemisphere continents, formed at 750–550 Ma. The earliest well-documented supercontinent is Rodinia, which formed about 1.3–1.0 Ga, fragmented at 750–600 Ma, and appears to have included most of the continents in a configuration quite different from Pangea.

26 To prepare for the survival of living systems on planet Earth, it is important to understand the nature and causes of interactions between Earth systems and between Earth and extraterrestrial systems.

Suggestions for further reading

Kearey, P. and Vine, F. J. (1996). *Global Tectonics* (second edition). Cambridge, MA, Blackwell Scient., 348 pp.

Klein, G. D. (1994). Pangea: Paleoclimate, Tectonics, and Sedimentation during Accretion, Zenith, and Breakup of a Supercontinent. *Geol. Soc. America, Spec. Paper 288.*

Moores, E. M. and Twiss, R. J. (1995). *Tectonics.* New York, W. H. Freeman, 415 pp.

Storey, B. C. (1995). The role of mantle plumes in continental breakup: Case histories from Gondwanaland. *Nature,* **377,** 301–308.

Windley, B. F. (1995). *The Evolving Continents* (third Edition). New York, J. Wiley, 526 pp.

Chapter 2

The Earth's crust

Introduction

The Earth's crust is the upper rigid part of the lithosphere, the base of which is defined by a prominent seismic discontinuity, the Mohorovicic discontinuity or Moho. There are three crustal divisions – oceanic, transitional, and continental – of which oceanic and continental crust greatly dominate (Table 2.1). Typically, oceanic crust ranges from 3–15 km thick and comprises fifty-nine per cent of the crust by area and twenty-one per cent by volume. Islands, island-arcs, and continental margins are examples of transitional crust that have thicknesses of 15 to 30 km. Continental crust ranges from 30–70 km thick and comprises seventy-nine per cent of the crust by volume but only forty-one per cent by area. With the exception of the upper continental crust, most of both the continental and oceanic crust remains relatively inaccessible. Our knowledge of oceanic crust comes largely from ophiolites, which are thought to represent tectonic fragments of oceanic crust that are preserved in the continents. Our view of the lower continental crust is based chiefly on a few uplifted slices of this crust in collisional orogens and from xenoliths brought to the surface in young volcanics.

The crust can be further subdivided into **crustal types**, which are segments of the crust exhibiting similar geological and geophysical characteristics. There are thirteen major crustal types, listed in Table 2.1 with some of their physical properties. The first two columns of the table summarize the area and volume abundances, and column 3 describes tectonic stability in terms of earthquake and volcanic activity and recent deformation.

To understand better the evolution of the Earth, we must understand the origin and evolution of the crust. This chapter reviews what is known about the physical and chemical properties of the crust, based on geophysics and chemically-analysed samples. Of the geophysical data base, seismic-wave velocities and heat flow provide the most robust constraints on the nature and composition of the crust.

Crustal types

Oceanic crust

Ocean ridges

Ocean ridges are widespread linear rift systems in oceanic crust where new lithosphere is formed as the flanking oceanic plates move away from each other. They are topographic highs on the sea floor and are tectonically unstable. A medial rift valley generally occurs near their crests in which new oceanic crust is produced by intrusion and extrusion of basaltic magmas. Iceland is the only known example of a surface exposure of a modern ocean ridge system, and it is complicated by the presence of a mantle plume beneath the ridge. The world-wide ocean-ridge system is interconnected from ocean to ocean and is more than 70 000 km long (Figure 1.1). Ridge crests are cut by numerous transform faults, which may offset ridge segments by thousands of kilometres.

From geophysical and geochemical studies of ocean ridges, it is clear that both structure and composition vary along ridge axes (Solomon and Toomey, 1992). Ridge axes can be divided into distinct segments 10–100 km in length bounded by transform faults or topographic boundaries. In each segment, magmas from the upper mantle accumulate either by enhanced magma production rates or enhanced upwelling. In general, there is a good correlation between spreading rate and the supply of magma from the upwelling asthenosphere. Furthermore, within each segment the characteristics of deformation, magma emplacement, and hydrothermal circulation vary with distance from the magma centre. Also, both the forms of segmentation and the seismic crustal structure differ between fast (> 70 mm/y half rate) and slow (< 7 mm/y) spreading ridges.

From our geophysical and petrological data base, the following observations are important in understanding the evolution of slow- and fast-spreading ocean-ridge systems:

Table 2.1 Geophysical features of the crust

Crustal type	Area (%)	Volume (%)	Stability	Heat flow (mW/m²)	Bouguer anomaly (mgal)	Poisson's ratio
Continental						
1 Shield	6	11	S	40	−20 to −30	0.29
2 Platform	18	35	S, I	49	−10 to −50	0.27
3 Orogen						
PA	8	13	S	60	−100 to −200	0.26
MC	6	14	I, U	70	−200 to −300	0.26
4 Continental margin arc	2	4	I, U	50–70	−50 to −100	0.25
Transitional						
5 Rift	1	1	U	60–80	−200 to −300	0.30
6 Island arc	1	1	I, U	50–75	−50 to +100	0.25
7 Submarine plateau	3	3	S, I	50–60	−100 to +50	0.30
8 Inland-sea basin	1	1	S	50	0 to +200	0.27
Oceanic						
9 Ocean ridge	10	2	U	100–200	+200 to +250	0.22
10 Ocean basin	38	12	S	50	+250 to +350	0.29
11 Marginal-sea basin	4	2	U, I	50–150	+50 to +100	0.25
12 Volcanic island	< 1	< 1	I, U	60–80	+250	0.25
13 Trench	2	1	U	45	−100 to −150	0.29
Average continent	40	77		55	−100	0.27
Average ocean	54	17		67 (95)*	+250	0.29

Stability key: S, stable; I, intermediate stability; U, unstable
Poisson's ratio = $0.5\{1 - 1/[(v_p/v_s)^2 - 1]\}$; where v_p and v_s = p and s wave velocities
PA, Paleozoic orogen; MC, Mesozoic–Cenozoic orogen
* calculated oceanic heat flow

1 The lower crust (oceanic layer, see Figure 2.2 later) is thin and poorly developed at slow-spreading ridges. The Moho is a sharp, tectonic boundary and may be a detachment surface (Dilek and Eddy, 1992). In contrast, at fast-spreading ridges the Moho is a transition zone up to 1 km thick.

2 In general, rough topography occurs on slow-spreading ridges, while smooth topography is more common on fast-spreading ridges. Fast-spreading ridges also commonly lack well-developed axial rifts.

3 Huge, long-lived axial magma chambers capable of producing a thick gabbroic lower crust are confined to fast-spreading ridges.

4 At slow-spreading ridges, like the Mid-Atlantic and Southwest Indian ridges, permanent magma chambers are often absent, and only ephemeral intrusions (chiefly dykes and sills) are emplaced in the medial rift.

Ocean basins

Ocean basins compose more of the Earth's surface (38 per cent) than any other crustal type (Table 2.1). Because the oceanic crust is thin, however, they compose only 12 per cent by volume. They are tectonically stable and characterized by a thin, deepsea sediment cover (approximately 0.3 km thick) and linear magnetic anomalies that are produced at ocean ridges during reversed and normal polarity intervals. The sediment layer thickens near continents and arcs from which detrital sediments are supplied. Although ocean basins are rather flat, they contain topographic features such as abyssal hills, seamounts, guyots, and volcanic islands. Abyssal hills, which are the most abundant topographic feature on the ocean floor, are sediment-covered hills ranging in width from 1–10 km and in relief from 500–1000 m.

Volcanic islands

Volcanic islands (such as the Hawaiian Islands) occur in ocean basins or on or near ocean ridges (e.g., St Paul Rocks and Ascension Island in the Atlantic Ocean). They are large volcanoes erupted on the sea floor whose tops have emerged above sea level. If they are below sea level, they are called seamounts. Volcanic islands and seamounts range in tectonic stability from intermediate or unstable in areas where volcanism is active (like Hawaii and Reunion), to stable in areas of extinct volcanism (such as Easter Island). Volcanic islands range in size from < 1 to about 10^4 km². Guyots are flat-topped seamounts produced by erosion at sea level followed by submersion, probably due to sinking of the sea floor. Coral reefs grow on some guyots as they sink, producing atolls. Most volcanic islands have developed over mantle plumes, which are the magma sources.

Trenches

Oceanic **trenches** mark the beginning of subduction zones and are associated with intense earthquake activity. Trenches parallel arc systems and range in depth from 5–8 km, representing the deepest parts of the oceans. They contain relatively small amounts of sediment deposited chiefly by turbidity currents and derived chiefly from nearby arcs or continental areas.

Back-arc basins

Back-arc basins are segments of oceanic crust between island-arcs (as the Philippine Sea) or between island-arcs and continents (as the Japan Sea and Sea of Okhotsk, Figure 1.1). They are most abundant in the western Pacific, and are characterized by a horst-graben topography (similar to the Basin and Range province) with major faults subparalleling adjoining arc systems. The thickness of sediment cover is variable, and sediments are derived chiefly from continental or arc areas. Back-arc basins are classified into tectonically active and inactive types. Active basins, like the Lau-Havre and Mariana troughs in the Southwest Pacific, have thin sedimentary cover, rugged horst-graben topography, and high heat flow. Inactive basins such as the Tasman and West Philippine basins, have variable sediment thicknesses and generally low heat flow.

Transitional crust

Submarine plateaux

Submarine plateaux are large flat-topped plateaux on the sea floor comprised largely of mafic volcanic and intrusive rocks. They are generally capped with a thin veneer of deepsea sediments and typically rise 2 km or more above the sea floor. Next to basalts and associated intrusive rocks produced at ocean ridges, submarine plateaux are the largest volumes of mafic igneous rocks at the Earth's surface. The magmas in these plateaux, together with their continental equivalents known as flood basalts, appear to be produced at hotspots caused by mantle plumes. Such plumes deliver 5–10 per cent of the Earth's heat to the surface (Coffin and Eldholm, 1994). Some of the largest submarine plateaux, such as Ontong-Java in the South Pacific and Kerguelen in the southern Indian Ocean (see Figure 3.8 later), which together cover an area nearly half the size of the conterminous United States, were erupted in the Mid-Cretaceous, perhaps during a period of superplume activity. Most submarine plateaux are 15–30 km thick, although some may exceed 30 km.

Generally grouped with submarine plateaux are **submarine ridges**, also known as **aseismic ridges**. These are elongated, steep-sided volcanic ridges, such as the Walvis and Rio Grande ridges in the South Atlantic (Figure 1.24). Many aseismic ridges are produced as the oceanic lithosphere moves over a hotspot (mantle plume),

and thus the ridge is the track of plate motion erupted on the sea floor.

Arcs

Arcs occur above active subduction zones where one plate dives beneath another. They are of two types: island arcs develop on oceanic crust, and continental-margin arcs develop on continental or transitional crust. Island arcs commonly occur as arcuate chains of volcanic islands, such as the Mariana, Kermadec and Lesser Antilles arcs (Figure 1.1). Most large volcanic chains, such as the Andes, Cascades, and Japanese chains, are continental-margin arcs. Some arcs, such as the Aleutian Islands, continue from continental margins into oceanic crust. Modern arcs are characterized by variable, but often intense earthquake activity and volcanism, and by variable heat flow, gravity, crustal thickness, and other physical properties. Arcs are composed dominantly of young volcanic and plutonic rocks and derivative sediments.

Continental rifts

Continental rifts are fault-bounded valleys ranging in width from 30–75 km and in length from tens to thousands of kilometres. They are characterized by a tensional tectonic setting in which the rate of extension is less than a few millimetres a year. Shallow magma bodies (< 10 km deep) have been detected by seismic studies beneath some rifts such as the Rio Grande rift in New Mexico. The longest modern rift system is the East African system, which extends over 6500 km from the western part of Asia Minor to south-eastern Africa (Figure 1.1). Continental rifts range from simple grabens (a down-dropped block between two normal faults) or half-grabens (a tilted block with one or more normal faults on one side) to complex graben systems (a down-dropped or tilted block cut by many normal faults, many or most of which are buried by sediments). The Basin and Range province in western North America is a multiple rift system composed of a complex series of alternating grabens and horsts. **Aulacogens** are rifts that die out towards the interior of continents; many appear to have formed at triple junctions during fragmentation of a supercontinent and represent 'failed arms' of RRR triple junctions. Young rifts (< 30 Ma) are tectonically unstable, and earthquakes, although quite frequent, are generally of low magnitude.

Inland-sea basins

Inland-sea basins are partially to completely surrounded by tectonically stable continental crust. Examples include the Caspian and Black Seas in Asia and the Gulf of Mexico in North America (Figure 1.1). Earthquake activity is negligible or absent. Inland-sea basins contain thick successions (10–20 km) of clastic sediments and

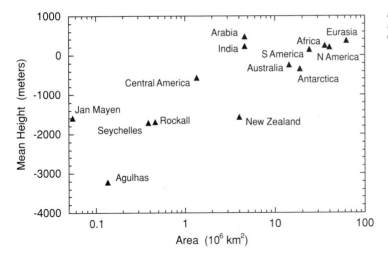

Figure 2.1 Mean continental heights as a function of continental area. Data from Cogley (1984).

both mud and salt diapirs are common. Some, such as the Caspian and Black Seas, are the remnants of large oceans that closed in the geologic past.

Continental crust

Precambrian shields

Precambrian shields are stable parts of the continents composed of Precambrian rocks with little or no sediment cover. Rocks in shields may range in age from 0.5 to > 3.5 Ga. Metamorphic and plutonic rock types dominate, and temperature–pressure regimes recorded in exposed rocks suggest burial depths ranging from as shallow as 5 km to as deep as 40 km or more. Shield areas, in general, exhibit very little relief and have remained tectonically stable for long periods of time. They comprise about 11 per cent of the total crust by volume, with the largest shields occurring in Africa, Canada, and Antarctica.

Platforms

Platforms are also stable parts of the crust with little relief. They are composed of Precambrian basement similar to that exposed in shields overlain by 1–3 km of relatively undeformed sedimentary rocks. Shields and the Precambrian basement of platforms are collectively referred to as cratons. A **craton** is an isostatically positive portion of the continent that is tectonically stable relative to adjacent orogens. As we shall see below, cratons are composed of ancient orogens. Sedimentary rocks on platforms range in age from Precambrian to Cenozoic and reach thicknesses up to 5 km, as for instance in the Williston basin in the north-central United States. Platforms compose most of the crust in terms of volume (35 per cent) and most of the continental crust in terms of both area and volume.

Collisional orogens

Collisional orogens are long, curvilinear belts of compressive deformation produced by the collision of continents. Giant thrust sheets and nappes are found in many orogens. Collisional orogens range from several thousands to tens of thousands of kilometres in length, and are composed of a variety of rock types. They are expressed at the Earth's surface as mountain ranges with varying degrees of relief, depending on their age. Older collisional orogens, such as the Appalachian orogen in eastern North America and the Variscan orogen in central Europe, are deeply eroded with only moderate relief, whereas young orogens such the Alps and Himalayas, are among the highest mountain chains on Earth. Tectonic activity decreases with age of deformation in orogens. Orogens older than Paleozoic are deeply eroded and are now part of Precambrian cratons. Large plateaux, which are uplifted crustal blocks that have escaped major deformation, are associated with some orogens, such as the Tibet plateau in the Himalayan orogen.

Continent size

It has been known for some time that there is an overall positive correlation between the area of continents and microcontinents, and their average elevation above sea level (Figure 2.1) (Cogley, 1984). Small continental fragments, such as Agulhas and Seychelles, lie two or more kilometres below sea level. Arabia and India are higher than predicted by the area–height relationship, a feature that is probably related to their tectonic histories. The height of a continent depends on the uplift rate and the rates of erosion and subsidence. Collision between continents resulting in crustal thickening is probably the leading cause of continental uplift in response to isostacy. The anomalous heights of Arabia and India undoubtedly reflect the collision of these microcontinents with Asia between 70 and 50 Ma, and in the case of India, this collision is continuing today. The six large continents have undergone numerous collisions in the last 1 Gy, each thickening the continental crust in collisional zones leading to a greater average continent elevation. Hotspot activity can also elevate continents, and it is possible that the high elevation of Jan Mayen may be related to the Iceland hotspot. Also, the modal height of Africa, which is higher than the other large continents by 200 m, may reflect the numerous hotspots beneath North and West Africa (Figure 1.24).

The other five small continents in Figure 2.1 are mostly fragments that have been rifted from supercontinents. Since the break up of supercontinents leads to a rise in sea level (Chapter 6), it is not surprising that these micro-continents are completely flooded. Not until they collide with other continents will they emerge above sea level.

Seismic crustal structure

The Moho

The Mohorovicic discontinuity or Moho is the outer-most seismic discontinuity in the Earth and defines the base of the crust (Jarchow and Thompson, 1989). It ranges in depth from about 3 km at ocean ridges to 70 km in collisional orogens, and is marked by a rapid increase in seismic P-wave velocity from < 7.6 km/sec to ≥ 8 km/sec. Because the crust is different in composition from the mantle, the Moho is striking evidence for a differentiated Earth. Detailed seismic refraction and reflection studies indicate that the Moho is not a simple boundary worldwide. In some crust, such as collisional orogens, the Moho is often offset by complex thrust faults. The Himalayan orogen is a superb example where a 20-km offset in the Moho is recognized beneath the Indus suture (Hirn et al., 1984). This offset was produced as crustal slices were thrust on top of each other during the Himalayan collision. In crust undergoing extension, such as continental rifts, a sharp seismic discontinuity is often missing, and seismic velocities change gradually from crustal to mantle values. In some collisional orogens, the Moho may not always represent the base of the crust. In these orogens, thick mafic crustal roots may invert to eclogite (a high density mafic rock) and Vp increases to mantle values, yet the rocks are still part of the crust (Griffin and Reilly, 1987). The petrologic base of the crust where eclogite rests on ultramafic mantle rocks may not show a seismic discontinuity, since both rock types have similar velocities. This has given rise to two types of Mohos: the seismic Moho (defined by a jump in seismic velocities), and the petrologic Moho (defined by the base of eclogitic lower crust).

Crustal layers

Crustal models based on seismic data indicate that oceanic crust can be broadly divided into three layers, which are, in order of increasing depth: the **sediment layer** (0–1 km thick), the **basement layer** (0.7–2.0 km

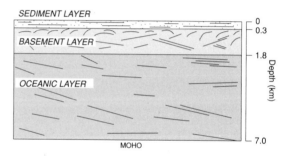

Figure 2.2 Standard cross-section of oceanic crust based on seismic reflection data. Solid lines are examples of major reflectors.

thick), and the oceanic layer (3–7 km thick) (Figure 2.2). Models for the continental crust show a greater range in both number and thicknesses of layers. Although two- or three-layer models for continental crust are most common (Christensen and Mooney, 1995), one-layer models and models with more than three layers are proposed in some regions. With the exception of continental borderlands and island-arcs, continents range in thickness from about 35–40 km (mean = 39 km), while the average thickness of the oceanic crust is only 5–7 km.

P-wave velocities in crustal sediment layers range from 2–4 km/sec, depending on degree of compaction, water content, and rock type. P-wave velocities in the middle oceanic crustal layer are about 5 km/sec, while those in the middle continental crustal layer are about 6.5 km/sec (Table 2.2). Lower crustal layers in both oceans and continents are characterized by P-wave velocities of 6.5–6.9 km/sec. The average crustal P-wave velocity (excluding the contribution of the sediment layer) in both oceanic and continental areas is 6.45 km/sec. The seismic velocity at the top of the mantle, known as the **Pn velocity**, increases both with mean crustal velocity and with depth to Moho in continental areas (Woollard, 1968).

Seismic-wave velocities increase with depth in the continental crust from 6.0–6.2 km/sec at depths of < 10 km to 6.6 km/sec at 25 km depth. Lower crustal velocities range from 6.8–7.2 km/sec, and in some cases show a bimodal distribution. In some continental crust there is evidence of a small discontinuity at mid-crustal depths, referred to as the **Conrad discontinuity** (Litak and Brown, 1989). When identified, the Conrad discontinuity varies in depth and character from region to region, suggesting that, unlike the Moho, it is not a fundamental property of the continental crust and it is diverse in origin. Seismic reflections also occur at mid-crustal depths in some extended crust like the Rio Grande rift in New Mexico.

Crustal types

Oceanic crust

Crustal structure in ocean basins is rather uniform, not deviating greatly in either velocity or layer thickness

Table 2.2 Crustal seismic-wave velocities

Crustal type	v_{Pn} (km/sec)	v_P Upper crust (km/sec)	v_P Middle crust (km/sec)	v_P Lower crust (km/sec)	Mean thickness (km)	Mean v_P (km/sec)
Continental						
1 Shield	8.1	6.2 (13)	6.5 (13)	7.0 (14)	40	6.5
2 Platform	8.1	6.2 (15)	6.5 (13)	7.0 (14)	42	6.5
3 Orogen						
PA	8.1	6.1 (10)	6.4 (12)	6.8 (13)	35	6.4
MC	8.0	6.0 (16)	6.3 (11)	6.7 (25)	52	6.3
4 Continental margin arc	7.9	6.2 (18)	6.6 (10)	7.0 (10)	38	6.4
Transitional						
5 Rift	7.8	6.0 (11)	6.5 (6)	6.9 (11)	28	6.5
6 Island arc	7.8	6.2 (6)	6.6 (6)	7.2 (7)	19	6.4
7 Submarine plateau	8.0	6.3 (5)	6.5 (10)	6.8 (10)	25	6.5
8 Inland-sea basin	8.1		2–5 (7)	6.8 (15)	22 [25]	6.0
Oceanic						
9 Ocean ridge	< 7.5		5.0 (1)	6.5 (4)	5 [6]	6.4
10 Ocean basin	8.2	2–4 (0.5)	5.1 (1.5)	6.8 (5)	7 [11]	6.4
11 Marginal-sea basin	7.5–8.0	2–4 (1)	5.3 (3)	6.6 (5)	9 [13]	6.3
12 Volcanic island	7.5–8.0		6.6 (6)	7.0 (7)	13	6.5
13 Trench	8.0	2–4 (3)	5.1 (1.5)	6.8 (5)	8 [14]	6.4
Average Continent	8.1	6.2 (13)	6.5 (13)	7.0 (14)	41	6.5
Average Ocean	8.1	2–4 (0.5)	5.1 (1.5)	6.8 (5)	7 [10]	6.4

Crustal layer thicknesses (in km) given in parenthesis
PA, Paleozoic orogen; MC, Mesozoic–Cenozoic orogen
Numbers in brackets [ ] are depths to the Moho below sea level

distribution from that shown in Figure 2.2 (Solomon and Toomey, 1992). Total crustal thickness ranges from 6–8 km, and Pn velocities are generally uniform in the range of 8.1–8.2 km/sec. Oceanic crust does not thicken with age (McClain and Atallah, 1986), and the width of the Moho may vary between 0 and 2 km. The sediment layer averages about 0.3 km in thickness and exhibits strong seismic reflecting zones with variable orientations (Figure 2.2), some of which are probably produced by cherty layers, as suggested by cores retrieved by the Ocean Drilling Project. The most prominent reflecting horizon is the interface between the sediment layer and basement layer, which is characterized by a rough topography. The thickness of the basement layer averages about 1.5 km, seismic-wave velocity increases rapidly with depth, and significant seismic anisotropy has been described in some areas (Stephen, 1985). There are also numerous reflective horizons in this layer. Such anisotropy is probably caused by preferred orientation of large fractures in the basement layer which were formed during the early stages of crustal development by near-ocean ridge extensional deformation. The oceanic layer is generally rather uniform in both thickness (4–6 km) and velocity (6.7–6.9 km/sec)

Beneath ocean ridges, crustal thickness ranges from 3–6 km, most of which is accounted for by the oceanic layer (Figure 2.3) (Solomon and Toomey, 1992). The sediment and basement layers thin or disappear entirely on most ridges. Seismic reflections indicate magma chambers beneath ridges at depths of 1–3 km. Unlike other

oceanic areas, the velocities in the oceanic layer are quite variable, ranging from 6.4–6.9 km/sec. Such a range of velocities is probably produced by seismic anisotropy and by varying but, on the whole, high temperatures beneath ridges. Anomalous mantle (Vp < 7.8 km/sec) occurs beneath ridge axes, reflecting high temperatures. Surface-wave data indicate that the lithosphere increases in thickness from < 10 km beneath ocean ridges to 50–65 km at a crustal age of 50 My. Anisotropy in S-wave velocities in the oceanic mantle lithosphere is often pronounced, with the fast wave travelling normally to ocean-ridge axes. Such anisotropy appears to be caused by alignment of olivine c axes in this direction (Raitt et al., 1969).

The crust of back-arc basins is slightly thicker (10–15 km) than that of ocean basins, due principally to a thicker sediment layer in marginal seas. Most seismic models show two to four layers above the oceanic layer, which exhibits velocities of 6.6–6.9 km/sec. The upper crustal layers commonly range in velocity from about 2 km/sec at the top (unconsolidated sediments) to about 5 km/sec above the oceanic layer. An intermediate velocity layer (6.0–6.2 km/sec) occurs just above the lower crustal layer in the Caribbean and New Caledonia basins. Crustal thickness in volcanic islands ranges from 10–20 km, with upper crustal velocities ranging from 4.7–5.3 km/sec and lower crustal velocities from 6.4–7.2 km/sec. Upper mantle velocities range from 7.8–8.5 km/sec. Crustal structure of trenches is not well known. Data suggest a wide range in crustal thickness beneath trenches

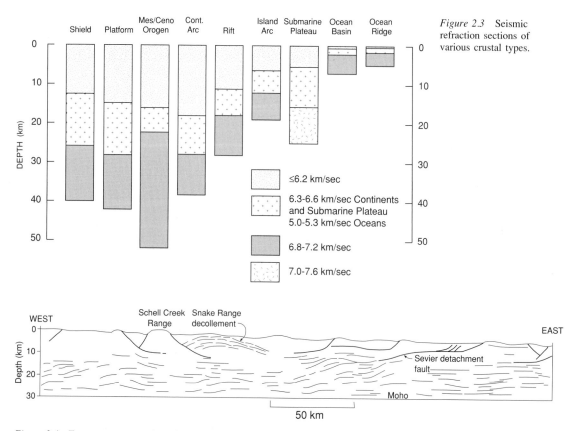

Figure 2.3 Seismic refraction sections of various crustal types.

Figure 2.4 East-west cross-section of the Great Basin from Long Valley, Nevada, to the Pavant Range in Utah, showing major seismic reflectors. Data from Gans (1987).

(3–12 km), with most thickening and thinning occurring in the oceanic layer, which ranges from 4–8 km thick. The sediment layer in trenches ranges from 2–4 km thick and exhibits variable Vp (3.8–5.5 km/sec). Most trenches have a sediment layer < 3 km thick.

Transitional crust

Submarine plateaux

The seismic structure of submarine plateaux is poorly known. On the basis of existing seismic and gravity data it would appear that the largest plateaux, such as Ontong-Java and Kerguelen, range from 20–30 km thick (Coffin and Eldholm, 1994) (Figure 2.3). Lower crustal P-wave velocities are anomalously high (7.0–7.6 km/sec), and seismic reflection data from Kerguelen indicate a typical pelagic sediment layer 2–3 km thick.

Continental rifts

Continental rifts are characterized by thin crust (typically 20–30 km thick) and low Pn velocities (< 7.8 km/sec). In these regions it is the lower crustal layer which is thinned (Figure 2.3), ranging from only 4–14 km thick. This reflects the ductile behaviour of the lower crust during extension. Although most earthquake foci in rifts are < 20 km deep, some occur as deep as 25–30 km. In young crust which is being extended, such as the Basin and Range province in the western United States, the lower crust is highly reflective in contrast to a relatively transparent upper crust (Figure 2.4) (Mooney and Meissner, 1992). Also, the reflection Moho is nearly flat, due presumably to removal of a crustal root during extension. Basin and Range province normal faults cannot be traced through the lower crust and do not appear to offset the Moho.

Arcs

Resolution of seismic data is poor in arc systems, and considerable uncertainty exists regarding arc crustal structure. Crustal thickness ranges from 5 km in the Lesser Antilles to 35 km in Japan, averaging about 22 km, and Pn velocities range from normal (8.0–8.2 km/sec) to low (< 7.8 km/sec). All values given in Table 2.2 have large standard deviations and many arcs cannot be fitted to a simple two- or three-layer crustal model. There is evid-

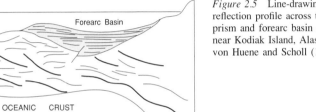

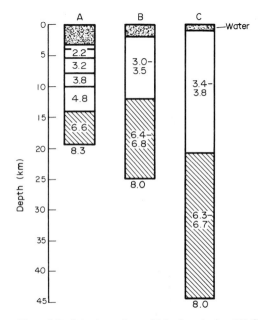

Figure 2.5 Line-drawing of a seismic reflection profile across the accretionary prism and forearc basin of the Aleutian arc near Kodiak Island, Alaska. Modified after von Huene and Scholl (1991).

ence in some island arcs for an intermediate-velocity layer (5.0–6.0 km/sec) of varying thickness (Figure 2.3).

Seismic reflection profiles in arcs are extremely complex as shown by a profile across the forearc region of the Aleutian arc in southern Alaska (Figure 2.5). Although reflections are deformed by steep faults that dip toward the continent (dark lines), they can be traced beneath the forearc basin, where they appear to plunge into the continent. The strong continuous reflection near the middle of the section (R) may be the contact of deformed ocean sediments with metamorphic basement. The sediments appear to be underplated beneath the arc along this boundary. The strong reflectors in the trench represent trench turbidites and oceanic sediments, and those in the forearc basin are chiefly volcanigenic turbidites.

Inland-sea basins

Inland-sea basins show a considerable range in crustal thickness and layer distributions (Figure 2.6). Crustal thickness ranges from about 15 km in the Gulf of Mexico to 45 km in the Caspian Sea basin. In general, the sedimentary layer or layers (Vp = 2–5 km/sec) rest directly on the lower crust (Vp = 6.3–6.7 km/sec), with little or no upper crust. Differences in crustal thickness among inland-sea basins are accounted for by differences in thickness of both sedimentary and lower crustal layers. Increasing velocities in sediment layers with depth, as for instance shown by the Gulf of Mexico, appear to reflect an increasing degree of compaction and diagenesis of sediments.

Continental crust

Shields and platforms

Shields and platforms have similar upper- and lower-layer thicknesses and velocities (Figure 2.3). The difference in their mean thickness (Table 2.2) reflects primarily the presence of the sediment layer in the platforms. Upper-layer thicknesses range from about 10–25 km and lower-layers from 16–30 km. Velocities in both layers are

Figure 2.6 Seismic sections of inland-sea basins. (A) Gulf of Mexico, (B) Black Sea, (C) Caspian Sea. P-wave velocities in km/sec. After Menard (1967).

rather uniform, generally ranging from 6.0–6.3 km/sec in the upper layer and 6.8–7.0 km/sec in the lower layer. Pn velocities are typically in the range of 8.1–8.2 km/sec, rarely reaching 8.6 km/sec. Shield areas with high-grade metamorphic rocks dominating at the surface may be characterized by a one-layer crust (with Vp = 6.5–6.6 km/sec) as exemplified, for instance, by the Grenville province in eastern North America. Relief up to 10 km per 100 km distance has been reported on the Moho in some platform and shield areas, and refinements in the interpretation of seismic and gravity data indicate the presence of significant lateral variations in crustal layer and velocity distributions. Also, results suggest the existence of a high-velocity layer (~ 7.2 km/sec) in lower crust of Proterozoic age. Seismic reflection studies show an increase in the number of reflections

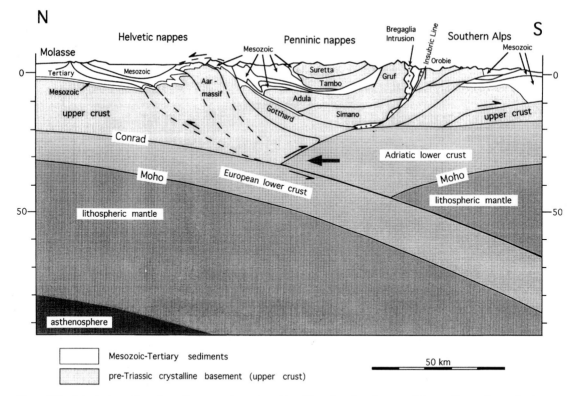

Figure 2.7   A tectonic north-south profile across the eastern Alps. The subsurface features of the profile are from seismic reflection data. Courtesy of Adrian Pfiffner.

with depth and generally weak, but laterally continuous Moho reflections (Mooney and Meissner, 1992).

Collisional orogens

Crustal thickness in collisional orogens is extremely variable, ranging from about 30 km in some Precambrian orogens to 70 km beneath the Himalayas. In general, thickness decreases with age. Average layer thicknesses and velocities of the upper two layers of Phanerozoic orogens are similar to platforms (Table 2.2; Figure 2.3), and average Phanerozoic orogen crustal thickness is about 46 km. In areas of very thick crust, such as beneath the Andes and Himalayas, the thickening occurs primarily in the lower crustal layer. Pn velocities in post-Paleozoic orogenic areas range from normal (8.0–8.2 km/sec) to low (< 7.8 km/sec). The velocity contrast between the lower crust and upper mantle is also commonly less beneath young orogenic areas (0.5–1.5 km/sec) than beneath platforms and shields (1–2 km/sec).

Phanerozoic orogens show considerable heterogeneity in terms of seismic reflection profiles. The Caledonian and Variscan orogens in Europe appear to have lost their crustal roots and show well-developed, chiefly subhorizontal reflectors in the lower crust (Mooney and Meissner, 1992). This loss of crustal roots may have

occurred during post-tectonic extensional collapse. Often, however, major thrust sheets, and in some instances sutures, are still preserved in the reflection profiles (Figure 2.7). In the southern Appalachians, the reflection data reveal a large decollement at mid-crustal depths separating Precambrian basement from overlying allochthonous rocks of the Appalachian orogen (Cook et al., 1981). In very young orogens like the Alps, lower crustal reflectors dip to the centre of a thick root, often with an irregular and offset Moho. At mid-crustal depths, however, reflectors typically show a complex pattern that is suggestive of crustal wedging and interfingering of nappes (Figure 2.7).

Crustal gravity

Shield and platform areas are characterized by broad Bouguer gravity anomalies of –10 to –50 mgal and occasional sharp anomalies of local importance (Table 2.1). Anomalies with widths of hundreds of kilometres reflect inhomogeneities in the lower crust or upper mantle, and those of smaller size reflect near-surface rock types or fault zones. Gravity anomalies can be used to trace structural trends or rock units in Precambrian shields beneath the sedimentary cover in platform areas. One example is the Keweenawan basalts (1 Gy in age), which can be traced south-west from Lake Superior beneath the sedimentary cover as the Mid-Continent Gravity High (Figure 2.8). Tectonic contacts between crustal provinces in shield and platform areas (such as between the Grenville and Superior Provinces in Canada) are also often expressed by gravity anomalies.

Phanerozoic orogens have large negative Bouguer anomalies of –200 to –300 mgal reflecting their thickened roots (Table 2.1). Young orogenic belts near continental borderlands and island-arc areas exhibit smaller anomalies (–50 to +100 mgal). Large negative anomalies in the Basin and Range province and in most continental rifts reflect thinning of the lower crust and the presence of shallow, low-density upper mantle. Small positive gravity anomalies superimposed on regional negative anomalies occur in the centres of some rifts and appear to be caused by near-surface intrusions of mafic magma (Daggett et al., 1986).

Bouguer anomalies in oceanic volcanic islands range from large and positive (+250 mgal) to small and variable (–30 to +45 mgal). Negative anomalies observed over trenches reflect the descent of less dense lithosphere into the upper mantle. Some gravity profiles suggest thinning of the crust on the oceanward side of trenches. Bouguer anomalies range from +250 to +350 in ocean basins, from +200 to +250 on ridges, and from 0 to +200 over marginal and inland-sea basins. In the North Atlantic, the Bouguer anomaly increases away from the Mid-Atlantic ridge due to cooling of the lithosphere as a function of age (Figure 2.9). In contrast, the free-air anomalies reflect topography on the ocean floor. The minimum in Bouguer anomaly over ocean ridges suggests they are isostatically compensated by hot, low-density upper mantle beneath them. Smaller Bouguer anomalies in marginal and inland seas result from thicker sediment layers in these basins.

Isostatic gravity anomalies are small over most crust, indicating a close approach to isostatic equilibrium. The major exceptions are arc–trench systems where lithosphere is being consumed by the asthenosphere. Thick roots exist under some mountains (e.g., the Sierra Nevada) but not under others (e.g., the Rocky Mountains), indicating that thickened crust is not always the dominant isostatic compensation mechanism. In any case, isostatic compensation must occur on a time scale of less than 50 My, because otherwise continents would be eroded to sea level in 25–50 My.

Crustal magnetism

Crustal magnetic anomalies are produced by magnetized rocks in the crust and, like gravity anomalies, provide a means of tracing major rock units and structural discontinuities beneath sediment cover in platform areas. For the most part, they reflect the distribution of magnetite in upper crustal rock types (Shive et al., 1992). In both shield and platform areas, magnetic anomalies

exhibit broad 'swirling' patterns. Broad anomalies are related to deeper crustal inhomogeneities and shallow, narrow anomalies to near-surface faults or mafic and ultramafic intrusions. In Archean shield areas granitic-gneiss terranes usually correlate with magnetic highs and mafic volcanic belts with lows (Figure 2.10). Structural discontinuities, such as major faults and crustal-province boundaries, also produce magnetic anomalies. Magnetic anomaly patterns in Phanerozoic orogens and island-arcs are more complex and show a wide range of amplitudes. In extended, relatively hot crust, such as the Basin and Range province, magnetic anomalies have small amplitudes, indicating that the lower crust and/or upper mantle are above the Curie point of magnetite (578 °C).

As discussed in Chapter 1, oceanic crust is characterized by linear magnetic stripes that roughly parallel ridge crests (Figure 1.3). They are produced by ocean-ridge basalts magnetized in normal and reversed magnetic fields. The anomalies are typically 5–50 km wide, hundreds of kilometres long and range from about 400–700 gammas in amplitude (relative to sea level). Anomalies are less distinct in back-arc basins, in part due to high temperatures beneath the basins, above or close to the magnetite Curie temperature. In the Gulf of Mexico, the thick sediment cover appears to mask or partially mask magnetic anomalies.

Crustal electrical conductivity

Although the electrical conductivity distribution in the continental crust is not well understood (Jones, 1992), there is a strong indication that it increases sharply in many places at mid-crustal depths. Data also suggest that crustal conductivity decreases with crustal age, with the cores of Precambrian shields having lower conductivities than their warmer margins. One or more high-conductivity layers are recognized at 20–40 km depth in some continental crust, and in some instances highly conducting layers correlate with low-velocity zones in the crust, such as is found in eastern New Mexico (Mitchell and Landisman, 1971). Linear crustal conductivity anomalies have been described from several continental regions, as for example the North American Central Plains (NACP) anomaly, which extends for a distance of nearly 2000 km from southeastern Wyoming to the edge of the Canadian shield in Saskatchewan (Camfield and Gough, 1977). This anomaly lines up with a major Early Proterozoic suture exposed in the Trans-Hudson orogen in the Canadian shield and with the Cheyenne belt in southeastern Wyoming. It is possible that NACP is produced by dehydration reactions in the suture zone.

Several origins have been considered for the high-conductivity layers in the middle and lower crust, of which high salinity pore waters, partial melting, and carbon films on grain boundaries are most widely cited (Jones, 1992). Experimental data, combined with geo-

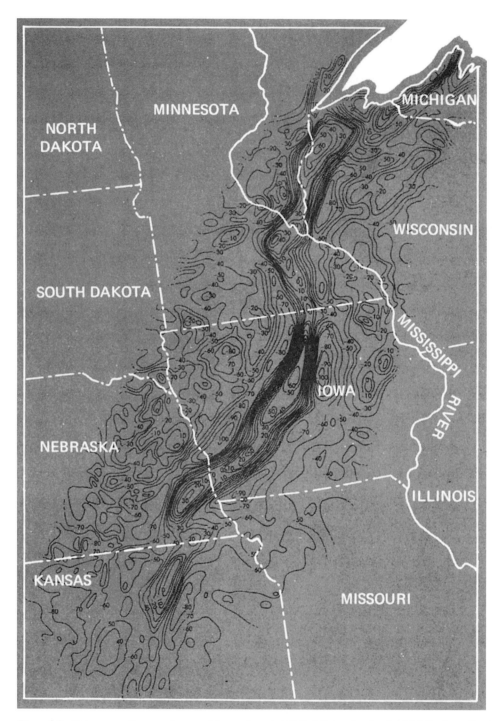

Figure 2.8 Bouguer gravity anomaly map of the north-central United States showing the Mid-Continent Gravity High. From Bouguer Anomaly Map of the United States, courtesy of the US Geological Survey. Contours in milligals.

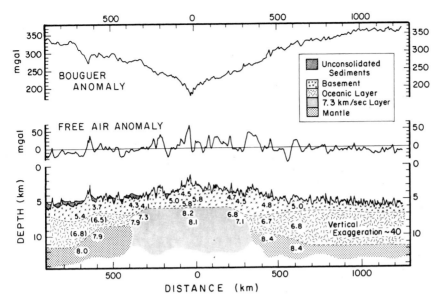

Figure 2.9 Gravity anomaly profiles and seismic structure across the northern Mid-Atlantic ridge. P-wave velocities in km/sec. After Talwani et al. (1965).

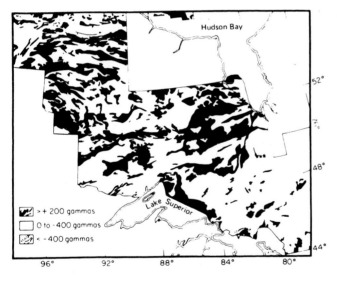

Figure 2.10 Magnetic anomaly pattern in the southern Canadian shield. Magnetic lows reflect greenstone belts and highs (in black) are granitoid terrains. After Morley et al. (1967). Courtesy of the Geological Association of Canada.

thermal results, indicate that partial melting is unlikely in all but young continental crust in active orogenic areas. Aqueous fluids with high salinity can produce highly conducting crustal layers (Hoak and Hutton, 1986). A possible source of saline water at shallow crustal depths (< 15 km) is from dehydration reactions accompanying metamorphism. However, if an aqueous phase is present in the lower crust it would react with the rocks to form hydrous minerals, which are not found in lower crustal xenoliths. Also, studies of high-grade metamorphic rocks from the lower crust suggest that these rocks formed in a dry environment or in the presence of CO_2-rich fluids. It would therefore seem that water is not responsible for lower-crustal high conductivities. The origin of lower-crustal high conductivity zones is uncertain and is a subject of continued discussion and debate among geologists and geophysicists.

Heat flow

Heat-flow distribution

Surface heat flow in continental and oceanic crust is controlled by many factors, and for any given segment of crust it decreases with mean crustal age. One of the most important sources of heat loss from the Earth is hydrothermal circulation associated with ocean ridges. Model calculations suggest that approximately 25 per cent of total global heat flow can be accounted for by hydrothermal transport (Davies, 1980). Heat flow on continents and islands varies as a function of the age of the last magmatic event, the distribution of heat-producing elements, and erosion level. Considering all sources of heat loss, the total heat loss from the Earth is about 42×10^{12} W, 12×10^{12} W from the continents and 30×10^{12} W from the oceans (Sclater et al., 1980). The equivalent heat flows are 55 and 95 mW/m^2, respectively, for a worldwide average heat flow of 81 mW/m^2 (Table 2.1). The difference between the average measured oceanic heat flow (67 mW/m^2) and the calculated value (95 mW/m^2) is due to heat losses at ocean ridges by hydrothermal circulation. Models indicate that 88 per cent of the Earth's heat flow is lost from the mantle: 66 per cent is lost at ocean ridges and by subduction, 10 per cent by conduction from the subcontinental lithosphere, and 12 per cent from mantle plumes (Davies and Richards, 1992). This leaves 12 per cent lost from the continents by radioactive decay of heat-producing elements.

Heat flows of major crustal types are tabulated in Table 2.1. Shield areas exhibit the lowest and least variable continental heat-flow values, generally in the range of 35–42 mW/m^2, averaging about 40 mW/m^2. Platforms are more variable, usually falling between 35–60 mW/m^2 and averaging about 49 mW/m^2. The difference between shield and platform heat flows reflects the greater thickness of crust (and hence more radiogenic heat) in platform areas. Young orogens, arcs, continental rifts, and oceanic islands exhibit high and variable heat flow in the range of 50–80 mW/m^2. The high heat-flow in some arcs and volcanic islands reflects recent volcanic activity in these areas. Heat flow in ocean basins generally falls between 35–60 mW/m^2, averaging about 50 mW/m^2. Ocean ridges, on the other hand, are characterized by extremely variable heat flow, ranging from < 100 to > 200 mW/m^2, with heat flow decreasing with increasing distance from ocean ridges (Figure 2.13). Back-arc basins are also characterized by high heat flow (60–80 mW/m^2), whereas inland-sea basins exhibit variable heat flow (30–75 mW/m^2), reflecting, in part, variable Cenozoic sedimentation rates in the basins.

A **geotherm**, which is the temperature distribution with depth in the Earth beneath a given surface location, is dependent upon surface and mantle heat flow and the distribution of thermal conductivity and radioactivity with depth. Geotherms in four crustal types are given in Figure 2.11. All of the geotherms have a nearly constant gradient in the upper crust, but diverge at depth resulting in up to 500 °C difference in crustal types at Moho depths. The geotherms will converge again at depths > 300 km where convective heat transfer in the mantle dominates.

Heat production and heat flow in the continents

Heat-flow values are significantly affected by the age and intensity of the last orogenic event, the distribution

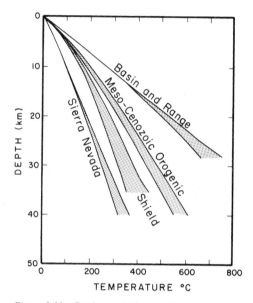

Figure 2.11 Geotherms in four continental crustal types. Shaded areas represent uncertainty ranges for each geotherm. After Blackwell (1971) and Chapman (1986).

of radioactive elements in the crust, and the amount of heat coming from the mantle (Chapman and Furlong, 1992). Continental surface heat flow (q_o) is linearly related to average radiogenic heat production (A_o) of near-surface basement rocks by the following equation:

$$q_o = q_r + A_o D. \tag{2.1}$$

q_r, commonly referred to as the **reduced heat flow**, is the intercept value for rocks with zero heat production; and D, the slope of the line relating q_o to A_o, has units of depth and is commonly referred to as the **characteristic depth**. The characteristic depth is interpreted to reflect the thickness of the upper crustal layer enriched in heat-producing elements, and ranges from about 3–14 km. The reduced heat flow is the heat coming from below this layer and includes radiogenic heat from both the lower crust and mantle.

Examples of the linear relationship between q_o and A_o for several crustal types are given in Figure 2.12. The Appalachian orogen is typical of pre-Mesozoic orogens and most platform areas. Studies of elevated fragments of lower continental crust, such as the Pikwitonei province in Canada (Fountain et al., 1987), show that heat production and surface heat flow also yield a linear relationship, indicating that the q–A relationship is independent of crustal level exposed at the surface. The chief factors distinguishing individual crustal types are the reduced heat-flow values (q_r) and, to a lesser extent, the range of surface heat flow and heat productivity. Thermal modelling indicates that the q–A relationship can best be explained by a decrease in radiogenic heat production with depth in the crust. Uplifted blocks of middle crust have moderate levels of radiogenic elements

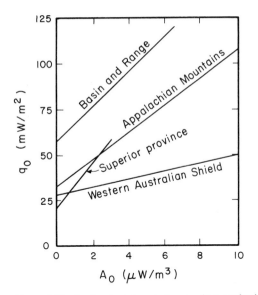

Figure 2.12 Surface heat flow (q_o) versus heat production of near-surface rocks (A_o) for several crustal types.

(U, Th, K), so most of the drop in radiogenic heat production must occur in the lower crust, in agreement with a largely mafic composition as inferred from xenolith data (Rudnick and Fountain, 1995).

Reduced heat flow decreases with the average age of the continental crust (Table 2.3). No relationship, however, appears to exist between the characteristic depth and mean crustal age. The reduced heat flow decays within 200–300 My to a value between 20–30 mW/m^2, which is about a factor of three faster than surface heat flow decays. Corresponding equilibrium q values for oceans are 30–40 mW/m^2. Two interpretations of the change in D between reduced heat flow provinces have been proposed (Jessop and Lewis, 1978): D values reflect differences in crustal erosion level or differences in degree of chemical fractionation of the crust. Because the crustal level exposed at the surface as deduced from metamorphic mineral assemblages does not correlate with D, nor does D correlate with mean crustal age, the first

interpretation seems unlikely. The alternate interpretation necessitates variable intensities of fractionation of U, Th and K into the upper crust, which is not related to the age of the crust. This could occur by partial melting in the lower crust producing granitoid magmas (in which U, Th, and K are concentrated), which rise to shallow levels. High-grade (granulite facies) metamorphism in the lower crust also liberates fluids in which U, Th, and K can be readily dissolved, and thus transported to higher crustal levels.

A relatively low heat flow is observed in all Archean cratons (such as the Superior province and Baltic shield, Table 2.3) compared with post-Archean cratons. This may be caused by either

1   a greater concentration of U, Th, and K in post-Archean cratons, or
2   a thick lithospheric root beneath the Archean cratons which effectively insulates the crust from asthenospheric heat (Nyblade and Pollack, 1993).

As we shall see in Chapter 4, the second factor is probably the principal cause of the low heat flow from Archean crust.

Heat flow and crustal thickness

For continental crust < 40 km thick, average heat flow is inversely correlated with crustal thickness by the expression,

$$q_o = -2.48 \, (t - 3.48) + 61.2, \qquad (2.2)$$

where q_o is given in mW/m^2 and t in km (Bodri and Bodri, 1985). The correlation coefficient r = −0.89. Although various crustal types fall on different parts of the regression line and reflect different temperature regimes at depth, all reflect similar rates of thinning of about 4 km for each 10 mW/m^2 increase in heat flow. The origin of the inverse q–t relationship is not well understood. One possibility is that thickening of the crust occurs primarily by underplating of mafic rocks with relatively low radiogenic heat production. Thus, the thicker the crust, the greater the amount of uplift and erosion of the upper heat-producing layer. Seismic velocity data lend support to this model.

Table 2.3   Reduced heat-flow provinces

		Age (Ma)	q_o (mW/m^2)	q_r (mW/m^2)	D (km)
1	Basin and Range (Nevada)	0–65	77	63	9.4
2	Eastern Australia	0.65	72	57	11.0
3	Appalachian Mountains	400–100	49	28	7.5
4	United Kingdom	1000–300	69	24	16.0
5	Western Australian shield	> 2500	39	30	4.5
6	India shield	> 1800	71	38	15.0
7	Superior province (Canada)	> 2500	34	22	14.0
8	Baltic shield	> 1800	33	22	8.5

q_o and q_r are average surface and reduced heat flow respectively
D is the characteristic depth

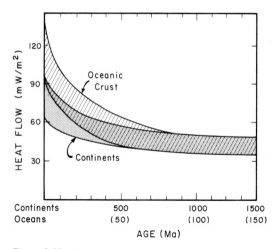

Figure 2.13 Age dependence of average heat flow for continental and oceanic crust. Numbers in parentheses on horizontal axis refer to ages in oceanic areas. After Morgan (1985).

Age dependence of heat flow

Average surface heat flow decreases with the average age of crustal rocks in both oceanic and continental areas (Sclater et al., 1980; Morgan, 1985). Continental heat flow drops with age to an approximately constant value of 40–50 mW/m^2 in about 1 Gy and, in oceanic areas, similar constant heat-flow values are reached in only 50–100 My (Figure 2.13). Continental heat flow can be considered in terms of three components, all of which decay with time (Vitorello and Pollack, 1980). Radiogenic heat in the upper crust contributes about 40 per cent of the total heat flow in continental crust of all ages. The absolute amount of heat from this source decreases with time in response to erosion of the upper crust. A second component of heat, contributing about 30 per cent to the heat flow in Cenozoic terranes, is residual heat of igneous activity associated with orogeny. These two components decay rapidly in a few hundred million years. The third and generally minor component comes from convective heat from within the mantle.

In oceanic areas, surface heat flow (q_o) falls off with the square root of crustal age (t), according to

$$q_o = 11.3 \; t^{(-1/2)} \qquad (2.3)$$

to approximately 120 My (Sclater et al., 1980). As the lithosphere cools it contracts, and the depth to the ocean floor increases going away from ocean ridges. The depth of the ocean floor (d) in crust with an age up to about 70 My can be approximated by

$$d \text{ (metres)} = 2500 + 350 \; t^{(1/2)} \qquad (2.4)$$

The age of the oceanic crust increases as a function of distance from ocean ridges and as a function of decreasing mean elevation as a consequence of seafloor spreading. The decrease in heat flow with distance to an active ocean ridge reflects cooling of new crust which was formed at the ridge by injection of magma. Models assuming a lithospheric plate 50–100 km thick with basal temperatures in the range of 550–1500 °C are consistent with observed heat-flow distributions adjacent to ocean ridges, which decay to equilibrium levels of about 40 mW/m^2 in 50–100 My.

Why should continental and oceanic heat flow decay to similar equilibrium values of 40–50 mW/m^2, when these two types of crust have had such different origins and histories? One model which accommodates this similarity is summarized in Table 2.4. The model assumes a 10-km thick oceanic crust and a 40-km thick continental crust, with an exponential distribution of radiogenic heat sources in the continental crust. Equilibrium reduced heat flow values (q_r) are selected from the range of observed values in both crustal types. Both models have similar temperatures by depths of 100 km, the average thickness of the lithosphere. The results clearly imply that more mantle heat is entering the base of the oceanic lithosphere (44 mW/m^2) than of the continental lithosphere (12 mW/m^2). Hence, when a plate carrying a Precambrian shield moves over oceanic mantle, the surface heat flow should rise until mantle convective systems readjust so that they are again liberating most heat beneath oceanic areas. This may account for the relatively high surface heat flow of the India shield (Table 2.3), which has moved across oceanic crust in the last 70 My.

Exhumation and cratonization

Introduction

In order to understand how continental crust evolves, we need to understand how collisional orogens evolve,

Table 2.4 Heat flow model for ocean basin and Precambrian shield

	Ocean basin $q_o = 50$ mW/m^2			Precambrian shield $q_o = 38$ mW/m^2	
Thickness (km)	A (mW/m^3)	q (mW/m^2)	*Thickness* (km)	A (mW/m^3)	q (mW/m^2)
10	0.42	4	35	0.7*	25
90	0.02	2	65	0.02	1
Mantle	0.02	44	Mantle	0.02	12

*A = $A_o e^{-(x/D)}$ where A_o = 2.5 mW/m^3 and D = 10 km

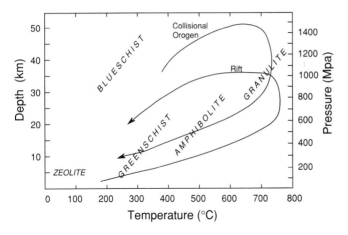

Figure 2.14 Pressure–temperature diagram showing the distribution of metamorphic facies and typical P–T–t paths of a collisional orogen and a continental rift.

since continents are made of collisional orogens of various ages. During continental collisions, large segments of continental crust are highly deformed and thickened and later they are uplifted and eroded to form cratons. The process of uplift and erosion is known as crustal **exhumation**. Crustal history is imprinted in rocks as pressure–temperature–time trajectories, generally referred to as crustal **P–T–t paths**. Not only collisional orogens, but all crustal types record P–T–t paths, and to understand continental growth and development, it is important to reconstruct and understand the meanings of these P–T–t paths.

Burial of rocks in the Earth's crust results in progressive metamorphism as the rocks are subjected to increasing pressure and temperature. Given that metamorphic reactions are more rapid with rising than with falling temperature, metamorphic mineral assemblages may record the highest P–T regime to which rocks have been subjected. Later exhumation makes it possible to study directly rocks that were once buried at various depths in the crust but are now exposed at the Earth's surface. Such rocks contain metamorphic mineral assemblages 'arrested' at some burial depth or metamorphic grade. Minor changes in mineralogy known as retrograde metamorphism may occur during uplift, but such changes commonly can be identified without great difficulty by studying textural features of the rocks. Progressive metamorphism is accompanied by losses of H_2O and other volatile constituents, with some high-grade rocks being almost anhydrous.

Zones of increasing metamorphic grade can be classified into **metamorphic facies**, which represent limited ranges of burial depth, temperature, and water content in the crust. Five major facies of regional metamorphism are recognized (Figure 2.14). The *zeolite facies* is characterized by the development of zeolites in sediments and volcanics and reflects temperatures generally less than 200 °C and burial depths up to 5 km. The *greenschist facies* is characterized by the development of chlorite, actinolite, epidote, and albite in mafic volcanics and of

muscovite and biotite in pelitic rocks. *Blueschist-facies* assemblages form at high pressures (> 800 Mpa) yet low temperatures (< 400 °C) in subduction zones, and typical minerals are glaucophane, lawsonite and jadeite. The *amphibolite facies* is characterized by kyanite, staurolite, and sillimanite in metapelites and plagioclase and hornblende in mafic rocks. The highest grade rocks occur in the *granulite facies*, which is characterized by a sparsity or absence of hydrous minerals and the appearance of pyroxenes. Using experimental petrologic and oxygen isotope data, it is possible to estimate the temperature and pressure at which metamorphic mineral assemblages crystallize, and from these results, burial depths of metamorphic terrains exposed at the Earth's surface can be estimated.

Unravelling P–T–t histories

To begin to understand complex P–T–t histories, many data sources must be used. It is important to know the sequence in which metamorphic minerals have grown and how their growth is related to deformation (Brown, 1993). This can be established from petrographic studies of metamorphic rocks in which textural relationships between mineral growth and deformational fabric are preserved. Particularly important are such features as mineral inclusions in porphyroblasts, replacement textures of one mineral by another, and mineral zoning, all of which reflect changing conditions along the P–T–t path. By relating mineral growth relationships to experimentally determined P–T stability fields of metamorphic minerals, it is possible in some instances to pin down the pressure and temperature at which a given mineral assemblage grew. This is known as **geothermobarometry**. In turn, local P–T conditions may be related to broad regional fabrics in crustal rocks, thus tying microscopic growth data to orogenic development.

To add time to these events, it is necessary to date isotopically minerals that form along different segments of the P–T–t path and become isotopically closed at

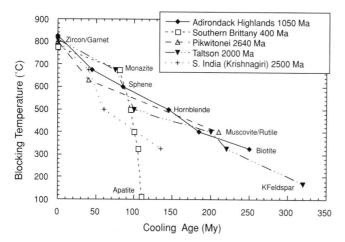

Figure 2.15 Cooling histories of several orogens. Ages of maximum temperatures given in the explanation and equated to zero age on the cooling age axis. Blocking temperature is the temperature at which the daughter isotope is trapped in a host mineral. Data from Harley and Black (1987), Dallmeyer and Brown (1992), Mezger et al. (1991), Kontak and Reynolds (1994).

given temperatures. For instance, the $^{40}Ar/^{39}$ method can be used to date hornblende, which closes to argon diffusion at about 500 °C, muscovite, that closes at 350–400 °C, K-feldspar, that closes at 150–350 °C, and fission tacks, which anneal in apatite at about 150 °C (McDougall and Harrison, 1988; Crowley and Kuhlman, 1988). Diffusion experiments have confirmed these mineral closure temperatures. Thus, to unravel P–T–t histories in the crust, it is critically necessary to combine results from field, petrographic, experimental, and isotopic studies. Even then it may not be possible to untangle a complex history in which a crustal segment has undergone multiple deformation and metamorphism.

Some typical P–T–t paths

Varying tectonic histories yield different P–T–t paths (Chapman and Furlong, 1992). As examples, let us consider collisional orogens and continental rifts. P–T–t paths of collisional orogens are typically clockwise in P–T–t space and their general features are reasonably well understood from classic studies in the Appalachians, Caledonides, Alps, and Himalayas (Brown, 1993). This type of P–T–t path results from rapid crustal thickening so that the maximum pressure is reached before the maximum temperature (Figure 2.14). Hence, the metamorphic peak generally post-dates early deformation in the orogen. This evolutionary path commonly leads to dehydration melting of the lower crust producing granitic magmas, which abound in collisional orogens.

During the development of a continental rift by crustal extension, the crust is heated from below by mantle upwelling before crustal thickening occurs, and thus the maximum temperature is reached before the maximum pressure. This produces a counterclockwise P–T–t path (Figure 2.14). The metamorphic peak usually predates or is synchronous with early deformation in these cases. Again, heating of the lower crust can lead to dehydration melting as a consequence of the metamorphism, producing felsic magmas. Any crustal environment in which

heating precedes thickening of the crust results in counterclockwise P–T–t paths. In addition to rifts, extended margins of platforms and magmatically underplated crust, such as beneath continental flood basalts, have counterclockwise P–T–t paths. In some instances, such as the northern Appalachians, clockwise and counterclockwise P–T–t paths can occur in adjacent segments of the same orogen.

Cratonization

Although cratons have long been recognized as an important part of the continental crust, their origin and evolution is still not well-understood. Most investigators agree that cratons are the end product of collisional orogenesis, and thus they are the building blocks of continents. Just how orogens evolve into cratons and how long it takes, however, are not well known. Although studies of collisional orogens show that most are characterized by clockwise P–T–t paths (Thompson and Ridley, 1987; Brown, 1993), the uplift/exhumation segments of the P–T paths are poorly constrained (Martignole, 1992). In terms of craton development, it is the < 500 °C portion of the P–T–t path that is most important.

Using a variety of radiogenic isotopic systems and estimated closure temperatures in various minerals, it is possible to track the cooling histories of crustal segments and, when coupled with geothermobarometry, also the uplift/exhumation histories. Results suggest a wide variation in cooling and uplift rates with most orogens having cooling rates of < 2 °C/My, whereas a few (like Southern Brittany) cool at rates of > 10 °C/My (Figure 2.15). In most cases, it would appear to take a minimum of 300 My to make a craton. Some terranes, such as Enderbyland in Antarctica, have had very long, perhaps exceedingly complex cooling histories lasting for more the 2 Gy. Many orogens, such as the Grenville orogen in eastern Canada, have been exhumed as indicated by unconformably overlying sediments, reheated during subsequent burial and then re-exhumed (Heizler, 1993).

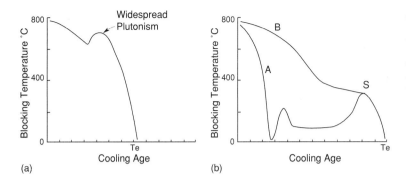

Figure 2.16 Two possible cooling scenarios in cratons. (a) Overprinting of a post-tectonic granite intrusion, and (b) a complex, multiple-event cooling history. Te, final exhumation age; S, suturing age of terranes A and B.

In some instances, post-orogenic thermal events such as plutonism and metamorphism have thermally overprinted earlier segments of an orogen's cooling history, such that only the very early high temperature history (> 500 °C) and, perhaps, the latest exhumation record (< 300 °C) are preserved. Fission track ages suggest that final uplift and exhumation of some orogens, like the 1.9-Ga Trans-Hudson orogen in central Canada, may be related to the early stages of supercontinent fragmentation.

An important, yet poorly-understood aspect of cratonization is that of how terranes which amalgamated during a continent–continent collision evolve into a craton. Does each terrane maintain its own identity and have its own cooling and uplift history? Or, alternatively, do terranes anneal to each other at an early stage and then the entire orogen cools and is elevated as a unit? What is the effect of widespread post-tectonic (anorogenic) plutonism? Does it overprint and erase important segments of the orogen cooling history? It is well-known that crustal cooling curves are not always equivalent to exhumation curves (Thompson and Ridley, 1987). In fact, some granulite-grade blocks appear to have undergone long periods of isobaric (constant depth) cooling before exhumation. Also, discrete thermal events can completely or partially reset thermochronometers without an obvious geologic rock record, and this can lead to erroneous conclusions regarding average cooling rates (Heizler, 1993).

Post-tectonic plutonism, which follows major deformation or multiple deformation of an orogen, can lead to a complex cooling history. Widespread post-tectonic plutonism can perturb the cooling curve of a crustal segment, prolonging the cooling history (Figure 2.16a). In an even more complex scenario a crustal domain can be exhumed, reburied as sediments accumulate in an overlying basin, age for hundreds of millions of years at about the same crustal level, and finally be re-exhumed (A, Figure 2.16b). In this example, all of the < 400 °C thermal history is lost by the overprinting of the final thermal event. A second terrane B could be sutured to A during this event (S in Figure 2.16b), and both domains exhumed together. It is clear from these examples that much or all of a complex thermal history can be erased by the last thermal event, producing an apparent gap in a cratonization cooling curve.

Processes in the continental crust

Rheology

The behaviour of the continental crust under stress depends chiefly on the temperature and the duration of the stresses. The hotter the crust, the more it behaves like a ductile solid deforming by plastic flow, whereas if it is cool, it behaves as an elastic solid deforming by brittle fracture and frictional gliding (Ranalli, 1991; Rutter and Brodie, 1992). The distribution of strength with depth in the crust varies with tectonic setting, strain rate, thickness and composition of the crust, and the geotherm. The **brittle–ductile transition** corresponding to an average surface heat flow of 50 mW/m^2 is at about 30 km depth, which also corresponds to the depth limit of most shallow earthquakes. Even in the lower crust, however, if stress is applied rapidly it may deform by fracture, and likewise if pore fluids are present in the upper crust, weakening it, and stresses are applied slowly, the upper crust may deform plastically. In regions of low heat flow, such as shields and platforms, brittle fracture may extend into the lower crust or even into the upper mantle because mafic and ultramafic rocks can be very resistant to plastic failure at these depths, and thus brittle faulting is the only way they can deform. Lithologic changes at these depths, the most important of which is at the Moho, are also likely to be rheological discontinuities.

Examples of two rheological profiles into the crust and subcontinental lithosphere are shown in Figure 2.17. The brittle–ductile transition occurs at about 20 km depth in the rift, whereas in the cooler and stronger Proterozoic shield it occurs at about 30 km. In both cases, the strength of the ductile lower crust decreases with increasing depth, reaching a minimum at the Moho. The rapid increase in strength beneath the Moho reflects chiefly the increase in olivine, which is stronger than pyroxenes and feldspars. The rheological base of the lithosphere, which is generally taken as a strength equal to about 1 Mpa, occurs at 55 km beneath the rift and 120 km beneath the Proterozoic shield. In general, the brittle–ductile transition occurs at relatively shallow depths in warm and young crust (10–20 km), whereas in cool and old crust it occurs at greater depths (20–30 km).

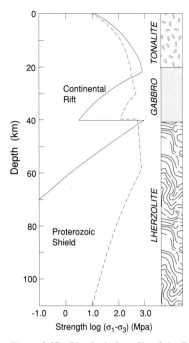

Figure 2.17 Rheological profile of the East African rift and the Proterozoic shield in East Africa. Strength expressed as the difference between maximum and minimum compressive stresses (σ_1 and σ_3, respectively). Strain rate = 10–15 s^{-1} and model compositions are tonalite, gabbro, and lherzolite. Diagram after Ranalli (1991).

The role of fluids and crustal melts

Fluid transport in the crust is an important process affecting both the rheology and chemical evolution of the crust. Because crustal fluids are mostly inaccessible for direct observation, this process is poorly understood and difficult to study. Studies of fluid inclusions trapped in metamorphic and igneous minerals indicate that shallow crustal fluids are chiefly water, whereas deep crustal fluids are mixtures of H_2O and CO_2, and both contain various dissolved species (Bohlen, 1991; Wickham, 1992). Fluids are reactive with silicate melts and in the lower crust they can promote melting, as well as changing the chemical and isotopic composition of rocks.

In the lower crust, only small amounts of fluid can be generated by the breakdown of hydrous minerals such as biotite and hornblende. Hence, the only major source of fluids in the lower crust is from the mantle. Studies of xenoliths suggest that the mantle lithosphere provides a potentially large source for CO_2 in the lower crust, and may be the principal source for CO_2 which is important in the production of deep crustal granulites.

The formation of granitic melts in the lower crust and their transfer to shallower depths is a fundamental process leading to the chemical differentiation of the continents. This is particularly important in arcs and collisional orogens. The melt-producing capacity of a source rock

in the lower crust is determined chiefly by its chemical composition, but also depends on temperature regime and the fluid content (Brown et al., 1995). Orogens that include a large volume of juvenile volcanics and sediments are more fertile (high melt-producing capacity) than those which include chiefly older basement rocks from which fluids and melts have been extracted (Vielzeuf et al., 1990). A fertile lower crust can generate a range of granitic melt compositions and leave behind a residue of granulites. Segregation of melt from source rocks can occur by several different processes, and just how much and how fast melt is segregated is not well known. It depends, however, on whether deformation occurs concurrently with melt segregation. Experiments indicate that melt segregation is enhanced by increased fluid pressures and fracturing of surrounding rocks. Modelling suggests that shear-induced compaction can drive melt into veins that transfer it rapidly to shallow crustal levels (Rutter and Neumann, 1995).

Crustal composition

Approaches

Several approaches have been used to estimate the chemical and mineralogical composition of the crust. One of the earliest methods used to estimate the composition of the upper continental crust was based on chemical analysis of glacial clays, which were assumed to be representative of the composition of large portions of the upper continental crust. Estimates of total continental composition were based on mixing average basalt and granite compositions in ratios generally ranging from 1:1 to 1:3 (Taylor and McLennan, 1985), or by weighting the compositions of various igneous, metamorphic, and sedimentary rocks according to their inferred abundances in the crust (Ronov and Yaroshevsky, 1969). Probably the most accurate estimates of the composition of the upper continental crust come from extensive sampling of rocks exhumed from varying depths in Precambrian shields and from the composition of Phanerozoic shales (Taylor and McLennan, 1985; Condie, 1993). Because the lower continental crust is not accessible for sampling, indirect approaches must be used. These include:

1   measurement of seismic-wave velocities of crustal rocks in the laboratory at appropriate temperatures and pressures, and comparing these with observed velocity distributions in the crust
2   sampling and analysing rocks from blocks of continental crust exhumed from middle to lower crustal depths
3   analysing xenoliths of lower crustal rocks brought to the surface during volcanic eruptions.

The composition of oceanic crust is estimated from the composition of rocks in ophiolites and from shallow cores

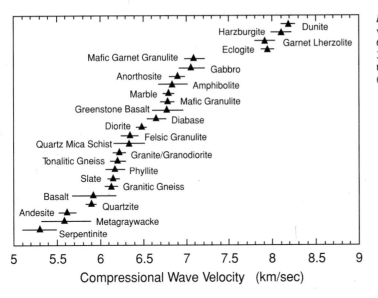

Figure 2.18 Average compressional wave velocities and standard deviations at 600 Mpa (20 km depth equivalent) and 300 °C (average heat flow) for major rock types. Data from Christensen and Mooney (1995).

into the sediment and basement layers of oceanic crust retrieved by the Ocean Drilling Project. Results are again constrained by seismic velocity distributions in the oceanic crust.

Before discussing the chemical composition of the crust, the major sources of data will be reviewed.

Seismic-wave velocities

Because seismic-wave velocities are related to rock density and density is related to rock composition, the measurement of these velocities provides an important constraint on the composition of both the oceanic and continental crust (Rudnick and Fountain, 1995). Poisson's ratio, which is the ratio of P- to S-wave velocity, is more diagnostic of crustal composition than either P- or S-wave data alone (Zandt and Ammon, 1995) (Tables 2.1 and 2.2). Because pressure and temperature influence seismic velocities, it is necessary to correct laboratory measurements to a common pressure and temperature for comparisons. These effects tend to cancel each other, so the magnitude of this correction is not large (< 5 per cent). Another variable in measuring seismic velocities is anisotropy, which is variation in velocity in different directions in a rock. Metamorphic rocks, in particular, can show significant anisotropy, reaching values as high a 17 per cent in phyllites and schists (Christensen and Mooney, 1995). Typical igneous rocks, granulites, and eclogites, however, have anisotropies of < 5 per cent.

Figure 2.18 shows average compressional wave velocities (at 600 Mpa and 300 °C) in a variety of crustal rocks. Velocities under 6 km/sec are limited to serpentinite, metagraywacke, andesite, and basalt. Many rocks of diverse origins have velocities between 6 km/sec and 6.5 km/sec, including slates, granites, altered basalts, and felsic granulites. With the exception of marble and anorthosite, which are probably minor components of

the crust based on exposed blocks of lower crust and xenoliths, all rocks with velocities of 6.5–7.0 km/sec are mafic in composition and include amphibolites and mafic granulites without garnet (Holbrook et al., 1992; Christensen and Mooney, 1995). Rocks with average velocities of 7.0–7.5 km/sec include gabbro, hornblendite, and mafic garnet granulite, and velocities above 7.5 km/sec are limited to non-serpentinized ultramafic rocks and eclogite (a very high-pressure mafic rock). It is important to note also that the order of increasing velocity in Figure 2.18 is not a simple function of increasing metamorphic grade. For instance, low, medium, and high grade metamorphic rocks all fall in the range of 6.0–7.5 km/sec.

As discussed below, rock types in the upper continental crust are reasonably well known from studies in Precambrian shields. The distribution of rock types in the lower crust, however, still remains uncertain. Platform lower crust, although having relatively high S-wave velocities, shows similar Poisson's ratios to collisional orogens (Figure 2.19b; Tables 2.1 and 2.2). The lower crust of continental rifts, however, shows distinctly lower velocities, a feature that would appear to reflect hotter temperatures in the lower crust. Two observations are immediately apparent from the measured rock velocities as summarized in Figure 2.19a:

1 the velocity distribution in the lower crust indicates compositional heterogeneity
2 metapelitic rocks overlap in velocity with mafic and felsic igneous and metamorphic rocks.

It is also interesting that, with the exception of rifts, mean lower crustal velocities are strikingly similar to mafic rock velocities. However, because of the overlap in velocities of rocks of different compositions and origins, it is not possible to assign unique rock compositions to the lower crust from seismic velocity data alone.

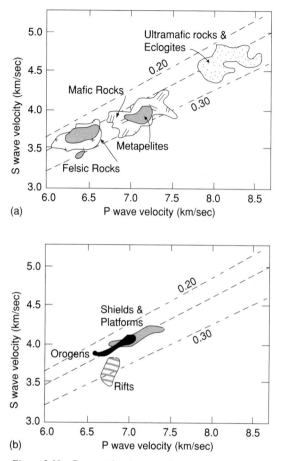

(a)

(b)

Figure 2.19 Compressional versus shear-wave velocity diagrams showing, (a) fields of various crustal and upper mantle rocks, and (b) lower crust of various crustal types. Velocities normalized to 20 km depth and room temperature. Dashed lines are Poisson's ratio $(0.5\{1 - 1/[(Vp/Vs)^2 - 1]\})$. Modified after Rudnick and Fountain (1995).

When coupled with xenolith data, however, the seismic velocity distributions suggest that the lower continental crust is composed largely of mafic granulites, gabbros, and amphibolites (50–65 per cent), with up to 10 per cent metapelites, and the remainder of intermediate to felsic granulites (Rudnick and Fountain, 1995).

In common rock types, Poisson's ratio (s), varies from about 0.2 to 0.35 and is particularly sensitive to composition. Increasing silica content lowers s and increasing Fe and Mg increases it (Zandt and Ammon, 1995). The average value of s in the continental crust shows a good correlation with crustal type (Figure 2.20; Table 2.1). Precambrian shield s values are consistently high, averaging 0.29, and platforms average about 0.27. The lower s in platforms and Paleozoic orogens appears to reflect the silica-rich sediments that add 4–5 km of crustal thickness to the average shield (Table 2.1). In Mesozoic–Cenozoic orogens, however, s is even lower but more variable, reflecting some combination of lithologic and thermal differences in the young orogenic crust. The high ratios in continental-margin arcs may reflect the importance of mafic rocks in the root zones of these arcs, although again the variation in s is great.

The origin of the Moho continues to be a subject of widespread interest (Jarchow and Thompson, 1989). Because the oceanic Moho is exposed in many ophiolites, it is better known than the continental Moho. From seismic velocity distributions and from ophiolite studies, the oceanic Moho is probably a complex transition zone from 0–3 km thick, between mixed mafic and ultramafic igneous cumulates in the crust and harzburgites (orthopyroxene-olivine rocks) in the upper mantle. It would appear that large tectonic lenses of differing lithologies occur at the oceanic Moho, which are the products of ductile deformation along the boundary. The continental Moho is considerably more complex and varies in nature with crustal type and crustal age. Experimental, geophysical, and xenolith data, however, do not favour a gabbro–eclogite transition to explain the

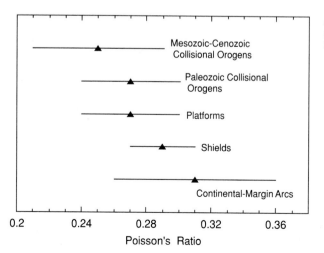

Figure 2.20 Mean values and one standard deviation of Poisson's ratio in various crustal types. Data from Zandt and Ammon (1995).

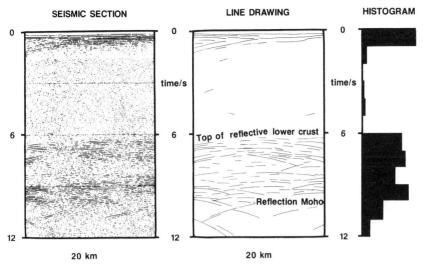

SEISMIC SECTION   LINE DRAWING   HISTOGRAM

Figure 2.21 Seismic
reflection profile southwest
of England. Also shown is a
line-drawing of the data and
a histogram of reflection
depths summed over 1 sec
travel-time intervals. After
Klemperer (1987).

continental Moho. Also, the absence of a correlation between surface heat flow and crustal thickness does not favour a garnet granulite–eclogite phase change at the Moho. Although eclogite is thermodynamically stable throughout most or all of the continental crust, it will not form below about 500 °C due to very small reaction rates, and hence mafic granulite is the stable mafic mineral assemblage in the lower crust. Beneath platforms and shields, the Moho is only weakly (or not at all) reflective, suggesting the existence of a relatively thick transition zone (> 3 km) composed of mixed mafic granulites, eclogites, and lherzolites, with no strong reflecting surfaces. Xenolith populations in the Aleutian arc indicate that the lower crust is composed of cumulates of arc magmas and fragments of older oceanic crust overlying ultramafic cumulates in the mantle (Kay and Kay, 1985). In modern rift systems, magmatic underplating results in a continually changing Moho, but again with mafic rocks in the crust and ultramafic rocks in the mantle.

Seismic reflections in the lower continental crust

Many explanations have been suggested for the strong seismic reflectors found in much of the lower continental crust (Figure 2.21) (Nelson, 1991; Mooney and Meissner, 1992). The most important possible causes fall into one of three categories:

1 layers in which fluids are concentrated
2 strain-banding developed from ductile deformation
3 lithologic layering.

The jury is still out on the role of deep crustal fluids. Some investigators argue that physical conditions in the lower crust allow up to 4 per cent of saline pore waters, and that the high electrical conductivity of the lower crust supports such a model. In this model, the seismic reflections are produced by layers with strong porosity contrasts. However, textural and mineralogical evidence from deep crustal rocks exposed at the surface and from xenoliths do not have high porosities, and thus contradict this idea.

Deformation bands hold more promise, at least for some of the lower crustal reflections, in that shear zones exposed at the surface can be traced to known seismic reflectors at depth in shallow crust (Mooney and Meissner, 1992). Some lower crustal reflection patterns in Precambrian cratons preserve structures that date from ancient collisional events, as for instance in the TransHudson orogen of Early Proterozoic age in Canada. At least in these cases, it would appear that the reflections are caused by tectonic boundaries, or syntectonic igneous intrusions. In extended crust, like that occurring in rifts, ductile shearing in the lower crust may enhance metamorphic or igneous layering.

The most probable cause of many lower crustal reflections is lithologic layering, caused by mafic sills, compositional layering in mafic intrusions, or metamorphic fabrics. Supporting this conclusion is the fact that some shallow reflectors in the crust which have been traced to the surface are caused by such layering (Percival et al., 1989). Furthermore, a bimodal distribution of acoustic impedance in the lower crust favours layered sequences of rocks, especially interlayered mafic and felsic units (Goff et al., 1994). Also, models of reflectivity in the Ivrea zone (a fragment of mafic lower crust faulted to the surface in the Alps) show that lower crustal reflections are expected when mafic rocks are interlayered with felsic rocks (Holliger et al., 1993).

The fact that seismic reflectivity in the lower crust is widespread and occurs in crustal types with differing heat flow characteristics favours a single origin for most reflectors. From studies of exhumed lower crust and lower crustal xenoliths, it would seem that most lower crustal reflections are caused by mafic intrusions, and in some instances the reflections have been enhanced by later ductile deformation.

Sampling of Precambrian shields

Widespread sampling of metamorphic terranes exposed in Precambrian shields, and especially in the Canadian shield, has provided an extensive sample base to estimate both the chemical and lithologic composition of the upper part of the Precambrian continental crust (Shaw et al., 1986; Condie, 1993). Both individual and composite samples have been analysed. Results indicate that although the upper crust is lithologically heterogeneous, granitoids of granodiorite to tonalite composition dominate and the weighted average composition is that of granodiorite.

Use of fine-grained terrigenous sediments

Fine-grained terrigenous sediments may represent well-mixed samples of the upper continental crust, and thus provide a means of estimating upper crustal composition (Taylor and McLennan, 1985). However, in order to use sediments to estimate crustal composition it is necessary to evaluate losses and gains of elements during weathering, erosion, deposition, and diagenesis. Elements, such as rare earths (REE), Th, and Sc that are relatively insoluble in natural waters and have short residence times in seawater ($< 10^3$y) may be transferred almost totally into terrigenous clastic sediments. The remarkable uniformity of REE in pelites and loess compared to the great variability observed in igneous source rocks attests to the efficiency of mixing during erosion and deposition. Studies of REE and element ratios such as La/Sc, La/Yb, and Cr/Th indicate that they remain relatively unaffected by weathering and diagenesis. REE distributions are especially constant in shales and resemble REE distributions in weighted averages from Precambrian shields. With some notable exceptions (Condie, 1993), estimates of the average composition of the upper continental crust using the composition of shales are in remarkable agreement with the weighted chemical averages determined from rocks exposed in Precambrian shields.

Exhumed crustal blocks

Several blocks of middle to lower continental crust have been recognized in Precambrian shields or collisional orogens, the best known of which is the Kapuskasing uplift in southern Canada (Percival et al., 1992; Percival and West, 1994) and the Ivrea Complex in Italy (Sinigoi et al., 1994). Four different mechanisms have been suggested as bringing these deep crustal sections to the surface:

1 large thrust sheets formed during continent–continent collisions

2 transpressional faulting
3 broad tilting of a large segment of crust
4 asteroid impacts.

However, tectonic settings at the times of formation of rocks within the uplifted blocks appear to be collisional orogens, island arcs, or continental rifts. Common to all sections that have been studied are high grade metamorphic rocks that formed at depths of 20–25 km, with a few, like the Kohistan arc in Pakistan, coming from depths as great as 40–50 km. Metamorphic temperatures recorded in the blocks are typically in the range of 700–850 °C. All blocks consist chiefly of felsic components at shallow structural levels and mixed mafic, intermediate, and felsic components at deeper levels. Commonly, lithologic and metamorphic features in uplifted blocks are persistent over lateral distances of over 1000 km, as for instance evidenced by three deep crustal exposures in the Superior province in southern Canada (Percival et al., 1992).

Examples of five sections of middle to lower continental crust are given in Figure 2.22. Each section is a schematic diagram illustrating the relative abundances of major rock types and the base of each section is a major thrust fault. The greatest depths exposed in each section are in the order of 25–35 km. Each column has a lower granulite zone with mafic granulites dominating in three sections and felsic granulites in the other two. The sections show considerable compositional variation at all metamorphic grades attesting to the heterogeneity of the continental crust at all depths. Mafic and ultramafic bodies and anorthosites occur at deep levels in some sections and probably represent layered igneous sheets intruded into the lower crust (Figure 2.23). Volcanic and sedimentary rocks are also buried to great depths in some sections. Intermediate and upper crustal levels are characterized by large volumes of granitoids.

More than anything else, the crustal sections indicate considerable variations in lithologic and chemical composition both laterally and vertically in the continental crust. The only large-scale progressive change in the sections is an increase in metamorphic grade with depth. Although there is no evidence for a Conrad discontinuity in the sections, rapid changes in lithology may be responsible for seismic discontinuities of a more local extent. Again, it should be emphasized that many uplifted blocks probably do not sample the lower crust, but only the middle crust (~ 25 km). Today these blocks are underlain by 35–40 km of crust, probably composed in large part of mafic granulites. The crust in these areas may have thickened during continental collision (to 60–70 km), thus burying upper crustal rocks to granulite grade (35–40 km). Uplift and erosion of this crust brought these felsic granulites to the surface with a possible mafic granulite root still intact. Thus, the differences between the generally felsic to intermediate compositions of uplifted crustal blocks and the mafic compositions of lower crustal xenolith suites (see below), may be due in part to different levels of sampling in the crust.

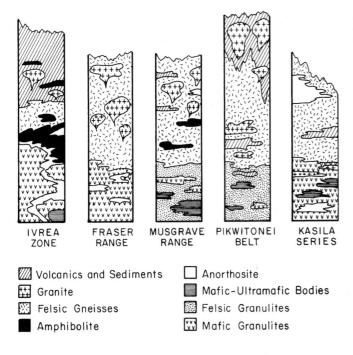

Figure 2.22 Generalized cross-sections of continental crust based on exhumed sections of deep crustal rocks. Modified after Fountain and Salisbury (1981).

IVREA ZONE   FRASER RANGE   MUSGRAVE RANGE   PIKWITONEI BELT   KASILA SERIES

- ▨ Volcanics and Sediments
- ⊞ Granite
- ▨ Felsic Gneisses
- ■ Amphibolite
- ☐ Anorthosite
- ▊ Mafic-Ultramafic Bodies
- ▦ Felsic Granulites
- ▨ Mafic Granulites

Crustal xenoliths

Crustal xenoliths are fragments of the crust brought to the Earth's surface by volcanic eruptions. If one can determine the depth from which xenoliths come by thermobarometry, and estimate the relative abundances of various xenolith populations in the crust, it should be possible to reconstruct a crustal cross-section (Kay and Kay, 1981). Although metamorphic xenoliths can be broadly ordered in terms of their crustal depths, metamorphic mineral assemblages in many lower crustal xenoliths are not definitive in determining precise depths (Rudnick, 1992). Even more difficult is the problem of estimating the relative abundances of xenolith types in the crust. Some lithologies may be oversampled and others undersampled by ascending volcanic magmas. Hence, it is generally not possible to come up with a unique crustal section from xenolith data alone.

Xenolith-bearing volcanics and kimberlites occur in many different tectonic settings, thus giving a wide lateral sampling of the continents. Lower crustal xenoliths from arc volcanics are chiefly mafic in composition and xenoliths of sediments are rare to absent. These results suggest that the root zones of modern arcs are composed chiefly of mafic rocks. Xenoliths from volcanics on continental crust are compositionally diverse and have complex thermal and deformational histories (Rudnick, 1992). Metasedimentary xenoliths, however, are minor compared to meta-igneous xenoliths. In general, xenoliths of mafic granulite are more abundant than felsic granulite, suggesting that a mafic lower crust is also important in cratons (Rudnick and Taylor, 1987). Most of these xenoliths appear to be basaltic melts and their cumulates

that were intruded into or underplated beneath the crust. Most granulite-grade xenoliths reflect equilibration depths in the crust of > 20 km and some more than 40 km. A smaller number of metasedimentary and gneissic xenoliths recording similar depths in many xenolith suites seems to require interlayering of felsic and mafic rocks in the lower crust. Where isotopic ages of xenoliths can be estimated, they range from the age of the host crust to considerably younger. For instance, mafic lower crustal xenoliths from the Four Corners volcanic field in the Colorado Plateau appear to be about 1.7 Ga, the same age as the Precambrian basement in this area (Wendlandt et al., 1993). In contrast, mafic xenoliths from Kilbourne Hole in southern New Mexico in the Rio Grande rift yield mid-Tertiary ages, yet are housed in crust about 1.65 Ga. In the latter case, it would appear that these xenoliths come from basalts underplated beneath the crust as the Rio Grande rift extended.

An estimate of crustal composition

Continental crust

The average chemical composition of the upper continental crust is reasonably well-known from widespread sampling of Precambrian shields, geochemical studies of shales, and exposed crustal sections (Taylor and McLennan, 1985; Condie, 1993). An average composition from Condie (1993) is similar to granodiorite (Table 2.5), although there are differences related to the age of the crust as discussed in Chapter 5. The composi-

Figure 2.23 Layered mafic granulites from the Ivrea Zone in the Alps. Similar rocks may comprise large volumes of the lower continental crust. Courtesy of K. R. Mehnert.

tion of the lower continental crust is much less con-strained. Uplifted crustal blocks, xenolith populations, and seismic velocity and Poisson's ratio data suggest that a large part of the lower crust is mafic in overall composition. In our estimate, we accept the middle and lower crustal estimates of Rudnick and Fountain (1995) based on all of the data sources described above. If the upper continental crust is felsic in composition and the lower crust is mafic as most data suggest, how do these two layers form and how do they persist over geologic time? This intriguing question will be considered in Chapter 5.

The estimate of total continental crust composition in Table 2.5 is a mixture of upper, middle, and lower crustal averages in equal amounts. The composition is similar to other published total crustal compositions indicating an overall intermediate composition (Taylor and McLennan, 1985; Wedepohl, 1995; Rudnick and Foun-tain, 1995). **Incompatible elements**, which are elements that are strongly partitioned into the liquid phase upon

Table 2.5 Average chemical composition of continental and oceanic crust

	Continental crust				Oceanic crust
	Upper	Middle	Lower	Total	
SiO$_2$	66.3	60.6	52.3	59.7	50.5
TiO$_2$	0.7	0.8	0.54	0.68	1.6
Al$_2$O$_3$	14.9	15.5	16.6	15.7	15.3
FeOT	4.68	6.4	8.4	6.5	10.4
MgO	2.46	3.4	7.1	4.3	7.6
MnO	0.07	0.1	0.1	0.09	0.2
CaO	3.55	5.1	9.4	6.0	11.3
Na$_2$O	3.43	3.2	2.6	3.1	2.7
K$_2$O	2.85	2.0	0.6	1.8	0.2
P$_2$O$_5$	0.12	0.1	0.1	0.11	0.2
Rb	87	62	11	53	1
Sr	269	281	348	299	90
Ba	626	402	259	429	7
Th	9.1	6.1	1.2	5.5	0.1
Pb	18	15.3	4.2	13	0.3
U	2.4	1.6	0.2	1.4	0.05
Zr	162	125	68	118	74
Hf	4.4	4.0	1.9	3.4	2.1
Nb	10.3	8	5	7.8	2.3
Ta	0.82	0.6	0.6	0.7	0.13
Y	25	22	16	21	28
La	29	17	8	18	2.5
Ce	59.4	45	20	42	7.5
Sm	4.83	4.4	2.8	4.0	2.6
Eu	1.05	1.5	1.1	1.2	1.0
Yb	2.02	2.3	1.5	1.9	3.1
V	86	118	196	133	275
Cr	112	150	215	159	250
Co	18	25	38	27	47
Ni	60	70	88	73	150

Major elements in weight percentages of the oxide and trace elements in ppm
Lower/Middle crust from Rudnick and Fountain (1995); Upper crust from Condie (1993); Oceanic crust (NMORB) from Sun
  and McDonough (1989) and miscellaneous sources

melting, are known to be concentrated chiefly in the continental crust. During melting in the mantle, these elements will be enriched in the magmas, and thus be transferred upward into the crust as the magmas rise. Relative to primitive mantle composition, 35–65 per cent of the most incompatible elements (such as Rb, Th, U, K, and Ba) are contained in the continents, whereas continents contain < 10 per cent of the least incompatible elements (like Y, Yb, and Ti).

Oceanic crust

Because fragments of oceanic crust are preserved on the continents as ophiolites, there is direct access to sampling for chemical analysis. The chief problem with equating the composition of ophiolites to average oceanic crust, however, is that some or most ophiolites appear to have formed in back-arc basins and, in varying degrees, have geochemical signatures of arc systems (Chapter 3). Other sources of data for estimating the composition of oceanic crust are dredge samples from

the ocean floor and drill cores retrieved from the Ocean Drilling Project which penetrated the basement layer. Studies of ophiolites and P-wave velocity measurements are consistent with basement and oceanic layers being composed largely of mafic rocks metamorphosed to the greenschist or amphibolite facies. The sediment layer is composed of pelagic sediments of variable composition and extent, and contributes < 5 per cent to the bulk composition of the oceanic crust.

An estimate of the composition of oceanic crust is also given in Table 2.5. It is based on the average composition of normal ocean-ridge basalts, excluding data from back-arc basins, and pelagic sediments are ignored in the estimate. Although ophiolites contain minor amounts of ultramafic rock and felsic rock, they are much less variable in lithologic and chemical composition than crustal sections of continental crust, suggesting that the oceanic crust is rather uniform in composition. Because of the relatively small volume of oceanic crust compared with continental crust (Table 2.1), and because oceanic basalts come from a mantle source that is de-

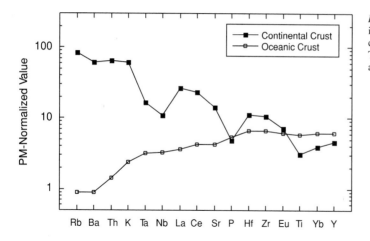

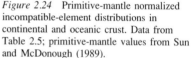

Figure 2.24 Primitive-mantle normalized incompatible-element distributions in continental and oceanic crust. Data from Table 2.5; primitive-mantle values from Sun and McDonough (1989).

pleted in incompatible elements (Chapter 4), the oceanic crust contains very little of the Earth's inventory of these elements.

Complementary compositions of continental and oceanic crust

The average composition of the continental and oceanic crusts relative to the primitive mantle composition form surprisingly complementary patterns (Figure 2.24). In continental crust the maximum concentrations, which reach values of 50–100 times primitive mantle values, are for the most incompatible elements K, Rb, Th, and Ba. There same elements only reach < 3 times primitive mantle values in the oceanic crust. The patterns cross at P, and the least incompatible elements Ti, Yb and Y are more enriched in oceanic than in continental crust. The relative depletions in Ta-Nb, P, and Ti are important features of the continental crust and will be discussed more fully in Chapter 5. The complementary crustal element patterns can be explained if most of the continental crust is extracted from the upper mantle first, leaving an upper mantle depleted in very incompatible elements. The oceanic crust is then continuously produced from this depleted upper mantle throughout geologic time (Hofmann, 1988).

Crustal provinces and terranes

Introduction

Stockwell (1965) suggested that the Canadian shield can be subdivided into structural provinces based on differences in structural trends and styles of folding. Structural trends are defined by foliation, fold axes, bedding and sometimes by geophysical anomalies. Boundaries between the provinces are drawn where one trend cross-cuts another, along either unconformities or structural-metamorphic breaks. Large numbers of isotopic dates from the Canadian shield indicate that structural prov-

inces are broadly coincident with age provinces. Similar relationships have been described on other continents and lead to the concept of a crustal province as discussed below.

Terranes are fault-bounded crustal blocks which have distinct lithologic and stratigraphic successions and which have geologic histories different from neighbouring terranes (Schermer et al., 1984). Most terranes have collided with continental crust, either along transcurrent faults or at subduction zones, and have been sutured to continents. Many terranes contain faunal populations and paleomagnetic evidence indicating they have been displaced great distances from their sources prior to continental collision. For instance, Wrangellia, which collided with western North America in the Late Cretaceous, had travelled many thousands of kilometres from what is now the South Pacific. Results suggest that as much as 30 per cent of North America was formed by terrane accretion in the last 300 My and that terrane accretion has been an important process in the growth of continents.

Terranes form in a variety of tectonic settings, including island arcs, submarine plateaux, volcanic islands, and microcontinents. Numerous potential terranes exist in the oceans today and are particularly abundant in the Pacific basin (Figure 2.25). Continental crust may be fragmented and dispersed by rifting or strike-slip faulting. In western North America, dispersion is occurring along transform faults such as the San Andreas and Fairweather faults, and in New Zealand movement along the Alpine transform fault is fragmenting the Campbell Plateau from the Lord Howe Rise (Figure 1.1). Baja California and California west of the San Andreas fault were rifted from North America about 3 Ma, and today this region is a potential terrane moving northward, perhaps on a collision course with Alaska. Terranes may continue to fragment and disperse after collision with continents, as did Wrangellia which is now distributed in pieces from Oregon to Alaska. The 1.9-Ga Trans-Hudson orogen in Canada and the 1.75–1.65 Yavapai orogen in the southwestern United States are examples of Proterozoic orogens composed of terranes, and the

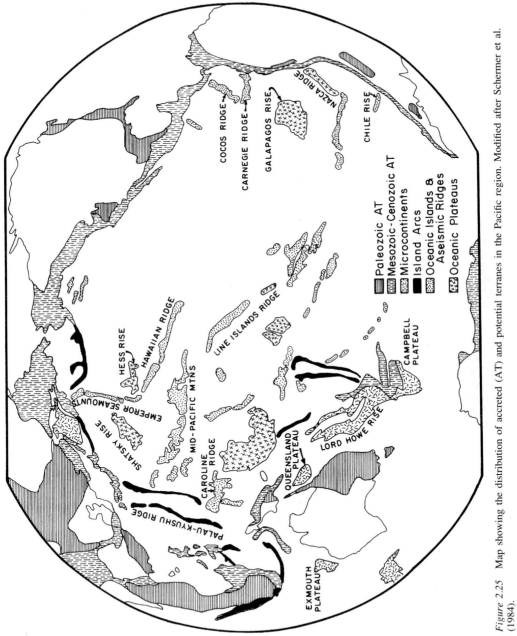

Figure 2.25 Map showing the distribution of accreted (AT) and potential terranes in the Pacific region. Modified after Schermer et al. (1984).

COCOS RIDGE

CARNEGIE RIDGE

GALAPAGOS RISE

NAZCA RIDGE

CHILE RISE

Paleozoic AT
Mesozoic-Cenozoic AT
Microcontinents
Island Arcs
Oceanic Islands &
Aseismic Ridges
Oceanic Plateaus

CAMPBELL
PLATEAU

LORD HOWE RISE

QUEENSLAND
PLATEAU

EXMOUTH
PLATEAU

PALAU-KYUSHU RIDGE

CAROLINE
RIDGE

SHATSKY RISE

EMPEROR SEAMOUNTS

MID-PACIFIC MTNS

HAWAIIAN RIDGE

HESS RISE

LINE ISLANDS RIDGE

Alps, Himalayas, and American Cordillera are Phanerozoic examples of orogens composed of terranes (Figure 2.26). Most crustal provinces, and indeed all collisional and accretionary orogens, are composed of terranes and, in turn, cratons are composed of exhumed orogens. In fact, terranes could be considered the basic building blocks of continents, and terrane collision a major means by which continents grow in size.

A **crustal province** is an orogen, active or exhumed, composed of terranes, and it records a similar range of isotopic ages and exhibits a similar post-amalgamation deformational history. Structural trends within provinces range from linear to exceedingly complex swirling patterns, reflecting polyphase deformation superimposed on differing terrane structural patterns. Exhumed crustal provinces which have undergone numerous episodes of deformation and metamorphism are old orogens, sometimes referred to as mobile belts. It is important to note that crustal provinces need not be coincident with crustal types. For instance, a crustal province may extend from a Precambrian shield into an orogen or across more than one orogen. Isotopic dating using multiple isotopic systems is critical to defining and unravelling the complex, polydeformational histories of crustal provinces.

The definition of crustal provinces is not always unambiguous. Most crustal provinces contain rocks of a wide range in age and record more than one period of deformation, metamorphism, and plutonism. For instance, the TransHudson orogen in North America (Figure 2.26) includes rocks ranging from about 3.0 to 1.7 Ga and records several periods of complex deformation and regional metamorphism. Likewise, the Grenville province records a polydeformational history with rocks ranging in age from 2.7 to 1.0 Ga. Some parts of crustal provinces are new mantle-derived crust, known as **juvenile crust**, while other parts represent reworked older crust. **Reworking**, also known as **overprinting** or **reactivation**, describes crust that has been deformed, metamorphosed and partially melted more than once. It is possible, in some instances, to map reworked crust within crustal provinces and these are sometimes referred to as relict-age subprovinces. In some areas and especially in central and southern Africa, provinces cross-cut one another, and older provinces have superimposed deformation and metamorphism from the younger provinces. There is increasing evidence that crustal reworking results from continental collisions, and large segments of Phanerozoic crust appear to have been reactivated by such collisions. For instance, much of central Asia at least as far north as the Baikal rift was affected by the India–Tibet collision beginning about 50 Ma. Widespread faulting and magmatism at present crustal levels suggest that deeper crustal levels may be extensively reactivated. In Phanerozoic collisional orogens where deeper crustal levels are exposed, such as the Appalachian and Variscan orogens, there is isotopic evidence for widespread reactivation.

One of the most important approaches to extracting multiple ages from crustal provinces is dating single zircons by the U–Pb method using an ion probe. Figure 2.27 shows an example for a trondhjemitic gneiss (Na-rich granitoid) from southern Africa. The scatter of data on the Concordia diagram shows complex Pb-loss from the zircons, and even from within a single zircon. Note, for instance, the complex Pb-loss from zircon grain 4. The most concordant domains can be fitted to a discordia line intersecting Concordia at 3505 ± 24 Ma, which is interpreted as the igneous crystallization age of this rock (Kroner et al., 1989). Three spots analysed on the clear prismatic zircon grain 6 have a near concordant age of 3453 ± 8 Ma. This records a period of intense deformation and high-grade metamorphism in which new metamorphic zircons formed in the gneiss. Grain 20 has a slightly discordant age of 3166 ± 4 Ma and comes from a later granitic vein that crosses the rock. Other discordant data points in Figure 2.27 cannot be fitted to regression lines and reflect Pb-loss at various times in the past, perhaps some as young as 3 Ga. When combined with other single zircon ages from surrounding gneisses, major orogenic-plutonic events are recorded at 3580, 3500, 3450, 3200, and 3000 Ma in this very small geographic area of Swaziland in southern Africa.

Crustal province and terrane boundaries

Contacts between crustal provinces or between terranes are generally major shear zones, only some of which are the actual sutures between formerly-colliding crustal blocks. Portions of province boundaries may be rapid changes in metamorphic grade or intrusive contacts. Boundaries between terranes or provinces may be parallel or at steep angles to the structural trends within juxtaposed blocks. Some boundary shear zones exhibit transcurrent motions and others pass from flat-lying to steep structures and may have thrust or transcurrent offsets. Magnetic and gravity anomalies also generally occur at provincial boundaries, reflecting juxtaposition of rocks of differing densities, magnetic susceptibilities, and crustal thicknesses. The Grenville Front, which marks the boundary between the Proterozoic Grenville and Archean Superior provinces in eastern Canada, is an example of a well-known crustal province boundary (Figure 2.26). Locally, the Grenville Front, which formed about 1 Ga, ranges in width from a few to nearly a hundred kilometres and includes a large amount of reworked Archean crust (Culotta et al., 1990). It also produces a major negative gravity anomaly. Seismic reflection data show that the Grenville Front dips to the east and probably extends to the Moho (Figure 2.28). K–Ar biotite ages are reset at rather low temperatures (200 °C) and gradually decrease from 2.7 Ga to about 1.0 Ga in an easterly direction across the Grenville Front. This Front, however, is not a suture, but a major foreland thrust associated with the collision of crustal provinces. The actual suture has not been identified, but may be the Carthage–Colton shear zone some 250 km east of the Grenville Front (Figure 2.28).

Shear zones between crustal provinces range up to

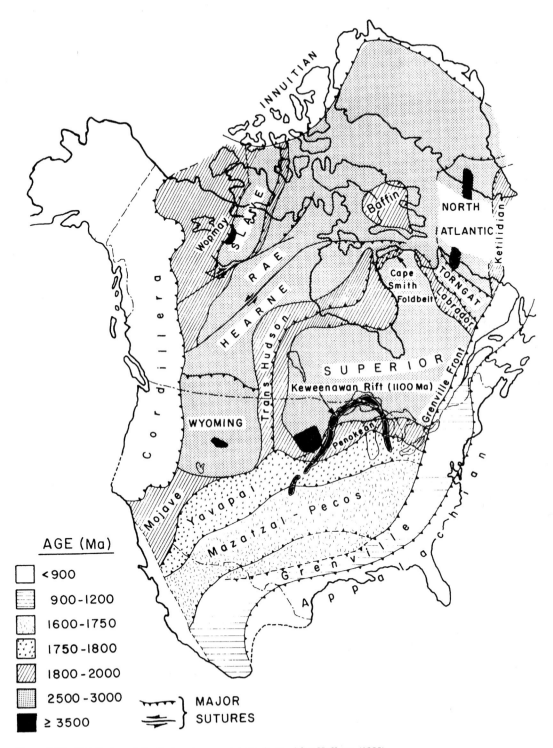

Figure 2.26   Distribution of North American crustal provinces. After Hoffman (1988).

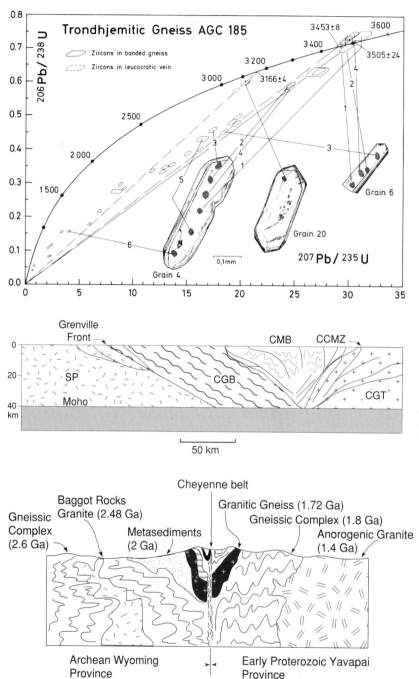

Figure 2.27 U–Pb Concordia diagram showing ion probe analyses of zircons from a trondhjemitic gneiss in northeast Swaziland, southern Africa. From Kroner et al. (1989). Concordia is the bold solid line defined by concordant $^{206}Pb/^{238}U$ and $^{207}Pb/^{235}U$ ages.

Figure 2.28 Diagrammatic cross-section of the 1-Ga Grenville province in eastern Canada. SP, Archean Superior province; CGB, Central gneiss belt; CMB, Central metasedimentary belt; CGT, Central granulite terrane; CCMZ, Carthage–Colton shear zone.

Figure 2.29 Schematic cross-section of the Cheyenne belt, a major shear zone separating the Archean Wyoming province and the Early Proterozoic Yavapai province in southeast Wyoming.

tens of kilometres in width, as illustrated by the Cheyenne belt in the Medicine Bow Mountains in southeastern Wyoming (Figure 2.29). The Cheyenne belt is a near vertical shear zone separating Archean gneisses of the Wyoming province from Early Proterozoic juvenile crust of the Yavapai province on the south. The timing of collision along this boundary is constrained by zircon ages from pre- and post-tectonic plutons at about 1.75 Ga (Condie, 1992b). This boundary is complicated by the fact that deformed metasedimentary rocks (about 2.0 Gy in age) rest unconformably on the Archean gneisses and are also cut by the shear zone. The shear zone, which is up to several kilometres wide, is composed chiefly of mylonitized quartzo-feldspathic gneisses.

The United Plates of America

North America provides an example of the birth and
growth of a continent through geologic time. Field, geophysical, and Nd and U–Pb isotopic data from the
Canadian shield and from borehole samples in platform
sediment, indicate that North America is an amalgamation of plates, referred to by Paul Hoffman as the 'United
Plates of America' (Nelson and DePaolo, 1985; Patchett
and Arndt, 1986; Hoffman, 1988). The Archean crust
includes at least six separate provinces joined by Early
Proterozoic orogenic belts (Figure 2.26). The assembly
of constituent Archean provinces took only about 100
My between 1950 and 1850 Ma. Apparent from the
map is the large amount of crust formed in the Late
Archean, constituting about 30 per cent of the continent.
Approximately 35 per cent of the continent appears to
have formed in the Early Proterozoic, 9 per cent in the
mid- to late-Proterozoic, and about 26 per cent in the
Phanerozoic.

The systematic asymmetry of stratigraphic sections,
structure, metamorphism, and igneous rocks in North
American orogens is consistent with an origin by subduction and collision. Such asymmetry is particularly
well-displayed along the TransHudson, Labrador, and
Penokean orogenic belts. In these belts, zones of foreland
deformation are dominated by thrusts and recumbent
folds, whereas hinterlands typically show transcurrent
faults. Both features are characteristic of subduction
zones. Some Proterozoic orogens have large accretionary prisms, whereas others do not. For instance, the Rae
and Hearne provinces involve only suturing of Archean
crust, while the TransHudson province is a collisional
orogen up to 500 km wide. The Penokean, Yavapai and
Mazatzal provinces are accretionary orogens added to
North America at 1.9, 1.75 and 1.65 Ga, respectively,
and the Grenville province was added by one or more
collisions at 1200–900 Ma. The Cordilleran and Appalachian provinces represent collages of accretionary
terranes sutured along transform faults or large thrusts
during the Phanerozoic.

Summary statements

1   Crustal types are segments of the crust exhibiting
similar geological and geophysical characteristics.
There are thirteen recognized crustal types: shield,
platform, orogen, continental-margin arc, rift, island arc, submarine plateau, inland-sea basin, ocean
ridge, ocean basin, marginal-sea basin, volcanic
island, and trench.

2   Seismic velocity results indicate that both oceanic
and continental crust are layered and velocities
increase with depth. Oceanic crust is composed of
sediment, basement, and oceanic layers with increasing depth.

3   A positive correlation between the area of continents, their thickness, and their average elevation

above sea level may reflect the growth of continents by continental collisions.

4   Electrical conductivity commonly increases with
depth in the continental crust, a feature that may be
caused by pore water, partial melting, or carbon
films on grain boundaries in lower crustal rocks.

5   About 66 per cent of the Earth's heat is lost at by
plate tectonics, 10 per cent by conduction from the
subcontinental lithosphere, 12 per cent from mantle
plumes, and another 12 per cent is lost from the
continents by radioactive decay of heat-producing
elements.

6   Surface heat flow from the continents is linearly
related to crustal near-surface heat productivity. The
zero intercept, the reduced heat flow, is the heat
coming from the lower crust and mantle, and the
slope D (the characteristic depth) is the thickness of the upper layer enriched in radiogenic heat-
producing elements.

7   The average heat flow of the crust decreases with
mean crustal age. In the continents, heat flow decreases to 40–50 mW/m^2 in about 1 Gy and to
similar values in the oceans in 50–100 My. Reduced heat flow in the continents also decreases
with mean crustal age.

8   Magnetic anomalies in continental areas reflect near-
surface rock types and structural discontinuities.
Expect in regions of thin crust and high heat flow,
magnetic grain generally parallels orogenic structures. Oceanic crust is characterized by linear magnetic anomalies paralleling ocean ridge systems.

9   Pressure–temperature–time (P–T–t) paths are important in tracking the history of continental crust.
They are defined by textural relationships between
mineral growth and deformational fabrics, by relating mineral growth to experimentally determined
P–T stability fields, and by isotopic ages of minerals that form and close along different segments of
P–T–t paths.

10 P–T–t paths of collisional orogens are typically
clockwise and they result from rapid crustal thickening so that the maximum pressure is reached
before the maximum temperature. Extension of
continental crust produces counterclockwise P–T–t
paths, in which the maximum temperature predates
or is synchronous with early deformation.

11 Cratons are the end-product of collisional orogenesis
and are composed of eroded and uplifted orogens.
Cooling rates of cratons range from < 2 °C/My to
> 10 °C/My, and P–T–t paths are complex such that
much or all of a thermal history can be erased by
the last thermal event.

12 The transition between elastic and ductile behaviour in the continental crust occurs at relatively shallow depth in warm and young crust (10–20 km),
whereas in cool and old crust it occurs at greater
depths (20–30 km).

13 The formation of granitic melts in the lower crust
and their transfer to shallower depths is a funda-

mental process leading the chemical differentiation of the continents.

14   Fluid transport in the crust is an important process affecting both the rheology and chemical evolution of the crust. Shallow crustal fluids are chiefly water, whereas deep crustal fluids are mixtures of H_2O and CO_2.

15   The composition of the continental crust is constrained by seismic velocity distributions, the composition of Precambrian shields, the composition of fine-grained sediments, exposed deep crustal sections, and the composition of crustal xenoliths in volcanic rocks.

16   Because of the overlap in seismic velocities of rocks of different compositions and origins, it is not possible to assign unique rock compositions to the lower crust from seismic velocity data alone. When coupled with xenolith data, however, the seismic velocity distributions suggest that the lower continental crust is composed largely of mafic granulites, gabbros, and amphibolites (50–65 per cent), with up to 10 per cent metapelites, and the remainder intermediate to felsic igneous rocks.

17   From studies of exhumed lower crust and lower crustal xenoliths, most lower crustal reflections appear to result from mafic intrusions and, in some instances, the reflections have been enhanced by later ductile deformation.

18   The upper continental crust has an average composition of granodiorite, the lower crust is mafic in composition, and total crust is intermediate. Oceanic crust is mafic in composition.

19   Complementary crustal element distributions in continental and oceanic crust can be explained if most of the continental crust is extracted from the upper mantle first, leaving an upper mantle depleted in very incompatible elements. The oceanic crust is then continuously produced from this depleted upper mantle throughout geologic time.

20   Terranes are fault-bounded crustal blocks which have distinct lithologic and stratigraphic successions and which also have geologic histories different from neighbouring terranes. They form in a variety of tectonic settings including island arcs, submarine plateaux, volcanic islands, and microcontinents.

21   Terranes are the basic building blocks of continents, and terrane collision is a major means by which continents grow in size.

22   A crustal province is an orogen, active or exhumed, composed of terranes, and it records a similar range of isotopic ages and exhibits a similar post-amalgamation deformational history. Cratons are composed of exhumed orogens.

23   Terrane and crustal province boundaries are shear zones, some of which represent sutures between formerly colliding blocks.

Suggestions for further reading

Dawson, J. B., Carswell, D. A., Hall, J. and Wedepohl, K. H. (1986). The nature of the lower continental crust. *Geol. Soc. London, Special Publ. No. 24.* Blackwell Scient., 394 pp.

Dewey, J. F. et al., editors (1991). *Allochthonous Terranes.* Cambridge, Cambridge Univ. Press, 199 pp.

Fountain, D. M., Arculus, R. and Kay, R. W. (1992). *Continental Lower Crust.* Amsterdam, Elsevier, 486 pp.

Meissner, R. (1986). *The Continental Crust, A Geophysical Approach.* New York, Academic Press, 426 pp.

Rudnick, R. L. and Fountain, D. M. (1995). Nature and composition of the continental crust: A lower crustal perspective. *Revs. Geophys. Space Phys.*, **33**, 267–309.

Sclater, J. G., Jaupart, C. and Galson, D. (1980). The heat flow through oceanic and continental crust and the heat loss of the Earth. *Revs. Geophys. Space Phys.*, **18**, 269–311.

Taylor, S. R. and McLennan, S. M. (1985). *The Continental Crust: Its Composition and Evolution.* Oxford, Blackwell Scient., 312 pp.

Chapter 3

Tectonic settings

Introduction

Characteristic features of tectonic settings include lithologic assemblages (both supracrustal rocks and intrusive rocks), deformational styles and histories, metamorphism and P–T–t paths, as well as mineral and energy deposits. Rock assemblages that form in modern plate tectonic settings are known as **petrotectonic assemblages**. Such assemblages include both supracrustal rocks (sediments and volcanics) and shallow intrusive rocks. From studying modern plate settings, it is possible to learn more about what to look for in ancient rocks in order to identify, or at least constrain, the tectonic setting in which they formed. From the results, it may also be possible to evaluate the intriguing question of just how for back in time plate tectonics has operated.

Although sediments forming in modern plate settings can be sampled and studied, it is more difficult to study young plutonic and metamorphic rocks and deep-seated deformation. One approach is to examine deep canyons in mountain ranges where relatively young (< 50 Ma) deep-seated rocks are exposed, to which a tectonic setting can be assigned with some degree of confidence. Seismic reflection profiles provide a means of studying deep-seated deformation. These studies have been particularly useful in modern arc systems. As discussed in the previous chapter, it is also possible to study recently elevated blocks that expose lower crustal rocks as well as xenolith populations brought up in young volcanoes. Finally, it is possible to simulate P–T conditions that exist in modern plates in laboratory and computer models, from which constraints can be placed on geologic processes occurring at depth.

In this chapter some of the processes that leave tectonic imprints in rocks will be examined. It is only when results from a variety of studies converge on the same interpretation, that ancient tectonic settings can be identified with confidence. Even then, care must be used as one or more tectonic settings which no longer exist may

have left imprints in the geologic record similar to those of modern plate settings. This is especially true for the Archean.

Ocean ridges

Ocean-ridge basalts

Experimental petrologic results indicate that **ocean-ridge basalts (MORB)** are produced by 15–30 per cent partial melting of the upper mantle at depths of 50–85 km (Elthon and Scarfe, 1984). Melting at these depths produces olivine tholeiite magma. Seismic and geochemical results show that this magma collects in shallow chambers (< 35 km) (Forsyth, 1996), where it undergoes fractional crystallization to produce tholeiites, quartz tholeiites and minor amounts of more evolved liquids including plagiogranites. A residue of olivine and pyroxenes is left behind in the mantle and may be represented by the sheared harzburgites preserved in some ophiolites.

Depletion in **large ion lithophile (LIL) elements** (K, Rb, Ba, Th, U etc.) in **NMORB** (normal MORB) indicates a mantle source that has been depleted in these elements by earlier magmatic events. Low $^{87}Sr/^{86}Sr$ and $^{206}Pb/^{204}Pb$ and high $^{143}Nd/^{144}Nd$ isotopic ratios also demand an NMORB source that is depleted relative to chondrites. Incompatible element contents and isotopic ratios vary in MORB along ocean ridges and from ocean to ocean as illustrated by the distribution of Nb and Zr in modern MORB from the North Atlantic basin (Figure 3.1). Although some of this variation can reflect differences in degree of melting of the source, fractional crystallization, or magma mixing, the large differences between some sites require variation in the composition of the depleted mantle source.

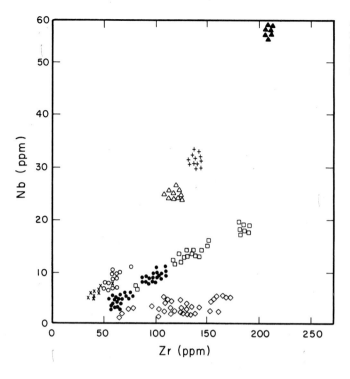

Figure 3.1 Zr and Nb distribution in ocean ridge basalts from the North Atlantic. Each symbol is for a different geographic location of the Mid-Atlantic ridge. Data from Tarney et al. (1979).

Ophiolites

General features

Ophiolites are tectonically emplaced successions of mafic and ultramafic rocks that are considered to represent fragments of oceanic or back-arc basin crust (Coleman, 1977; Moores, 1982). An ideal ophiolite includes from bottom to top the following units (Figure 3.2):

1 ultramafic tectonite (generally harzburgites)
2 layered cumulate gabbros and ultramafic rocks
3 non-cumulate gabbros, diorites and plagiogranites
4 sheeted diabase dykes
5 pillowed basalts.

Overlying this succession in many ophiolites are abyssal or/and pelagic sediments or arc-related volcaniclastic sediments. Due to faulting or other causes, the idealized ophiolite succession is rarely found in the geologic record. Instead, one or more of the ophiolite units are missing or they have been dismembered by faulting and occur as blocks in a tectonic melange.

Some ophiolites are in fault contact with underlying shallow marine cratonic sediments, while others occur as tectonic slivers in accretionary prisms with graywackes and other arc-related rocks. These ophiolites appear to be emplaced along passive and active continental margins, respectively. The basal ophiolite melange (Figure 3.2) consists of a chaotic mixture of diverse rocks in a highly-sheared matrix. Clast lithologies include ophiolite-derived materials, pelagic and abyssal sediments, graywackes, and various metamorphic and volcanic rocks. Matrices are commonly sheared serpentinite. Ophiolite melanges are of tectonic origin formed during ophiolite emplacement.

Sheared and serpentinized ultramafic rock is an important component in the lower part of most ophiolites. These ultramafic tectonites are composed chiefly of harzburgite with pronounced foliation and, generally, compositional banding. Lenses of dunite and chromite occur within the harzburgites. Overlying the tectonites are cumulate ultramafic and gabbroic rocks that have formed by fractional crystallization. These rocks have cumulus textures and well-developed compositional banding. Some ophiolites contain non-cumulate gabbro, diorite, and plagiogranite in the upper part of the non-cumulate zone. Plagiogranites are tonalites composed of quartz and sodic plagioclase with minor mafic silicates (Table 3.1). They typically have granophyric intergrowths and may be intrusive into layered gabbros (Coleman, 1977).

Above the non-cumulate unit in an idealized ophiolite is a **sheeted dyke complex** (Figure 3.2). Dikes may cross-cut or be gradational with the non-cumulate rocks. Although dominantly diabases, dykes range from diorite to pyroxenite in composition, and dyke thickness is variable, commonly from 1–3 m. One-way chilled margins are common in sheeted dykes, a feature generally interpreted to reflect vertical intrusion in an oceanic axial rift zone, where one dyke is intruded in the centre of another as the lithosphere spreads. The transition from sheeted

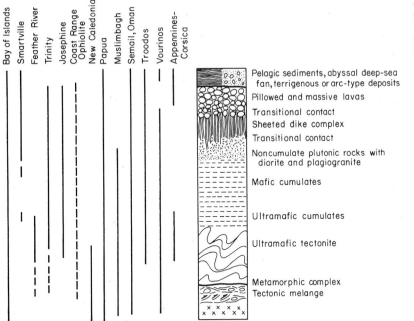

Figure 3.2 An idealized ophiolite succession compared with various exposed ophiolites. Modified after Moores (1982).

Table 3.1 Characteristic features of granitoids

	M-Type	I-Type	S-Type	A-type
Composition	Plagiogranite	Variable, tonalite to granite	Leucogranites with narrow silica range	High-K granites and syenites
Characteristic minerals	hb, biot, cpx	hb, biot, mgt, sph	biot, cord, ilm, musc, gar, monazite	biot, mgt, alkaline hb
Xenoliths	Mafic	Mixed	Chiefly metasediments	Mixed
Molecular ratio Al_2O_3/Na_2O+K_2O+CaO	< 0.6	0.5–1.1	> 1.1	0.9–1.1
Relation to deformation	Pre-tectonic	Pre- to syn-tectonic	Syn- to post-tectonic	Post-tectonic
Ba, Ti, P depletion	Minor	Moderate	Moderate	Strong
Subduction zone component	Variable	Strong	Variable	Absent
$\delta^{18}O$	5.5–6	8–10	> 10	8–10
Tectonic setting	Ophiolite	Arc or orogen	Orogen	Rift, orogen, arc
Source	FXL of MORB	FXL of andesite or partial melting of lower crust	Partial melting of sediments	Partial melting of lower crust

hb, hornblende; biot, biotite; cpx, clinopyroxene; mgt, magnetite; sph, sphene; cord, cordierite; musc, muscovite; gar, garnet
Subduction zone component (Nb–Ta depletion relative to Th and light REE)
FXL, fractional crystallization; MORB, ocean ridge basalt

dykes into pillow basalts generally occurs over an interval of 50–100 m where screens of basalt between dykes become more abundant. The uppermost unit of ophiolites is ocean-ridge basalt occurring as pillowed flows or hyaloclastic breccias. Thickness of this unit varies from a few metres to 2 km, and pillows form a honeycomb network with individual pillows ranging up to 1 m across. A few dykes cut the pillowed basalt unit.

Many ophiolites are overlain by sediments reflecting pelagic, abyssal, or arc depositional environments. Pelagic sediments include radiolarian cherts, red fossiliferous limestones, metalliferous sediments, and abyssal

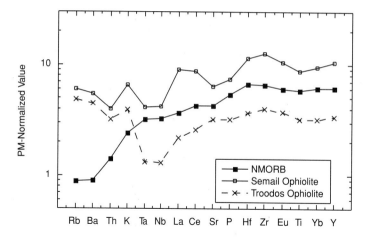

Figure 3.3 Primitive-mantle normalized incompatible-element distributions in ocean ridge basalts and ophiolites. Primitive-mantle values from Sun and McDonough (1989) and other data from Bednarz and Schmincke (1994) and Alabaster et al. (1982).

sediments. **Abyssal sediments** are chiefly pelites and silt-stones which were deposited on abyssal plains and often show evidence of both volcanic and continental provenance. Modern abyssal plains are most frequent adjacent to passive continental margins, and hence have significant input of sediment from continental shelves. Some ophiolites are overlain by graywackes and volcaniclastic sediments of arc provenance.

Tectonic setting and emplacement of ophiolites

Ophiolites have been described from three oceanic tectonic settings: mid-ocean ridges, back-arc basins and, in some instances, like the Metchosin Complex in British Columbia, immature island arcs (Coleman, 1977; Massey, 1986). Incompatible element distributions in most ophiolitic basalts, as for example the Semail ophiolite in Oman and the Troodos Complex in Cyprus, show a **subduction geochemical component** (relative depletion in Ta and Nb; Figure 3.3) suggesting that they are fragments of back-arc oceanic crust. Relatively few ophiolites have basalts showing NMORB element distributions.

Ophiolites are emplaced in arcs or collisional orogens by three major mechanisms (Figure 3.4) (Dewey and Kidd, 1977; Cawood and Suhr, 1992):

1 Obduction or thrusting of oceanic lithosphere onto a passive continental margin during a continental collision
2 splitting of the upper part of a descending slab and obduction of a thrust sheet onto a former arc
3 addition of a slab of oceanic crust to an accretionary prism in an arc system.

Although gravity sliding also has been proposed as a way of emplacing ophiolites, this requires both an elevated source area and a slope sufficient to allow sliding of 5-km thick slabs of oceanic crust onto the continent, a highly unlikely situation.

The metamorphic complex at the base of ophiolites (Figure 3.2) may play a major role in ophiolite emplacement (Spray, 1984). These metamorphic soles, as they are often called, have many features in common:

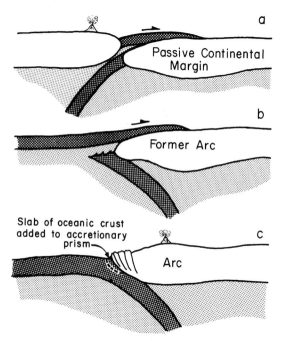

Figure 3.4 Three mechanisms of ophiolite emplacement, (a) obduction at a passive continental margin; (b) obduction of oceanic lithosphere; and (c) transfer of a slab of oceanic lithosphere to an accretionary prism.

1 thicknesses generally range from 10–500 m
2 they extend laterally for tens to hundreds of kilometres
3 most show a sharp decrease in metamorphic grade from top to bottom
4 they are highly deformed and have a pronounced tectonic foliation
5 they are composed of mixed mafic volcanics, serpentinite, and metasediments.

These data are consistent with an origin for soles of successive underplating and welding onto the base of an

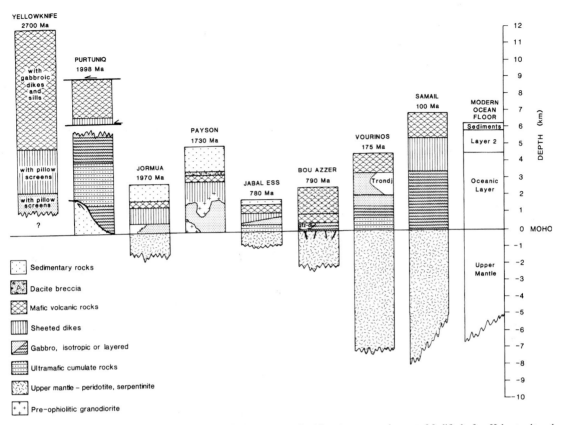

Figure 3.5 Simplified stratigraphic sections of ophiolites compared with average oceanic crust. Modified after Helmstaedt and Scott (1992).

ophiolite as it moves from the upper mantle to the surface. In some cases, like the Josephine ophiolite in northwest California, serpentinization of basal ultramafic rocks may have occurred prior to ophiolite emplacement, suggesting that the paleo-Moho was a serpentinization boundary (Coulton et al., 1995). The heat source for sole metamorphism is probably a combination of residual heat from the mantle and shear heating during ophiolite emplacement. If this interpretation is correct, ophiolites may represent tectonically decoupled slices of hot oceanic lithosphere.

Compared with seismic sections of oceanic or arc crust, most ophiolites are considerably thinner (chiefly < 5 km compared with 7–10 km for oceanic crust and 10–15 km for arcs) (Figure 3.5). In some highly-deformed ophiolites, the thickness may be controlled by tectonic thinning during emplacement. Most ophiolites, however, are not severely deformed and may have been thickened during emplacement (Moores, 1982). So where in the oceanic crust do most ophiolites actually come from? Perhaps ophiolites are decoupled from the mantle at the asthenosphere–lithosphere boundary in the vicinity of spreading centres where oceanic lithosphere is thin. In a closing ocean basin, lithosphere decoupling may occur when an ocean ridge enters a subduction zone or during a continent–continent collision. If decoupling occurs only at the asthenosphere–lithosphere boundary, it may not be possible to obduct thicker segments of oceanic lithosphere (up to 150 km thick at the time of subduction), thus accounting for the absence of ophiolites much thicker than 5 km.

Formation of ophiolites

Although details continue to be controversial, the overall mechanism by which a complete ophiolite succession forms is reasonably well understood. As pressure decreases in rising asthenosphere beneath ocean ridges, garnet lherzolite partially melts to produce basaltic magma (Figure 3.6). These magmas collect in shallow chambers (3–6 km deep) and undergo fractional crystallization. Layered ultramafic and gabbroic rocks accumulate in the magma chamber forming the cumulate section of ophiolites. The tectonized harzburgites represent residue left after melting, which is highly deformed and sheared due to lateral advective motion of the asthenosphere, and/or to deformation during emplacement. Dykes are ejected from the magma chamber forming the sheeted dyke complex and many are erupted

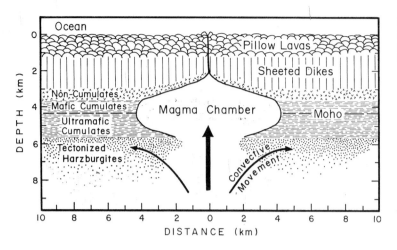

Figure 3.6 Schematic cross-section of oceanic crust near an ocean ridge showing possible relationships to an ophiolite succession.

forming pillow lavas in the axial rift. Fractional crystallization in the magma chamber also gives rise locally to diorites and plagiogranites, which are intruded at the top of the layered cumulate series.

Precambrian ophiolites

Ophiolites and associated deep-sea sediments are first recognized in the geologic record about 2 Ga. One of the oldest recognized ophiolites which has all of the essential components in the correct stratigraphic order is the Jormua complex in northern Finland (Figure 3.5). Although older ophiolites have been reported (de Wit et al., 1987; Helmstaedt et al., 1986; Helmstaedt and Scott, 1992), they lack a convincing sheeted dyke complex and tectonized harzburgites. The thickest, most laterally extensive Early Proterozoic ophiolite is the Purtuniq ophiolite in the Cape Smith orogen in Canada with an age of about 2 Ga (Figure 3.5) (Scott et al., 1992). Although few well-documented ophiolites have been reported older than about 1000 Ma, numerous occurrences in the age range of 1000–600 Ma are reported from the Pan-African provinces in Africa and South America and from areas around the North Atlantic. Estimated thicknesses of most Late Proterozoic ophiolites range up to 8 km, but most are less than 5 km. They are bounded by thrust faults and appear to have been emplaced by obduction. Although the range in metamorphic grade and degree of deformation is considerable, many Proterozoic ophiolites preserve primary textures and structures. Most, however, represent only partial ophiolite successions.

As with Phanerozoic ophiolites, many Proterozoic ophiolites carry a weak-to-strong subduction zone geochemical imprint. This suggests that they represent fragments of arc-related oceanic crust from either back-arc or intra-arc basins. Why there are no reported occurrences of ophiolites older than 2 Ga is an important question which will be considered in Chapter 5.

Tectonic settings related to mantle plumes

Submarine plateaux and aseismic ridges

Submarine plateaux, which are composed chiefly of basalt flows erupted beneath the oceans, are the largest topographic features of the sea floor. **Aseismic ridges** are extinct volcanic ridges on the sea floor. About 10 per cent of the ocean floors are covered by submarine plateaux and aseismic ridges, and more than 100 are known (Figure 3.7), many of which are in the Western Pacific (Coffin and Eldholm, 1994). These features rise thousands of metres above the sea floor and some, such as the Seychelles Bank in the Indian Ocean (SEYC, Figure 3.7), rise above sea level. Some have granitic basement (such as the Seychelles Bank, the Lord Howe Rise north of New Zealand [LORD], and the Agulhas Plateau south of Africa [AGUL]), suggesting that they are rifted fragments of continental crust. Others, such as the Cocos and Galapagos ridges west of South America (COCO, GALA), are of volcanic origin, related to hotspot activity. Some aseismic ridges, such as the Palau-Kyushu ridge south of Japan, are extinct oceanic arcs. Others, such as the Walvis and Rio Grande ridges in the South Atlantic (WALV, RIOG), are hotspot tracks.

Together with flood basalts on the continents, submarine plateaux are thought to be the products of magmas erupted from mantle plumes (Coffin and Eldholm, 1994; Carlson, 1991). The largest submarine plateau, which straddles the equator in the western Pacific, is the Ontong-Java plateau (ONTO) erupted from a mantle plume about 120 Ma in the South Pacific. This plateau is capped by seamounts and is covered by a veneer of pelagic sediments (limestone, chert, and radiolarite), in places over 1 km thick (Berger et al., 1992). Because only the tops of submarine plateaux are available for sampling, to learn more about their rock assemblages older plateaux, preserved on the continents as terranes must be

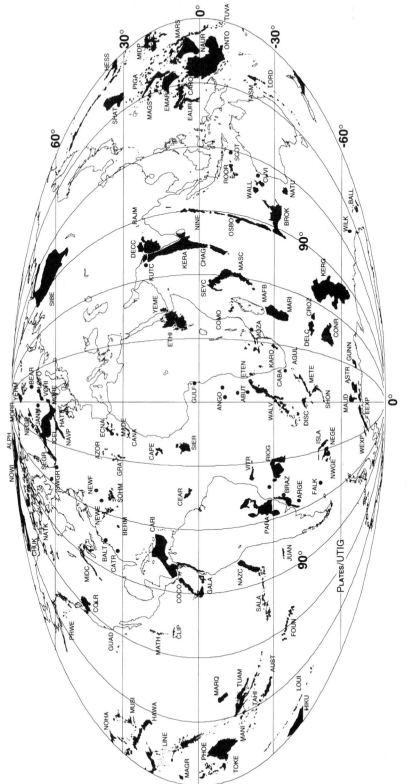

Figure 3.7 Distribution of Phanerozoic submarine plateaux, aseismic ridges, and flood basalts. After Coffin and Eidholm (1994).

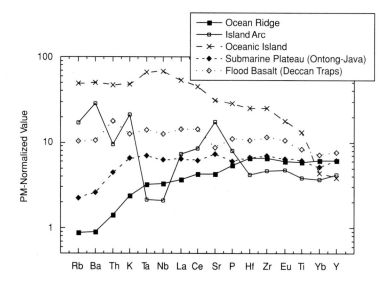

Figure 3.8 Primitive-mantle normalized incompatible-element distributions in various basalts. Primitive-mantle values and NMORB from Sun and McDonough (1989).

studied. Studies of the Wrangellia terrane, which is composed largely of a submarine plateau accreted to western North America in the Cretaceous, show that pillow basalts and hyaloclastic breccias compose most of the succession, with only a thin capping of pelagic limestones, shales, and cherts. Some submarine plateaux (like Kerguelen in the southern Indian Ocean [KERG]) emerge above sea level and are capped with subaerial basalt flows and associated pyroclastic volcanics. Submarine plateau basalts are largely tholeiites, with only minor amounts of alkali basalt, and most show incompatible elements only slightly enriched compared with NMORB (Figure 3.8).

Continental flood basalts

Flood basalts are thick successions of basalt erupted on the continents over short periods of time, like the Columbia River basalts in the northwestern United States (COLR, Figure 3.7) and the Deccan traps in India (DECC). They are composed chiefly of the tholeiitic basalt flows, and like submarine plateau basalts, they appear to be derived from mantle plumes or, in some cases, from melting of the subcontinental lithosphere caused by mantle plumes (Carlson, 1991). Thick successions of flood basalts compose important plateaux on the continents. One of the characteristic features of both submarine plateau basalts and flood basalts is the rapidity with which large volumes of basaltic magma are erupted. The Deccan traps, which were erupted at the Cretaceous-Tertiary boundary 65 Ma, include a preserved volume of basalt of about 1.5×10^6 km^3 erupted in ≤ 1 My, probably more than 80 per cent of these flows $< 500\,000$ y. The Columbia River basalts formed at 17.5–6 Ma, at which time 0.17×10^6 km^3 of magma were erupted. Eruption rates of flood basalts and submarine plateau basalts range from about 0.5 to > 1 km^3/y, considerably greater than rates typical of ocean ridges or

volcanic islands like Hawaii (with rates of 0.02–0.05 km^3/y) (Carlson, 1991; White and McKenzie, 1995).

Continental flood basalts tend to be lower in Mg and Fe and other compatible elements than MORB or island basalts. They typically show Fe enrichment fractionation trends, but in some instances have lower Ti than other oceanic basalts. Incompatible element distributions vary widely in flood basalts, often within the same volcanic field. Those with relatively high contents of LIL elements, like the Deccan traps (Figure 3.8), are either contaminated by continental crust, or come from enriched sources in the subcontinental lithosphere. In both cases a subduction geochemical component, as shown by the Deccan basalts, may be transferred to the magmas, either by magma contamination or, in the case of the lithosphere, directly from the source.

Volcanic Islands

As exemplified by Hawaii, oceanic volcanic islands are some of the largest mountains on Earth, often rising many kilometres above the sea floor. During the subaerial stages of eruption, volcanoes typically evolve from a shield-building stage (like Mauna Loa and Kilauea), through a caldera-filling stage (like Mauna Kea), terminating in a highly-eroded shield volcano with small, often alkaline magma eruptions (like most of the extinct volcanoes in Hawaii). Volcanic islands are composed chiefly of tholeiitic basalts, with only minor amounts of alkali basalts and their derivatives erupted during terminal volcanism. Dredging of seamounts and submarine slopes of islands reveals a dominance of hyaloclastic volcanics, in contrast to the abundance of flows found along oceanic ridges (Bonati, 1967). This difference indicates that island-seamount magmas are considerably more viscous than MORB magmas and readily fragment upon eruption into seawater (Kokelaar, 1986). In striking contrast to subduction-related basalts, many island basalts show a relative enrichment in Nb and Ta (Figure 3.8), reflecting a mantle-plume source enriched in these elements.

As discussed in Chapter 1, volcanic chains on the sea floor may result from the lithosphere moving over rising mantle plumes. Many aseismic ridges represent such volcanic chains. As shown by the Hawaiian–Emperor island-seamount chain (Figures 1.26 and 3.7), volcanic chains may show a systematic change in magma composition with time. Changes also occur at single volcanoes. For instance, in Hawaii the oldest rocks from a given volcano are olivine tholeiites, followed by an increasing abundance of Fe-rich quartz tholeiites and terminating with minor volumes of alkali basalt and its differentiates. Incompatible elements become more enriched during the change from tholeiite to alkali basalt. Such a sequence of events can be interpreted in terms of a hotspot model, in which early phases of magmatic activity reflect extensive melting in or above a mantle plume. As an oceanic plate moves over a mantle plume, large volumes of tholeiitic basalt are erupted, forming one or more oceanic islands or seamounts. As the island approaches the edge of the plume, isotherms drop and the depth and degree of melting decrease, resulting in the production of small volumes of alkali basalt. Thus, the bulk of oceanic islands comprises tholeiitic basalts (chiefly submarine), with only a frosting of alkali basalts erupted, before volcanoes move off the hotspot and become extinct.

Giant mafic dyke swarms

Vast swarms of mafic dykes have been intruded into the continents, covering areas of tens to hundreds of thousands of square kilometres (Ernst et al., 1995). These swarms, which range up to 500 km in width and over 3000 km in length, occur on all Precambrian shields and include many thousands of dykes. Individual dykes range from 10–50 m in width, with some up to 200 m. Most dykes have rather consistent strikes and dips, and individual dykes have been traced for up to 1000 km. Dyke widths tend to be greater where depth of erosion is least, suggesting that dykes widen in the upper crust, and dyke spacing ranges generally from 0.5–3 km with dykes branching both vertically and along strike. Major swarms appear to have been intruded episodically, with important ages of intrusion at 2500, 2390–2370, 2150–2100, 1900–1850, 1270–1250 and 1150–1100 Ma. Structural studies indicate that most swarms are intruded at right angles to the minimum compression direction and, except near the source, most are intruded laterally, not vertically. Some swarms, such as the giant Mackenzie swarm in Canada (Figure 3.9), appear to radiate from a point, commonly interpreted as a plume source for the magmas (Baragar et al., 1996). The Mackenzie swarm is also an example of a swarm associated with flood basalts (Coppermine River basalts, Cb in Figure 3.9) and a layered intrusion (Muskox intrusion, M). U–Pb dating of baddeleyite shows that most swarms are emplaced in very short periods of time, often less 2–3 My (Tarney,

1992). Giant dyke swarms are typically comprised of tholeiites, although norite swarms are also important in the Proterozoic.

Like the Mackenzie swarm, many giant dyke swarms appear to be associated with mantle plumes, and their emplacement is accompanied by significant extension (over 30 per cent) of the continental crust (Tarney, 1992; Baragar et al., 1996). During the early stages of plume activity and rifting, dykes may develop radial patterns from an underlying plume source (Figure 3.10, stage 1). As with the Red Sea and Gulf of Aden rifts associated with the Afar hotspot, an ocean basin may open between two of the radiating dyke swarms, leaving the third swarm as part of an aulacogen (stage 2) (Fahrig, 1987). Later, during closure of the ocean basin and continental collision, dykes are consumed or highly deformed. The swarms intruded in the aulacogen, however, are more likely to be preserved if they are on the descending plate (stage 3). Hence, major dyke swarms which radiate from a point, such as the Mackenzie swarm, may indicate the presence of a former ocean basin. Some giant dyke swarms may have been emplaced during supercontinent fragmentation (Chapter 5), and thus their ages and distributions can be helpful in reconstructing supercontinents.

Cratons and passive margins

Rock assemblages deposited in cratonic basins and passive margins are mature clastic sediments, chiefly quartz arenites and shales, and shallow marine carbonates. In Late Archean–Early Proterozoic successions, banded iron formation may also be important (Klein and Beukes, 1992; Eriksson and Fedo, 1994). Because passive continental margins began life as continental rifts, rift assemblages generally underlie passive continental margin successions. When back-arc basins develop between passive margins and arcs, like the Sea of Japan, cratonic sediments may interfinger with arc sediments in these basins. Cratonic sandstones are relatively pure quartz sands reflecting intense weathering, low relief in source areas, and prolonged transport across subdued continental surfaces. Commonly associated marine carbonates are deposited as blankets and as reefs around the basin margins. Transgression and regression successions in large cratonic basins reflect the rise and fall of sea level, respectively.

Depositional systems in cratonic and passive margin basins vary depending on the relative roles of fluvial, aeolian, deltaic, wave, storm and tidal processes. Spatial and temporal distribution of sediments is controlled by regional uplift, the amount of continent covered by shallow seas, and climate (Klein, 1982). If tectonic uplift is important during deposition, continental shelves are narrow and sedimentation is dominated by wave and storm systems. However, if uplift is confined chiefly to craton margins, sediment yield increases into the craton and fluvial and deltaic systems may dominate. For trans-

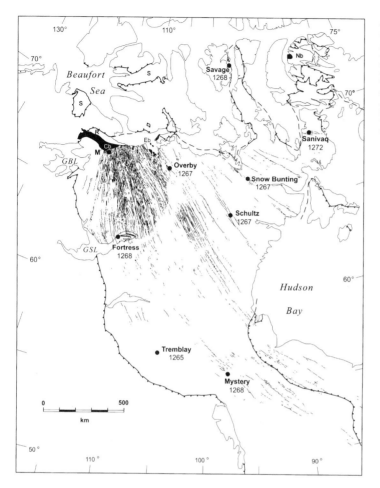

Figure 3.9 The giant Mackenzie dyke swarm in northern Canada. This dyke swarm was intruded at 1267 Ma, probably in response to a mantle plume located at that time north of Coronation Gulf. The Muskox intrusion (M) and the Coppermine River flood basalts (Cb) are part of the same event. GBL, Great Bear Lake; GSL, Great Slave Lake. Courtesy of A. N. LeCheminant.

gressive marine clastic sequences, shallow seas are extensive and subtidal, and storm-dominated and wave-dominated environments are important. During regression, fluvial and aeolian depositional systems become dominant.

The rates of subsidence and uplift in cratons are a function of the time interval over which they are measured. Current rates are of the order of a few centimetres per year, whereas data from older successions suggest rates one to two orders of magnitude slower. In general, Phanerozoic rates of uplift appear to have been 0.1–1 cm/y over periods of 10^4–10^5 years and over areas of 10^4–10^6 km^2. One of the most significant observations in cratonic basins is that they exhibit the same exponential subsidence as ocean basins (Sleep et al., 1980). Their subsidence can be considered in terms of two stages: in the first stage the subsidence rate varies greatly, whereas the second stage subsidence is widespread. After about 50 My, the depth of subsidence decreases exponentially to a constant value.

Several models have been suggested to explain cratonic subsidence (Bott, 1979; Sleep et al., 1980). Sediment loading, lithosphere stretching, and thermal doming fol-

lowed by contraction are the most widely-cited mechanisms. Although the accumulation of sediments in a depression loads the lithosphere and causes further subsidence, calculations indicate that the contribution of sediment loading to subsidence must be minor compared to other effects. Subsidence at passive margins may result from thinning of continental crust by progressive creep of the ductile lower crust towards the suboceanic upper mantle. As the crust thins, sediments accumulate in overlying basins. Alternatively, the lithosphere may be domed by upwelling asthenosphere or a mantle plume, and uplifted crust is eroded. Thermal contraction following doming results in platform basins or a series of marginal basins around an opening ocean, which fill with sediments. Most investigators now agree that subsidence along passive continental margins is due to the combined effects of thermal contraction of continental and adjacent oceanic lithosphere together with sediment loading.

In no other tectonic setting do such large volumes of detrital quartz accumulate as in cratonic and passive-margin basins. Where does all the quartz come from and what processes concentrate it into these basins? Although

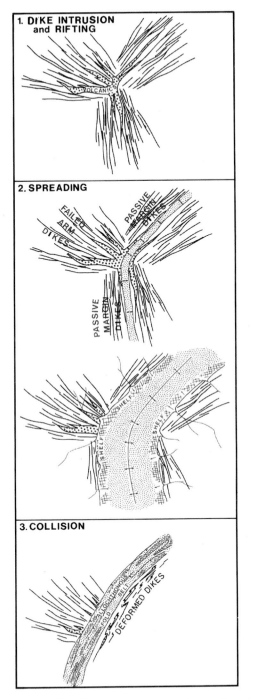

Figure 3.10 Three stages in the tectonic evolution of mafic dyke swarms. From Fahrig (1987).

most of the detrital quartz in cratonic sandstones is probably recycled, we are still faced with the problem of what served as the primary source. A felsic volcanic source for detrital quartz requires weathering of vast amounts of volcanic rock, since quartz phenocrysts rarely comprise more than 10–20 per cent of such rocks. Not only is there no apparent evidence for such large amounts of felsic volcanics in the geologic record, but quartz phenocrysts are generally smaller than the average grain size of most quartz arenites, and thus such a source seems unlikely. Vein quartz and silicified rocks are very minor in the crust and so are also unlikely to serve as major sources for detrital quartz. Most detrital quartz occurs as monocrystalline or polycrystalline grains or as quartzite rock fragments, and hence cannot represent reworked chert. It would appear therefore that the only suitable primary source for large quantities of detrital quartz are granitoids or felsic gneisses. Although relatively pure quartz sands can be produced in one cycle by intense chemical weathering, high-energy environments like tidal flats and sand dunes can selectively remove feldspars and produce relative pure quartz without intense weathering (Chandler, 1988).

Igneous rocks are rare in cratonic and passive margin basins, and when found are small intrusive bodies, dykes, sills or volcanic pipes, generally of alkaline compositions. **Kimberlites** are ultramafic breccias found in cratonic areas, and they are significant in that they contain xenoliths of ultramafic rocks from the upper mantle (Chapter 4) as well as diamonds and other high-pressure minerals which indicate depths of origin for kimberlitic magmas of up to 300 km (Pasteris, 1984). They range in age from Archean to Tertiary and most occur as pipes or dykes < 1 km^2 in cross-sectional area. Kimberlite magmas appear to have been produced by partial melting of enriched sources in the subcontinental lithosphere.

Continental rifts

General features

Continental rifts are fault-bounded basins produced by extension of continental crust. They may be single, like the rifts in East Africa, or multiple, as in the case of the Basin and Range province in the western United States (Figure 1.1). Also included in this category are aulacogens. **Aulacogens** are failed or less-active arms of triple junctions, such as the Ethiopian and Benue rifts in Africa. Spreading began in the Red Sea and Gulf of Aden 15–25 Ma and may be beginning today in the Ethiopian aulacogen. The Benue rift became a failed arm in the Late Cretaceous as the other two rift segments opened as part of the Atlantic basin. Rifts are of different origins and occur in different regional tectonic settings. Although the immediate stress environment of rifts is extensional, the regional stress environment may be compressional, extensional or nearly neutral. Rifts that form in cratons, such as the East African rift system, are commonly associated with domal uplifts, although the timing of doming relative to rifting may vary (Mohr, 1982). Geophysical data indicate that both the crust and lithosphere are thinned beneath rifts, and that most or all of the crustal thinning occurs in the ductile lower crust over a much broader area than represented by the surface expression of the rift (Thompson and Gibson, 1994). The amount of extension in young rifts can be estimated from offset on faults and from gravity profiles. Results range from as little as 10 km for the Baikal rift to > 50 km for the Rio Grande rift. Some rifts, such as the Kenya rift, have a prominent gravity anomaly running along the axis generally interpreted to reflect emplacement of dykes or sills at shallow depth. Magma chambers have been described in rifts based on seismic data, of which the best-documented cases are magma chambers in central New Mexico beneath the Rio Grande rift and in Iceland (Sanford and Einarsson, 1982). Although in some instances tectonic and volcanic activity in rifts are related in space and time, in many cases they are not. Zones of major eruption seldom coincide with the main rift faults.

Rock assemblages

Continental rifts (hereafter referred to as rifts) are characterized by immature terrigenous clastic sediments and bimodal volcanics (Wilson, 1993). **Bimodal volcanics** are basalts (tholeiites) and felsic volcanics, as found for instance in the Rio Grande rift, or alkali basalts and phonolites, as occur in the East African rift. Few, if any, igneous rocks of intermediate composition occur in rifts. Rift felsic volcanics are emplaced usually as ash flow tuffs or glass domes. In older rifts that are uplifted, the plutonic equivalents of these rocks are exposed, as in the Oslo rift in Norway. Granitic rocks range from granite to monzodiorite in composition with granites and syenites usually dominating (Williams, 1982). Most are A-type granitoids (Table 3.1). Rifts range from those with abundant volcanics, like the East African rift, to those with only minor volcanics, like the Baikal rift. Alkali basalts, basanites, and tephrites dominate in the East African rift, followed in abundance by phonolites and trachytes. In the northern Rio Grande rift, tholeiites and basaltic andesites are most important, followed by rhyolites. Farther south in the Rio Grande rift, alkali basalts dominate. The Kenya and Oslo rifts show a decrease in magma alkalinity with time, a feature that probably reflects a secular decrease in depth of magma generation. Many rifts show an increase in magma alkalinity from the centre to the edge of the rift, suggesting a deepening of magma sources in this direction. Geochemical and isotopic studies of rift basalts show that they are derived either from mantle plumes or from the subcontinental lithosphere, or both (Bradshaw et al., 1993; Thompson and Gibson, 1994). Similar data suggest that associated felsic volcanics and granitoids are of crustal origin.

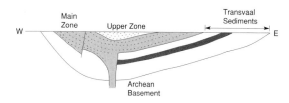

Figure 3.11   Idealized cross-section of the 2-Gy-old Bushveld Complex in South Africa. Length of section about 150 km. After Hunter (1975).

Layered igneous intrusions are also a common rock type in exhumed rifts. These intrusions are large mafic to ultramafic bodies which exhibit internal layering formed by the accumulation of crystals during fractional crystallization of basaltic or komatiitic parent magmas. Cyclic layering in large mafic intrusions is generally interpreted to reflect episodic injections of new magma into fractionating magma chambers. The largest well-studied intrusion is the vast Bushveld Complex in South Africa, which is over 8 km thick, covers a minimum area 66 000 km^2, and was intruded 2 Ga (Figure 3.11). Other large bodies with their corresponding ages of intrusion include the Great Dyke in Zimbabwe (2.5 Ga), the Widgiemooltha Complex in Western Australia (2.4 Ga), the Muskox intrusion in northern Canada (1.27 Ga) and the Duluth Complex in Minnesota (1.1 Ga). Large layered intrusions are major sources of metals, such as Cr, Ni, Cu, and Fe and, in the case of the Bushveld Complex, also Pt.

Rift sediments are chiefly arkoses, feldspathic sandstones, and conglomerates derived from rapidly-uplifted fault blocks, in which granitoids are important components. Evaporites also are deposited in many rifts. If a rift is inundated with seawater, as exemplified by the Rhine graben in Germany, marine sandstones, shales and carbonates may also be deposited.

Rift development and evolution

As a starting point for understanding rift development and evolution, rifts can be classified into two categories, depending on the mechanisms of rifting (Sengor and Burke, 1978; Ruppel, 1995). One group of rifts, referred to as **active rifts**, is produced by doming and cracking of the lithosphere, where doming results from upwelling asthenosphere or rising mantle plumes (Figure 3.12). **Passive rifts**, on the other hand, are produced by stresses in moving lithospheric plates or drag at the base of the lithosphere. Active rifts contain relatively large volumes of volcanic rock, while in passive rifts immature clastic sediments exceed volcanics in abundance. Active rifts are also characterized by early uplift and basement-stripping resulting from crustal expansion due to a deep heat source. In general, uplift in passive rifts is confined to the stretched and faulted near-surface region and to the shoulder of the rift zone, whereas in active rifts, uplift commonly extends for hundreds of kilometres

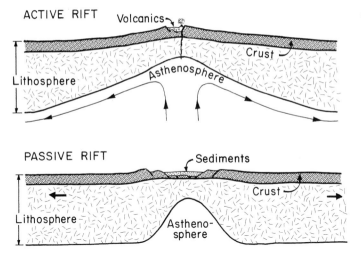

beyond the rift zone proper. Lithospheric thinning is laterally confined to the rift zone in passive rifts, while in active rifts the zone of thinning is several times wider than the rift width (Thompson and Gibson, 1994).

Although active rifts may form in a number of tectonic settings, they commonly exhibit similar overall patterns of development. The major stages are as follows (after Mohr, 1982):

1   Development of a broad, shallow depression prior to doming or volcanism.
2   Diapiric asthenosphere or a mantle plume is forcefully injected into the base of the lithosphere. During injection, diapirs undergo adiabatic decompression leading to partial melting and the onset of basaltic magmatism.
3   Buoyant isostatic uplift of heated lithosphere leads to doming.
4   Doming and extensional forces cause crustal attenuation and thinning of the lithosphere.
5   Episodic dyke injection and volcanism alternate with faulting, and the duration of episodes and volumes of magma erupted decrease with time.
6   Rift valleys develop and may be associated with voluminous felsic magmatism, both extrusive and intrusive.
7   An active rift may be aborted at a relatively early stage of its development, such as the Late Proterozoic Keweenawan rift in the north-central United States, or it may continue to open and evolve into an ocean basin, such as the Red Sea.

There are three recognized types of both active and passive rifts (Figure 3.13). Active rifts are represented by ocean ridges, continental rifts and aulacogens, and back-arc basins. As continental rifts and back-arc rifts continue to open, they can evolve into ocean ridges, and all three types of passive rift can evolve into active rifts. A possible sequence in the break up of a continent by continued opening of a continental rift is shown in

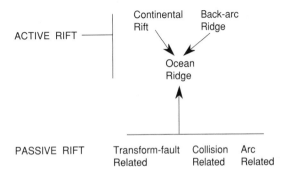

Figure 3.13 Possible paths of rift evolution.

Figure 3.14. This history is based on the opening of the Red Sea beginning about 30 Ma. A mantle plume or asthenosphere upwelling causes the lithosphere to thin and fracture into a series of grabens, which collect clastic sediments from intervening horsts (A and B), and basaltic magma is injected into the axial portion of the graben system. Eventually, new oceanic crust (and lithosphere) is produced as the continent separates and an ocean ridge forms (C). Remnants of basaltic flows and sills and clastic sediments reflecting early stages of rupturing may be preserved in grabens on retreating continental margins (D).

Passive rifts develop along faulted continental margins, in zones of continental collision, and in arc systems (Ingersoll, 1988). Examples of rifts associated with faulted continental margins are those found adjacent to the San Andreas and related faults in California and the rifts in western Turkey associated with the Anatolian transform fault. Rifts can be generated along collision boundaries by irregularities in continental margins and by transcurrent and normal faulting caused by nonperpendicular collision. The Rhine graben in Germany is an example of a rift that developed at a steep angle to a collisional boundary, and the term impactogen has been applied to this type of rift. The Rhine graben appears to

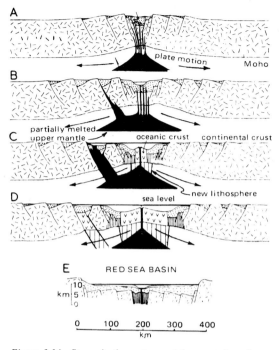

Figure 3.14 Stages in the rupture and fragmentation of a continent and development of a new ocean ridge.

have been produced as Africa collided with an irregularity along the European continental margin. An aborted attempt to subduct the irregularity resulted in depression and rifting. Numerous studies confirm that stresses associated with the Tibet–India collision were transmitted as transcurrent and normal faulting into the Eurasian plate, forming such rifts as the Baikal rift in Siberia and the Shanshi graben system in China.

As rifts evolve, their rock assemblages change. Oceanic rifts are represented by ophiolites. Arc-related rifts contain arc volcanics and graywackes, and moderate-to-large back-arc basins are characterized by mixed assemblages including some combination of ophiolites and deepsea sediments, arc-derived graywackes, and cratonic

sediments. Continental rifts and aulacogens contain arkoses, feldspathic sandstones, conglomerates, and bimodal volcanics. Passive rifts contain a variety of immature sediments and, in some instances, minor volcanic rocks.

Mechanisms of rifting

Two major mechanisms are proposed for continental rifting as mentioned above: upwelling or diapirism of the mantle, and stresses produced by moving plates (McKenzie, 1978; Wernicke, 1985). Both involve extension of the lithosphere. Just how this extension occurs and the relative roles of brittle and ductile deformation continue to be subjects of disagreement among geologists. In general, two models have been proposed for rifting in the crust. One model consists of multiple horsts and grabens, which form during brittle extension of the upper crust. These blocks sink differentially into a plastically-deforming middle and lower crust, where they become isostatically adjusted. The second, and currently more popular, model involves curved, downward-flattening listric faults which exhibit moderate to extreme rotation of hanging-wall blocks. Seismic reflection profiling in the Basin and Range province strongly favours the existence of listric faults at mid-crustal levels (Figure 2.4). These faults appear to merge with or be truncated by a flat decollement or detachment fault that develops during extension (Figure 3.15). **Detachment faults** are major flat-lying shear zones separating disrupted, tilted blocks of an upper plate from ductilely-deformed mylonitic rocks of a lower plate. When exposed at the surface, as in the Whipple Mountains in southeast California, these are known as **core complexes** (Lister and Davis, 1989). Up to 100 per cent crustal extension may occur with the combined offsets of listric and detachment faults. In addition to curving listric faults, sets of near-planar faults may also accommodate extension as these faults rotate and flatten like a stack of tilted books in a bookcase (Rehrig, 1986).

In detail, detachment faults are not simple or single surfaces of displacement. They are generally multiple surfaces where exposed, with slivers of brecciated rock

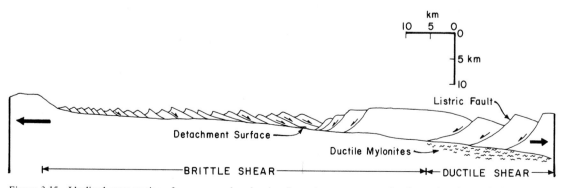

Figure 3.15   Idealized cross-section of a core complex showing the major components related to a detachment fault.

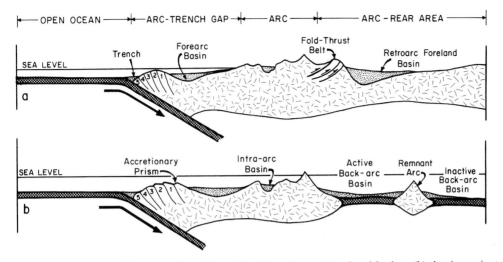

Figure 3.16 Diagrammatic cross-sections of a continental margin arc (a) and an island arc (b) showing major tectonic divisions. Numbers in accretionary prism indicate relative age of slabs (1 the oldest).

between surfaces. Detachment faults commonly bound upper-plate rocks (often Tertiary volcanics and sediments) which display chaotic tilting generally in one direction with consistently-orientated rotational axes. Upper-plate extensional structures initiated at higher angles have been progressively flattened, whereas detachment faults are always nearly horizontal. The observations are generally interpreted to indicate that a detachment fault is the lower decoupling surface upon which the upper plate fragments and is transported (Figure 3.15).

Arc systems

Subduction-related rock assemblages

Numerous geologic environments are associated with subduction. In an idealized arc system, three zones are recognized: the arc–trench gap, the arc, and the arc–rear area (Hamilton, 1988). From the ocean side landwards, a continental-margin arc is characterized by the trench; an accretionary prism with overlying forearc basins; the volcanic arc with intra-arc basins; a fold-thrust belt; and a retroarc foreland basin (Figure 3.16a). An oceanic arc (b), differs from a continental-margin arc primarily in the arc–rear area where it includes some combination of active and inactive back-arc basins and, in some instances, remnant arcs. Each of these arc environments is characterized by different rock assemblages as described below.

Trench

Trenches are formed where lithospheric slabs begin to descend into the mantle. Trench sediments are dominantly fine-grained graywacke turbidites with minor pelagic components. Turbidity currents generally enter trenches at submarine canyons and flow along trench axes. Sediments can be transported along trenches for up to 3000 km, as for example in the Sunda trench south of Sumatra where detritus from the Himalayas enters the trench on the north from the Bengal submarine fan (Moore et al., 1982). Although most detrital sediment is clay or silt, sand and coarser sediments may be deposited in proximal facies. Down-axis changes in facies suggest that trenches are filled by a succession of radiating submarine fans. This longitudinal filling is supplemented by lateral infilling from side canyons. If a trench forms near a continent, continental sources generally dominate, with volcanic arc sources of minor importance. Most trench sediments, however, are derived from local arc volcanic and plutonic sources. Seismic reflection studies of trench sediments show hummocky topography in places suggestive of buried debris flows or submarine slumps. Trench sediments may be preserved in the geologic record if they are added to the arc as faulted slabs in the accretionary prism.

Accretionary Prism

The **accretionary prism** (or subduction zone complex) consists of a series of steeply-inclined, fault-bounded wedges of sediment and volcanic rocks above a descending slab. These wedges represent oceanic crust and trench sediments which have been accreted to the front of the arc. Individual wedges in the accretionary prism decrease in age as the trench is approached (Figure 3.16). Accretionary prisms are intensely deformed producing **melange**, which is a mappable body of rock characterized by the lack of continuous bedding and the inclusion of fragments of rocks of all sizes (up to more than a kilometre across) contained in a fine-grained, deformed matrix. Both sedimentary and tectonic processes can give rise to melange. **Olistostromes** are melanges produced

Figure 3.17 Franciscan melange near San Simeon, California. Large fragments of greenstone (metabasalt) (top) and graywacke (bottom) are enclosed in a sheared matrix of serpentine and chlorite. Courtesy of Darrel Cowan.

by gravitational sliding and accumulate as semi-fluid bodies that do not have bedding, but include associated turbidites. Melange clasts may be exotic (derived from another environment) or native (reworked from the immediate environment) and are generally matrix supported (Figure 3.17). Clast lithologies include graywacke, mudstone, chert, basalt (greenstone) and other ophiolite lithologies, arc volcanics, and rare granitoids. Melanges are commonly folded and may contain more than one cleavage or foliation. Sheared matrices are usually composed of serpentine and fine-grained rock and mineral fragments.

Although melanges are typical of accretionary prisms, they are formed by several different processes and occur in different tectonic environments (Cowan, 1986). Tectonic melanges are produced by compressive forces along the upper part of descending slabs at shallow depths. Fragmentation and mixing of largely lithified rocks may occur along a migrating megashear zone subparallel to a subducting slab. Fragments of oceanic crust and trench sediment are scraped off the descending plate and accreted to the overriding plate. Olistostromes may be produced by gravitational slumping or debris flows on oversteepened trench walls or along the margins of a forearc basin. Debris flows, in which clay minerals and water form a single fluid-possessing cohesion, are probably the most important transport mechanism of olistostromes. Still other melanges form along major shear zones, and they are also characteristic of collisional sutures.

Forearc basins

Forearc basins are marine depositional basins on the trench sides of arcs (Figure 3.16), and they vary in size and abundance with the evolutionary stage of an arc. In continental-margin arcs, such as the Sunda arc in Indonesia, forearc basins range to 700 km in strike length. They overlie the accretionary prism, which may be exposed as submarine hills within and between forearc basins. Sediments in forearc basins, which are chiefly turbidites with sources in the adjacent arc system, range up to many kilometres in thickness. Hemipelagic sediments are also of importance in some basins, such as in the Mariana arc. Olistostromes can form in forearc basins by sliding and slumping from locally-steepened slopes. Forearc-basin clastic sediments may record progressive unroofing of adjoining arcs, as for instance shown by the Great Valley Sequence (Jur-Cret) in California (Dickinson and Seely, 1986). Early sediments in this sequence are chiefly volcanic detritus from active volcanoes and later sediments reflect progressive unroofing of the Sierra Nevada batholith. Volcanism is rare in modern forearc regions and neither volcanic nor intrusive rocks are common in older forearc successions.

Arcs

Volcanic **arcs** range from entirely subaerial, such as the Andean and Middle America arcs, to mostly or completely submarine, such as many of the immature oceanic arcs in the Southwest Pacific. Other arcs, such as the Aleutians, change from subaerial to partly submarine along strike. Subaerial arcs include flows and associated pyroclastic rocks, which often occur in large stratovolcanoes. Submarine arcs are built of pillowed basalt flows and large volumes of hyaloclastic tuffs and breccias. Volcanism begins rather abruptly in arc systems at a volcanic front. Both tholeiitic and calc-alkaline magmas characterize arcs, with basalts and basaltic andesites dominating in oceanic arcs, and andesites and dacites often dominating in continental-margin arcs. Felsic magmas are generally emplaced as batholiths, although felsic volcanics are common in most continental-margin arcs.

Back-arc basins

Active **back-arc basins** occur over descending slabs behind arc systems (Figure 3.16) and commonly have high heat flow, relatively thin lithosphere and, in many instances, an active ocean ridge which is enlarging the size of the basin (Jolivet et al., 1989; Fryer, 1996).

Sediments are varied, depending on basin size and nearness to an arc. Proximal to arcs and remnant arcs, volcaniclastic sediments generally dominate, whereas in more distal regions, pelagic, hemipelagic, and biogenic sediments are more widespread (Klein, 1986). During the early stages of basin opening, thick epiclastic deposits, largely representing gravity flows, are important. With continued opening of a back-arc basin, these deposits pass laterally into turbidites, which are succeeded distally by pelagic and biogenic sediments (Leitch, 1984). Discrete layers of air-fall tuff may be widely distributed in back-arc basins. Early stages of basin opening are accompanied by diverse magmatic activity including felsic volcanism, whereas later evolutionary stages are characterized by an active ocean ridge. As previously mentioned, many ophiolites carry a subduction zone geochemical signature, and thus appear to have formed in back-arc basins.

Subaqueous ash flows may erupt or flow into back-arc basins and form in three principal ways (Fisher, 1984). The occurrence of felsic welded ash flow tuffs in some ancient back-arc successions suggests that hot ash flows enter water without mixing and retain enough heat to weld (Figure 3.18a). Alternatively, submarine eruptions may eject large amounts of ash into the sea, which fall onto the sea floor forming a dense, water-rich debris flow, (b). In the closing stages of eruption, small turbidity currents are deposited as thin-graded beds above there debris flows. In addition to direct eruption, slumping of unstable slopes composed of pyroclastic debris can produce ash turbidites, (c). Submarine fans composed of volcanic debris are also found in back-arc basins, and turbidites often comprise a large portion of these fans (Carey and Sigurdsson, 1984). Fan sediments come from multiple volcanic sources along the arc, and massive influxes of sediment are related to arc volcanism and/or to uplift of the arc (Klein, 1986). Results from modern arcs suggest that rates of arc uplift must be > 400 m/My to generate the requisite sediment yield to form submarine fans.

Processes controlling back-arc basin deposition include latitude-dependent biogenic productivity and oceanic circulation, climate, volume of arc volcanism, rate of arc uplift, and regional wind patterns. Because of the highly-varied nature of modern back-arc sediments and the lack of a direct link between sediment type and tectonic setting, one cannot assign a distinct sediment assemblage to these basins. It is only when a relatively complete stratigraphic succession is preserved and detailed sedimentological and geochemical data are available that ancient back-arc successions can be identified. Inactive back-arc basins, such as the western part of the Philippine plate, have a thick pelagic sediment blanket and lack evidence for recent seafloor spreading.

Remnant arcs

Remnant arcs are submarine aseismic ridges that are extinct portions of arcs which have been rifted away by

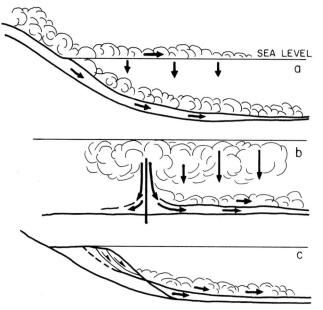

SEA LEVEL

a

b

c

Figure 3.18   Mechanisms for the origin of subaqueous ash flows (from Fisher, 1984). (a) Hot ash flow erupted on land flowing into water; (b) ash flow forms from column collapse; (c) ash turbidites develop from slumping of hyaloclastic debris.

the opening of a back-arc basin (Figure 3.16b) (Fryer, 1996). They are composed chiefly of subaqueous mafic volcanic rocks similar to those formed in submarine arcs. Once isolated by rifting, remnant arcs subside and are blanketed progressively by deep-water pelagic and biogenic deposits and distal ash showers.

Retroarc foreland basins

Retroarc foreland basins form behind continental-margin arc systems (Figure 3.16a), and are filled largely with clastic terrigenous sediments derived from a fold-thrust belt behind the arc. A key element in foreland basin development is the syntectonic character of the sediments (Graham et al., 1986). The greatest thickness of foreland basin sediments borders the fold-thrust belt, reflecting enhanced subsidence caused by thrust-sheet loading and deposition of sediments. Another characteristic feature of retroarc foreland basins is that the proximal basin margin progressively becomes involved with the propagating fold-thrust belt (Figure 3.19). Sediments shed from the rising fold-thrust belt are eroded and redeposited in the foreland basin, only to be recycled again with basinward propagation of this belt. Proximal regions of foreland basins are characterized by coarse, arkosic alluvial-fan sediments, and distal facies by fine-grained sediments and variable amounts of marine carbonates. Progressive unroofing in the fold-thrust belt should lead to an 'inverse' stratigraphic sampling of the source in foreland basin sediments as illustrated in Figure 3.19. Such a pattern is well-developed in the Cretaceous foreland basin deposits in eastern Utah (Lawton, 1986). In this basin, early stages of uplift and erosion resulted in deposition of Paleozoic carbonate-rich clastic

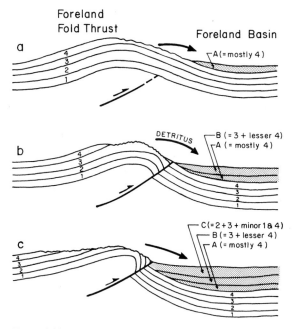

Foreland
Fold Thrust

a

Foreland Basin

A(=mostly 4)

b

DETRITUS

B (= 3 + lesser 4)
A (= mostly 4)

4
3
2
1

c

C(=2+3 + minor 1 & 4)
B (= 3 + lesser 4)
A (= mostly 4)

4
3
2
1

Figure 3.19   Progressive unroofing of an advancing foreland thrust sheet. After Graham et al. (1986).

sediments followed later by quartz-feldspar-rich detritus from the elevated Precambrian basement. Foreland basin successions also typically show upward coarsening and thickening terrigenous sediments, a feature which reflects progressive propagation of the fold-thrust belt into the basin.

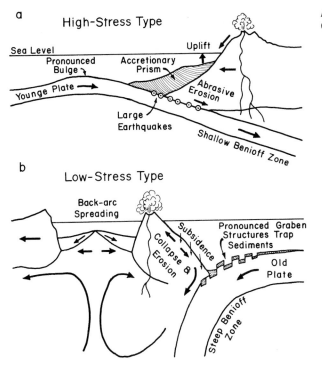

Figure 3.20   Idealized cross-sections of (a) high- and (b) low-stress subduction zones. After Uyeda (1983).

High- and low-stress subduction zones

Uyeda (1983) suggested that subduction zones are of two major types, each representing an end member in a continuum of types (Figure 3.20). The relatively **high-stress type**, as exemplified by the Peru–Chile arc, is characterized by a pronounced bulge in the descending slab, a large accretionary prism, relatively large shallow earthquakes, buoyant subduction (producing a shallow dipping slab), a relatively young descending slab, and a wide range in composition of calc-alkaline and tholeiitic igneous rocks (a). The **low-stress type**, of which the Mariana arc is an example, has little or no accretionary prism, few large earthquakes, a steep dip of the descending plate which is relatively old, a dominance of basaltic igneous rocks, and a back-arc basin (b). In the high-stress type, the descending and overriding plates are more strongly coupled than in the low-stress type, explaining the importance of large earthquakes and growth of the accretionary prism. This stronger coupling, in turn, appears to result from buoyant subduction. In the low-stress type, the overriding plate is retreating from the descending plate resulting in the opening of a back-arc basin (Scholz and Campos, 1995). In the high-stress type, however, the overriding plate is either retreating very slowly compared with the descending plate, or perhaps converging against the descending plate. Thus, the two major factors contributing to differences in subduction zones appear to be:

1 relative motions of descending and overriding plates
2 the age and temperature of the descending plate.

Arc processes

Seismic reflection profiling and geological studies of uplifted and eroded arc systems have led to a greater understanding of arc evolution and of the development of accretionary prisms and forearc basins. Widths of modern arc–trench gaps (75–250 km) are proportional to the ages of the oldest igneous rocks exposed in adjacent arcs (Dickinson, 1973). As examples, the arc–trench gap width in the Solomon Islands is about 50 km, with the oldest igneous rocks about 25 Ma, and the arc–trench gap width in northern Japan (Honshu) is about 225 km, with the oldest igneous rocks about 125 Ma. The correlation suggests progressive growth in width of arc–trench gaps with time. Such growth appears to reflect some combination of outward migration of the subduction zone by accretionary processes and inward migration of the zone of maximum magmatic activity. **Subduction zone accretion** involves the addition of sediments and volcanics to the margin of an arc in the accretionary prism (von Huene and Scholl, 1991). Seismic profiling suggests that accretionary prisms are composed of sediment wedges separated by high-angle thrust faults produced by offscraping of oceanic sediments (Figures 2.5 and 3.21). During accretion, oceanic sediments and fragments of oceanic crust and mantle are scraped off and added to the accretionary prism (Scholl et al., 1980). This offscraping results in outward growth of the prism, and also controls the location and evolutionary patterns of overlying forearc basins. Approximately half of modern arcs are growing by offscraping accretion. Reflection profiles suggest that deformational patterns are

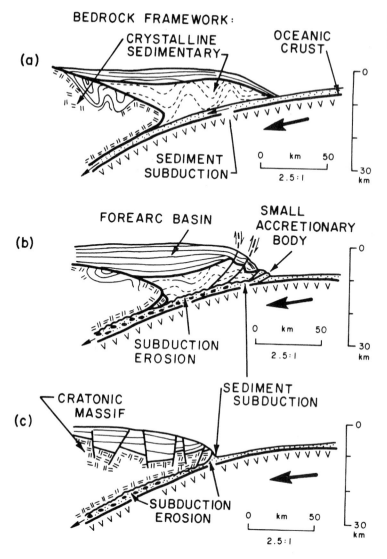

Figure 3.21 Sediment subduction and sediment erosion at a convergent plate boundary. After Scholl et al. (1980).

considerably more complex than simple thrust wedges. Deformation may include large-scale structural mixing and infolding of forearc basin sediments. Geological evidence for such mixing comes from exposed accretionary prisms, exemplified by parts of the Franciscan Complex in California. Fluids also play an important role in facilitating mixing and metasomatism in accretionary prisms (Tarney et al., 1991).

In addition to accretion to the landward side of the trench, material can be underplated beneath the arc by a process known as **duplex accretion**. A duplex is an imbricate package of isolated thrust slices bounded on top by a thrust and below by a low-angle detachment fault (Sample and Fisher, 1986). During the transfer of displacement from an upper to a lower detachment horizon, slices of footwall are accreted to the hanging wall (accretionary prism) and rotated by bending of the fron-

tal ramp (Figure 3.22). Observations from seismic reflection profiles, as well as exposed accretionary prisms, indicate that duplex accretion occurs at greater depths than offscraping accretion. Although some arcs, such as the Middle America and Sunda arcs, appear to have grown appreciably by accretionary processes others, such as the New Hebrides arc, have little if any accretionary prism. In these latter arcs, either very little sediment is deposited in the trench or most sediments are subducted (Figure 3.21, a). One possible way to subduct sediments is in grabens in the descending slab, a mechanism supported by the distribution of seismic reflectors in descending plates. Interestingly, if sediments are subducted in large amounts beneath arcs, they cannot contribute substantially to arc magma production as constrained by isotopic and trace element distributions in modern arc volcanics.

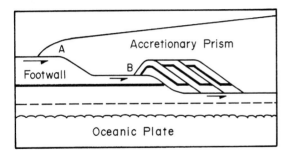

Figure 3.22 Idealized cross-sections showing duplex accretion. After Sample and Fisher (1986). A, zone of offscraping and B, zone of duplex accretion.

Subduction erosion is another process proposed for arcs with insignificant accretionary prisms. It involves mechanical plucking and abrasion along the top of a descending slab, which causes a trench's landward slope to retreat shoreward (Figure 3.21). Subduction erosion may occur either along the top of the descending slab (b), or at the leading edge of the overriding plate (c). Evidence commonly cited for subduction erosion includes:

1 an inland shift of the volcanic front, as has occurred in the Andes in the last 100 My
2 truncated seaward trends and seismic reflectors in accretionary prisms and forearc basins
3 not enough sediment in trenches to account for the amount delivered by rivers
4 evidence for crustal thinning such as tilting of unconformities towards the trench, which is most easily accounted for by subsidence of the accretionary prism.

All of these can be explained by erosion along the top of the descending slab. Subduction erosion rates have been estimated along parts of the Japan and Chile trenches at 25–50 km³/My for each kilometre of shoreline (Scholl et al., 1980).

Accretion, mixing, subduction erosion, and sediment subduction are all potentially important processes in subduction zones and any one of them may dominate at a given place and evolutionary stage. Studies of modern arcs indicate that about half of the ocean floor sediment arriving at trenches is subducted and does not contribute to growth of accretionary prisms either by offscraping or duplex accretion (von Huene and Scholl, 1991). At arcs with significant accretionary prisms, 70–80 per cent of incoming sediment is subducted, and at arcs without accretionary prisms, all of the sediment is subducted. The combined average rates of subduction erosion (0.9 km³/y) and sediment subduction (0.7 km³/y) suggest that, on average, 1.6 km³ of sediment are subducted each year.

High-pressure metamorphism

Blueschist-facies metamorphism is important in subduction zones, where high-pressure, relatively low-temperature mineral assemblages form. Glaucophane and lawsonite, both of which have a bluish colour, are common minerals in this setting. In subduction zones, crustal fragments can be carried to great depths (> 50 km), yet remain at rather low temperatures, usually < 500 °C. A major unsolved question is how these rocks return to the surface. One possibility is by the continual underplating of the accretionary prism with low-density sediments, resulting in fast, buoyant uplift during which high-density pieces of the slab are dragged to the surface (Cloos, 1993). Another possibility is that blueschists are thrust upwards during later collisional tectonics.

One of the most intriguing fields of research at present is seeing just how far crustal fragments are subducted before returning to the surface. Recent discoveries of coesite (high-pressure silica phase) and diamond inclusions in pyroxenes and garnet from eclogites from high-pressure metamorphic rocks in eastern China record astounding pressures of 4.3 Gpa (about 150 km burial depth) at 740 °C (Schreyer, 1995). Several other localities have reported coesite-bearing assemblages recording pressures of 2.5–3 Gpa. Also, several new high-pressure hydrous minerals have been identified in these assemblages, indicating that some water is recycled into the mantle and that not all water is lost by dehydration to the mantle wedge. Perhaps the most exciting aspect of these findings is that for the first time there is direct evidence that crustal rocks (both felsic and mafic) can be recycled into the mantle.

Igneous rocks

The close relationship between active volcanism in arcs and descending plates implies a genetic connection between the two. Subduction-zone related volcanism starts abruptly at the **volcanic front**, which roughly parallels oceanic trenches and begins 200–300 km inland from trench axes adjacent to the arc–trench gap (Figure 3.23). It occurs where the subduction zone is 125–150 km deep, and the volume of magma erupted decreases in the direction of subduction zone dip. The onset of volcanism at the volcanic front probably reflects the onset of melting above the descending slab, and the decrease in volume of erupted magma behind the volcanic front may be caused by either a longer vertical distance for magmas to travel or a decrease in the amount of water liberated from the slab as a function of depth.

The common volcanic rocks in most island arcs are basalts and basaltic andesites, whereas andesites and more felsic volcanics also become important in continental-margin arcs. Whereas basalts and andesites are erupted chiefly as flows, felsic magmas are commonly plinian eruptions in which much of the ejecta are in the form of ash and dust. These eruptions give rise to ash flows and associated pyroclastic (or hyaloclastic) deposits. Arc volcanic rocks are generally porphyritic containing up to 50 per cent phenocrysts in which plagioclase dominates.

Arc volcanoes are typically steep-sided stratovolcanoes composed of varying proportions of lavas and fragmental

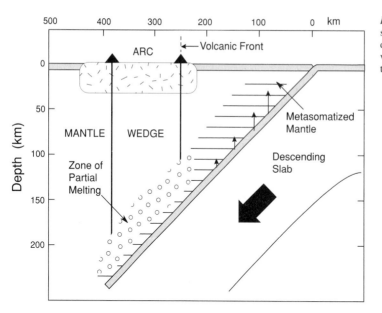

Figure 3.23 Cross-section of a subduction zone showing shallow devolatilization of descending slab (short vertical lines) and magma production in the mantle wedge.

materials (Figure 3.24). Their eruptions range from mildly explosive to violently explosive, and contrast strikingly with the eruptions of oceanic-island and continental-rift volcanoes. Large amounts of water are given off during eruptions. Rapid removal of magmas may result in structural collapse of the walls of stratovolcanoes, producing calderas such as Crater Lake in Oregon. The final stages of eruption in some volcanic centres are characterized by felsic ash flows that may travel great distances. Seismic shadow-zone studies indicate that modern magma reservoirs in subduction-zone areas are commonly 50–100 km deep. The migration of earthquake hypocentres from depths of up to 200 km over periods of a few months prior to eruption reflects the ascent of magmas at rates of 1–2 km/day. The height of stratovolcanoes varies with magma composition (McBirney, 1976). The higher a volcano, the more felsic its composition. This appears to be due to increased fractional crystallization in areas of thickened crust in which volcanoes have higher base elevations due to isostasy.

The cores of arc systems comprise granitic batholiths as evidenced in deeply eroded arcs. Such batholiths, which are composed of numerous plutons, range in composition from diorite to granite, with granodiorite often dominating.

In contrast to oceanic basalts, arc basalts are commonly quartz normative, with high Al_2O_3 (16–20 per cent) and low TiO_2 (< 1 per cent) contents. Igneous rocks of the tholeiite and calc-alkaline series are typical of both island arcs and continental-margin arcs. $^{87}Sr/^{86}Sr$ ratios in volcanics from island arcs are low (0.702–0.705) while those from continental-margin arcs are variable, reflecting variable contributions of continental crust to the magmas. Arc basalts also exhibit a subduction-zone component (depletion in Nb and Ta relative to neighbouring incompatible elements on a primitive-mantle normalized graph; Figure 3.8) (Hawkesworth et al., 1994). Arc granitoids are chiefly I-types, typically meta-aluminous, with tonalite or granodiorite dominating (Table 3.1).

Compositional variation of arc magmas

Igneous rocks from arcs vary in composition in both space and time. It has long been recognized that some arc volcanics show increasing LIL elements and $^{87}Sr/^{86}Sr$ ratios with increasing depth to subduction zone. This change in composition is known as **compositional polarity** and is well documented in Japan (Sakuyama and Nesbitt, 1986). Because many or perhaps most arcs do not show compositional polarity (Arculus and Johnson, 1978), the linear relationship between LIL element content and depth to subduction zone proposed by Dickinson (1970) should not be used to determine polarity of ancient subduction zones unless supported by other data. Arc volcanics also exhibit lateral compositional changes as evidenced, for instance, by the $^{87}Sr/^{86}Sr$ ratio in the Sunda–Banda arc. Progressing eastwards in this arc, the $^{87}Sr/^{86}Sr$ ratio increases, especially east of Flores. Some oceanic arcs seem to show a geochemical evolution with time. The best documented case is in Fiji where the earliest volcanics (30–25 Ma) are relatively depleted in LIL elements, whereas later volcanics show moderate LIL-element enrichment and a strong subduction geochemical component (Gill, 1987).

Both experimental and geochemical data show that most arc basalts are produced by partial melting of the mantle wedge in response to the introduction of volatiles (principally water) from the breakdown of hydrous minerals in descending slabs (Pearce and Peate, 1995; Poli and Schmidt, 1995) (Figure 3.23). However, other pro-

Figure 3.24   Eruption of Mount St Helens, southwest Washington, in May, 1980.

cesses, like fractional crystallization, assimilation of crust, and contamination by subducted sediment, also affect magma composition. Trace element and isotope distributions cannot distinguish between a subducted sediment contribution to arc magmas and continental assimilation. If arc magmas reside in the continental crust during fractional crystallization for significant amounts of time, they may become contaminated with incompatible elements from the surrounding crust. The very high Sr and Pb isotope ratios and low Nd isotope ratios in some felsic volcanics and granitic batholiths from continental-margin arcs, such as those in the Andes, suggest that these magmas were produced either by partial melting of older continental crust or by significant contamination with this crust. Incompatible elements are also strongly enriched in these rocks, a feature which is consistent with a major crustal component in their source.

One unsolved problem related to the production of arc magmas is that of how the subduction zone geochemical component is acquired by the mantle wedge. Whatever process is involved, it requires decoupling of Ta–Nb, and in some cases Ti, from LIL elements and REE (McCulloch and Gamble, 1991). Liberation of saline aqueous fluids from oceanic crust may carry LIL elements, which are soluble in such fluids, into the overlying mantle wedge, metasomatizing the wedge, and leaving Nb and Ta behind (Saunders et al., 1980; Keppler, 1996). The net result is relative enrichment in Nb–Ta in the descending slab and corresponding depletion in the mantle wedge. Magmas produced in the mantle wedge should therefore inherit this subduction zone signature. A potential problem with this model is that hydrous secondary minerals in descending oceanic crust (e.g., chlorite, biotite, amphiboles, talc, etc.) breakdown and liberate water at or above 125 km (Figure 3.23). Only phlogopite may persist to greater depths. Yet the volcanic front appears at subduction-zone depths of 125–150 km. Devolatilization of descending slabs by 125 km should add a subduction component only to the mantle wedge above this segment of the slab, which is shallower than the depth of partial melting. How, then, do magmas acquire a subduction-zone component in the mantle wedge? Two possibilities have been suggested, although neither have been fully evaluated:

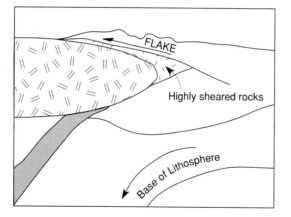

Figure 3.25 Schematic cross-section showing the emplacement of an allochthon (flake) during a continent–continent collision.

1   asthenosphere convection extends into the 'corner' of the mantle wedge and carries ultramafic rocks with a subduction component to greater depths (Figure 3.23)
2   frictional drag of the descending plate drags mantle with a subduction component to greater depths.

Orogens

Two types of orogens

Two types of orogens are recognized in the continental crust (Windley, 1992). The first type, **collisional orogens**, involve collision of two or more continental provinces; when the direction of colliding plates is orthogonal, the crust is greatly thickened, and older crustal provinces are extensively reworked by thrusting, metamorphism, and partial melting of the crust. Very little juvenile crust is produced or tectonically 'captured' during collisional orogeny. In contrast, **accretionary orogens** involve collision and suturing of largely juvenile crustal blocks (ophiolites, island arcs, submarine plateaux, etc.) to continental crust. Accretionary orogens contain very little reworked older crust.

Collisional Orogens

During continental collisions, major thrusts and nappes are directed toward the converging plate as the crust in a collisional zone thickens by ductile deformation, and perhaps by underplating with mafic magmas. In some instances, sheet-like slabs, commonly referred to as flakes or **allochthons**, may be sheared from the top of the converging plate and thrust over the overriding plate (Figure 3.25). In the Eastern Alps, for instance, Paleozoic metamorphic rocks have been thrust more than 100 km north over the Bohemian Massif and a zone of highly-sheared Mesozoic metasediments lies within the thrust

zone (Pfiffner, 1992). The allochthon is less than 12 km thick and appears to represent part of the Carnics plate which was detached during the Mid-Tertiary (Figure 2.7). Seismicity along continent–continent collision boundaries suggests partial subduction of continental crust. Thickening of crust in collisional zones results in partial melting of the lower crust producing felsic magmas that are chiefly intruded as plutons, with some surface eruptions. Fractional crystallization of basalts may produce anorthosites in the lower crust, and losses of fluids from the thickened lower crust may leave behind granulite-facies mineral assemblages. Isostatic recovery of collisional orogens is marked by the development of continental rifting with fluvial sedimentation and bimodal volcanism, as evident in the Himalayas and Tibet (Dewey, 1988).

To some extent, each collisional belt has its own character. In some instances, plates lock together with relatively little horizontal transport, such as along the Caledonian suture in Scotland or the Kohistan suture in Pakistan. In other orogens, such as the Alps and Himalayas, allochthons are thrust for considerable distances and stacked one upon another. Hundreds of kilometres of shortening may occur during such a collision. In a few collisional belts, such as the Late Proterozoic Damara belt in Namibia, a considerable amount of deformation and thickening may occur in the overthrust plate. In the Caledonides and parts of the Himalayan belt, ophiolite obduction precedes continent–continent collision and does not seem to be an integral part of the collision process. In some cases, the collisional zone is oblique, with movement occurring along one or more transform faults. In still other examples, one continent will indent another, thrusting thickened crust over the indented block. Thickened crust tends to spread by gravitational forces, and spreading directions need not parallel the regional plate movement. Which block has the greatest effects of deformation and metamorphism depends on such factors as the age of the crust, its thermal regime, crustal anisotropy, and the nature of the subcrustal lithosphere. Older lithosphere generally has greater strength than younger lithosphere.

Accretionary orogens

Accretionary orogens, for example, the western Cordillera in Alaska and western Canada, develop as oceanic terranes such as island arcs, submarine plateaux, and ophiolites collide with a continental margin. Collisions may occur between oceanic terranes, producing a superterrane, before collision with a continent. In some instances, older continental blocks may also be involved in the collisions, such as those found in the 1.9-Ga Trans-Hudson orogen in eastern Canada and in the Ordovician Taconian orogen in the eastern United States. Seismic reflection profiles in western Canada show west-dipping seismic reflectors which are probably major thrusts, suggesting that accretionary orogens comprise stacks of thrusted terranes. Because accretionary orogens are made

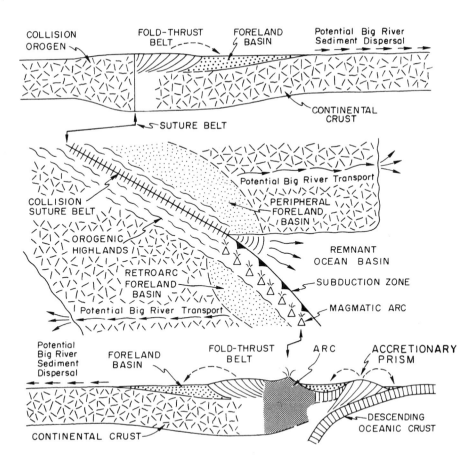

Figure 3.26 Cross-sections and plan view of an evolving collisional boundary. Modified after Dickinson and Suczek (1979). Dashed arrows indicate sediment transport directions.

chiefly of juvenile crust, they represent new crust added to the continents during collisions and, as we shall see in Chapter 5, represent one of the major processes of continental growth. As an example, during the Early Proterozoic, an aggregate area of new crust up to 1500 km wide and over 5000 km long was added to the southern margin of the Baltica–Laurentia supercontinent (Hoffman, 1988). One of the major questions addressed in later chapters is how this accreted mafic crust changes into continental crust.

Orgenic rock assemblages

It is difficult to assign any particular rock assemblage to collisional and accretionary orogens because rock assemblages change both with time and space as collision progresses. Also, it is necessary to sort out a myiad of older rock assemblages contained in the colliding blocks and representing every conceivable tectonic setting. Sediments accumulate in peripheral foreland and hinterland basins, which develop in response to uplift and erosion of a collisional zone. These basins and the sediments contained in them evolve in a manner similar to retroarc foreland basins. Classic examples of peripheral foreland basins developed adjacent to the Alps and the Himalayas during the Alpine–Himalayan collisions in the Tertiary. During the Alpine collision, up to 6 km of alluvial fan deposits known as molasse were deposited in foreland basins (Homewood et al., 1986). Individual alluvial fans up to 1 km thick and 40 km wide have been recognized in the Alps. Coarsening upward cycles and intraformational unconformities characterize collisional molasse deposits, both of which reflect uplift of an orogen and propagation of thrusts and nappes into foreland basins. Collision-derived sediment may also be shed longitudinally from an orogen and enter remnant ocean basins as turbidite fans (Figure 3.26). Sediments and volcanics may accumulate in fault-bounded basins in the thickened overriding continental plate during a continental collision. These are generally similar to rift assemblages with abundant felsic ash flow tuffs.

At deeper exposure levels (10–20 km) in collisional orogens, granitoids are common and appear to be produced by partial melting of the lower crust during a collision. Thickening of continental crust, both in descending and overriding plates, leads to the production of granulites at depths > 20 km as fluids escape upwards. Anorthosites may form as cumulates from fractional crystallization of basalt in the lower and middle crust. Basaltic magma also may underplate the crust and occurs as gabbro or mafic granulites in uplifted crustal sections (Chapter 2). Collisional granites include pre-,

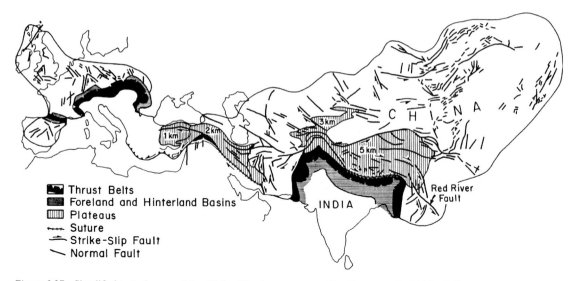

Figure 3.27 Simplified tectonic map of the Alpine–Himalayan orogen. After Dewey et al. (1986). Collisional plateau elevations given in kilometres. 1 km, Anatolia Plateau; 5 km, Tibet Plateau.

syn-, and post-collisional types (Pearce et al., 1984). Pre- and post-collisional granitoids are chiefly I-type granites, whereas syn-collisional granites are commonly leucogranites (in collisional orogens) and many exhibit features of S-type granites (Table 3.1). They are commonly peraluminous and often contain muscovite with or without biotite, and tourmaline also may be present (Harris et al., 1986). Silica contents of leucogranites exceed 70 per cent and most show a subduction geochemical component inherited from their lower crustal source. Supporting a crustal source for many collisional granitoids are their high $^{87}Sr/^{86}Sr$ ratios (> 0.725) and high $\delta^{18}O$ values (Vidal et al., 1984). Post-collisional granites, which generally post-date collision by 40–50 My, are sharply cross-cutting and are dominantly tonalite to granodiorite in composition.

Tectonic elements of a collisional orogen

Continental collision involves progressive compression of buoyant terranes within subduction zones. These terranes may vary in scale from seamounts or island arcs to large continents. The scale of colliding terranes dictates the style, duration, intensity and sequence of strain systems (Dewey et al., 1986). If colliding continental margins are irregular, the strain sequences are variable along great strike lengths. Prior to terminal collision, one or both continental margins may have had a long complex history of accretionary terrane assembly. Continental collisional boundaries are wide, complicated structural zones (Figure 3.27), where plate displacements are converted into complex and variable strains. Just how this conversion is accomplished remains a fundamental problem of plate tectonics. Horizontal strain alone cannot explain thickened crust in the

Tibetan Plateau where vertical stretching may play an important role.

Collisional orogens can be considered in terms of five tectonic components (Figures 3.27 and 3.28): thrust belts, foreland flexures, plateaux, widespread foreland/hinterland deformational zones, and zones of orogenic collapse. **Foreland** and **hinterland** refer to regions beyond major overthrust belts in the direction and away from the direction of principal orogenic vergence, respectively. Thrust belts develop where thinned continental crust is progressively restacked and thickened towards the foreland. If detachment occurs along a foreland thrust, rocks in the allochthon can shorten significantly independent of shortening in the basement. The innermost nappes and the suture zone are usually steepened and overturned during advanced stages of collision. The upper crust in collisional orogens is a high-strength layer that may be thrust hundreds of kilometres as relatively thin, stacked sheets that merge along a decollement surface. For instance, in the southern Appalachians a decollement has moved westwards over the foreland for at least 300 km. Foreland flexures are upwarps in the foreland lithosphere caused by progressive downbending of the lithosphere by advancing thrust sheets. The wavelength and amplitude of foreland flexures depend on rigidity and temperature distribution of the lithosphere.

Collisional plateaus, such as the Tibet and Anatolia plateaus (Figure 3.27), form in the near-hinterland area adjacent to the suture and may rise to 1–5 km above sea level. Although several models have been suggested for the formation of these plateaux, only three seem to be realistic (Harrison et al., 1992) (Figure 3.29):

1   underthrusting of continental crust and lithosphere, which buoyantly elevates the plateau

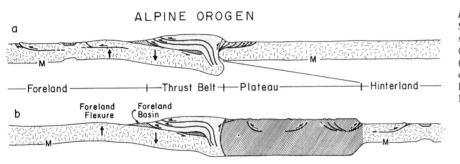

ALPINE OROGEN

—Foreland —— ——Thrust Belt —+—Plateau —————— +—Hinterland —

HIMALAYAN OROGEN

Figure 3.28
Schematic cross-sections of Alpine (a) and Himalayan (b) collisional orogens. After Dewey et al. (1986). M, Moho.

2 compressive shortening and thickening of both the crust and lithosphere, followed by isostatic rebound to form the plateau

3 either (1) or (2) followed by delamination of the thickened lithosphere.

Neither model (1) nor model (2) adequately explains the delayed uplift of Tibet, which began 20 Ma, whilst the India–Tibet collision began about 55 Ma. However, if mantle lithosphere was detached some 30 My after collision and sank into the mantle, it would be replaced with hotter asthenosphere, which would immediately cause isostatic uplift, elevating Tibet.

The widespread foreland/hinterland deformational zones are evidence that stresses generated by collision can affect continental areas thousands of kilometres from

a. Underthrusting of Continental Lithosphere

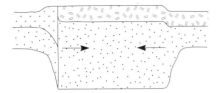

b. Compressional Thickening of the Lithosphere

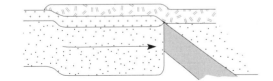

c. Delamination of the Lithosphere

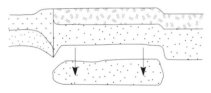

Figure 3.29 Schematic cross-sections of Tibet showing three tectonic models for the uplift of the Tibet Plateau.

the thrust belt (Figure 3.27). For instance, the Tien Shan Range in central Asia appears to have formed in response to stresses transmitted across the continent from continued post-collisional indentation of India into Eurasia. The relative amounts of deformation taken up by the foreland and hinterland vary considerably and depend on thermal age and anisotropy of the lithosphere in converging and overriding plates. The preferential hinterland deformation in Asia associated with the India–Tibet collision probably reflects warm, thin Tibetan lithosphere produced by pre-collisional subduction and accretion. Extensional collapse zones, such as the Aegean basin, are generally of local occurrence in foreland areas and may develop in response to rollback of a partially-subducted slab.

Sutures

Sutures are ductile shear zones produced by thrusting along converging plate boundaries, and they range from a few hundred metres to tens of kilometres wide (Coward et al., 1986). Rocks on the hinterland plate adjacent to sutures are chiefly arc-related volcanics and sediments from a former arc system on the overriding plate, whereas those on the foreland plate are commonly passive-margin sediments. Fragments of rocks from both plates as well as ophiolites occur in suture melanges. These fragments are tens to thousands of metres in size and are randomly mixed in a sheared, often serpentine-rich, matrix.

One of the problems in recognizing Precambrian collisional orogens is the absence of well-defined suture zones. However, at the crustal depths exposed in most reactivated Precambrian terranes, suture zones are difficult to tell from other shear zones. The classic example of a Cenozoic suture is the Indus suture in the Himalayas. Yet, in the Nanga Parbat area where deep levels of this suture are exposed, it is difficult to identify the suture because it looks like any other shear zone (Coward et al., 1982). Rocks on both sides of the suture are complexly-deformed amphibolites and gneisses indistinguishable from each other. If equivalent crustal levels of the Indus suture are compared with those of exhumed Precambrian orogens, there are striking similarities, and deep-seated shear zones in these orogens may or may

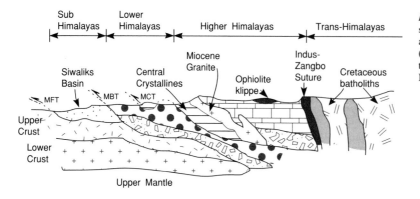

Figure 3.30 Schematic cross-section of the Himalayas in Nepal and Tibet. Modified after Windley (1983). MBT, Main Boundary thrust; MFT, Main Frontal thrust; MCT, Main Central thrust.

not be sutures. Without precise geochronology and detailed geologic mapping on both sides of shear zones, it is not possible to identify correctly which shear zones are sutures and which are not.

Foreland and hinterland basins

Peripheral foreland and hinterland basins are like retroarc foreland basins in terms of sediment provenance and tectonic evolution. Major peripheral foreland and hinterland basins developed in the Tertiary in response to the Alpine–Himalayan collisions (Figure 3.28) (Allen and Homewood, 1986). These basins exhibit similar stages of development which have been explained by thermal-mechanical models (Stockwell et al., 1986). In the case of foreland basins, during the first stage a passive-margin sedimentary assemblage is deposited on a stretched and rifted continental margin. Collision begins as a terrane is thrust against and over continental crust on the descending slab. Continental convergence results in thickening of this terrane and partial melting of the root zones to form syntectonic granite magmas. Topographic relief develops at this stage as the thrust belt rises above sea level and sediments derived from erosion of the highland begin to fill a foreland basin. The first sediments come from distal low-relief terranes and are largely fine-grained, giving rise to deepwater marine shales and siltstones exposed in the lowest stratigraphic levels of foreland basin successions. In contrast, in hinterland basins the early sediments are commonly alluvial-fan deposits, shed from the rising mountain range. Continued convergence causes thin-skinned thrust sheets to propagate into foreland basins and relief increases rapidly. Erosion rates increase as do grain size and feldspar content of derivative sediments in response to increased rates of tectonic uplift in the thrust belt. During this stage of development, thick alluvial fans may also be deposited in foreland basins.

The Himalayas

As an example of a young collisional mountain range, none can surpass the high Himalayas. The Himalayan story began some 80 Ma when India fragmented from

Gondwana and started on its collision course with eastern Asia. Collision began about 55 Ma and is still going on today. Prior to collision, Tibet was a continental-margin arc system with voluminous andesites and felsic ash flow tuffs, and northern India was a passive continental margin with a marine shelf-facies on the south, passing into a deep-water Tethyan facies on the north. As the collision began, folds and thrusts moved southwards onto the Indian plate (Searle et al., 1987). This resulted in thickening of the crust, high-pressure metamorphism, and partial melting of the root zones to produce migmatites and leucogranites. Thrusting continued on both sides of the Indus suture as India continued to converge on Tibet. By 40 Ma, deformation had progressed southwards across both the Lower and Higher Himalayan zones (Figure 3.30). Collapse of the Lower and Sub-Himalayas during the Miocene juxtaposed lower Paleozoic shelf sediments north of the Main Boundary thrust against metamorphic rocks and leucogranites south of the thrust. Continued convergence of the two continental plates led to oversteepening of structures in the Indus suture, and finally to backthrusting on to the Tibet plate as well as continued southward-directed thrusting. The Siwaliks foreland basin continued to move to the south in the Sub-Himalayas as thrust sheets advanced from the north.

The amount of crustal shortening recorded across the Himalayan orogen is almost 2500 km for a time-averaged compression rate of ~ 5 cm/y (Searle et al., 1987). However, the crust of Tibet is only about 70 km thick, which can account for only about 1000 km of shortening. The remainder appears to have been taken up by transcurrent faulting north of the collision zone (Tapponnier et al., 1986; Windley, 1995). The Tertiary geological record in Southeastern Asia required 1000–1500 km of cumulative strike-slip offset in which India has successively pushed Southeastern Asia followed by Tibet and China in an ESE direction. Most of the Mid-Tertiary displacement has occurred along the left-lateral Red River fault zone (Figure 3.27) accompanying the opening of the South China Sea.

Most models for the India–Tibet collision involve buoyant subduction of continental crust. Earthquake data from the Himalayas suggest a shallow (~ 3°) northward-

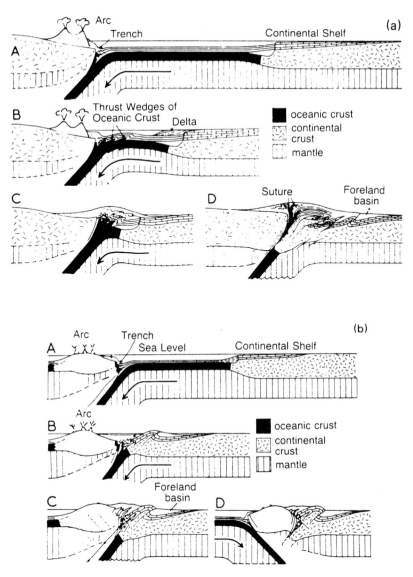

Figure 3.31 Schematic sequence of events in (a) a continent–continent and (b) continent–arc collision. After Dewey and Bird (1970).

dipping detachment zone extending beneath the foreland basins on the Indian plate (Figure 3.30). This detachment surface is generally interpreted as the top of the descending Indian plate and may have a surface expression as the Min Frontal thrust. Convergence since the Early Miocene has been taken up chiefly along the Main Central thrust and its splays by counterclockwise rotation of India beneath Tibet.

Idealized orogenic scenarios

A schematic sequence of events in an orogeny caused by a continent–continent collision is illustrated in Figure 3.31a. As closure of an ocean basin occurs (A), slices of oceanic crust may be thrust towards the converging continent (B). Upon collision, the buoyancy of the con-

verging continent prevents its subduction, and large nappes and thrusts from the convergence zone propagate outwards on to the descending plate (C) and (D). Remnants of the intervening ocean basin (ophiolites) may be incorporated in the suture melange together with deepsea sediments. A continent–continent collision results in the termination of a convergent plate boundary, and world-wide spreading rates must adjust to compensate for the loss of a subduction zone or a new subduction zone must develop elsewhere. There is some evidence that a new convergent plate boundary, defined by a shallow seismicity zone extending southeast from India, may be in the early stages of development to compensate for the India–Tibet collision.

A possible sequence of events in a continent–arc collision is illustrated in Figure 3.31b. This is representative

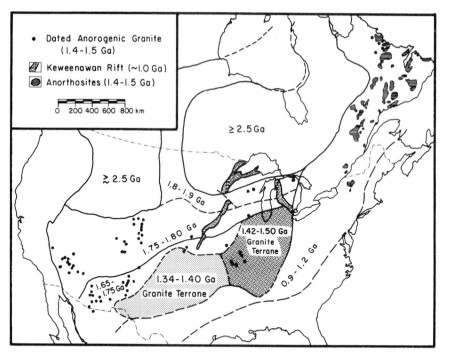

Figure 3.32 The Mid-Proterozoic anorogenic granite–anorthosite belt in North America. After Anderson (1983).

of accretionary orogens. When a plate carrying a continent reaches an island arc (B, C), the subduction zone becomes choked with thrusts and nappes directed towards the continent. Eventually the island arc is accreted to the edge of the continent (C, D). Spreading rates may adjust to compensate for the loss of the subduction zone, or the subduction zone may move to the ocean side of the arc, dipping in the opposite direction (D). The latter change, known as a subduction-zone flip, results in the accretion of an arc to a continent. Such a flip has occurred in the last 20 My as an arc on the southern edge of the Pacific plate was accreted to New Guinea on the northern edge of the Australian plate.

Uncertain tectonic settings

Anorogenic granites

General features

A wide belt of Proterozoic granites and associated anorthosites extends from southwestern North America to Labrador, across southern Greenland and into the Baltic shield in Scandinavia and Russia. The granites are massive and undeformed and appear to have been emplaced in an 'anorogenic' tectonic environment (Anderson and Morrison, 1992; Windley, 1993). Many have rapakivi textures and other features of A-type

granitoids (Table 3.1). Two important field observations for anorogenic granites are:

1 they occur chiefly in accretionary (juvenile) orogens
2 often there is a close spatial and temporal relation between granite magmatism and crustal extension.

Early Proterozoic anorogenic granites in the Laurentia–Baltica belt range in age from about 1.8 Ga–1.0 Ga. Most of those in North America are 1.5–1.4 Ga and tend to increase from 1.44–1.43 Ga in the southwestern United States to 1.48–1.46 Ga in the mid-continent area (Figure 3.32). Major subprovinces of 1.4–1.34 Ga and 1.5–1.42 Ga granites occur in the mid-continent region. The largest and oldest anorogenic granites in this Proterozoic belt occur in Finland and Russia dating from 1.8–1.65 Ga (Haapala and Ramo, 1990). Large anorthosite bodies are associated with some anorogenic granites, and most occur in the Grenville province and adjacent areas in eastern Canada (Figure 2.26). Although Middle Proterozoic anorogenic granites have received most attention, granites with similar field and geochemical characteristics are also known in the Archean and Phanerozoic, some of the youngest of which are in the American Cordillera.

Proterozoic anorogenic granites are A-type granites enriched in K and Fe and depleted in Ca, Mg, and Sr relative to I- and S-type granitoids (Table 3.1). They are subalkalic to marginally peraluminous and plot near the minimum in the Q-Ab-Or system at 5–10 kb pressure

(Anderson, 1983). Anorogenic granites are typically enriched in REE, Zr, and Hf and have striking depletions in Sr, P and Ti compared with most other granites. In addition, they appear to have been emplaced under relatively dry conditions at temperatures of 650–800 °C and depths of chiefly < 15 km. Rapakivi textures may have developed by volatile losses at shallow depths during emplacement. Another characteristic feature of anorogenic granites is that they crystallized over three orders of magnitude of oxygen fugacities as reflected by their Fe–Ti oxide mineralogy. Their relatively high initial $^{87}Sr/^{86}Sr$ ratios (0.705 ± 0.003) and negative or near zero E_{Nd} values are consistent with a lower crustal source, as are incompatible element and oxygen isotope distributions.

Associated anorthosites

Associated anorthosites, which are composed of more than 90 per plagioclase (An_{45-55}), are interlayered with gabbros and norites and exhibit cumulus textures and rhythmic layering. Most bodies range from 10^2 to 10^4 km^2 in surface area. Many are intruded into older granulite-facies terranes and some are highly fractured. Gravity studies indicate that most anorthosites are from 2–4 km thick and are sheet-like in shape, suggesting that they represent portions of layered igneous intrusions. Furthermore, the close association of granites and anorthosites suggests a genetic relationship. Geochemical and isotopic studies, however, indicate that the anorthosites and granites are not derived from the same parent magma by fractional crystallization or from the same source by partial melting. Data are compatible with an origin for the anorthosites as cumulates from fractional crystallization of high-Al_2O_3 tholeiitic magmas produced in the upper mantle (Emslie, 1978). The granitic magmas, on the other hand, appear to be the products of partial melting of lower crustal rocks of intermediate or mafic composition (Anderson and Morrison, 1992). The heat for lower crustal melting may be derived from crystallizing basalts in magma chambers in the lower crust. In this respect, the anorthosite-granite association is a product of bimodal magmatism.

Tectonic setting

The tectonic setting of anorogenic granites continues to baffle geologists. Unlike most other rock assemblages, young counterparts of anorogenic granites have not been recognized. The general lack of preservation of supracrustal rocks further hinders identifying the tectonic setting of these granites. The only well-documented outcrops of coeval supracrustal rocks and anorogenic granite are in the St Francois Mountains of Missouri where felsic ash flow tuffs and calderas appear to represent surface expressions of granitic magmas. Both continental rift and convergent-margin models have been proposed for

anorogenic granites, and both models have problems. The incompatible element distributions in most anorogenic granites are suggestive of a within-plate tectonic setting. If an extensive granite-rhyolite province of late Paleozoic to Jurassic age in Argentina represents an anorogenic granite province (Kay et al., 1989), this would favour a back-arc continental setting. The intrusion of anorogenic granites in the Mid-Proterozoic belt in Laurentia–Baltica described above usually follows the main deformational events in this belt by 60 to 100 My. This may reflect the time it takes to heat the lower crust to the point at which at which it begins to melt. Although our current data base seems to favour an extensional regime for anorogenic granites (± anorthosites), we cannot draw a strict parallel to modern continental rifts. It would appear that an event which transcended time from about 1.9 to 1.0 Ga is responsible for the Proterozoic anorogenic granite belt in Laurentia–Baltica. Perhaps this represents the movement of a supercontinent over one or more mantle plumes which heated the lower crust? In any case, it seems clear that Proterozoic-style anorogenic magmatism is not a one-time event, but that it also has occurred in the Late Archean and perhaps many times in the post-Archean.

Archean greenstones

Although at one time it was thought that greenstones were an Archean phenomenon, it is now clear that they have formed throughout geologic time (Condie, 1994). It is equally clear that all greenstones do not represent the same tectonic setting, nor do the proportions of preserved greenstones of a given age and tectonic setting necessarily reflect the original proportions of that tectonic setting. Greenstone has been used rather loosely in the literature, suggesting perhaps that the term should be abandoned. However, it is a useful term when referring to submarine mafic volcanic successions. In this text **greenstone** is defined as a supracrustal succession in which the combined submarine mafic volcanic and volcaniclastic sediment component exceeds 50 per cent. Thus, from a modern perspective, greenstones are volcanic-dominated successions which have formed in arcs, submarine plateaus, volcanic islands, and oceanic crust. It is now known that greenstones contain various packages of supracrustal rocks separated by unconformities or faults (Thurston and Chivers, 1990; Williams et al., 1992). **Greenstone belts** are linear-to irregular-shaped volcanic-rich successions that average 20–100 km in width and extend for distances of several hundred kilometres. They contain several to many greenstone assemblages or domains, and in this sense a greenstone belt could be equated to a terrane, or more specifically to an oceanic terrane. As an example, the largest preserved greenstone belt, the Late Archean Abitibi belt in eastern Canada, contains several greenstone domains, and can therefore be considered a terrane which was amalgamated at about 2.7 Ga (Figure 3.33). Some greenstone

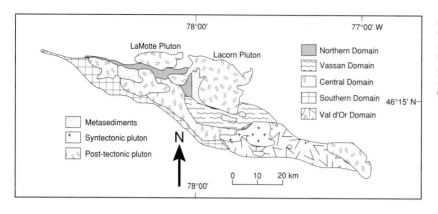

Figure 3.33 Schematic map of part of the Late Archean Abitibi greenstone belt in southeast Ontario, Canada showing tectonic domains. Modified after Descrochers et al. (1993).

belts may be superterranes. Subprovinces in the Archean Superior province, such as the Wawa-Abitibi and Wabigoon subprovinces (Figure 3.34), can be considered as superterranes and represent amalgamations of greenstone terranes of various oceanic settings. High-precision U–Pb zircon isotopic dating of Archean greenstone terranes indicates that they formed in short periods of time, generally < 50 My. In some areas, more than one volcanic-plutonic cycle may be recorded for a cumulative history of 200–300 My. Although Archean greenstones can be described in terms of terranes, their tectonic settings continue to be a subject of lively debate among Archean investigators. Let us review some of the principal features of Archean greenstones that are useful in constraining tectonic settings.

General features

Although field evidence clearly indicates that most greenstones are intruded by surrounding granitoids, there are some areas in which greenstone successions lie unconformably on older granitic basement (Condie, 1981). In a few greenstone belts, such as Kambalda in Southwestern Australia, volcanic rocks contain zircon xenocrysts from gneissic basement which is at least 700 My older than the host rocks. Thus, although most Archean greenstones are clearly juvenile oceanic terranes, some were erupted on or close to continental crust.

In Canadian Archean greenstones, four lithologic associations are recognized (Thurston and Chivers, 1990; Thurston, 1994). Most widespread are the basalt-komatiite (mafic plain) and mafic to felsic volcanic cycle associations, comprising most of the major greenstone belts in Canada. These two associations are also the most common associations found in Archean greenstones on other continents. Of more local importance is a calc-alkaline volcanic and fluvial sediment association, and a carbonate-quartz arenite association, the latter of which is volumetrically insignificant. In the Archean Superior province in eastern Canada, greenstone subprovinces alternate with metasedimentary subprovinces (Figure 3.34). Granitoids are more abundant than volcanic and sedimentary rocks in both subprovinces, with gneisses

and migmatites most abundant in the metasedimentary belts. U–Pb zircon ages indicate younging in volcanism and plutonism from the northwest (Sachigo subprovince) to the southeast (Wawa-Abitibi subprovinces). The oldest magmatic events in the northwest occurred at 3.0, 2.9–2.8, and 2.75–2.7 Ga followed by major deformation, metamorphism, and plutonism at about 2.7 Ga. In the south, magmatism occurred chiefly between 2.75 and 2.7 Ga. The near contemporaneity of magmatic and deformational events along the lengths of the volcanic subprovinces, coupled with structural and geochemical evidence, supports a subduction-dominated tectonic regime in which oceanic terranes were successively accreted from northwest to southeast.

Greenstone volcanics

Archean greenstones are structurally and stratigraphically complex. Although stratigraphic thicknesses up to 20 km have been reported, because of previously unrecognized tectonic duplication it is unlikely that any sections exceed 10 km (Condie, 1994). Most Archean greenstones are comprised chiefly of subaqueous basalts (Figure 3.35) and komatiites (ultramafic volcanics) with minor amounts of felsic tuff and layered chert. Many Archean volcanics are highly altered and silicified, probably from submarine hydrothermal fluids in a manner similar to that characteristic of modern submarine arcs and ocean ridges. Two general trends observed with increasing stratigraphic height in Late Archean greenstone successions are:

1   a decrease in the amount of komatiite
2   an increase in the ratio of volcaniclastics to flows and in the relative abundance of andesitic and felsic volcanics.

These changes reflect an evolution from voluminous submarine eruptions of basalt and komatiite, commonly referred to as a **mafic plain**, to more localized calc-alkaline and tholeiitic stratovolcanoes, which may become emergent with time, and intervening sedimentary basins.

Archean volcanoes were in some respects similar to modern submarine volcanoes in arc systems (Ayers and Thurston, 1985). Similarities include:

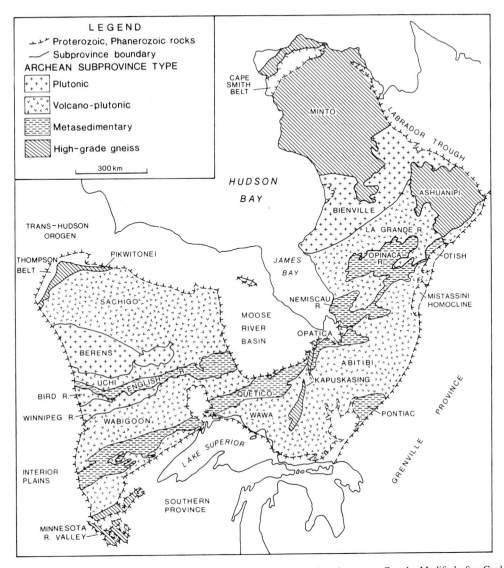

Figure 3.34 Generalized geologic map of the Archean Superior province in eastern Canada. Modified after Card and Ciesielski (1986).

1 a general upward change from basalts to calc-alkaline and tholeiitic volcanics
2 eventual emergence of subaqueous volcanoes to form islands
3 a linear alignment of volcanoes
4 flanking sedimentary aprons leading into basins between volcanoes.

Differences between modern arc volcanoes and Archean volcanoes include:

1 the occurrence of komatiites in many Archean volcanoes
2 the bimodal character of some Archean volcanics, especially in older greenstone successions

3 the paucity of Archean shoshonitic volcanics
4 thick, laterally-continuous submarine flows which collectively form large submarine mafic plains upon which Archean volcanoes grew.

Perhaps the most distinctive volcanic rock of the Archean is komatiite (Arndt, 1994). **Komatiites** are ultramafic lava flows (or hypabyssal intrusives) that exhibit a quench texture known as spinifex texture and contain > 18 per cent MgO. Although komatiites are common in some Archean greenstone successions, they are uncommon in the Proterozoic and very rare in the Phanerozoic. Spinifex texture is commonly preserved in the upper parts of komatiite flows. This texture is characterized by

Figure 3.35   Pillows in Archean basalts from the Abitibi greenstone belt in southern Ontario.

randomly-orientated skeletal crystals of olivine or pyro-xene (Figure 3.36), and forms by rapid cooling in the near-absence of crystal nuclei. Flows range from those with well-developed spinifex zones, which comprise nearly all of the upper half of a flow, to flows without spinifex textures but with polygonal cooling fractures. Archean komatiite and basalt flows range from about 2 m to > 200 m thick, are commonly pillowed, and may be associated with sills or volcaniclastic rocks of similar composition. Intrusive mafic and ultramafic igneous rocks are also found in most greenstone successions and geochemical studies indicate that they are closely related to enclosing volcanic rocks. Archean komatiitic and mafic volcanics are depleted in LIL elements and show variable amounts of light-REE depletion. While Archean mafic plain basalts have incompatible element distributions similar to modern submarine plateaux (Figure 3.8), Archean calc-alkaline basalts (associated with andesites and felsic volcanics) show a distinct subduction zone geochemical component (Figure 3.37) (Condie, 1989).

Andesites and felsic volcanic rocks in greenstone successions occur chiefly as volcaniclastic rocks and minor flows. Tuffs, breccias and agglomerates are common, and their distribution can be used to define volcanic centres. Felsic ash flow tuffs with well-preserved

eutaxitic textures (flattened shards and pumice fragments) have been described from Archean greenstones in Australia and Canada. Archean andesites are similar to modern andesites from volcanic arcs in incompatible element distributions. Alkaline volcanic rocks are rare in Archean greenstone successions and, when found, occur chiefly as volcaniclastics and associated hypabyssal intrusives.

Greenstone sediments

Four types of sediments are recognized in Archean greenstones (Mueller, 1991; Lowe, 1994a), which are, in order of decreasing importance, volcaniclastic, chemical, biochemical, and terrigenous sediments. Volcaniclastic sediments (including graywackes) range from mafic to felsic in composition and are common as turbidites. Chemical and biochemical sediments, principally chert, banded iron formation, and carbonate, are of minor importance but of widespread distribution, and terrigenous sediments such as shale, quartzite, arkose, and conglomerate are of only local significance. Although plutonic sources are of local importance in some graywackes, most greenstone detrital sediments are clearly derived from nearby volcanic sources. Layered chert and

Figure 3.36 Spinifex texture in an Archean komatiite from the Barberton greenstone in South Africa (× 6). Radiating crystals are serpentine pseudomorphs after olivine.

banded iron formation are the most important nonclastic sediments in greenstones. Relict detrital textures in some cherts, however, indicate that they are chertified clastic sediments or volcanics. Few if any Archean cherts appear to represent pelagic sediments. Most are very local in extent and are probably hydrothermal vent deposits. Low $\delta^{18}O$ values for many Archean cherts support a volcanigenic origin. The earliest evidence of life on Earth occurs in the form of microstructures in Archean cherts. Some cherts in Early Archean greenstones con-

tain abundant casts of gypsum crystals (and in some instances, halite) suggesting a shallow-water evaporite origin. Carbonates and barite are very minor components in some greenstone belts. The barites appear to represent hydrothermal deposits and the carbonates, either hydrothermal or evaporite-related deposits.

Five different sedimentary environments are recognized in Archean greenstones (Lowe, 1994a). In some greenstones, such as the Northern Volcanic Zone in the Abitibi belt, more than one sedimentary environment occurs in the same succession (Figure 3.38). The most widespread environment, especially in Early Archean greenstones, is the mafic plain environment. In this setting, large volumes of basalt and komatiite were erupted to form a mafic plain, which may have risen to shallow depths beneath sea level or even became emergent. Sediments in this setting include hyaloclastic sediments, chert, banded iron formation, carbonate and, locally, shallow-water evaporites and barite. These rocks, which commonly preserve primary textures such as mudcracks, oolites, and gypsum casts indicate shallow-water deposition. A second environment is a deep-water, nonvolcanic environment in which chemical and biochemical chert, banded iron formation, and carbonate are deposited. The third association, a graywacke-volcanic association very widespread in Late Archean greenstones, and often stratigraphically on top of the mafic plain succession, is composed chiefly of graywackes and interbedded calc-alkaline volcanics and may have been deposited in or near island arcs. Fluvial and shallow-marine detrital sediments probably deposited in pull-apart (rift) basins, and mature sediments (quartz arenites etc.) deposited in continental rifts, are the last two environments. These two environments are generally not volumetrically important in most Archean greenstones, although the pull-apart basin setting is widespread.

Structure and metamorphism

Archean greenstone terranes are remnants of basinal or platform sequences which have undergone multiple deformations, with compressive forces largely dominating

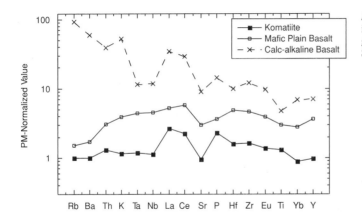

Figure 3.37 Primitive-mantle normalized incompatible-element distributions in Archean greenstone basalts. Primitive-mantle values from Sun and McDonough (1989).

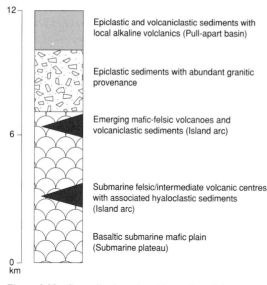

12

Epiclastic and volcaniclastic sediments with local alkaline volcanics (Pull-apart basin)

Epiclastic sediments with abundant granitic provenance

Emerging mafic-felsic volcanoes and volcaniclastic sediments (Island arc)

6

Submarine felsic/intermediate volcanic centres with associated hyaloclastic sediments (Island arc)

Basaltic submarine mafic plain (Submarine plateau)

0
km

Figure 3.38 Generalized stratigraphic section of the Northern Volcanic Zone in the Abitibi greenstone belt, Quebec, showing the major lithologic assemblages. Modified after Lowe (1994b).

in earlier stages of development changing to vertical forces during intrusion of major plutons. Thrust faults, recumbent folds and transcurrent faults are widespread in greenstones, and most successions have undergone two or three periods of major deformation and metamorphism (de Wit et al., 1987; Williams et al., 1992). Greenstones in the Kalgoorlie area of Southwestern Australia show repetitions of sections caused by multiple thrust faults, and contacts between stacked units are shear zones characterized by mylonites. In the Michipicoten greenstone in Canada, major recumbent folds facing in opposite directions developed synchronously with major wrench faults. In the Abitibi belt, transcurrent faults separate blocks with distinct structural-metamorphic histories that are allochthonous relative to each other.

Later stages of deformation in greenstones reflect vertical forces during which diapiric plutons were intruded. Metamorphic grade ranges from greenschist (or locally, zeolite) to amphibolite facies, and higher grades typically occur around margins of greenstone belts probably caused by intrusion of plutons. Metamorphic mineral assemblages are of the low-pressure type with implied temperature gradients of 30–50 °C/km.

Associated granitoids

Granitoids associated with Archean greenstones fall into one of three categories: gneissic complexes and batholiths, diapiric (syntectonic) plutons of variable composition, and late discordant granite plutons (Martin, 1994; Sylvester, 1994). Gneissic complexes and

batholiths, which are dominantly tonalite, trondhjemite, and granodiorite, the so-called **TTG suite**, compose most of the preserved Archean crust. They contain large infolded remnants of supracrustal rocks as well as numerous inclusions of surrounding greenstone belts. Contacts between TTG complexes and greenstones are generally intrusive and usually strongly deformed. Granitic plutons range from foliated to massive, discordant to concordant, and some are porphyritic. Geophysical studies indicate that most Archean plutons extend to depths of less than 15 km. Diapiric plutons have well developed concordant foliation around their margins and appear to have been deformed during forceful injection. Others, which are post-tectonic, are usually granite (*sensu strictu*) in composition and have discordant contacts and massive interiors. Most of these are post-tectonic (anorogenic?) plutons with A-type characteristics.

All but the post-tectonic granites are geochemically similar to felsic volcanics in greenstones. Together with their common syntectonic mode of emplacement, this fact suggests that most granitoids and felsic volcanics are genetically related. In areas where detailed U–Pb zircon chronology is available, such as in parts of the Superior Province, entire cycles of volcanism, sedimentation, deformation and plutonism occur in < 50 My, and late discordant granites are generally emplaced during the following 50 My. In some Early Archean greenstones (Pilbara, Barberton) plutonism can continue for up to 500 My after major volcanism.

Geochemically, the TTG suite and most diapiric plutons are I-type granitoids, and they are distinct from most of their post-Archean counterparts in that they are extremely depleted in heavy REE (Martin, 1994). Most show a subduction geochemical component and no significant Eu anomalies. Geochemical and experimental data favour an origin for these rocks as partial melts of garnet amphibolites. Archean post-tectonic and anorogenic granites are calc-alkaline, strongly peraluminous, and commonly have A- or S-type affinities (Table 3.1) (Sylvester, 1994). They appear to be the products of partial melting of igneous and sedimentary rocks in the lower crust.

Mineral and energy deposits

Introduction

Plate tectonics provides a basis for understanding the distribution and origin of mineral and energy resources in space and time (Rona, 1977; Sawkins, 1990). The occurrence of energy and mineral deposits can be related to plate tectonics in three ways:

1   geological processes driven by energy liberated at plate boundaries control the formation of energy and mineral deposits
2   deposits form in specific tectonic settings, which are controlled by plate tectonics

Table 3.2 Summary of major mineral and energy deposits by tectonic setting

Tectonic setting	Mineral deposit	Energy deposit
1 Oceanic settings		
(a) Ophiolite	Cyprus-type Cu-Fe massive sulphides; podiform chromite	
(b) Ocean ridge		Geothermal
(c) Back-arc basin		Hydrocarbons
2 Subduction zone		
(a) Arc	Hydrothermal: Au, Ag, Cu, Mo, Pb, Sb, Hg, Sn, W	Geothermal, hydrocarbons
	Porphyries: Cu, Mo, Sn	
	Massive sulphides: Cu, Pb, Zn	
(b) Foreland basin	Red bed U, V, Cu	Hydrocarbons, coal
(c) Forearc basin		Hydrocarbons
3 Orogens		
(a) Highlands	Sn–W granites	
	Gemstones	
	Deposits from older tectonic settings	
(b) Foreland/hinterland basin	Redbed U, V, Cu	Hydrocarbons, oil shale, coal
	Stratiform Pb–Zn–Ag	
4 Continental rifts	Stratiform Pb–Zn	Geothermal
	REE, Nb, U, Th, P, Cl, F, Ba, Sr associated with alkaline intrusives	
	Sn granites	
	Stratiform Cu	
	Evaporites	
5 Cratons, passive margins	Diamonds (kimberlites)	Hydrocarbons, coal
	Bauxite	
	Ni laterite	
	Evaporites	
	Clays	
6 Archean greenstones	Cu–Zn massive sulfides	
	Ni–Cu sulfides	
	Au quartz veins	

3 reconstruction of fragmented supercontinents can be used in exploration for new mineral and energy deposits.

In attempting to relate mineral and energy deposits to plate tectonics, it is important to know the relationship between the deposits and their host rocks. If deposits are syngenetic with host rocks, then they formed in the same tectonic setting. If, however, mineral and energy deposits are secondary in origin and entered the host rocks later (such as by hydrothermal fluids or migration of oil), they may have formed in a tectonic setting different from the host rocks. Major in-situ mineral and energy deposits in various tectonic settings are summarized in Table 3.2.

Mineral deposits

Ocean ridges

Employing deepsea photography and using small submersibles, numerous hydrothermal fields have been identified on ocean ridges (Rona et al., 1986; Von Damm, 1990). The submersibles allow direct observations and measurements on the sea floor. On the Galapagos and Juan De Fuca ridges, hydrothermal veins ranging in size from 500–1600 m^2 are spread along fissure systems for 500–2500 m. The TAG hydrothermal field on the Mid-Atlantic ridge (26 °N) occurs along a fault zone on the east wall of the medial rift valley, and vent waters have temperatures up to 300 °C. Animal communities living near the vents appear to be totally dependent upon energy derived from seawater-rock reactions and sulphur-oxidizing bacteria. Seafloor observations also indicate that hydrothermal vents are locally developed in areas of the youngest sea floor and occur along both fast- and slow-spreading ridge segments. Each hydrothermal field has multiple discharge sites, with sulphide chimneys rising up to 10 m above the sea floor. Minerals recovered from chimneys are chiefly Fe, Cu and Au sulphides with minor amounts of anhydrite and amorphous silica (Humphris et al., 1995). Hydrothermal waters are acid, rich in H_2S, and are major sources of Mn, Li, Ca, Ba, Si, Fe, Cu and Au. These metals are either deposited in the chimneys or with sediments near the vents.

The hydrothermal waters erupted along ocean ridges appear to be seawater which has circulated through hot oceanic crust with thermal gradients of > 150 °C/km.

The seawater circulates by convection through the upper part of the permeable crust (1–2 km deep), is heated at depth and discharges along active fissures in the medial rift. The discharging water leaches metals from the oceanic crust and alters the crust by additions of H_2O and elements such as Mg. Upon erupting on the sea floor, the hydrothermal waters are rapidly cooled and increase in pH resulting in deposition of sulphides and sulphates.

In considering ancient oceanic crust, two major mineral deposits are formed in ophiolites: Cyprus-type Cu–Fe massive sulphides and podiform chromite. Cyprus-type ores occur as stratiform deposits in pillowed basalt layers. These ores are exhalative deposits formed by hydrothermal vents along ocean ridges as described above. Podiform chromite deposits are formed in ultramafic cumulates and relict textures suggest that they are the products of fractional crystallization.

Arc Systems

Metalliferous mineral deposits are important in both continental-margin arcs and island arcs. Base metals (Zn, Cu, Mo, Pb), precious metals (Ag, Au) and other metals (Sn, W, Sb, Hg) are found in hydrothermal veins and lodes formed in arc systems. Following erosion, placer deposits of these metals are important in some geographic areas. Veins and lodes are commonly associated with volcanics or granitic plutons, where they represent late-stage fluids derived from differentiated arc magmas. Cu, Mo and Sn porphyry deposits are formed also in arc systems. These large-volume, low-grade disseminated deposits occur in altered porphyritic granites and are important sources of Cu and Mo in the Southwestern United States and in the Andes. Kuroko-type massive sulphides (Cu, Pb, Zn) are important in intra-arc and back-arc basin successions where they have formed as exhalatives on the sea floor. Redbed U, V and Cu deposits occur in some retroarc foreland basins such as in the Mesozoic redbeds of the Colorado Plateau.

Zonation of metallic mineral deposits has been reported in Late Tertiary rocks from the Andes (Sillitoe, 1976). In the direction of the dipping lithospheric slab, the major metallic mineral zones encountered are: contact metasomatic Fe deposits, Cu, Au and Ag veins, porphyry Cu–Mo deposits, Pb–Zn–Ag vein and contact metasomatic deposits, and Sn and Mo vein and porphyry deposits. Zonation is believed to result from progressive liberation of metals from the descending slab, with Sn coming from an extreme depth of about 300 km. Geochemical and isotopic data support the general concept that metal deposits associated with subduction are derived from some combination of the descending slab and the overlying mantle wedge. Metals move upward in magmas or in fluids and are concentrated in late hydrothermal or magmatic phases.

Orogens

Metalliferous deposits are abundant at collisional boundaries where a variety of tectonic settings exist, depending on location and stage of evolution. In addition, older mineral deposits associated with ophiolites, arc, craton, and continental rift assemblages occur in collisional zones. Sn and W deposits are associated with collisional leucogranites in the Himalayas and in the Variscan orogen, and at deeper crustal levels, Fe and Ti deposits occur associated with anorthosites. Many gemstones (ruby, sapphire) also are found in high-grade metamorphic rocks or in syntectonic nepheline syenites from collisional orogens. In peripheral foreland basins, stratiform Pb–Zn–Cu sulphides and redbed U–V–Cu deposits may be of economic importance.

Continental rifts

Pb–Zn–Ag stratiform deposits occur in ancient continental rift sediments (Sawkins, 1990). These deposits, which are not associated with igneous rocks, occur in marine carbonates and are probably deposited from brines which migrate to the edge of rift basins. REE, Nb, U, Th, Ba, P, Sr and halogens are concentrated in carbonatites and other alkaline igneous rocks that occur in some continental rifts. Granites intruded during the late stages of rifting are often associated with Sr and fluorite. Stratiform Cu deposits occur in rift-related shales and sandstones as exemplified by deposits in the Zambian Copper Belt. The Cu appears to be derived from associated basalts. Evaporites are important non-metallic deposits found in some rifts.

Some of the major occurrences of Cr, Ni, Cu and Pt are found in Proterozoic layered igneous intrusions. Chromite occurs as primary cumulates within the ultramafic parts of these intrusions and Cu and Ni generally occur as late-stage hydrothermal replacements. Pt occurs in a great variety of cumulus minerals, the most famous occurrence of which is the Merensky Reef in the Bushveld Complex in South Africa. In addition, some layered intrusions have magmatic ore deposits of Sn (in late granites) and Ti or V-rich magnetite.

Cratons and passive continental margins

Few if any metallic deposits are known to form on modern cratons or passive continental margins. Among the non-metallic deposits formed in cratonic areas are diamonds from kimberlite pipes and associated placer deposits, bauxite, Ni-laterite deposits, and evaporites.

Important Early Proterozoic mineral deposits are the placer deposits of Au and U that occur in quartz arenites and conglomerates in cratonic successions. The largest deposits are in the Witwatersrand (2900 Ma) and Huronian (2300 Ma) Supergroups (Pretorius, 1976). Detrital Au and uraninite appear to have been concentrated by fluvial and deltaic processes in shallow-water, high-energy environments. Sources for the Au and U are older greenstone–granite terranes. Other important Proterozoic sedimentary mineral deposits include banded iron formation and manganese rich sediments. Banded iron formation, described more fully in Chapter 6, reached the peak of its development at about 2500 Ma. It occurs

Table 3.3 Oil reserves in Devonian and younger reservoirs (in percentages)

	Intracratonic basin	Passive margin	Foreland basin
Cenozoic	5	8	18*
Mesozoic	43	5	6
Paleozoic	13	1	1
Total	61	14	25

* 14 per cent in the Persian Gulf and 4 per cent elsewhere.

as alternating quartz-rich and magnetite- (or hematite-) rich laminae and some was deposited in basins which are hundreds of kilometres across. Most banded iron formation is not associated with volcanic rocks and appears to have been deposited in shallow cratonic basins. Manganese-rich sediments occur also in cratonic successions associated with carbonates.

Archean greenstones

Some of the world's major reserves of Cu, Zn and Au occur in Archean greenstone belts (Groves and Barley, 1994). Those greenstones ≥ 3.5 Ga contain only minor mineral deposits, including Cu–Mo porphyry and stockwork deposits and small occurrences of barite and banded iron formation. Late Archean greenstones contain major Cu–Zn massive sulphides associated with submarine felsic volcanics and Ni–Cu sulphides associated with komatiites. The latter sulphides appear to have formed as cumulates by fractional crystallization of immiscible sulphide melts associated with komatiite magmas. Minor occurrences of banded iron formation occur in most Late Archean greenstones. One of the most important deposits in greenstones is Au which occurs in quartz veins and in late disseminated sulphides commonly associated with hydrothermal chert or carbonate deposits.

Energy Deposits

Several requirements must be met in any tectonic setting for the production and accumulation of hydrocarbons such as oil or natural gas. First of all, the preservation of organic matter requires restricted seawater circulation to inhibit oxidation and decomposition. High geothermal gradients are needed to convert organic matter into oil and gas and, finally, tectonic conditions must be such as to create traps for the hydrocarbons to accumulate. Several tectonic settings potentially meet these requirements (Table 3.3).

Both oil and gas are formed in forearc and back-arc basins, which can trap and preserve organic matter and where geothermal heat facilitates conversion of organic matter into hydrocarbons. Later deformation, generally accompanying continental collisions, creates a variety of structural and stratigraphic traps in which hydrocarbons can accumulate. An important site of hydrocarbon

formation is in foreland basins. The immense accumulations in the Persian Gulf area formed in a peripheral foreland basin associated with the Arabia–Iran collision in the Tertiary.

The majority of the oil and gas reserves in the world have formed either in intracratonic basins or passive continental margins (Table 3.3). During the early stages of the opening of a continental rift, seawater can move into the rift valley and if evaporation exceeds inflow evaporites are deposited. This environment is also characterized by restricted water circulation in which organic matter is preserved. As the rift continues to open, water circulation becomes unrestricted and accumulation of organic matter and evaporite deposition cease. High geothermal gradients beneath the opening rift and increasing pressure due to burial of sediments facilitates the conversion of organic matter into oil and gas. At a later stage of opening, salt in the evaporite succession, because of its gravitational instability, may rise as salt domes and trap oil and gas. Oil and gas may also be trapped in structural or stratigraphic traps as they move upward in response to increasing pressures and temperatures at depth. Supporting this model are data from wells in the Red Sea, which represents an early stage of ocean development. These wells encounter hydrocarbons associated with high geotherms and with rock salt up to 5 km thick. Also, around the Atlantic basin there is a close geographic and stratigraphic relationship between hydrocarbons and evaporite accumulation.

Hydrocarbon production in intracratonic basins may also be related to plate tectonic processes (Rona, 1977). Increases in seafloor spreading rates and ocean-ridge lengths may cause marine transgression, which results in deposition in intracratonic basins. Decreasing spreading rates causes regression, which results in basins with limited circulation, and hence organic matter and evaporites accumulate. Unconformities also develop during this stage. Burial and heating of organic matter in intracratonic basins facilitates hydrocarbon production, and salt domes and unconformities may provide major traps for accumulation.

Coal is formed in two major tectonic settings: cratonic basins and foreland basins. For coal to form, plant remains must be rapidly buried before they decay, and such rapid burial occurs in swamps with high plant productivity. Widespread transgression in the Cretaceous was particularly suitable for coal swamps in cratonic areas. Swamps in foreland basins are generally part of large lake basins, and the widespread Late Paleozoic coals of central Europe appear to have formed in such environments. Oil shales can accumulate under similar conditions, as exemplified by the Tertiary oil shales in the Early Tertiary foreland basins of eastern Utah.

Geothermal energy sources occur along ocean ridges, with Iceland representing the only subaerial expression of a ridge. Other geothermal sources occur associated with hotspots (such as Yellowstone Park), arc systems (such as the Taupo area in New Zealand) and continental rifts (such as the Jemez site in New Mexico).

Summary statements

1 Depletion in LIL elements (such as K, Rb, Ba, Th, U) in NMORB (normal ocean ridge basalt) indicates a mantle source which has been depleted in these elements by earlier magmatic events.

2 Ophiolites are tectonically-emplaced successions of mafic and ultramafic rocks that appear to represent fragments of oceanic or back-arc crust and upper mantle. In ascending statigraphic order, complete ophiolites include ultramafic tectonite, cumulate gabbros and ultramafic rocks, non-cumulate igneous rocks, sheeted diabase dykes, and pillow basalts. Ophiolites are emplaced by obduction at collisional plate boundaries or as fault slices in accretionary prisms of arcs.

3 Mantle-plume related tectonic settings include submarine plateaux and aseismic ridges, flood basalts, volcanic islands, and giant mafic dyke swarms. Submarine plateaux and flood basalts compose chiefly basalt flows erupted in short time intervals (≤ 1 My).

4 Sediments deposited in cratonic and passive margin basins are chiefly mature clastic sediments (quartz arenites and shales) and shallow marine carbonates. Subsidence in these basins is due to the combined effects of thermal contraction of the lithosphere and sediment loading. Kimberlites are ultramafic breccias emplaced in cratonic areas, and they contain xenoliths of ultramafic rocks from the upper mantle as well as diamonds and other high-pressure minerals that indicate magma sources of up to 300 km deep.

5 Continental rifts and aulacogens are characterized by immature, terrigenous sediments (arkose, conglomerate), bimodal igneous rocks, and mafic layered intrusions. Active rifts, characterized by early and widespread uplift and abundant volcanics, develop in response to rising mantle plumes or upwelling asthenosphere. Passive rifts, characterized by a narrow zone of uplift and few volcanics, develop in response to stresses in lithospheric plates.

6 Extension during continental rifting is accommodated by successive listric faults which merge with a large subhorizontal detachment fault at mid-crustal levels. When exposed at the surface, these features are known as a core complex.

7 Seven environments are recognized at convergent margins. These are, with corresponding petrotectonic assemblages: (1) the trench (graywacke turbidites); (2) the accretionary prism (melange, fragments of ophiolite and sediment); (3) forearc basins (turbidites, hemipelagic sediments); (4) the arc (flows, fragmental deposits); (5) back-arc basins (arc and/or pelagic sediments); (6) the remnant arc (mafic volcanics); and (7) retroarc foreland basins (immature clastic sediments).

8 Subduction zone processes include accretion – scraping off oceanic sediments and fragments of oceanic crust and mantle and adding them to the accretionary prism; mixing and infolding of forearc basin sediments into the accretionary prism; and subduction erosion – mechanical plucking and abrasion along the top of a descending slab.

9 High-stress subduction zones, which have large accretionary prisms, relatively large earthquakes, and a shallow dip, develop in response to buoyant subduction. Low-stress subduction zones have little if any accretionary prism, few large earthquakes, and a very steep dip.

10 Coesite (high-P silica) and diamond inclusions in pyroxenes and garnet from eclogites indicate that crust can be subducted to depths of at least 150 km, and that both felsic and mafic crust can be recycled into the mantle.

11 Arc magmatism begins at a volcanic front, which overlies the subduction zone at 125–150 km depth. Arc basalts are produced by partial melting of the mantle wedge in response to increased volatile content caused by devolatilization of the descending slab, and the abrupt onset of volcanism at the volcanic front reflects the onset of melting above the slab, controlled by volatiles and temperature.

12 Arc igneous rocks exhibit a subduction geochemical component (depletion in Nb and Ta relative to neighbouring incompatible elements compared with primitive mantle), the latter of which are enriched in the mantle wedge by volatiles derived from the descending slab.

13 Two types of orogens are recognized. Collisional orogens, involving collision of two or more continental provinces, are characterized by extensive reworking of older crust, and very little juvenile crust. In contrast, accretionary orogens involve collision and suturing of largely juvenile crustal blocks to continental crust, and they contain very little reworked older crust.

14 Collisional orogens vary greatly in character, ranging from those with large overthrusts to those in which compression is taken up partly by shortening of the overriding plate. Collisional rock assemblages are highly variable and change both with time and space as collision progresses. Important components at deep crustal levels are syntectonic granites, granulites, and anorthosites. In some orogens, collision is oblique producing a major transcurrent component.

15 An idealized collisional orogen includes: (1) a thrust belt where continental crust is restacked and propagates toward the foreland; (2) a foreland flexure caused by down-bending of the lithosphere by advancing thrust sheets; (3) plateaux produced by shortening and thickening of the lithosphere in the overriding plate or/and by buoyant subduction; (4) widespread foreland/hinterland deformation zones involving both transcurrent and normal faults; and (5) extensional collapse zones.

16 Sutures are tectonic melanges containing a wide variety of rock fragments in a sheared matrix. They

are major shear zones separating terranes or continental provinces.

17 Foreland and hinterland basins develop during plate collisions in response to sediments derived from advancing thrust sheets (foreland basins) or uplift of mountains (hinterland basins). Sediments are immature terrigenous sediments reflecting progressively deeper crustal levels with time.

18 Anorogenic granites are massive, relatively undeformed and appear to have been emplaced in an extensional tectonic regime. They occur chiefly in accretionary (juvenile) orogens, are sometimes associated with large anorthosite bodies, and often there is a close spatial and temporal relation between granite magmatism and crustal extension. Anorogenic granites are produced by partial melting of the lower crust, and associated anorthosites are derived by fractional crystallization of associated mafic magmas which may have been the heat sources for crustal melting.

19 Archean greenstones are assemblages dominated by submarine mafic, volcanic and volcaniclastic sediments which represent mostly oceanic tectonic settings. Greenstone belts are terranes that contain several to many greenstone assemblages or domains.

20 Komatiites are ultramafic lavas that exhibit a quench texture known as spinifex texture and contain > 18 per cent MgO. Although komatiites are common in some Archean greenstones, they are uncommon in the Proterozoic and very rare in the Phanerozoic.

21 Archean volcanic/sedimentary associations include: (1) the mafic plain with large volumes of basalt and komatiite, and small amounts hyaloclastic sediments, chert, banded iron formation, carbonate and, locally, shallow-water evaporites and barite; (2) a deepwater, non-volcanic association in which chemical and biochemical chert, banded iron formation, and carbonate dominate; (3) a graywacke–volcanic association comprised chiefly of graywackes and calc-alkaline volcanics probably deposited in or near island arcs; (4) fluvial and shallow-marine detrital sediments probably deposited in pull-apart basins; and (5) mature sediments deposited in continental rifts.

22 Mineral and energy deposits are closely related to tectonic setting. Sulphides are being deposited along modern ocean ridges by hydrothermal waters. Arc systems are major sites of formation of Zn, Cu, Pb, Mo, Ag, Au, Sn, W, Sb and Hg deposits, and continental rifts are important sites of formation of REE, Th, U, P, Sr, Ba and Nb deposits. Sn, W, Fe, Ti, Pb, Cu, Zn, U and V occur in collisional orogens. Most hydrocarbons and coal are formed in cratonic or foreland basins.

Suggestions for further reading

Allen, P. A. and Homewood, P., editors (1986). *Foreland Basins*. Oxford, Blackwell Scient., 453 pp.

Carlson, R. W. (1991). Physical and chemical evidence on the cause and source characteristics of flood basalt volcanism. *Austral. J. Earth Sc.*, **38**, 525–544.

Condie, K. C., editor (1994). *Archean Crustal Evolution*. Amsterdam, Elsevier, 528 pp.

Coward, M. P., editor (1995). Early Precambrian Processes. *Geol. Soc. London, Spec. Publ., No. 95*, 308 pp.

Coward, M. P. and Ries, A. C. (1986). *Collision Tectonics*. Oxford, Blackwell Scient., 420 pp.

Gass, I. G., Lippard, S. J. and Shelton, A. W., editors (1984). *Ophiolites and Oceanic Lithosphere*. Oxford, Blackwell Scient., 413 pp.

Leitch, E. C. and Scheibner, E., editors (1987). Terrane accretion and orogenic belts. *Geol. Soc. America–Amer. Geophys. Union, Geodynam. Series, Vol. 19*, 354 pp.

Moores, E. M. and Twiss, R. J. (1995). *Tectonics*. New York, W. H. Freeman, 415 pp.

Olsen, K. H., editor (1995). *Continental Rifts: Evolution, Structure, Tectonics*. Amsterdam, Elsevier, 520 pp.

Raymond, L. A. (1984). Melanges: Their nature, origin and significance. *Geol. Soc. America, Spec. Paper 198*, 176 pp.

Smellie, J. L. (1994). Volcanism associated with extension at consuming plate margins. *Geol. Soc. London, Spec. Publ. No. 81*, 272 pp.

Treloar, P. J. (1993). Himalayan Tectonics. *Geol. Soc. London, Spec. Publ. No. 74*, 640 pp.

Wezel, F. C. (1986). *The Origin of Arcs*. Amsterdam, Elsevier, 568 pp.

Chapter 4

The Earth's mantle and core

Introduction

The Earth's core and mantle play important roles in the evolution of the crust and provide the thermal and mechanical driving forces for plate tectonics. Heat liberated by the core is transferred into the mantle where most of it (> 90 per cent) is convected through the mantle to the base of the lithosphere. The remainder is transferred upwards by mantle plumes generated in the core–mantle boundary layer. The mantle is also the graveyard for descending lithospheric slabs, and the fate of these slabs in the mantle is the subject of ongoing discussion and controversy. Do they collect at the 660-km discontinuity in the upper mantle or do they descend to the bottom of the mantle? Just what happens to these slabs controls the type of convection in the mantle. If they do not penetrate the 660-km discontinuity, the upper mantle may convect separately from the lower mantle, whereas if they sink to the base of the mantle, whole-mantle convection is probable. A related question is that of how and where mantle plumes are generated and what role they play in mantle–crust evolution?

Another exciting mantle topic is that of the origin and growth of the lithosphere, and whether its role in plate tectonics has changed with time. The Archean lithosphere, for instance, is considerably thicker than post-Archean lithosphere, and geochemical data from xenoliths suggest it had quite a different origin. Still another hot topic is the origin of isotopic differences in basalts, which reflect different compositions and ages of mantle sources. How did these sources form and survive for billions of years in a convecting mantle? How did the core form and do core processes affect mantle processes? These are some of the questions addressed in this chapter.

Seismic structure of the mantle

Upper mantle

From studies of spectral amplitudes and travel times of body waves, it is possible to refine details of the structure of the mantle. The P- and S-wave velocity structure beneath several crustal types is shown in Figures 4.1 and 4.2. Cratons are typically underlain by a high-velocity lid (V_P > 7.9 km/sec) overlying, in turn, the low-velocity zone (LVZ) beginning at a depth as shallow as 60–70 km and extending to depths of 100–300 km. A prominent minimum in Q, a measure of the attenuation of seismic-wave energy, coincides with the LVZ (Figure 4.1). S-wave low-velocity zones may extend to 400 km (Figure 4.2). The top of the LVZ is generally assumed to mark the base of the lithosphere and averages 50–100 km deep beneath oceans. Beneath ocean ridges and most continental rifts, the LVZ may extend nearly to the Moho at depths as shallow as 25 km. The LVZ is minor or not detected beneath most Precambrian shields and may be non-existent beneath Archean shields. The thickest lithosphere also occurs beneath shields (150–200 km), and beneath Archean shields it may be > 300 km thick.

At depths < 200 km, seismic-wave velocities vary with crustal type, whereas at greater depths the velocities are rather uniform regardless of crustal type. From the base of the LVZ to the 410-km discontinuity and from the 410-km discontinuity to the 660-km discontinuity, P-wave velocities increase only slightly (Figure 4.1). In some velocity profiles, a minor discontinuity is observed between 500 and 550 km. Unlike P-waves, S-wave velocities show significant lateral variation in the upper mantle, indicating compositional heterogeneity and/or anisotropy. S-wave velocities beneath ocean ridges are strongly correlated with spreading rates at shallow depths (< 100 km), and very slow velocities are limited to depths of < 50 km (Zhang and Tanimoto, 1993). Also, some hotspots are associated with very slow S-wave velocities.

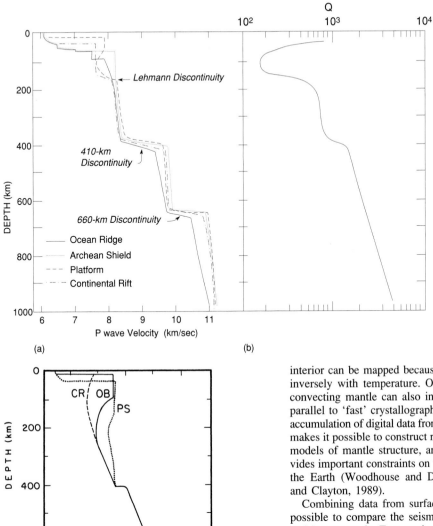

Figure 4.1 (a) P-wave velocity distribution in the mantle (after Walck, 1985); (b) Average distribution of Q, the specific attenuation factor, in the mantle (excluding regions beneath Precambrian shields). Q varies inversely with the amount of seismic wave attenuation, thus low values of Q show the greatest attenuation.

Figure 4.2 S-wave velocity distribution in the upper mantle beneath three crustal types: CR, continental rift; OB, ocean basin: and PS, Proterozoic shield or platform. After Grand and Helmberger (1984).

Lower mantle

In the lower mantle, seismic-wave velocities continue to increase with depth and no further discontinuities are recognized (Figure 1.2). However, seismic tomographic studies have shown that various 'velocity domains' exist in both the upper and lower mantle. **Seismic tomography**, like its medical analogue, combines information from a large number of criss-crossing waves to construct three-dimensional images of the interior of the Earth. Relatively hot and cool regions of the Earth's interior can be mapped because seismic velocities vary inversely with temperature. Orientation of minerals in convecting mantle can also increase seismic velocities parallel to 'fast' crystallographic axes of minerals. The accumulation of digital data from global seismic networks makes it possible to construct reliable three-dimensional models of mantle structure, and such information provides important constraints on the style of convection in the Earth (Woodhouse and Dziewonski, 1984; Hager and Clayton, 1989).

Combining data from surface and body waves, it is possible to compare the seismic structure of the upper and lower mantle. Two vertical cross-sections of tomographic data are illustrated in Figure 4.3. Results for the upper mantle (25–660 km) are based on surface-wave data and those for the lower mantle (660–2900 km) on travel-time residuals of P-waves, which accounts for the mismatch at the 660-km discontinuity. The tomographic sections clearly show that large inhomogeneities exist in the mantle and that they extend to great depths. Noteworthy in the equatorial cross-section are the deep high-velocity roots beneath the Brazilian and African shields, which appear to merge below 400 km. This high-velocity region continues into the lower mantle, perhaps at a steeper angle. Other Precambrian shields, such as the Canadian and Baltic shields, also have deep high-velocity roots. In both tomographic sections, low-velocity zones underlie ocean ridges, although their depth of penetration into the deep mantle varies. For instance, the root zone of the Mid-Atlantic ridge seems to end at the 410-km discontinuity, while that of the East Pacific rise may extend to the base of the mantle (offset by a

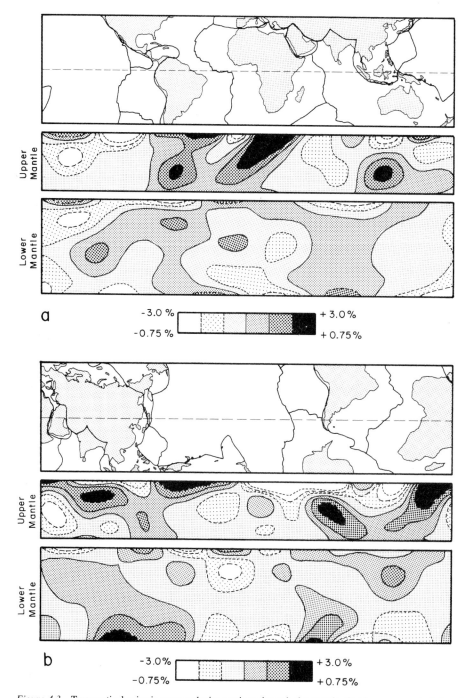

Figure 4.3 Two vertical seismic-wave velocity sections through the mantle. Sections are great circles shown on the maps. The upper mantle (25–660 km) has an 8:1 vertical exaggeration and the lower mantle (660–2900 km) a 4:1 vertical exaggeration. Upper-mantle distributions based on S-waves and lower mantle on P-waves. Deviations in percentages from average velocities, + refers to fast and − to slow. After Woodhouse and Dziewonski (1984).

Precambrian shield high-velocity root). The dipping high-velocity roots beneath Japan, the southern tip of South America, and Indonesia appear to penetrate the 660-km discontinuity and probably reflect descending slabs of lithosphere. The high-velocity regions in the very deep mantle beneath Asia and South America are probably caused by lithospheric slabs that have sunk to the base of the mantle. Although the origin and significance of all of the anomalies shown on these sections are not yet understood, the fact that many anomalies cross the 660-km discontinuity tends to favour whole-mantle rather than layered-mantle convection.

Geoid anomalies

Seismic velocity distributions in the Earth reflect the density distribution which, in turn, influences the topographic shape of the Earth's surface, known as the geoid. The **geoid** is a gravitational equipotential surface defined in oceanic areas by sea level. The Earth's surface deviates by up to 100 m from that of an ideal rotating Earth producing what are known geoid anomalies. Geodynamic theory indicates that a negative correlation should exist between density variation in the Earth and geoid elevation in a convecting Earth (Hager et al., 1985). Relatively hot upwellings in the mantle should cause uplift of the Earth's surface and of the core–mantle boundary. This dynamically-maintained topography has a strong affect on the gravity field which, in turn, reflects the viscosity distribution and presence or absence of stratification in the mantle. The calculated geoid of Earth is shown in Figure 4.4a. Geoid highs over Africa and the central Pacific reflect hotter (less dense) than average mantle in these regions. Hotspots caused by rising mantle plumes are also frequent in these regions, as discussed in Chapter 1 (Figure 1.25) and, as discussed later, geochemical anomalies also correlate with geoidal topography. Geoid lows overlie much of Asia and North America, as well as South America and Antarctica and reflect relatively cool mantle temperatures.

The calculated topography of the core–mantle boundary is shown in Figure 4.4b. The similarity between the shapes of the geoid and the core–mantle boundary is striking. Furthermore, both the core–mantle boundary and the Earth's surface are upwarped in regions of low seismic velocities and downwarped in regions of high velocities (compare with Figure 4.3).

The correlation of deviations in the geoid, the shape of the core–mantle boundary, hotspot density, lower mantle seismic velocities, and geochemical anomalies (discussed later) is consistent with the same cause: large-scale horizontal variations in temperature in the deep mantle. Thus, regions of warm temperature in the deep mantle are inferred beneath Africa and the Central Pacific, whereas cool temperatures characterize the regions beneath Asia, Antarctica, and South America (Figure 4.3). What is responsible for heating the mantle beneath Africa and the Pacific and cooling it beneath Asia and

South America? The relatively hot regions may be large **mantle upwellings** bringing hot, deep mantle material to shallower levels. Such hot upwellings would have lower seismic velocities and could elevate the Earth's surface and the core–mantle boundary, especially if heat is liberated from the core beneath the upwellings (Hager and Clayton, 1987). Release of core heat could also produce more mantle plumes accounting for the high density of hotspots in the upwellings and, as we shall see later, if the plumes carry distinct geochemical signatures, they could be responsible for geochemical anomalies in oceanic basalts. What about the relatively cool regions of the lower mantle? These regions might contain previously subducted slabs sinking to the base of the mantle (Richards and Engebretson, 1992). During the break up of Pangea, which began about 160 Ma, a large volume of lithosphere was subducted into the mantle. Reconstructions of the break-up history indicate that most of this lithosphere should have descended beneath the present locations of Asia, Antarctica, and South America (Anderson, 1989; Richards and Engebretson, 1992), and this is precisely where there is now evidence of geoid lows and relatively high seismic velocities. Thus, relatively cool sinking slabs from the break up of Pangea may be responsible for the geoid lows and high seismic velocities in the deep mantle.

Electrical conductivity of the mantle

The electrical conductivity distribution in the mantle is not well-known due to poor resolution of data. Especially large variations occur in conductivity for depths less than 500 km (Figure 4.5). In most continental areas, the conductivity of the lower crust is in the range of 10^{-2} to 10^{-3} S/m, and it increases rapidly to about 1 S/m at the base of the upper mantle and less rapidly to about 10 S/m at the core–mantle interface (Figure 4.5). The deep oceanic mantle in the Pacific region may be more conductive than the mantle beneath most continents at similar depths (Roberts, 1986). Such a pattern is consistent with the distribution of seismic velocities and supports the existence of relative hot mantle in this region. Where the seismic LVZ occurs at shallow depths, such as beneath the Basin and Range province, ocean ridges, and young continental rift systems, a region of shallow, high electrical conductivity also exists.

The cause of high conductivity layers in the upper mantle appears to be related to relatively high temperatures and in some cases, partial melting. For instance, an increase in depth of high conductivity zones beneath trenches reflects depression of isotherms produced by subduction (Honkura, 1978). A shallow depth of high-conductivity layers above descending slabs, such as beneath the Philippine and Japan Seas, is related both to higher temperatures and to partial melting above descending slabs. The shallowing of conductivity anomalies beneath ocean ridges and continental rift systems probably reflects upwelling asthenosphere from the LVZ

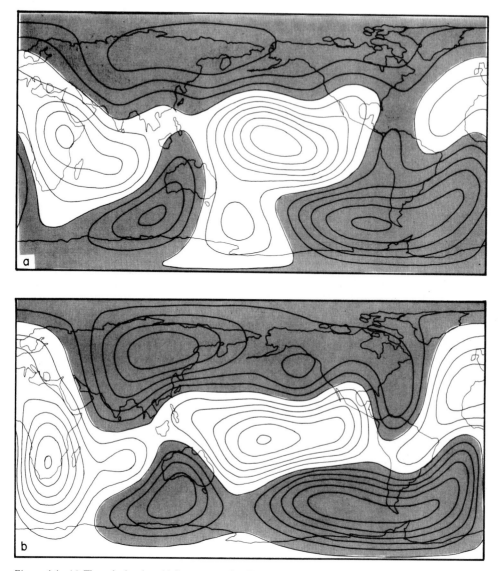

Figure 4.4   (a) The calculated geoid for a convecting Earth. Density contrast assumed equal to 2.3 gm/cm³, contour interval 200 m, and degrees 1–6. Geoid lows shaded. After Hager and Clayton (1987). (b) Calculated topography at the core–mantle boundary. Density contrast across the boundary assumed to be 4.5 gm/cm³, contour interval 400 m, and degrees 1–6. Geoid lows shaded. After Hager and Clayton (1987).

or mantle plumes which have thinned the lithosphere. In both cases, partial melting of the mantle may contribute to the high electrical conductivities (Tarits, 1986).

Temperature distribution in the mantle

Estimates of temperatures in the lower crust and mantle can be made from surface heat-flow measurements, models of heat production, and thermal conductivity distributions with depth. Convection is the dominant mode of heat transfer in the asthenosphere and mesosphere where an adiabatic gradient is maintained, and thus temperature increases at a very slow rate with depth (Figure 1.2). On the other hand, conduction is the main way heat is lost from the lithosphere, and temperatures change rapidly with depth and with tectonic setting (Figure 4.6). In this respect, the lithosphere is both a mechanical and a thermal boundary layer in the Earth.

Although major differences in temperature distribution exist in the upper mantle, it is necessary that all geotherms converge at depths of ≤ 400 km or large

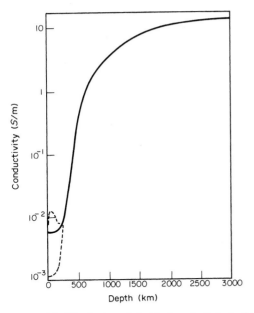

Figure 4.5 Distribution of electrical conductivity in the mantle. The dashed lines indicate the combined range of uncertainty and real variations at depths less than 250 km. After Rikitake (1973).

intersect the adiabat at greater depths (Figure 4.6). Beneath ocean ridges (A, Figure 4.6) geotherms are very steep and intersect the **mantle solidus** (i.e., the temperature at which the mantle begins to melt) at depths generally < 50 km. Thus, ocean ridge basaltic magma is produced at shallow depths beneath ridges. With increasing distance from a ridge axis as the lithosphere cools, geotherms decrease at a rate inversely proportional to lithosphere spreading rates (Bottinga and Allegre, 1973). The decrease results in progressively greater depths of intersection of the geotherm with the mantle solidus, and hence to a deepening of the lower thermal boundary of the oceanic lithosphere as it ages. After about 100 My, the thermal lithosphere is about 100 km thick, in good agreement with thicknesses estimated from surface-wave studies (Figure 4.7). On average, neutral buoyancy of the oceanic lithosphere is reached at about 20 My, and further cooling leads to negative buoyancy and to subduction where the mean age is about 120 My. Rarely are plates > 100 km thick when they subduct.

Geotherms beneath Archean shields are not steep enough to intersect the mantle solidus, whereas most other continental geotherms intersect a slightly hydrous mantle solidus at depths of 150–200 km (Figure 4.6). Subduction geotherms vary with the age of the onset of subduction. In most cases, however, they are not steep enough for the descending slab to melt. The temperature distribution beneath a descending slab 10 My after the onset of subduction is shown in Figure 4.8. Note that isotherms are bent downward as the cool slab descends, and the phase changes at 150 km (gabbro–eclogite) and 410 km (olivine–wadsleyite) occur at shallower depths in the slab than in surrounding mantle in response to cooler temperatures in the slab. As slabs descend into the mantle, they warm up by heat transfer from surrounding mantle, adiabatic compression, frictional heating along the upper surface of the slab, and exothermic phase changes in the slab (Toksoz et al., 1971).

unobserved gravity anomalies would exist between continental and oceanic areas. Heat flow distribution, heat production calculations, and seafloor spreading rates suggest that most geotherms to about 50 km deep range from 10–30 °C/km. With the exception of beneath ocean ridge axes, geotherms must decrease rapidly between 50–200 km to avoid large amounts of melting in the upper mantle, which are not allowed by seismic or gravity data. Oceanic geotherms intersect the mantle adiabat at depths of < 200 km, whereas continental geotherms

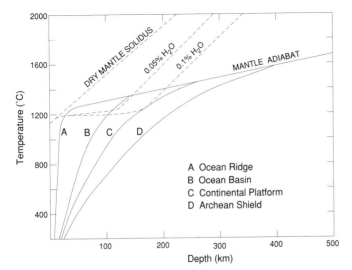

Figure 4.6 Typical lithosphere geotherms. Note that geotherms intersect the mantle adiabat at variable depths. Also shown is the dry mantle solidus and mantle solidus with 0.1 and 0.05 per cent H_2O.

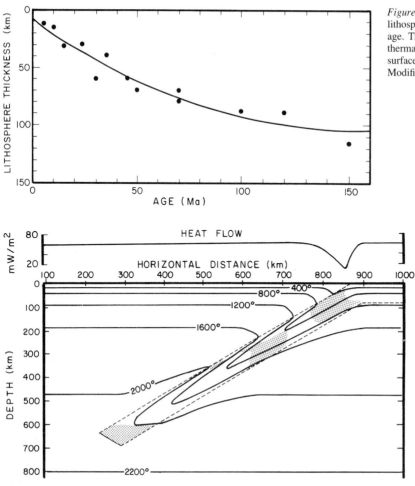

Figure 4.7 Average oceanic lithosphere thickness as a function of age. The base of the lithosphere is a thermal boundary estimated from surface wave dispersion studies. Modified after Sacks (1983).

Figure 4.8 Temperature distribution in a descending slab 10 My after initiation of subduction for a spreading rate of 8 cm/y. Shaded areas are the gabbro–eclogite, olivine–wadsleyite, and 660-km discontinuity phase changes, in order of increasing depth. Modified after Toksoz et al. (1971).

Composition of the mantle

Introduction

The mineralogical and chemical composition of the mantle can be approximated from the combined results of seismic velocity distributions in the Earth, high-pressure–temperature experimental studies, and geochemical and isotopic studies of meteorites and ultramafic rocks. It is now possible to attain static pressures in the laboratory of 10 Gpa to more than 50 Gpa, which allows direct investigation of the physical and chemical properties of minerals which may occur throughout most of the Earth.

Several lines of evidence indicate that ultramafic rocks compose large parts of the upper mantle. These rocks are composed of > 70 per cent of Fe- and Mg-rich minerals such as pyroxenes, olivine, and garnet. Geochemical, isotopic, and seismic studies all agree that the mantle is heterogeneous. Scales and causes of heterogeneities, however, are still not well understood and published models for the mantle range from a layered mantle to a well-stirred, but inhomogeneous ('plum pudding') mantle. As we shall see later in the chapter, the distribution of radiogenic isotopes in basalts clearly indicates the existence of distinct mantle reservoirs which have been in existence for more than 10^9 years. The sizes, shapes and locations of these reservoirs, however, are not well-constrained by the isotopic data and it is only through increased resolution of seismic velocity distributions that these reservoirs can be more accurately characterized.

Seismic velocity constraints

Laboratory measurements of V_p at pressures up to 5 Gpa provide valuable data regarding mineral assemblages in the upper mantle (Christensen, 1966; Christensen and

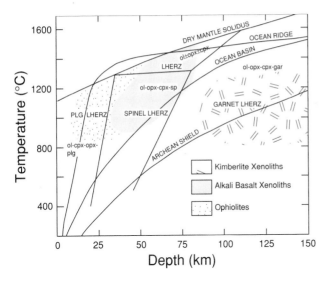

Figure 4.9 Stability fields of ultramafic mineral assemblages in pressure–temperature space. Also shown are lithosphere geotherms and the dry mantle solidus. sp, spinel; plg, plagioclase; gar, garnet; ol, olivine; cpx, clinopyroxene; opx, orthopyroxene; LHERZ, lherzolite.

Mooney, 1995). Dunite, pyroxenite, harzburgite, various lherzolites, and eclogite or some mixture of these rock types are consistent with observed P-wave velocities in the upper mantle (Figure 2.18). Higher than normal temperatures and/or some combination of garnet granulite or serpentinite may account for regions with anomalously low velocities ($V_P < 8.0$ km/sec). Measured velocities of highly-serpentinized ultramafic rocks (5.1–6.5 km/sec), however, are too low for most upper mantle. The high temperatures characteristic of the upper mantle also exceed the stability of serpentine, thus greatly limiting the extent of even slightly-serpentinized lherzolite in the mantle. Serpentinization of ultramafic rocks occurs at low temperatures and pressures (100–400 °C; < 25 km), probably during tectonic emplacement in the crust.

Upper mantle mineral assemblages

Ultramafic xenoliths are common in alkali basalts and in kimberlite pipes and provide an important constraint on mineral assemblages in the upper mantle. Experimental studies of clinopyroxenes coexisting with olivine, orthopyroxene and an Al-rich phase (plagioclase, spinel, or garnet) in xenoliths indicate that Al, Mg, and Ca content of clinopyroxene is sensitive to the temperature and pressure at which it last crystallized in the mantle (MacGregor, 1974). Hence, clinopyroxene compositions in ultramafic rocks provide information on minimum depths for mantle xenoliths. Results sometimes define a P–T curve which may be a 'paleogeotherm' through the lithosphere (Boyd, 1973; 1989). Supporting such an interpretation is the fact these P–T paths are similar in shape to calculated modern geotherms.

From measured seismic velocities (Figure 2.18), mantle xenoliths, and high-pressure experimental data, possible upper-mantle mineral assemblages are shown in

Figure 4.9. Also shown is the mantle solidus and ocean ridge, ocean basin, and Archean shield geotherms. Xenoliths from kimberlites indicate sources with depths > 100 km, and ultramafic xenoliths from alkali basalts record depths of 50–100 km. The data are consistent with an upper mantle lithosphere beneath both the oceans and post-Archean continents composed chiefly of **spinel lherzolite** (olivine-cpx-opx-spinel) and **lherzolite** (olivine-cpx-opx), whereas in the thick Archean lithosphere and in the asthenosphere **garnet lherzolite** (olivine-cpx-opx-garnet) is the dominant rock type. Although measured seismic-wave velocities also allow **dunite** (all olivine), **pyroxenite** (all pyroxenes), **eclogite** (garnet-cpx), and **harzburgite** (olivine-opx) in the upper mantle, the fact that the first three of these rocks are not abundant in xenolith populations or ophiolites suggests they are not important constituents in the upper mantle. Harzburgites, however, are widespread in most ophiolites and may be the most important rock type in the shallow oceanic upper mantle. As suggested by the steep ocean-ridge geotherm in Figure 4.9, some ophiolite ultramafics are dominantly **plagioclase lherzolites** (olivine-cpx-opx-plag).

Ultramafic rocks from ophiolites record source depths of 10–75 km, and thus verify the interpretation that they are slices of the oceanic upper mantle. Ophiolites interpreted as fast-spreading ocean ridges (> 1 cm/y) commonly contain harzburgites from which > 20 per cent basaltic melt has been extracted. Foliations and lineations in these rocks are gently dipping and may actually reflect subhorizontal asthenosphere flow. Ophiolites representative of slow-spreading ridges (< 1 cm/y) record less melt extraction (15 per cent) and a residual mineral assemblage in which plagioclase lherzolite dominates. In these rocks, foliations and lineations are steep and may reflect mantle diapirs that have risen beneath ocean ridges.

Table 4.1 Average chemical composition of the mantle

	Primitive mantle	Post-Archean lithosphere	Archean lithosphere	Depleted mantle
SiO_2	46.0	44.1	46.6	43.6
TiO_2	0.18	0.09	0.04	0.134
Al_2O_3	4.06	2.20	1.46	1.18
FeOT	7.54	8.19	6.24	8.22
MgO	37.8	41.2	44.1	45.2
CaO	3.21	2.20	0.79	1.13
Na_2O	0.33	0.21	0.09	0.02
K_2O	0.03	0.028	0.08	0.008
P_2O_5	0.02	0.03	0.04	0.015
Mg/Mg+Fe	90	90	93	91
Rb	0.64	0.38	1.5	0.12
Sr	21	20	27	13.8
Ba	7.0	17	25	1.4
Th	0.085	0.22	0.27	0.018
U	0.02	0.04	0.05	0.003
Zr	11.2	8.0	7.3	9.4
Hf	0.31	0.17	0.17	0.26
Nb	0.71	2.7	1.9	0.33
Ta	0.04	0.23	0.10	0.014
Y	4.6	3.1	0.63	2.7
La	0.69	0.77	3.0	0.33
Ce	1.78	2.08	6.3	0.83
Eu	0.17	0.10	0.11	0.11
Yb	0.49	0.27	0.062	0.30
Co	104	111	115	87
Ni	2080	2140	2120	1730

Major elements in weight percentages of the oxide and trace elements in ppm
Mg number = Mg/Mg+Fe mole ratio × 100
Depleted mantle calculated from NMORB; Archean lithosphere from garnet lherzolite xenoliths and post-Archean lithosphere
    from spinel lherzolite xenoliths
Data from Hofmann (1988), Sun and McDonough (1989), McDonough (1990), Boyd (1989), and miscellaneous sources

Chemical composition of the mantle

Several approaches have been used to estimate the composition of the mantle:

1 using the compositions of ultramafic rocks from ophiolites and xenoliths
2 using theoretical compositions calculated from geochemical models
3 using compositions of various meteorite mixtures
4 using results from high-pressure and high-temperature experimental studies.

Estimates of the average composition of the primitive mantle, the mantle lithosphere, and depleted mantle are given in Table 4.1. All four estimates have a combined total of more than 90 per cent of MgO, SiO_2, and FeO, and no other oxide exceeds 4 per cent. In a plot of modal olivine versus Mg number of olivine, ophiolite ultramafics and spinel lherzolite xenoliths from post-Archean lithosphere define a distinct trend, while those from the Archean lithosphere define a population with a relatively high Mg number but no apparent trend (Figure 4.10). The trends for oceanic and post-Archean suites

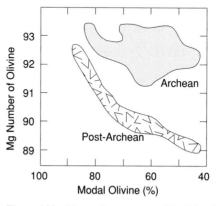

Figure 4.10 Mg number versus modal olivine in post-Archean mantle xenoliths and ophiolite ultramafics and in Archean lithosphere xenoliths from South Africa. After Boyd (1989). Mg number = Mg/Mg+Fe molecular ratio.

are readily interpreted as restite trends, where **restite** is the material remaining in the mantle after extraction of variable amounts of basaltic magma (Boyd, 1989). In the case of ophiolites, they represent the restite remain-

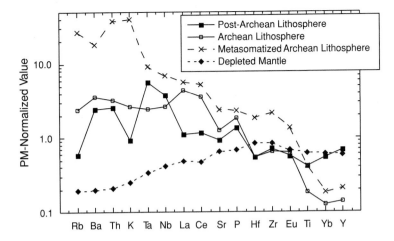

ing after extraction of ocean-ridge basalts. The xenoliths from the post-Archean lithosphere may be restites from extraction of plume-related basalts (flood basalts), in which case they come from rocks underplated beneath the Archean continental lithosphere. The high Mg numbers of olivines from Archean xenoliths suggest they are restites from the extraction of komatiite magmas. Because garnet, the most dense mineral in the upper mantle, is preferentially removed during melting, the restites in all cases are less dense than a primitive mantle of garnet lherzolite mineralogy, and thus they would tend to rise and remain part of the lithosphere.

The Earth's **primitive mantle** composition (Table 4.1) is calculated from geochemical modelling of mantle–crust evolution, and represents the average composition of the silicate part of the Earth just after planetary accretion (Sun and McDonough, 1989). Compared with primitive mantle, incompatible-element distributions in the mantle lithosphere and depleted mantle are quite distinct (Figure 4.11; Table 4.1). The striking depletion in the most incompatible elements (Rb, Ba, Th, etc.) in depleted mantle, as represented by ophiolite ultramafics and mantle compositions calculated from ocean-ridge basalts, reflects removal of basaltic liquids enriched in these elements, perhaps early in Earth history (Hofmann, 1988). In striking contrast to depleted mantle, lithosphere mantle shows prominent enrichment in the most incompatible elements. The spinel lherzolites from post-Archean lithosphere show a positive Nb–Ta anomaly, suggestive that they represent plume material (as discussed later) plastered on the bottom of the lithosphere (McDonough, 1990). Archean garnet lherzolites commonly show textural and mineralogical evidence for metasomatism (i.e., modal metasomatism), such as veinlets of amphibole, micas, and other secondary minerals (Waters and Erlank, 1988). The very high content of the most incompatible elements in modally metasomatized xenoliths (Figure 4.11) probably records metasomatic additions of these elements to the lithosphere and, in some cases, like in the xenoliths from Kimberley in South Africa (Hawkesworth et al., 1990), these addi-

tions occurred after the Archean. Another striking feature of the Archean mantle lithosphere is the relative depletion in the least incompatible elements, such as heavy rare earths (like Yb, Figure 4.11). Because garnet is the only mantle mineral that concentrates heavy rare earths, when the Archean upper mantle was melted garnet could not have remained in the restite or heavy rare earths would not have been lost from the restite. Why then is modal garnet found in these rocks today? It would seem that they must have been buried deep in the Archean lithosphere (> 75 km, Figure 4.9), where garnet was again stabilized, and it formed as the mineral assemblage recrystallized.

The lithosphere

Introduction

Although the lithosphere can be loosely thought of as the outer rigid layer of the Earth, more precise definitions in terms of thermal and mechanical characteristics are useful. For the oceanic lithosphere, where thickness is controlled by cooling, it can be defined as the outer shell of the Earth with a conductive temperature gradient overlying the convecting adiabatic interior (White, 1988). This is known as the **thermal lithosphere**. Thus, asthenosphere can be converted to oceanic lithosphere simply by cooling. The progressive thickening of the oceanic lithosphere continues until about 70 My, and afterwards it remains relatively constant in thickness until subduction (Figure 4.7). Convective erosion at the base of the oceanic lithosphere may be responsible for maintaining this constant depth. The thickness of the rigid part of the outer layer of the Earth that readily bends under a load, known as the **elastic lithosphere**, is less than that of the thermal lithosphere. The base of the oceanic elastic lithosphere varies with composition and temperature, increasing from about 2 km at ocean ridges to 50 km just before subduction. It corresponds roughly to the 500–600 °C isotherm. In continental lithosphere,

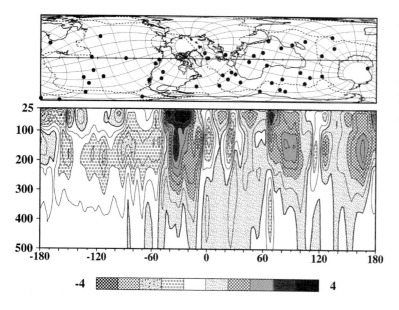

Figure 4.12 S-wave velocity distribution in the upper mantle along a great circle passing through Hawaii and Iceland. Darker shades indicate faster velocities. Map shows location of great circle and major hotspots (black dots). From Zhang and Tanimoto (1993). Depth (km) shown on vertical axis.

the elastic thickness is less than crustal thickness, often by 10–15 km.

Thickness of continental lithosphere

The thermal continental lithosphere varies considerably in thickness depending on its age and mechanism of formation. Shear-wave tomographic studies of the upper mantle have been most definitive in estimating thickness of continental lithosphere (Grand, 1987; Polet and Anderson, 1995). Most post-Archean lithosphere is 100–200 km thick, while lithosphere beneath Archean shields is commonly > 300 km thick. Rheological models suggest thicknesses in these same ranges (Ranalli, 1991). In a shear-wave tomographic cross-section around the globe, high-velocity roots underlie Archean crust, as for example in northern Canada, central and southern Africa, and Antarctica (Figure 4.12). The base of the lithosphere in these and other areas overlain by Archean crust can nearly reach the 410-km discontinuity. Under Proterozoic shields, however, lithospheric thicknesses rarely exceed 200 km. Consideration of elongation directions of Archean cratons relative to directions of modern plate motions suggests that the thick Archean lithosphere does not aid or hinder plate motions (Stoddard and Abbott, 1996). As expected, hotspots (plumes), such as Hawaii and Iceland, are associated with slow velocities between 50 and 200 km deep (Figure 4.12). These slow-velocity anomalies are clearly different from those associated with ocean ridges (Figure 4.27), which bottom out at no more than 100 km. This feature supports the view that passive upwelling mantle is responsible for ocean ridges, while active upwelling mantle (plumes) forms hotspots. Thermal and geochemical modelling has shown that the lithosphere can be thinned by as much as 50 km by extension over mantle plumes (White and McKenzie, 1995). Isotopic and geochemical data from mantle xenoliths

indicate that the mantle lithosphere beneath Archean shields formed during the Archean and that it is chemically distinct from post-Archean lithosphere. Because of the buoyant nature of the depleted Archean lithosphere, it tends to ride high compared with adjacent Proterozoic lithosphere, as evidenced by the extensive platform sediment cover on Proterozoic cratons compared with Archean cratons (Hoffman, 1990). As discussed in the next chapter, the thick Archean lithosphere may have a unique origin by underplating of buoyantly-subducted slabs. The thick roots of Archean lithosphere often survive later tectonic events and thermal events, such as continental collisions and supercontinent rifting. However, mantle plumes or extensive later reactivation can remove the thick lithosphere keels, as for instance is the case with the Archean Wyoming province in North America.

Seismic anisotropy

P-wave velocity measurements at various orientations to rock fabrics show that differences in mineral alignment can produce significant anisotropy (Ave'Lallemant and Carter, 1970; Kumazawa et al., 1971). Differences in V_P of more than 15 per cent occur in some ultramafic samples and are related primarily to the orientation of olivine grains. Seismic-wave anisotropy in the mantle lithosphere beneath ocean basins may be produced by recrystallization of olivine and pyroxene accompanying seafloor spreading with the [100] axes of olivine and [001] axes of orthopyroxene oriented normal to ridge axes (the higher V_P direction) (Raitt et al., 1969; Estey and Douglas, 1986). Supporting evidence for alignment of these minerals comes from studies of ophiolites and upper-mantle xenoliths, and flow patterns in the oceanic upper mantle can be studied by structural mapping of olivine orientations (Nicholas, 1986). The mechanism of

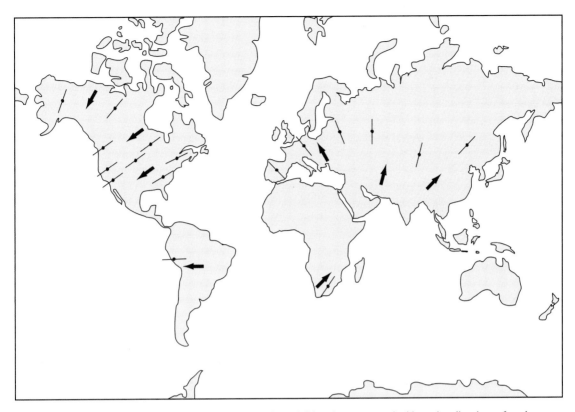

Figure 4.13 Fast S-wave velocity directions in the subcontinental lithosphere compared with motion directions of modern plates (bold arrows). After Silver and Chan (1991).

mineral alignment requires upper-mantle shear flow, which aligns minerals by dislocation glide. The crystallographic glide systems have a threshold temperature necessary for recrystallization of about 900 °C, which yields a thermally-defined lithosphere depth similar to that deduced from seismic data (~ 100 km). Mineral alignment is actively maintained below this boundary in the LVZ by creep, and it is preserved in a fossil state in the overlying lithosphere.

One of the exciting discoveries of the last decade is that the subcontinental lithosphere exhibits seismic anisotropy of S-waves parallel to the surface of the Earth. This is evidenced by **S-wave splitting**, where the incident wave is polarized into two orthogonal directions travelling at different velocities (Silver and Chan, 1991). As with the oceanic lithosphere, this seismic anisotropy also appears to be caused by strain-induced preferred orientation of anisotropic crystals such as olivine. Seismic and thermal modelling indicate that the continental anisotropy occurs within the lithosphere at depths of 150–400 km. The major problem in the subcontinental lithosphere has been to determine how and when such alignment has occurred in tectonically-stable cratons. Was it produced during assembly of the craton in the Precambrian or is it a recent feature caused by deformation of the base of the lithosphere as it moves about?

In most continental sites, the azimuth of the fast S-wave is closely aligned with the direction of absolute plate motion for the last 100 My (Silver and Chan, 1991; Vinnik et al., 1995) (Figure 4.13). This coincidence suggests that the anisotropy is not a Precambrian feature, but results from resistive drag along the base of the lithosphere. Supporting this interpretation is the fact that seismic anisotropy does not correlate with single terranes in Precambrian crustal provinces. These provinces were assembled in the Precambrian by terrane collisions, and if anisotropy was acquired at this time, the azimuths should vary from terrane to terrane according to their pre-assembly deformational histories. Instead, the seismic anisotropies show a uniform direction across cratons, aligned parallel to modern plate motions (Figure 4.13).

Thermal structure of Precambrian continental lithosphere

It has long been known that heat flow from Archean cratons is less than that from Proterozoic cratons (Figure 4.14). Two explanations for this relationship have been proposed (Nyblade and Pollack, 1993):

1   there is a greater heat production in Proterozoic crust than in Archean crust

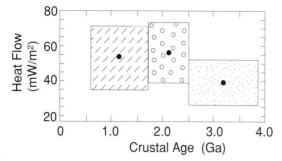

Figure 4.14   Heat flow versus age of Precambrian lithosphere. Width of each box shows the age range and the height is one standard deviation of the mean heat flow. From Nyblade and Pollack (1993).

2   it is due to a thick lithospheric root beneath the Archean lithosphere which is depleted in radiogenic elements.

As we shall see in Chapter 5, the Proterozoic upper continental crust, relative to its Archean counterpart, appears to be enriched in K, U, and Th, whose isotopes produce most of the heat in the Earth. Using recent estimates of the concentration of these elements in the crust (Condie, 1993), only part of the difference between heat flow from Proterozoic and Archean lithosphere can be explained. Thus, it would appear that the thick root beneath the Archean cratons must also be depleted in radiogenic elements and contribute to the difference in heat flows.

Age of subcontinental lithosphere

It is important in terms crust–mantle evolution to know if the thick lithospheric roots beneath Archean cratons formed in the Archean in association with the overlying crust, or if they were added later by underplating. Although in theory mantle xenoliths can be used to isotopically date the lithosphere, because later deformation and metasomatism may reset isotopic clocks, ages obtained from xenoliths are generally too young. In fact, some xenoliths give the isotopic age of eruption of the host magma. What are really needed to determine the original age of the subcontinental lithosphere are minerals that did not recrystallize during later events, or an isotopic system that was not affected by later events. At this point diamonds and Os isotopes enter the picture. Diamonds, which form at depths > 150 km, are very resistant to recrystallization at lithosphere temperatures, and sometimes trap silicate phases as they grow, shielding these minerals from later recrystallization (Richardson, 1990). Pyroxene and garnet inclusions in diamonds, which range from about 50–300 microns in size, have been successfully dated by the Sm–Nd isotopic method, and appear to record the age of the original ultramafic rock. Often more than one age is recorded by diamond inclusions from the same kimberlite pipe,

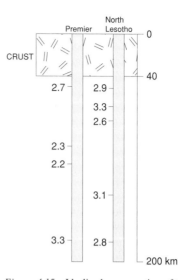

Figure 4.15   Idealized cross-section of the Archean lithosphere in South Africa constructed from mantle xenoliths from two kimberlite pipes. Ages are Re-depletion model ages (in Ga) from Carlson et al. (1994).

as for instance the Premier pipe in South Africa. Diamonds in this pipe with gar–opx inclusions have Nd and Sr mineral isochron ages > 3.0 Ga, suggesting that the mantle lithosphere formed in the Early Archean at the same time as the overlying crust formed (Richardson et al., 1993). Those diamonds with gar-cpx-opx inclusions record an age of 1.92 Ga and those with gar-cpx (eclogitic), an age of 1.15 Ga, an age only slightly older than the age of kimberlite emplacement. The younger ages clearly indicate multiple events in the South African lithosphere. Diamond-inclusion ages from lithosphere xenoliths from Archean cratons in South Africa, Siberia, and Western Australia all indicate that the lithosphere in these regions is also Archean in age.

The Re–Os isotopic system differs from the Sm–Nd and Rb–Sr systems in that Re is very incompatible in the mantle but Os is very compatible. In contrast, in the other isotopic systems both parent and daughter elements are incompatible. Hence, during the early magmatic event that left the Archean mantle lithosphere as a restite, Re was completely extracted from the rock while Os was unaffected (Carlson et al., 1994). When Re was extracted from the rock, the Os isotopic composition was 'frozen' into the system, and hence by analysing a mantle xenolith later brought to the surface, we can date the Re depletion event. Calculated Os isotopic ages of xenoliths from two kimberlite pipes in South Africa are shown in Figure 4.15, plotted at depths of origin inferred from major element distributions. In the Premier and North Lesotho pipes, ages range from 3.3 to 2.2 Ga, similar to the ranges found in xenoliths from pipes in the Archean Siberian craton. Results support the diamond-inclusion ages indicating an Archean age for the thick Archean mantle keels. It is not yet clear if the range in ages from a given pipe really records the range in formation ages of the lithosphere or a series of metasomatic remobilization events of lithosphere that occurred about 3 Ga.

The maximum isotopic ages obtained for the mantle keels in the Siberia and South African cratons are similar to the oldest isotopic ages obtained from the overlying crust (Carlson et al., 1994; Pearson et al., 1995). This suggests that substantial portions of the mantle keels beneath the continents formed at the same time as the overlying crust and that they have remained firmly attached to the crust ever since.

Origin of the mantle lithosphere

As previously mentioned, the origin of the oceanic lithosphere is reasonably well understood from thermal modelling of seafloor spreading and from mantle rocks preserved in ophiolites. The oceanic lithosphere begins life at ocean ridges as a restite left when ocean-ridge basaltic magma is extracted, and this residue is largely composed of harzburgites. As the lithosphere spreads and cools, it thickens (Figure 4.7) by the transfer of mass from the convecting asthenosphere to the lithosphere.

The continental lithosphere has had a much more complicated history, perhaps involving more than one growth mechanism and changes in growth mechanisms at the end of the Archean. The post-Archean subcontinental lithosphere may chiefly represent the remnants of 'spent' mantle plumes, that is, mantle plumes which rose to the base of the continental lithosphere, partially melted producing flood basalts, with the restite remaining behind as a lithospheric underplate. Seismic reflection results, however, suggest that at least some of the continental lithosphere represents remnants of partially-subducted oceanic lithosphere. In northern Scotland, for instance, dipping reflectors in the lower lithosphere are thought to represent fragments of now eclogitic oceanic crust, a relic of pre-Caledonian oceanic subduction (Warner et al., 1996). Isotopic data from xenoliths also indicate that some asthenosphere is added directly to the subcontinental lithosphere. Although basal plume accretion may also have been important in the formation of the thick Archean lithospheric keels (Campbell and Griffiths, 1992), as discussed in Chapter 5, buoyant subduction must also have been important in the Archean, and thus oceanic plates may have been plastered beneath the continents contributing to lithospheric thickening.

The low-velocity zone (LVZ)

The low-velocity zone (LVZ) in the upper mantle is characterized by low seismic-wave velocities, high seismic energy attenuation, and high electrical conductivity. The bottom of the LVZ, sometimes referred to as the **Lehmann discontinuity**, has been identified from the study of surface wave and S-wave data in some continental areas (Figure 4.1) (Gaherty and Jordan, 1995). This discontinuity, which occurs at depths of 180–220 km, appears to be thermally controlled and at least in part

reflects a change from an anisotropic lithosphere to a more isotropic asthenosphere.

Because of the dramatic drop in S-wave velocity and increase in attenuation of seismic energy, it would appear that partial melting must contribute to producing the LVZ. The probable importance of incipient melting is attested to by the high surface heat flow observed when the LVZ reaches shallow depths, such as beneath ocean ridges and in continental rifts. Experimental results show that incipient melting in the LVZ requires a minor amount of water to depress silicate melting points (Wyllie, 1971). With only 0.05–0.1 per cent water in the mantle, partial melting of garnet lherzolite occurs in the appropriate depth range for the LVZ as shown by the geotherm–mantle solidus intersections in Figure 4.6. The source of water in the upper mantle may be from the breakdown of minor phases that contain water, such as hornblende, mica, titanoclinohumite, or other hydrated silicates. The theory of elastic wave velocities in two-phase materials indicates that only 1 per cent melt is required to produce the lowest S-wave velocities measured in the LVZ (Anderson et al., 1971). If, however, melt fractions are interconnected by a network of tubes along grain boundaries, the amount of melting may exceed 5 per cent (Marko, 1980). The downward termination of the LVZ appears to reflect the depth at which geotherms pass below the mantle solidus as illustrated in Figure 4.6. Also possibly contributing to the base of the LVZ is a rapid decrease in the amount of water available (perhaps free water enters high-pressure silicate phases at this depth). The width or even the existence of the LVZ depends on the steepness of the geotherms. With steep geotherms like those characteristic of ocean ridges and continental rifts, the range of penetration of the mantle solidus is large, and hence the LVZ should be relatively wide (Figure 4.6, A and B). The gentle geotherms in continental platforms, which show a narrow range of intersection with the hydrated mantle solidus, produce a thin or poorly defined LVZ, (C). Beneath Archean shields, geotherms do not intersect the mantle solidus, and hence there is no LVZ, (D).

The LVZ plays a major role in plate tectonics, providing a relatively low-viscosity region upon which lithospheric plates can slide with very little friction. The fact that the LVZ is absent or poorly developed beneath Proterozoic shields, and is probably absent beneath Archean shields, suggests that the roots of shields may actually drag at the base of the lithosphere as it moves. Model calculations, however, indicate that this drag is small compared to the forces pulling plates into the mantle at subduction zones.

The transition zone

The 410-km discontinuity

The transition zone is that part of the upper mantle where two major seismic discontinuities occur: one at 410 km

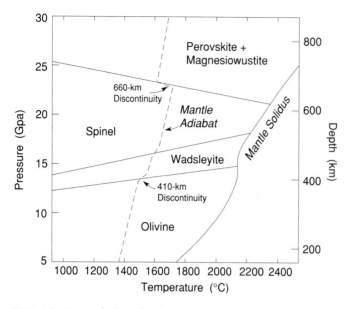

Figure 4.16 Summary of phase relations for Mg$_2$SiO$_4$ in the mantle from high-pressure and high-temperature experimental studies. The dashed line is the mantle adiabat. After Christensen (1995).

Table 4.2 Summary of mantle mineral assemblages for average garnet lherzolite from high-pressure studies

	Depth (km)	*Mineral assemblage (minerals in vol %)*	*Density contrast* (%)	*Slope of reaction* (Mpa/°C)
	< 410	Olivine (58)		
		Opx (11)		
		Cpx (18)		
		Garnet (13)		
	350–450	Opx-cpx → majorite garnet	6	+1.5
410-km discontinuity	410	Olivine (α phase) → wadsleyite (β phase)	6	+5.5
	410–550	Wadsleyite (58)		
		Maj garnet (30)		
		Cpx (9)		
		Opx (3)		
	500–550	Wadsleyite → spinel (γ phase)	2	+3.0
	550–660	Spinel (58)		
		Maj garnet (37)		
		Ca-perovskite (5)		
		Ca-garnet → Ca-perovskite		
660-km discontinuity	660	Spinel → perovskite + magnesiowustite	8	–3.0
	660–2900	Maj Garnet → perovskite		+
		Perovskite (77)		
		Magnesiowustite (15)		
		Ca-perovskite (8)		
		Silica (?)		

Data from Ita and Stixrude (1992) and Christensen (1995)

and the other at 660 km (Figure 4.1). High-pressure experimental studies document the breakdown of Mg-rich olivine to a high-pressure phase known as **wadsleyite** (beta phase) at about 14 Gpa, which is equivalent to 410 km burial depth in the Earth (Figure 4.16). There is no change in chemical composition accompanying this phase change or other phase changes described below. Mantle olivine (Fo$_{90}$) completely transforms to wadsleyite over a < 300 Mpa pressure range at appropriate temperatures for the 410-km discontinuity (~ 1000 °C) (Ita and

Stixrude, 1992). This pressure range is in good agreement with the < 10 km width of the 410-km discontinuity deduced from seismic data (Vidale et al., 1995). If olivine composes 40–60 per cent of the rock, as it does in garnet lherzolites, the olivine–wadsleyite phase change may account for the approximately 6 per cent increase in density observed at this discontinuity (Table 4.2). Measurements of elastic moduli of olivine at high pressures suggest that 40 per cent olivine explains the velocity contrast better than 60 per cent olivine (Duffy et al., 1995). Because garnet lherzolites typically have 50–60 per cent olivine, modal olivine must decrease with depth in the upper mantle to meet this constraint.

Experimental data furthermore indicate that wadsleyite should transform to a more densely-packed spinel structured phase (gamma phase) at equivalent burial depths of 500–550 km. This mineral, hereafter referred to as **spinel**, has the same composition as Mg-rich olivine, but the crystallographic structure of spinel. The small density change (~ 2 per cent) associated with this transition, however, does not generally produce a resolvable seismic discontinuity.

High-pressure experimental data also indicate that at depths of 350–450 km, both clino- and orthopyroxenes are transformed into a garnet-structured mineral known as **majorite garnet**, involving a density increase of about 6 per cent (Christensen, 1995). This transition has been petrographically observed as pyroxene exsolution laminae in garnet in mantle xenoliths derived from the Archean lithosphere at depths of 300–400 km (Haggerty and Sautter, 1990). It is probable that an increase in velocity gradient sometimes observed beginning at 350 km and leading up to the 410-km discontinuity is caused by these pyroxene transformations. At a slightly higher temperature, Ca-garnet begins to transform to Ca-perovskite (a mineral with Ca-garnet composition but with perovskite structure). All of the above phase changes have positive slopes in P–T space, and thus the reactions are exothermic (Table 4.2).

The 660-km discontinuity

One of the most important questions related to the style of mantle convection in the Earth is the nature of the 660-km discontinuity (Figure 4.1). If descending slabs cannot readily penetrate this boundary or if the boundary represents a compositional change, two-layer mantle convection is favoured with the 660-km discontinuity representing the base of the upper layer. Large increases in both seismic-wave velocity (5–7 per cent) and density (8 per cent) occur at this boundary. High-frequency seismic waves reflected at the boundary suggest that it has a width of only about 5 km, but has up to 20 km of relief over distances of hundreds to thousands of kilometres (Wood, 1995). There also may be a positive correlation between relief on the 410-km and on the 660-km discontinuities.

As with the 410-km discontinuity, it appears that a phase change in Mg_2SiO_4 is responsible for the 660-km discontinuity (Christensen, 1995). High-pressure experimental results indicate that spinel transforms to a mixture of perovskite and magnesiowustite at a pressure of about 23 Gpa, and can account for both the seismic velocity and density increases at this boundary if the rock contains 50–60 per cent spinel:

$$(Mg,Fe)_2SiO_4 \rightarrow (Mg,Fe)SiO_3 + (Mg,Fe)O$$

spinel   perovskite   magnesiowustite

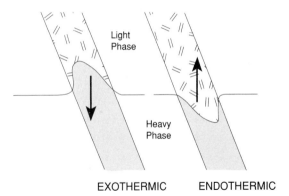

Figure 4.17 Deflection of phase boundaries for exothermic and endothermic reactions for subducting slabs.

Mg **perovskite** and **magnesiowustite** are extremely high-density minerals and appear to comprise most of the lower mantle. High-pressure experimental studies show that small amounts of water may be carried as deep as the 660-km discontinuity in hydrous phases that are stable to 23 Gpa (Ohtani et al., 1995).

Unlike the shallower-phase transitions, the spinel–perovskite transition has a negative slope in P–T space (Figure 4.16; Table 4.2), and thus the reaction is endothermic and may impede slabs from sinking into the deep mantle or impede plumes from rising into the upper mantle. The latent heat associated with phase transitions in descending slabs and rising plumes can deflect phase transitions to shallower depths for exothermic (positive P–T slope) reactions and to greater depths for endothermic (negative P–T slope) reactions (Liu, 1994) (Figure 4.17). For a descending slab in the exothermic case, like the olivine–wadsleyite transition, the elevated region of the denser phase exerts a strong downward pull on the slab or upward pull on a plume, thus helping to drive convection. In contrast, for an endothermic reaction, like the spinel–perovskite transition, the low-density phase is depressed, enhancing a slab's buoyancy and resisting further sinking of the slab. This same reaction may retard a rising plume. At a depth somewhat greater than the 660-km discontinuity, majorite garnet begins to transform to perovskite (Table 4.2), but unlike the spinel transition, the garnet transition is gradual and does not produce a seismic discontinuity. Because the garnet–perovskite transition has a positive slope, the net effect of both the garnet and spinel transitions may be near zero. Recent computer models by Davies (1995) suggest that stiff slabs can penetrate the boundary more readily than plume heads, and plume tails are the least able to penetrate it. Some geophysicists have suggested that slabs may locally accumulate at the 660-km discontinuity, culminating in occasional 'avalanches' of slabs into the lower mantle.

Seismic tomographic images of descending slabs provide an important constraint on the depth of penetration into the mantle. Data suggest that the degree of penetra-

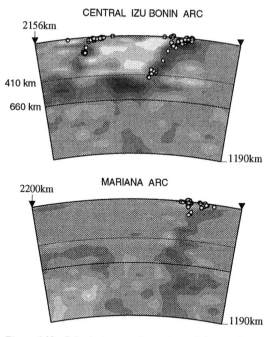

Figure 4.18 Seismic tomography sections of the mantle beneath the Central Izu-Bonin arc and the Mariana arc. White dots are earthquake hypocentres; darker shades are higher velocities, lighter shades are lower velocities. Courtesy of Rob van der Hilst.

tion is related to slab dip and migration of converging margins (Zhong and Gurnis, 1995). Slabs with steep dip angles and relatively stationary trenches, like the Mariana and Tonga slabs (van der Hilst, 1995), are more likely to descend into the lower mantle (Figure 4.18). In contrast, those with shallow dip angles, like the Izu-Bonin and Japan slabs, are commonly associated with rapid retrograde trench migration and these have greater difficulty in penetrating the 660-km discontinuity. Although some slabs may be delayed at the 660-km discontinuity, tomographic images suggest that all modern slabs eventually sink into the lower mantle. Thus, there is no evidence for layered convection in the Earth in terms of slab distributions in the mantle.

The lower mantle

General features

High-pressure experimental studies clearly suggest that Mg perovskite is the dominant phase in the lower mantle (Table 4.2). However, it is still not clear if the seismic properties of the lower mantle necessitate a change in major element composition (Wang et al., 1994). Results allow, but do not require, the Fe/Mg ratio of the lower mantle to be greater than that of the upper mantle. If this were the case, it would greatly limit the mass flux across the 660-km discontinuity to maintain such a

chemical difference, and thus favour layered convection. It is also possible that free silica could exist in the lower mantle. Stishovite (a high-P phase of silica) inverts to an even more dense silica polymorph with a CaCl$_2$ structure at about 50 Gpa (Kingma et al., 1995), and it is possible that this silica phase exists in the Earth at depths > 1200 km.

New interpretations of the isostatic rebound of continents following Pleistocene glaciation together with gravity data indicate that the viscosity of the mantle increases with depth by two orders of magnitude, with the largest jump at the 660-km discontinuity. This conclusion is in striking agreement with other geophysical and geochemical observations. For instance, although mantle plumes move upwards relatively quickly, it would be impossible for them to survive convective currents in the upper mantle unless they were anchored in a 'stiff' lower mantle. Also, only a mantle of relatively high viscosity at depth can account for the small number (two today) of mantle upwellings. Geochemical domains which appear to have remained isolated from each other for billions of years in the lower mantle (as discussed below) can also be accounted for in a stiff lower mantle that resists mixing.

The D″ layer

The D″ layer is a region of the mantle within a few hundred kilometres of the core where seismic velocity gradients are anomalously low (Young and Lay, 1987; Loper and Lay, 1995). Calculations also indicate that only a relative small temperature gradient (1–3 °C/km) is necessary to conduct heat from the core into the D″ layer. Because of diffraction of seismic waves by the core, the resolution in this layer is poor, and thus details of its structure are not well-known. However, results clearly indicate that D″ is a complex region that is both vertically and laterally heterogeneous (Kendall and Silver, 1996). Data seem to be equally consistent with either a sharp interface at 100–300 km above the core–mantle boundary and/or small-scale discontinuities that scatter seismic waves near the boundary (Figure 4.19). Estimates of the thickness of the D″ layer suggest that it ranges from 100 to about 500 km. Despite the poor resolution, large-scale lateral heterogeneities can be recognized in D″ (Loper and Lay, 1995). For instance, regions beneath circum-Pacific subduction zones have anomalously fast P- and S-waves, interpreted by many to represent lithospheric slabs which have sunk to the base of the mantle (Figure 4.3). It is noteworthy that slow velocities in D″ occur beneath the Central Pacific and correlate with both the surface and core–mantle boundary geoid anomalies (Figure 4.4) and a concentration of hotspots (see Figure 4.25).

There are three possible contributions to the complex seismic structures seen in D″: temperature variations, compositional changes, and mineralogical phase changes. Temperature variations appear to be caused chiefly by slabs sinking into D″ (a cooling effect which produces

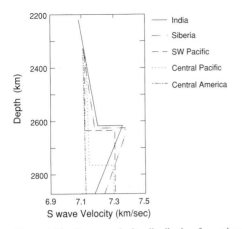

Figure 4.19   S-wave velocity distribution for various geographic regions in the lower mantle. Although most regions show a sharp upper boundary at the top of the D″ layer, the velocity profiles show a great deal of lateral heterogeneity in this region. Data after Knittle and Jeanloz (1991).

relatively fast velocities) and heat released from the core (causing slow velocities). Mixing of molten iron from the core with high-pressure silicates can lead to compositional changes with corresponding velocity changes. Experiments have shown, for instance, that when liquid iron comes in contact with silicate perovskite at high pressures, these substances vigorously react to produce a mixture of Mg perovskite, a high-pressure silica polymorph, wustite (FeO), and Fe silicide (FeSi). These experiments also suggest that liquid iron in the outer core will seep into D″ by capillary action to hundreds of metres above the core–mantle boundary and that the reactions will occur on timescales of less than 10^6 years. Phase changes, such as the possible breakdown of perovskite to magnesiowustite and silica, provide the best explanation for the sharp velocity increase seen at about 2600 km (Figure 4.19). Because the low seismic-wave velocities in D″ reflect high temperatures, and thus a lowering of mantle viscosity, this layer is commonly thought to be the source of mantle plumes. The lower viscosity will also enhance the flow of material into the base of newly-forming plumes, and the lateral flow into plumes will be balanced by slow subsidence of the overlying mantle. The results of Davies and Richards (1992) suggest that a plume could be fed for 100 My from a volume of D″ only a few tens of kilometres thick and 500–1000 km in diameter. These results are important for mantle dynamics because they suggest that plumes are fed from the lowermost mantle, whereas ocean ridges are fed from the uppermost mantle. Although chemical segregation may occur in the D″ layer, calculated temperatures are not high enough to melt perovskite or magnesiowustite phases.

The heterogeneous nature of the D″ layer is consistent with the presence of somewhat denser material, commonly referred to as 'dregs'. Slow upward convection of the mantle may pull dense phases such as wustite and FeSi upwards from the core–mantle boundary. Eventually, they begin to sink because of their greater density, and these may form dregs that accumulate near the base of D″. Modelling suggests that dregs should pile up in regions of mantle upwelling (the fast regions in Figure 4.3) and thin in regions of downwelling, with the possibility that parts of D″ could be swept clean of dregs beneath downwellings. This means that the dregs must be continually supplied by reactions and upwelling from the core–mantle boundary. Lateral variations in the thickness of D″ caused by lateral dreg movements could account for the large-scale seismic-wave velocity variations and the variations in thickness of D″. Small-scale compositional heterogeneity in D″ or convection including the dregs in this layer could account for small-scale variations and for scattering of seismic waves.

Mantle plumes

Theoretical and laboratory models have clarified the dynamics of plumes and have suggested numerous ways in which plumes may interact with the mantle. Also, there has been a great deal of interest in the possible effects of plumes on a variety of near-surface phenomena such as the evolution of the atmosphere and life, a subject that will be addressed in Chapter 6. Although the base of both the upper and the lower mantle have been suggested as sites of mantle-plume generation, five lines of evidence strongly suggest that plumes are produced just above the core–mantle boundary in the D″ layer (Campbell and Griffiths, 1992; Davies and Richards, 1992):

1 Computer modelling indicates that plume heads can only achieve a size (~ 1000 km diameter) required to form large volumes of flood basalt and large submarine plateaux, like the Ontong–Java plateau, if they come from the deep mantle.
2 The approximate fixed position of hotspots relative to each other in the same geographic region is difficult to explain if plumes originate in the upper mantle, but is consistent with a lower mantle source.
3 The amount of heat transferred to the base of the lithosphere by plumes, which is ≤ 12 per cent of the Earth's total heat flux, is comparable to the amount of heat estimated to be emerging from the core as it cools.
4 Periods of increased plume activity in the past seem to correlate with normal polarity epochs and decreased pole reversal activity in plume-related basalts (Larson, 1991a and 1991b). Correlation of plume activity with magnetic reversals implies that heat transfer across the core–mantle boundary starts a mantle plume in D″, and at the same time changes the pattern of convective flow in the outer core, which in turn affects the Earth's magnetic field.

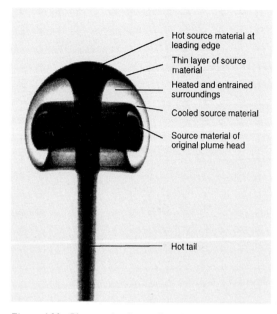

Figure 4.20 Photograph of a starting plume in a laboratory experiment showing the large head and narrow tail. The plume was produced by continuously injecting hotter, lower-viscosity dyed fluid into the base of a cool, higher-viscosity layer of the same fluid. Light regions in the head are entrained from surrounding undyed fluid. Courtesy of Ian Campbell.

5 By analogy with iron meteorites, the Earth's core should be enriched in Os with a high $^{187}Os/^{188}Os$ ratio compared to the mantle (Walker et al., 1995). Thus, plumes produced in the D″ layer could be contaminated with radiogenic Os from the core. Some plume-derived basalts have $^{187}Os/^{188}Os$ ratios up to 20 per cent higher than primitive or depleted mantle, suggesting core contamination and thus, a source near the core–mantle interface.

Theoretical and laboratory models show that a plume is composed of a head of buoyant material whose diameter is much larger than the following tail (Figure 4.20). This occurs because the ascent velocity of a new plume is limited by the higher viscosity of the surrounding mantle, and the head grows until its velocity matches the flux of material into its base (Davies and Richards, 1992). Because of the thermal buoyancy of the head, it entrains material from its surroundings as it grows and, depending on the rate of ascent, it may entrain up to 90 per cent of its starting mass. Streamlines in computer models show that most of the entrained fraction should come from the lower mantle (Hauri et al., 1994). Results of these studies also indicate that a vertical stationary plume source should give rise to a stationary plume trace at the surface (like the Hawaiian seamount chain), despite any horizontal flow in the surrounding mantle. If a plume is inclined > 60 ° from the vertical, however, it should break up, although entrainment may suppress this. Petrologic

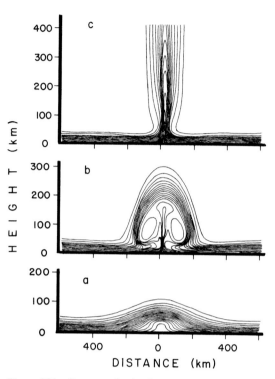

Figure 4.21 Computer-simulated generation of a mantle plume in the D″ layer, shown here at (a) 110 My, (b) 130 My, and (c) 140 My after thermal perturbation from the underlying core. Contour interval is 40 °C. Modified after Olson et al. (1987).

observations, some of which are discussed in Chapter 5, and plume models (Sleep, 1990) constrain the maximum temperature of plumes in the upper mantle to ≤ 250 °C greater than surrounding mantle.

A computer-simulated sequence of events for the generation of a plume in the D″ layer is shown in Figure 4.21. The plume begins to form in response to a pulse of heat lost from the core, and after 130 My (b), it is 50–100 km across surrounded by boundary zones of entrained mantle about 50 km thick. A successful plume, one that survives and rises out of the D″ layer, requires a temperature perturbation exceeding 200 °C. By 140 My after initiation (c), the plume has ascended from the D″ layer leaving its trailing tail. Plumes produced in such a manner would appear to travel upwards to the base of the lithosphere in 50 to 100 My, a time period much shorter than the convective overturn of the mantle (~ 10^9 y). Thus, convection should not disturb or destroy plumes on the their upward journey.

Although convection should not affect plumes, what about the endothermic phase change at the 660-km discontinuity? Davies (1995) has shown by computer models that for the temperature- and depth-dependent viscosity in the Earth, most plumes should pass through the spinel–perovskite phase boundary unimpeded. In some cases, however, plume tails may slow down due to the negative buoyancy effect caused by an upward displacement of the phase boundary within the plume. However, if the slope of the phase change in P–T space is more negative than –3 Mpa/ °C, plumes may not penetrate the phase boundary, at least not without a significant delay. The positive slope of the other phase transitions in the upper mantle (Table 4.2), however, should enhance the ascent rate of plumes.

As a plume head approaches the lithosphere it begins to flatten and eventually intersects the mantle solidus, resulting in the production of large volumes of basaltic magma in relatively short times (White and McKenzie, 1995). Plumes with flattened heads ≥ 1000 km diameter are plausible sources for flood basalts and large submarine plateaux as discussed in Chapter 3.

Mantle upwellings, plumes, and supercontinents

As discussed in Chapter 1, supercontinents appear to have aggregated and dispersed several times during geologic history, although our geologic record of supercontinent cycles is only well-documented for the last two cycles: Gondwana–Pangea and Rodinia. It now seems almost certain that the supercontinent cycle is closely tied to mantle processes, including both convection and perhaps mantle plumes. Pangea 200 Ma was centred approximately over the Atlantic–African geoid high (Figure 4.4a), and the other continents moved away from this high during break up. Because this high contains many of the Earth's hotspots and is characterized by low seismic velocities in the lower mantle, it is probably hotter than average, as discussed earlier in the chapter. Except for Africa, which still sits over the geoid high, continents seem to be moving towards geoid lows, which are also regions with relatively few hotspots and high lower-mantle velocities, all of which point to cooler mantle. These relationships suggest that supercontinents may affect the thermal state of the mantle, with the mantle beneath continents becoming hotter than normal, expanding and producing the geoid highs. This is followed by increased mantle-plume activity, which may fragment a supercontinent or at least contribute to dispersal of its cratons.

Gurnis (1988) proposed a computer model based on feedback between continental plates and mantle convection, whereby supercontinents insulate the mantle, leading to high mantle temperatures, and the perturbed mantle, in turn, fragments and disperses the supercontinents. Experimental studies support such a model (Guillou and Jaupart, 1995). As an example, beginning with a supercontinent with cold downwellings on each side, a hot upwelling generated beneath the supercontinent by its insulating effect fragments the supercontinent (Figure 4.22a). Immediately after break up, two smaller continental cratons begin to separate rapidly as the hot upwelling extends to the surface between the two plates,

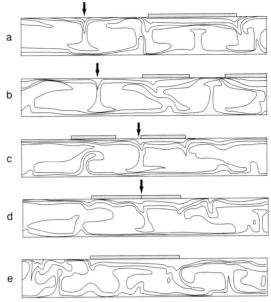

Figure 4.22 Computer model of the break up of a supercontinent: (a) hot upwelling beneath a supercontinent; (b) break up of the supercontinent and opening of an ocean basin; (c) closing of the ocean basin as continents move towards cool downwellings; (d) continental collision over a downwelling; and (e) a new hot upwelling forms under the supercontinent and the cycle begins again. The frame of reference is fixed with respect to the left corner of the plate in (a) and the right continent in (b) moves with respect to the left continent, which is stationary in this model. Contours are isotherms. After Gurnis (1988).

producing a thermal boundary layer (b). Both plates rapidly move towards the cool downwellings marked with a vertical arrow (c). Approximately 150 My after break up, the two continental fragments collide over a downwelling (d). Nearly 450 My after break up, a new thermal upwelling develops beneath the new supercontinent, and the cycle starts again (e).

One question not fully understood yet is the role that mantle plumes may play in the supercontinent cycle. Are they responsible for fragmenting supercontinents, or do they play a more passive role? Because plumes have the capacity to generate large quantities of magma, it should be possible to track the role of plumes in continental break up by the magmas they have left behind as flood basalts and giant dyke swarms. From studies of these features and the break up of Gondwana, it seems that plumes are not the ultimate driving force for the break up of this supercontinent in the early Paleozoic (Storey, 1995). Rather, fragmentation can generally be linked to changes in plate-boundary forces. When mantle plumes were involved, they appear to have assisted break up, and perhaps controlled its location. However, they may have been important in the splitting of smaller continental blocks during the dispersal of Gondwana.

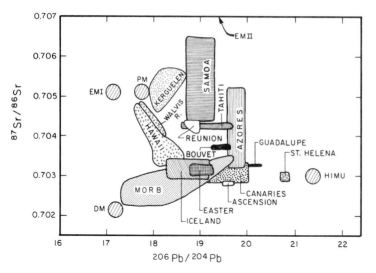

Figure 4.23 Sr and Pb isotope distributions in MORB and ocean island basalts. Modified after Zindler and Hart (1986). DM, depleted mantle; PM, primitive mantle; HIMU, high U/Pb component; EMI (EM1) and EMII (EM2), enriched mantle components.

Mantle geochemical components

Introduction

Radiogenic isotopes can be used as geochemical 'tracers' to track the geographic and age distribution of crustal or mantle reservoirs from which the elements that house the parent (P) and daughter (D) isotopes have been fractionated from each other. For instance, and process that fractionates U from Pb or Sm from Nd will result in a changing rate of growth of the daughter isotopes ^{207}Pb and ^{206}Pb (from ^{237}U and ^{235}U, respectively) and ^{143}Nd (from ^{147}Sm). If the element fractionation occurred geologically long ago, the change in P/D ratio will result in a measurable difference in the isotopic composition of the melt and its source. This is the basis for the isotopic dating of rocks. On the other hand, if the chemical change occurs during a recent event, such as partial melting in the mantle just prior to magma eruption, there is not enough time for the erupted lava to evolve a new isotopic signature distinct from the source material reflecting its new P/D ratio. Hence, young basalts derived from the mantle carry with them the isotopic composition of their mantle sources, and this is the basis of the **geochemical tracer** method.

When isotopic ratios from young oceanic basalts are plotted on isochron diagrams, the data often fall close to an isochron with a Precambrian age, and these are sometimes referred to as **mantle isochrons** (Brooks et al., 1976). The interpretation of these 'isochrons', however, is ambiguous. They could represent true ages of major fractionation events in the mantle or, alternatively, they could be mixing lines between endmember components in the mantle. However, the fact that most ages calculated from U–Pb and Rb–Sr isotopic data from young oceanic basalts fall between 1.8–1.6 Ga, and most Sm–Nd results yield ages of 2.0–1.8 Ga seem to favour their interpretation as ages of mantle events. Even if the linear arrays on isochron diagrams are mixing lines, at least one of the endmembers has to be Early Proterozoic in age.

Identifying mantle components

Summary

At least four, and perhaps as many as six isotopic endmembers may exist in the mantle from results available from oceanic basalts (Hart, 1988; Hart et al., 1992) (Figures 4.23 and 24). These are **depleted mantle (DM)**, the source of normal ocean ridge basalts (NMORB); **HIMU**, distinguished by its high $^{206}Pb/^{204}Pb$ ratio, which reflects a high U/Pb ratio ($\mu = {}^{238}U/^{204}Pb$) in the source; and two enriched sources (EM), which reflect long-term enrichment in light REE in the source. **EM1** has relatively low $^{206}Pb/^{204}Pb$ and moderate $^{87}Sr/^{86}Sr$ ratios compared with **EM2**, which has intermediate $^{206}Pb/^{204}Pb$ and high $^{87}Sr/^{86}Sr$ ratios (Figure 4.23). Both EM components have high $^{207}Pb/^{204}Pb$ relative to $^{206}Pb/^{204}Pb$ compared with MORB. A possible fifth component, **primitive mantle (PM)**, has been described above. However, when several isotopic systems are considered for the same suite of samples, conclusive evidence for a PM reservoir is not found. Another possible end member, **FOZO**, defined by isotopic mixing arrays in three dimensional space, is described more fully below.

The existence of at least four mantle endmembers is now well-documented. What remains to be verified is the origin and location of each endmember, and their mixing relations. As summarized below, much progress has been made on these questions by also using rare gas isotopic data and trace element ratios. Further, the hierarchy of mixing of components in basalts from single islands or island chains can provide useful information on the location of the components in the mantle.

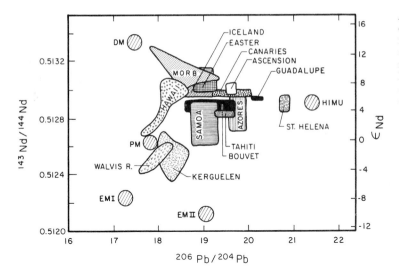

Figure 4.24 Nd and Pb isotope distributions in MORB and ocean island basalts. Modified after Zindler and Hart (1986). DM, depleted mantle; PM, primitive mantle; HIMU, high U/Pb component; EMI (EM1) and EMII (EM2), enriched mantle components.

DM

Depleted mantle (DM) is mantle that has undergone one or more periods of fractionation involving extraction of basaltic magmas. Depleted mantle is known to underlie ocean ridges and probably extends beneath ocean basins, although it is not the source of oceanic-island magmas. The depleted isotopic character (low $^{87}Sr/^{86}Sr$, $^{206}Pb/^{204}Pb$ and high $^{143}Nd/^{144}Nd$) and low LIL element contents of NMORB require the existence in the Earth of a widespread depleted mantle reservoir. Rare-gas isotopic compositions also require that this reservoir is highly depleted in rare gases compared with other mantle components. Although most of the geochemical variation within NMORB can be explained by magmatic processes, such as fractional crystallization, variations in isotopic ratios demand that the depleted mantle reservoir is heterogeneous, at least on scales of 10^2–10^3 km. This heterogeneity may be caused by small amounts of mixing with enriched mantle components.

HIMU

The extreme enrichment in ^{206}Pb and ^{208}Pb in some oceanic-island basalts (such as those from St Helena, Figures 4.23 and 24) requires the existence of a mantle source enriched in U+Th relative to Pb, and mantle isochrons suggest an age for this HIMU source of the order of 2.0–1.5 Ga. This reservoir is also enriched in radiogenic ^{187}Os (Hauri and Hart, 1993). Because HIMU has $^{87}Sr/^{86}Sr$ ratios similar to NMORB, however, it has been suggested that it may represent subducted oceanic crust in which the U+Th/Pb ratio was increased by preferential loss of Pb in volatiles escaping upward from descending slabs during subduction. Supporting a recycled oceanic crust origin for HIMU are relative enrichments in Ta and Nb in many oceanic-island basalts (Figure 3.8). As discussed in Chapter 3, devolatilized descending slabs should be relatively enriched in these

elements since neighbouring incompatible elements like Th, U, K, and Ba have been lost to the mantle wedge. Thus, the residual mafic part of the slab that sinks into the lower mantle and becomes incorporated in mantle plumes should be relatively enriched in Ta and Nb. This interpretation supports a mantle plume origin for the HIMU component in oceanic islands.

Enriched mantle

Enriched mantle components are mantle reservoirs enriched in Rb, Sm, U, and Th relative to Sr, Nd and Pb, respectively, compared with primitive mantle ratios of these elements. At least two enriched components are required (Zindler and Hart, 1986): EM1 with moderate $^{87}Sr/^{86}Sr$ ratios and low $^{206}Pb/^{204}Pb$ ratios, and EM2 with high $^{87}Sr/^{86}Sr$ ratios and moderate $^{206}Pb/^{204}Pb$ ratios. Both have low $^{143}Nd/^{144}Nd$ ratios (Figure 4.24). The two chief candidates for EM1 are strongly depleted oceanic mantle lithosphere and ancient sediments which have been recycled back into the mantle (Hart et al., 1992). EM2 has isotopic ratios closer to average upper continental crust or to modern subducted continental sediments (i.e., $^{87}Sr/^{86}Sr > 0.71$ and $^{143}Nd/^{144}Nd \sim 0.5121$). Subducted continental sediments are favoured for this endmember because EM2 commonly contributes to island arc volcanics in which continental sediments have been subducted, such as the Lesser Antilles and the Sunda arc. In contrast, it is notable that there are no known EM1-type arcs.

The Dupal anomaly

With few exceptions, most ocean islands showing EM-type mantle components occur in the Southern Hemisphere (Hart, 1988). Also, NMORB in the South Atlantic and Indian Ocean exhibit definite contributions from EM sources, unlike NMORB from other ocean-ridge sys-

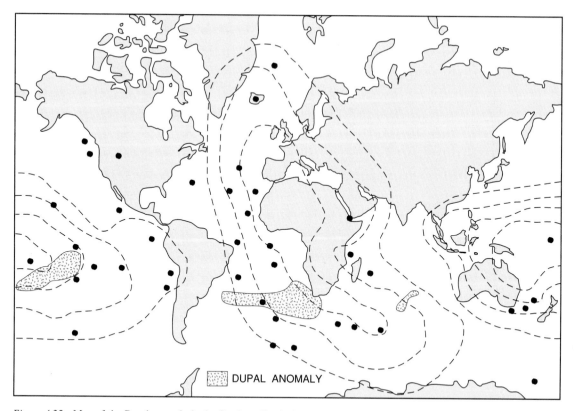

Figure 4.25 Map of the Dupal anomaly in the Southern Hemisphere (patterned areas) compared with the geoid anomalies and hotspot distributions. Dupal data from Hart (1988). Dashed lines are geoid anomalies, dots are hotspots.

tems. This band of EM enrichment in the Southern Hemisphere is known as the **Dupal anomaly** (Figure 4.25). Pb isotopic data indicate that the anomaly has existed for billions of years (Hart, 1984). One intriguing feature of this anomaly is the apparently complete lack of any EM signatures in hotspot basalts from polar regions. Both shallow and deep origins have been proposed for the Dupal anomaly. Favouring a deep source is the correlation of the Dupal anomaly with geophysical anomalies. The Dupal maxima occur over the slowest seismic velocities in the lower mantle, which also are highs in the geoid and have numerous hotspots at shallower depths, suggesting that the anomaly is an expression of material rising in plumes from the core–mantle boundary region (Figure 4.25). On the other hand, a shallow origin for Dupal is supported by EM components in South Atlantic and Indian Ocean NMORB and in island-arc basalts in the Southwest Pacific.

The best way out of this dilemma is for Dupal to have both deep and shallow sources. If EM2, which dominates the Dupal anomaly, is caused by modern continental sediments, the plume source could be from sediments which have been recycled through the deep mantle with sinking slabs, and the shallow source could be produced by subducted sediments that pass their geochemical signature to mantle wedges during plate devolatilization.

Plume contamination of DM near the South Atlantic and Indian Ocean ridges is a possible explanation for EM components in NMORB from these ridges.

Mixing regimes in the mantle

When viewed in Nd–Sr–Pb isotopic space, arrays defined by various oceanic islands tend to cluster along two or three component mixing lines with DM, HIMU, EM1 and EM2 at the corners of a tetrahedron (Figure 4.26). There is, however, a notable lack of mixing arrays joining EM1, EM2 and HIMU, which seems to rule out random mixing in the mantle (Hart et al., 1992; Hauri et al., 1993). The mixing arrays, in fact, seem to be systematically orientated originating from points along the EM1–HIMU and EM2–HIMU joins and converging in a region within the lower part of the tetrahedron, and not at DM. This convergence suggests the existence of another mantle component that appears in oceanic mantle plumes. This component has been called FOZO after focal zone, from the converging arrays, and like DM is also a depleted mantle component (Hart et al., 1992) (Figure 4.26). Helium isotope ratios (^{3}He/^{4}He) also progressively increase towards FOZO and, because ^{3}He is not formed by radiogenic decay but was incorporated in the Earth during planetary accretion, the high ^{3}He/^{4}He

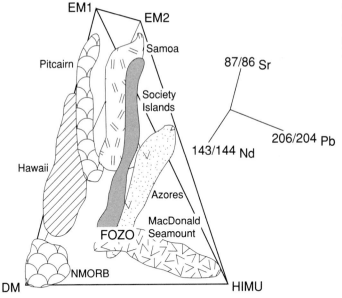

Figure 4.26 Three-dimensional plot of Sr, Nd, and Pb isotopic ratios from intraplate ocean basalts. The tetrahedron is defined by the mantle end members EM1, EM2, DM, and HIMU. Modified after Hauri et al. (1994).

ratios suggest that the FOZO component is relatively primitive.

The fact that mixing arrays do not emanate from the EM1–EM2 join nor from the EM1–EM2–HIMU face of the tetrahedron indicates there was some process that caused the juxtaposition of EM components with HIMU, which predates the mixing with FOZO. If, as previously suggested, EM1, EM2, and HIMU represent oceanic lithosphere, modern sediments, and recycled oceanic crust respectively, there is a possibility that these three components represent the sediment–basalt–ultramafic suite we see in ophiolites, as it is subducted into the mantle. The fact that hotspot sources show no mixing with DM is consistent with plume entrainment results previously discussed, indicating that most plume entrainment occurs in the lower, not the upper mantle where DM is located.

The isotopic mixing arrays do not tell us which components occur in the plume sources and which have been entrained in the mantle. A high $^3He/^4He$ ratio in FOZO in contrast to a low ratio in DM suggests that FOZO is in the lower mantle and does not mix with DM in the upper mantle. It is not possible from isotopic data alone to determine if the EM–HIMU components are the plume sources in the D″ layer, and FOZO is entrained in the lower mantle as the plumes rise, or the other way around (FOZO is the plume and EM–HIMU are entrained). When coupled with the geophysical data previously discussed, however, the former model seems most acceptable.

Scale of mantle heterogeneities

Although isotopic, rare gas, and trace element distributions in oceanic basalts demand a heterogeneous mantle, the nature, scale, preservation, and history of mantle heterogeneities remain problematic. Experimental studies on diffusion rates of cations in mantle minerals provide a basis to constrain the minimum scale of mantle heterogeneities. Results suggest that even in the presence of a melt where diffusion rates are high, heterogeneities larger than one kilometre in size can persist for several billions of years (Zindler and Hart, 1986; Kellogg, 1992). Geochemical heterogeneities in ultramafic portions of ophiolites occur on scales of centimetres to kilometres, indicating that small-scale heterogeneities survive in the upper mantle. Geochemical and isotopic variations in basalts from a single volcano require heterogeneity on the scale of several kilometres and variations in MORB along ocean ridges, or of arc basalts along single volcanic arcs, demand heterogeneities on scales of 10^2–10^3 km. At the large end of the scale is the Dupal anomaly (Figure 4.25), which requires global-scale anomalies that have survived for billions of years. Numerical stirring models show that if the viscosity of the lower mantle is 2–3 orders of magnitude greater than the upper mantle, mantle heterogeneities should survive for up to 2 Gy, as isotopic data from oceanic-island basalts suggest (Davies and Richards, 1992). These models, furthermore, allow heterogeneities of a variety of scales to survive for long periods of time.

An overview

By way of summary, perhaps the most significant chemical signatures recorded by the mantle are the complementary relationships between incompatible element enrichment in the continental crust (Figure 2.24) and the probable siderophile element (Fe, Ni, Co, Cr, etc.) enrichment of the core, and the corresponding depletion of these elements in the DM reservoir (Carlson, 1994). It would seem that the key events controlling compositional variations in the Earth are early core formation, followed by gradual or episodic extraction of continental crust to leave at least part of the mantle depleted in incompatible and siderophile elements. The DM and FOZO components may represent depleted mantle remaining from these events. The other mantle components would appear to reflect subducted inhomogeneities which were not sufficiently remixed into the convecting mantle to lose their geochemical and isotopic signatures. Many of these components, together with DM, appear to have been 'fossilized' in the subcontinental lithosphere by accretion of spent plume material to the base of the lithosphere throughout geologic time (Menzies, 1990). Geochemical evidence for layering in the mantle is equivocal and controversial. Although the data do not exclude chemical layering, the non-random occurrence of geochemical components in plume sources would seem to require exchange between the upper and lower mantle.

Convection in the mantle

The nature of convection

It is generally agreed that convection in the mantle is responsible for driving plate tectonics. Convection arises because of buoyancy differences, with lighter material rising and denser material sinking. In terms of quantitative laboratory models, **Rayleigh-Bernard convection** is best understood. This type of convection arises because of heating at the base and cooling at the surface of fluid. Lord Rayleigh was the first to show that the convective behaviour of a substance is dependent on a dimensionless number, now known as the **Rayleigh number**. For a simple homogeneous liquid to convect as it is heated at the base, the Rayleigh number must exceed 2000, and for the convection to be vigorous it must be the order of 10^5. Irregular turbulent convection begins when the Rayleigh number reaches about 10^6 and such convection probably exists in the Earth where the Rayleigh number is $> 10^6$. Another factor contributing to mantle convection is lateral motion of subducted slabs. Absolute plate motions indicate that slab migration is generally opposite to the direction of subduction at rates of 10–25 mm/y. As a result, the downward motions of slabs are generally steeper than their dips (Garfunkel et al., 1986). Calculations indicate that the mass flux in the upper mantle caused by lateral slab motions may be an important contribution to large-scale mantle convection.

A voluminous literature exists on models for convection in the Earth. In general, models fall into two categories:

1 layered convection
2 whole-mantle convection.

In the **layered convection** models, convection occurs separately below and above the 660-km discontinuity,

while **whole-mantle convection** involves the entire mantle. It is important to remember that these types of convection are two endmember scenarios, and that convection in the Earth may be somewhere in between (i.e., partially-layered convection). Although both models have a scientific following, as the resolution of seismic tomographic data has improved it appears certain that lithospheric slabs descend into the lower mantle. This seems to necessitate some style of whole-mantle convection. Classical pictures of convection in the Earth show convective upcurrents coming from the deep mantle beneath ocean ridges, and downcurrents returning at subduction zones. As discussed below, however, it seems clear that ocean ridges are shallow passive features, not related to deep convection in the Earth.

Passive ocean ridges

S-wave velocity distributions beneath ocean ridges are useful in distinguishing between shallow and deep sources for upwelling mantle. As shown in Figure 4.27a, ocean ridges are typically underlain by broad, low-velocity regions in the upper 100 km of the mantle. In all cases, the lowest velocities occur at depths < 50 km and, regardless of spreading rate, regions with velocities that are 1–2 per cent low are ~ 100 km deep (Zhang and Tanimoto, 1992). The widths of the low-velocity regions, however, correlate positively with spreading rates. This correlation may be due to dragging of the shallow asthenosphere away from ridges by the separating plates. In contrast to ocean ridges, hotspots have low-velocity roots that extend to much greater depths, with regions that are 1–2 per cent low extending to 200 km or more in depth, and low-velocity mantle, in general, continuing into the deep mantle (Figure 4.27b). These results support theoretical and laboratory plume models suggesting that plumes have deep roots.

The low-velocity anomalies under ocean ridges indicate that they are shallow features, with shallow roots probably caused by passive upwellings. Because plumes, on the other hand, are driven by density anomalies in the mantle, they should have low-velocity anomalies that extend much deeper into the mantle, as observed for Hawaii and the Azores (Figure 4.27b). Both situations occur along the Mid-Atlantic ridge, but only the hotspots on or near the ridge have deep, low-velocity roots, while the low-velocity roots of the ocean ridge are < 100 km deep. Thus, it would appear there is some degree of decoupling between active and passive asthenosphere at depths of 100–200 km. Just how this occurs and evolves with time is not well understood.

The layered convection model

The layered convection model involves separately convecting upper and lower mantle reservoirs. The upper mantle, above the 660-km discontinuity, is generally

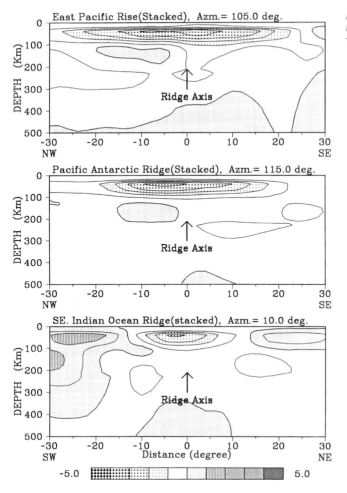

Figure 4.27 (a) S-wave velocity distribution across three ocean ridges. Note that very slow velocities are confined to very shallow depths.

equated with the geochemically depleted mantle reservoir (DM) and the lower mantle with the other mantle geochemical domains (Allegre, 1982; O'Nions, 1987). If DM is assumed to have formed chiefly by the extraction of continental crust from primitive mantle, model calculations based on Nd and Pb isotopes indicate that it must comprise between 35–50 per cent of the entire mantle.

One of the major lines of evidence used to support layered-mantle convection comes from the isotopic composition of rare gases in basalts (O'Nions, 1987; Allegre et al., 1995a). Non-radiogenic ^{3}He is enriched in many ocean-island basalts, and since these appear to be derived from mantle plumes, they probably come from an undegassed source in the deep mantle. Similar arguments can be made from Ar, Ne, and Xe isotopes. The simplest model to satisfy these constraints is a two-layer convective mantle with the upper layer (DM) strongly depleted in rare gases and LIL elements, and the lower layer, which contains enriched components (HIMU, EM), the site of generation of mantle plumes relatively enriched in ^{3}He.

Despite the rare gas data, layered convection has numerous difficulties, among the most robust of which are the following:

1 The aspect ratio (width/depth) of convection cells should be close to unity. This is not consistent with layered convection in which convection cells bottom out at 660 km, when horizontal measurements exceed 10^4 km as reflected by plate sizes.
2 Nd, Sr, and Pb isotopic data from oceanic basalts require not just two or three, but several ancient mantle sources as discussed above.
3 As we have seen, seismic-velocity studies indicate that descending lithosphere sinks into the lower mantle, a feature that would promote whole-mantle convection.
4 Mantle plumes appear to be derived from the lowermost mantle, another feature consistent with whole-mantle convection.

As pointed out above, however, layering in the mantle is not an all-or-none situation, and numerous factors influence the degree of layering in the mantle (Christensen,

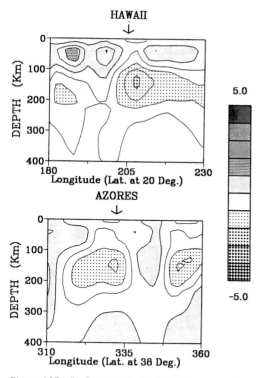

Figure 4.27 (b) S-wave velocity distribution across two hotspots, Hawaii and the Azores. See Figure 4.3 for explanation of patterns. Courtesy of Yu-Shen Zhang.

1995; Davies, 1995). Among the more important are the following:

1 *Temperature.* As temperature increases so does the Rayleigh number, and higher Rayleigh numbers tend to favour layered convection (Christensen and Yuen, 1985). This is caused by the fact that the sensitivity of an endothermic phase change, like that at the 660-km discontinuity, to retard descending slabs, increases with increasing Rayleigh number. As discussed in Chapter 5, this may have resulted in strongly layered convection in the Archean when mantle temperatures were higher.

2 *Internal heating.* Model calculations indicate that internal heating of a substance, as compared with bottom heating, increases the propensity for layered convection. Thus, the amount and distribution of radiogenic heat sources in the mantle can affect the type of convection.

3 *Exothermic phase changes.* As previously discussed, all of the phase changes recognized in the upper mantle, except the 660-km phase change, are exothermic, and exothermic phase changes enhance rather than retard the movement of descending slabs and rising plumes across the phase-change boundary. The possibility has already been mentioned that the spinel–perovskite and garnet–perovskite reactions with opposite P–T slopes may offset each other ther-

mally such that the net effect of the 660-km transition on slabs and plumes is near zero.

4 *Plate lengths.* Models of Zhong and Gurnis (1994) show that the mass flux between the upper and lower mantle increases strongly with total plate length. For a ratio of plate length to mantle thickness of 1, for instance, perfectly layered convection is predicted, whereas for a ratio of 5 perfect whole-mantle convection should exist.

5 *Mantle viscosity.* Penetration of slabs into the lower mantle is favoured by increasing viscosity with depth, a situation which is likely in the mantle.

6 *Slab dip angle.* As we have seen from tomographic profiles of subduction zones (Figure 4.18), shallow dip angles of descending slabs decrease the chances of penetrating the 660-km discontinuity without some holding time.

7 *Reaction rates.* Slow reaction rates of phase changes can temporarily retard descending slabs. This is due to the time it takes to heat up the coldest parts of the slab to temperatures necessary for the reaction to proceed.

8 *Change in chemical composition.* If, as some investigators suggest, there is an increase in the amount of Fe and perhaps Si at the 660-km discontinuity, some degree of layered convection seems necessary to preserve such a compositional boundary.

From the above considerations it appears that there is a real possibility that convection in the Earth involves partial layering, at least at certain times in the past. In fact, computer modelling by Tackley et al. (1994) suggests that descending slabs may be temporarily delayed at the 660-km discontinuity, where they accumulate and spread laterally. After some critical mass is reached, however, this plate graveyard overcomes the buoyancy of the phase boundary and sinks catastrophically into the lower mantle. This avalanche of plates, as it is commonly referred to, occurs only at Raleigh numbers > 10^6, and thus it may have been important early in the Earth's history. It should also be followed by a period of extensive plume generation after the dead slabs arrive in the D″ layer. At Raleigh numbers ≤ 10^6, a kind of 'leaky' two-layer convection is predicted, which is really time-delayed whole-mantle convection, and this may be the best approximation of what is going on in the mantle today.

Towards a convection model for the Earth

Figure 4.28 is a model for convection in the Earth proposed by Davies and Richards (1992), which accommodates the geophysical and geochemical data presented in this chapter. Although the model is for whole-mantle convection, it can easily accommodate delays in descending slabs at the 660-km discontinuity. As shown, ocean ridges are produced by shallow, passive upwellings of asthenosphere. The mantle increases in viscosity with depth leading to an approximate 10-fold decrease in

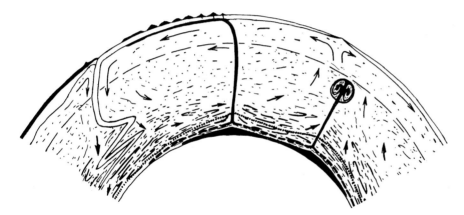

Figure 4.28   A convection model for the Earth. After Davies and Richards (1992).

convection rate in the lower mantle (as indicated by the short arrows which are vectors). Two plumes are shown rising from the D″ layer. One (on the left) has left a track of volcanoes on the surface of the ocean floor, beginning with a submarine plateau, which was derived from the plume head (long since disaggregated). The other plume is approaching the 660-km discontinuity. Compositional inhomogeneities (black squiggles) are more prevalent in the high viscosity region at the base of the mantle, and are few in number in the upper mantle. This greater homogeneity in the upper mantle (DM) reflects faster stirring here than at greater depths. Plumes are produced by buoyant masses of dead slabs (EM and HIMU sources) and surrounding lower mantle (FOZO) rising from the D″ layer. To explain the differences in ^{3}He/^{4}He ratios between DM and plume-FOZO sources, there must be either some remaining relatively undegassed (primitive) sources deep in the mantle, or a flux of primordial helium from the core.

Whether this model will withstand new data remains to be seen, but at least it is starting point that accommodates our present data base. As we shall see in the next chapter, during the early stages of Earth history, when the mantle was hotter, convective patterns and inhomogeneity distribution in the mantle may have been quite different from that shown in Figure 4.28.

The core

Introduction

Seismic velocity data indicate that the radius of the core is 3485 ± 3 km and that the outer core does not transmit S-waves (Jeanloz, 1990; Jacobs, 1992) (Figure 1.2). This latter observation is interpreted to mean that the outer core is in a liquid state. Supporting this interpretation are radio astronomical measurements of the Earth's normal modes of free oscillations. The inner core, with a radius of 1220 km, transmits S-waves at very low

velocities, suggesting that it is a solid near the melting point or partly molten. There is a sharp velocity discontinuity in P-wave velocity (0.8 km/sec; Figure 1.2) at the inner core boundary, and a low-velocity gradient at the base of the outer core. Results suggest that the top of the inner core attenuates seismic energy more than the deeper part of the inner core. Detailed analysis of travel-times of seismic waves reflected from and transmitted through the core indicate that the outer liquid core is relatively homogeneous and well-mixed, probably due to mixing by convection currents. Seismic data also suggest that relief on the core–mantle interface is limited to about 5 km. The viscosity of the outer core is poorly known with an estimated value of about 10^{-3} pa sec and in any case < 10^4 pa sec.

Three lines of evidence indicate that the core is composed chiefly of iron. First, the internal geomagnetic field must be produced by a dynamo mechanism, which is only possible in a liquid metal outer core. Second, the calculated density and measured body-wave velocities in the core are close to those of iron measured at appropriate pressures and temperatures. Third, iron is by far the most abundant element in the Solar System that has the seismic properties resembling those of the core.

Core temperature

An accurate knowledge of the temperature of the core is important for constraining both Earth's radioactivity budget and the generation of the magnetic field. Limits on the temperature profile in the core are determined by considering the solid–liquid interface at the inner core boundary from extrapolated phase equilibria and from high-pressure experimental studies of the melting of iron and iron alloys. Depending on the approach used, significant differences exist in the estimates of core temperature. For instance, estimates of the minimum temperature at the core–mantle boundary range from 2500 to 5000 °C. In order to generate the Earth's magnetic field, the higher temperatures seem necessary, and using the high-pressure melting point data of iron, a value of 4500 ± 500 °C seems a reasonable estimate for the temperature at core–mantle boundary (Jeanloz, 1990; Duba, 1992). Corresponding temperatures at the inner–outer core boundary and at the Earth's centre are 6200 and 6700 °C respectively, each with at least a 500 °C uncertainty (Figure 1.2).

The inner core

Although seismic data indicate that the inner core is solid, on geological timescales it may behave like a fluid, undergoing solid-state convection like the mantle. Support for this idea comes from seismic anisotropy in which the P-wave velocity is higher in the inner core along the Earth's rotational axis than in equatorial directions (Jeanloz, 1990). This anisotropy may be caused by alignment of iron crystals with the c axes of the crystals aligned parallel to the high velocities (N–S direction). The simplest explanation for the alignment of iron crystals in the inner core is solid-state convection.

Composition of the core

Although it is clear the core must be composed chiefly of iron, P-wave velocities and density of the outer core are about 10 per cent lower than that of iron (Jacobs, 1992). Thus, at least the outer core requires 5–15 per cent of one or more low atomic number elements to reduce its density. This also means there is not a single melting temperature at a given pressure, but that core must melt over an interval given by its solidus and liquidus. Just which alloying elements occur in the core is intimately linked with various models that have been proposed for the origin and evolution of the core. In addition to nickel, sulphur and oxygen are the two elements which have received most support from geochemical modelling as core contaminants. The presence of 5–10 per cent nickel in the core is supported by the composition of iron meteorites, which may represent fragments of core material from the asteroids (Chapter 7). Seismically, however, there is no reason for nickel to be in the core, since seismic velocities are essentially the same for iron and nickel. The common presence of iron sulphides in meteorites is also consistent with the presence of sulphur in the core, as is density modeling of Fe–S alloys at core pressures (Sherman, 1995). Also, a depletion of both sulphur and nickel in the mantle compared to cosmic abundances may reflect their preferential concentration in the core. If significant amounts of sulphur and nickel are in the core, they must have entered at low pressures as shown by experimental data, and thus would support core formation during the late stages of planetary accretion some 4.6 Ga (Newsom and Sims, 1991).

The possibility that oxygen is the dominant alloying agent in the core is based entirely on ultra-high-pressure experimental studies (Kato and Ringwood, 1989), and the relevant oxide phases are not stable at low pressures nor are they found in meteorites. Experimental results document that oxygen can alloy with molten iron at pressures generally in excess of 10 Gpa. Hence, in contrast to sulphur and nickel, alloying of oxygen would be expected well after the core began to form, and perhaps after planetary accretion was complete (Ito et al., 1995). The distinction between sulphur and oxygen as the primary low-atomic number dilutents in the core may be possible as the timing of core formation is better constrained.

It seems probable from comparing the seismically-deduced density of the inner core with measured densities of iron and iron alloys at high pressures that the inner core also cannot be composed of pure iron (Jephcoat and Olson, 1987). Like the outer core, it must contain a low-atomic number element, presumably sulphur or oxygen, but in smaller amounts of only 3–7 per cent. This would also be consistent with convection in the inner core, as mentioned above, since these elements would lower the effective viscosity enhancing the ability to convect.

Another approach to estimating the composition of the core is by using the composition of the bulk Earth and the primitive mantle as estimated from meteorite compositions, and determine core composition by difference. Results suggest that the total core contains about 7 per cent silicon and traces of both sulphur (2 per cent) and oxygen (4 per cent) (Allegre et al., 1995). A core of this composition must have formed at low pressures.

Age of the core

It is obviously not possible to date the core since samples have not been brought to the Earth's surface. However, there are indirect isotopic and geochemical arguments which can be used to constrain the timing of core formation. As discussed in Chapter 7, Pb isotopes provide an estimate of the age of Earth of 4.57–4.45 Ga, and it appears that at this time U and Pb were fractionated from each other. A significant proportion of the Earth's Pb (although still at trace concentrations) may have been incorporated in the core at this time, since it geochemically follows iron. If so, the Pb isotope age of the Earth suggests that separation of the core from the mantle occurred very early in Earth's history, perhaps beginning during the late stages of planetary accretion. ^{182}Hf–^{182}W isotope data from iron meteorites support early segregation of the core, probably in the first 50 My of planetary accretion (Harper and Jacobsen, 1996).

Also supporting early core formation is the lack of change in ratios of siderophile to lithophile elements in the mantle with time (Sims et al., 1990). If the core were to have grown gradually by blobs of liquid iron sinking through the mantle to the Earth's centre, the mantle should show progressive depletion of siderophile elements (like W, Mo, and Pb) relative to lithophile elements (like Ce, Rb, and Ba) with time. However, it does

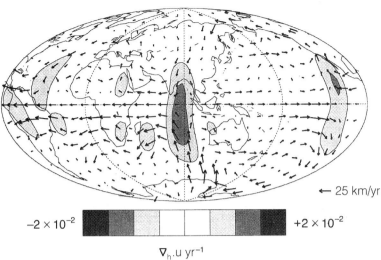

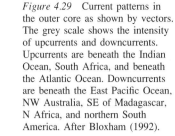

Figure 4.29 Current patterns in the outer core as shown by vectors. The grey scale shows the intensity of upcurrents and downcurrents. Upcurrents are beneath the Indian Ocean, South Africa, and beneath the Atlantic Ocean. Downcurrents are beneath the East Pacific Ocean, NW Australia, SE of Madagascar, N Africa, and northern South America. After Bloxham (1992).

not show such a trend. Even the oldest known basalts (~ 4 Ga), which carry a geochemical signature of their mantle source, have element ratios comparable to ratios in modern basalts.

Generation of the Earth's magnetic field

The geodynamo

It is well-known that the Earth's magnetic field is similar to that of a giant bar magnet aligned with the rotational axis of the planet. The magnetic lines of force trace curved paths, exiting near the South Pole and entering near the North Pole. Maps of the magnetic field at the surface show a secular variation that occurs over periods of decades to tens of thousands of years. Among the most striking changes in the last few thousand years are:

1 a decrease in the dipole component of the field
2 a westward drift of part of the field.

As we have seen earlier, the Earth's field can also reverse its polarity, and has done so many times in the geologic past. Most evidence strongly suggests that the Earth's magnetic field is generated in the fluid outer core by a dynamo-like action, although the details of how this occurs are very poorly understood (Jacobs, 1992).

Fluid motions in the outer core

In recent years it has been possible to map fluid motions in the outer core using the flow lines of the magnetic field (Bloxham, 1992). The maps show current movements at the core surface (as vectors) and regions of upwelling and downwelling known as core spots (Figure 4.29). There are four spots that contribute most strongly to the dipole component, which have been interpreted by Bloxham (1992) as the tops and bottoms of

two columns of liquid that appear to touch the inner core and run parallel to the Earth's rotational axis. Liquid iron may spiral down through the two columns creating a dynamo that concentrates magnetic flux within the columns. Magnetic field lines are 'frozen' in the iron liquid, and as it moves they are carried with the liquid, thus mapping current patterns. Results show two prominent cells of circulating fluid beneath the Atlantic basin, one south and one north of the Equator (Figure 4.29). Also, there is a region of intense upwelling near the Equator elongated in a N–S direction at 90 ° E longitude beneath the Indian Ocean, with a strong equatorial jet extending westward between the two circulating cells. This westward current may in fact explain the slow westward drift of the magnetic field in this area. The core spots appear to be produced by intense upward and downward flow of liquid. Evidence seems to be mounting that the fall in dipole component of the magnetic field is caused by the growth and propagation of the downward flux core spots beneath Africa and the Atlantic basin. This is supported by a correlation between the in tensity of the dipole component and the amount of lateral motion of these two spots in the last 300 years.

Fuelling the geodynamo

The energy for the geodynamo could be gravitational, chemical, or thermal, which in all cases is ultimately converted to heat that flows outward into the mantle (Jacobs, 1992). Remanent magnetism in rocks more than 3.5 Ga indicate that the Earth's geodynamo was in action by that time. Also, paleomagnetic studies show that the field intensity has never varied by more than a factor of two since the Archean, so the energy needed to drive the dynamo must have been available at about the same rate for at least 4 Gy.

There are two serious candidates for the geodynamo energy source:

1   thermal convection of the outer core
2   growth of the inner core.

In the first case, if the liquid outer core is stirred by thermal convection, most of the heat will be carried away either by convection or conduction and will not contribute to production of the magnetic field. However, thermal modelling indicates that up to 30 per cent of the geodynamo may be driven by thermal convection (Buffett et al., 1996). Growth of the inner core can supply gravitational energy in two ways:

1   latent heat of crystallization as iron crystallizes on the surface of the inner core
2   as metal accretes to the inner core, a lighter fraction is concentrated in the outer core and, because of its buoyancy, leads to compositionally-driven convection.

Calculations suggest that growth of the inner core is the most important stimulus for convection that drives the geodynamo in the outer core, and that the latter mechanism, i.e., compositionally-driven convection, dominates. Further, a large part of the gravitational energy released by growth of the inner core goes into electrical heat, which is released into the outer core and makes the compositionally-powered dynamo much more efficient.

What causes magnetic reversals?

Reversals in the Earth's magnetic field occur on different timescales and there are many aborted attempts to reverse when the field either immediately switched back or did not even reach the opposite polarity. At the other end of the timescale are superchrons, which record times when the field maintained the same polarity for periods of 20–50 My. It is very unlikely that these changes of polarity on very different timescales are the result of the same processes in the core (Jacobs, 1995). An incomplete reversal (or immediate swicthback), known as an excursion, probably results from some local instability at the core–mantle interface, and may not even be worldwide. Stable reversals, on the other hand, are always global in extent.

Reversals may be initiated either by changes at the core–mantle boundary or at the inner core boundary, and in both cases it appears that some physical or chemical process arising from an energy source independent of the source that powers the geodynamo actually initiates the reversal. For instance, instabilities may be generated at the core–mantle boundary by heat loss producing cooler, more dense blobs of molten iron that sink and destabilize the main convection in the outer core. Alternatively, hot plumes rising from the inner core boundary may have the same effect. Hollerbach and Jones (1993) suggest that reversals may be caused by oscillations of the magnetic field in the outer core. Their modelling concludes that although the field oscillates strongly in the outer core, only weak oscillations should occur near the core surface, and reversals may be caused when these

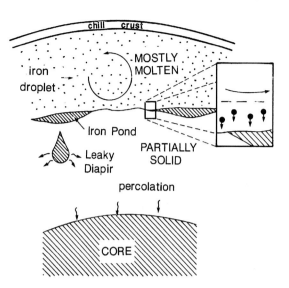

Figure 4.30 Possible core segregation processes. Molten metal droplets may have sunk rapidly in a silicate magma ocean, large blobs of molten iron may have sunk as diapirs, and alloying of oxygen with iron in the lower mantle may have allowed downward percolation of liquid alloy into the core. After Newsom and Sims (1991).

weak oscillations exceed some threshold value. This would occur only when the size ratio of the inner to outer core is greater than 0.25 and thus, before the inner core crystallized, there should have been no reversals. A period of rapid growth of the inner core may occur in response to cooling of the outer core, initiated by sudden heat loss at the core–mantle boundary. This heat loss could also give rise to numerous mantle plumes, sometimes referred to as 'superplumes', a topic we will return to in Chapter 6.

Origin of the core

Segregation of iron in the mantle

Many models have been proposed to explain how molten iron sinks to the Earth's centre as the melting point of metallic iron or iron alloys is reached in the mantle. Although molten iron could collect into layers and sink as small diapirs, the high surface tension of metallic liquids relative to solid silicates should cause melt droplets to collect at grain boundaries, rather than drain into layers (Newsom and Sims, 1991). However, at high pressures (> 6 Gpa) oxygen should be dissolved in liquid iron and, if so, it will lower the surface tension such that the liquid can collect into layers and sink as diapirs (Figure 4.30). This model obviously favours oxygen as the diluent in the core. If the Earth had a deep silicate magma ocean soon after accretion (see Chapter 7), droplets of liquid iron should exsolve from the silicate liquid and sink rapidly to the core (Figure 4.30). Alternatively,

if the magma ocean were relatively shallow, metal droplets could accumulate at the bottom of the magma ocean,
eventually coalescing into large blobs that could either
sink as diapirs or percolate downwards as an oxygen-
rich liquid.

In any case, the metal segregation in the Earth must
occur rapidly, probably during the late stages of planetary accretion. Indeed, it may have been a catastrophic
event that gave off a large amount of heat that partially
melted the mantle, or further melted the mantle if a
magma ocean already existed.

Siderophile element distribution in the mantle

Until very recently, it was generally assumed that the
concentrations of siderophile elements in the mantle are
higher than predicted for models of core segregation
(Newsom and Sims, 1991). Because of their very strong
tendency to follow iron, most of these elements should
have been scavenged by liquid iron as it sank to the
centre of the Earth to form the core. Yet the concentrations of many of these elements are similar to that of
primitive mantle. Murthy (1991), however, has shown
that this apparent depletion may not be real. In a molten
planet the size of the Earth, increasing pressure increases
the melting point of metal and silicates significantly, so
a large part of the Earth would be at very high temperatures (2000–4000 °C). Thus, the distribution of trace
elements between liquid and solid would be governed
by high-temperature distribution coefficients, and not
the low-temperature distribution coefficients that have
been used in all previous modelling. The siderophile
element distributions in the mantle calculated with high-
temperature distribution coefficients is similar to the
element distributions in mantle xenoliths. Thus, the problem of excess siderophile elements in the mantle goes
away when appropriate high-temperature modelling is
done.

Growth and evolution of the core

Although high-pressure phase equilibria in the Fe–O and
Fe–S systems is still not precisely known, it is informative using the available data to track the growth of the
core during the first 50 My of Earth history. During the
late stages of planetary accretion, steepening geotherms
in the Earth should intersect the melting curve of iron at
relatively shallow depths in the mantle. Continued heating raises the geotherms well above the melting point of
iron, producing widespread spherically-distributed layers of molten iron in the mantle which, because of their
gravitational instability, begin to sink to the centre of
the Earth to form the core. Alternatively, if a silicate
magma ocean existed at this time, iron droplets would
readily sink directly to the core (Figure 4.30). Some
time within the first 100 My of Earth history (about 4.5
Ga), a Mars-sized planet may have collided with the

Earth, forming the Moon, as discussed in Chapter 7.
During this collision, most of the mantle of the impactor
should have accreted to the Earth and its core probably
penetrated the Earth's core (Benz and Cameron, 1990),
possibly transferring iron to the Earth's core. As the
Earth began to cool, iron in the core began to crystallize
and the inner core began to form. Just when this process
started is presently unknown, although it was at the latest some time during the Early Archean.

As the core continues to cool in the future, the inner
core should continue to grow at the expense of the outer
core. Because the low-atomic numbered diluting element (S or O) is preferentially partitioned into the liquid
phase, the outer core will become progressively enriched
in this element with time, and thus the melting point of
the outer core should drop until the eutectic in the system is reached. The final liquid that crystallizes as the
outermost layer of the core, some time in the distant
geological future, should be this eutectic mixture.

Summary statements

1   Major seismic discontinuities are recognized in the
    mantle at 410 and 660 km and a seismic low velo-
    city zone, the LVZ, occurs between 50 and 300 km
    depth except beneath Archean shields where it
    appears to be absent.
2   Seismic tomography indicates the presence of large
    heterogeneities in the mantle, many of which
    extend across the 660-km discontinuity and some
    may extend to the core–mantle boundary. Both the
    Earth's surface and the core–mantle interface are
    upwarped in regions of low seismic velocities in
    the lower mantle.
3   Deviations in the geoid and the shape of the core–
    mantle boundary, hotspot density, lower mantle
    seismic velocities, and mantle geochemical anoma-
    lies suggest large-scale horizontal variations in tem-
    perature in the deep mantle. Relatively hot regions
    may be large upwellings bringing lower mantle
    material to shallower levels.
4   Unlike the asthenosphere and mesosphere, which
    cool by convection maintaining an adiabatic
    gradient, the lithosphere cools by conduction and
    temperatures change rapidly with depth and with
    tectonic setting.
5   Seismic-wave velocity measurements, high-pressure
    experimental studies, and studies of mantle xenoliths
    indicate that the mantle lithosphere and upper man-
    tle are composed chiefly of ultramafic rocks in which
    olivine is the dominant constituent.
6   Depletion in the most incompatible elements in
    oceanic basalt sources reflects removal of basaltic
    liquids enriched in these elements, perhaps early in
    Earth history. In contrast, lithosphere mantle shows
    prominent enrichment in the most incompatible
    elements, which were probably introduced by man-
    tle metasomatism.

7   Isotopic and geochemical data from mantle xenoliths indicate that the thick mantle lithosphere beneath Archean shields formed during the Archean and that it is chemically distinct from post-Archean lithosphere.

8   S-wave anisotropy in continental lithosphere correlates with directions of modern plate motions suggesting that the anisotropy is not a Precambrian feature, but results from resistive drag along the base of the lithosphere.

9   The oceanic lithosphere thickens as it spreads and cools by transfer of mass from the convecting asthenosphere. In contrast, post-Archean subcontinental lithosphere may grow by basal accretion of spent mantle plumes, and Archean lithosphere may include both spent plume material and partially subducted slabs.

10   The LVZ at the base of the lithosphere, which probably results from incipient melting, plays a major role in plate tectonics, providing a relatively low-viscosity region upon which lithospheric plates can slide with very little friction.

11   The 410-km discontinuity appears to be caused by the inversion of olivine to the high-pressure phase wadsleyite. Wadsleyite in turn inverts to spinel at 500–550 km.

12   The 660-km discontinuity is probably caused by the phase change of Mg spinel to Mg perovskite + magnesiowustite. However, because this reaction is endothermic and may impede slabs from immediately sinking into the deep mantle; there is no evidence from seismic tomography that slabs collect at the 660-km discontinuity.

13   The D″ layer is a region of the mantle within a few hundred kilometres of the core where seismic-velocity gradients are anomalously low and show extreme vertical and lateral heterogeneities. These heterogeneities appear to be caused by temperature variations, compositional changes, and mineralogical phase changes. Movement of dregs (denser material) could account for some of the large-scale seismic-wave velocity variations and variations in thickness of D″.

14   Theoretical and laboratory models show that mantle plumes are composed of a head of buoyant material whose diameter is much larger than the tail. Evidence strongly suggests that mantle plumes are produced in the D″ layer and may entrain up to 90 per cent of their starting mass, chiefly from the lower mantle. Most plumes should pass through the spinel–perovskite phase boundary at 660 km depth unimpeded.

15   Supercontinents appear to form over mantle downwellings, which change to upwellings partly in response to the insulation effect of supercontinents. This leads to fragmentation and dispersal of supercontinents, which travel towards mantle downwellings. The role of mantle plumes in supercontinent break up is not clearly understood.

16   Geochemical data from ocean basalts and theoretical models show that mantle heterogeneities on scales of a few metres to $> 10^4$ km can be preserved for times of the order of 10^9 years.

17   At least four, and perhaps as many as six isotopic components exist in the mantle. These are a depleted component (DM) occurring in the upper mantle; HIMU, distinguished by its high U/Pb ratio, may represent subducted oceanic crust; EM1, which has relatively low Pb and moderate Sr isotopic ratios and may represent depleted oceanic lithosphere or subducted ancient sediments; and EM2, which has intermediate Pb and high Sr isotopic ratios, may be subducted young continental sediments. If EM1, EM2, and HIMU represent oceanic lithosphere, young sediments, and recycled oceanic crust respectively, there is a possibility that these three components are the sediment–basalt–ultramafic suite we see in ophiolites.

18   The key events controlling compositional variations in the Earth are early core formation, followed by gradual or episodic extraction of continental crust to leave at least part of the mantle depleted in incompatible and siderophile elements (DM, FOZO). Other mantle components (HIMU, EM1, EM2) are in plumes and appear to reflect subducted inhomogeneities that were not sufficiently remixed into the convecting mantle to lose their geochemical and isotopic signatures.

19   The low-velocity anomalies under ocean ridges indicate that ridges are shallow passive features, not directly related to deep convection. Geophysical and geochemical features of the mantle are best explained by whole-mantle convection.

20   Because the outer core does not transmit S-waves, it must be in a liquid state, whereas the inner core is a solid near its melting point. Convection probably occurs in both the outer and inner core. Although the core must be composed chiefly of iron (with some nickel), low P-wave velocities require the presence of one or more low atomic number elements, probably sulphur or oxygen. Pb model ages and Hf-W isotopic ages indicate that the core formed in the first 100 My of Earth history, and probably before the end of planetary accretion.

21   Evidence suggests that the Earth's magnetic field is generated in the fluid outer core by a poorly understood dynamo-like action. Energy for the geodynamo appears to come from growth of the inner core. Reversals in the magnetic field may be caused by instabilities at the core–mantle boundary, which release blobs of molten iron that sink and destabilize convection in the outer core; hot plumes rising from the inner core boundary that have the same effect; or by occasional large oscillations of the magnetic field in the outer core.

22   The core formed as iron segregated in the mantle by settling of molten droplets in a silicate magma ocean, sinking of diapirs of molten iron, and/or by

downward percolation of iron–oxygen alloy liquids. Some of the metal in the core may have come from a Mars-sized planet that collided with the Earth during core formation. As the Earth began to cool, the inner core began to form as iron in the core crystallized and sank.

Suggestions for further reading

Campbell, I. H. and Griffiths, R. W. (1992). The changing nature of mantle hotspots through time: Implications for the chemical evolution of the mantle. *J. Geol.*, **92**, 497–523.

Christensen, U. R. (1995). Effects of phase transitions on mantle convection. *Ann. Rev. Earth Planet. Sci.*, **23**, 65–87.

Davies, G. F. and Richards, M. A. (1992). Mantle convection. *J. Geology*, **100**, 151–206.

Hauri, E. H., Whitehead, J. A. and Hart, S. R. (1994). Fluid dynamic and geochemical aspects of entrainment in mantle plumes. *J. Geophys. Res.*, **99**, 24 275–24 300.

Jacobs, J. A. (1992). *Deep Interior of the Earth*. London, Chapman & Hall, 167 pp.

Stacey, F. (1992). *Physics of the Earth* (third edition). Brisbane, Australia, Brookfield Press, 513 pp.

Zhang, Y. and Tanimoto, T. (1993). High-resolution global upper mantle structure and plate tectonics. *J. Geophys. Res.*, **98**, 9793–9823.

Chapter 5

Crustal and mantle evolution

Introduction

One of the unique features of the Earth is its crust. None of the other terrestrial planets seems to have a crust similar to that of the Earth, and the reason for this is related to plate tectonics, which in turn is related to the way the mantle cools and how it has evolved. Thus, most investigators now agree that the history of the Earth's crust and mantle are closely related, and that many of the features we find in the crust are controlled by processes in the mantle. In this chapter we will show just how closely the evolution of the crust is tied to the that of the mantle, and review how this dynamic system has evolved as planet Earth has cooled over the last 4.6 Ga. As a starting point, let us consider the thermal history of the Earth.

The Earth's thermal history

There are two lines of evidence indicating that the mantle was hotter during the Archean. The most compelling evidence is that the present-day heat loss from the Earth is approximately twice the amount of heat generated by radioactive decay, which requires that the excess heat come from cooling of the Earth. U, Th, and K isotopes provide most of the Earth's radiogenic heat (Chapter 2) and, from the decay rates of these isotopes, it would appear that heat production in the Archean was three to four times higher than today (Richter, 1988). A second argument for higher Archean mantle temperatures is the presence of high-Mg komatiites in Archean greenstones which, as discussed later, require higher mantle temperatures for their generation. Estimates of the average Archean mantle temperature at 3 Ga vary from about 100 to 300 °C higher than modern temperatures (Figure 5.1).

It is commonly thought that the most realistic approach to calculating the Earth's thermal history is by using parameterized convection models in which expres-

sions are solved that relate convective heat transport to the temperature difference between the surface and interior of a convecting cell in the mantle and to the viscosity distribution in that cell. Results for whole-mantle convection show that mantle temperature, heat production, and Rayleigh number (see Chapter 4) decrease with time, while mantle viscosity increases with time, all in accordance with a cooling Earth (McGovern and Schubert, 1989) (Figure 5.1). Early in the Earth's history, when the mantle was very hot, viscosity was low and convection very rapid, perhaps chaotic, as dictated by the high Rayleigh numbers. During this period of about 500 My the Earth cooled rapidly, followed by gradual cooling of about 100 °C/Gy to the present time.

The Earth's primitive crust

Origin of the first crust

With an ever-increasing data base, several outstanding questions related to the origin of the Earth's early crust can be addressed:

1 Was the first crust of local or global extent?
2 When and by what process did the first crust form?
3 What was the composition of the early crust?
4 When and how did oceanic and continental crustal types develop?

As discussed below, the oldest-preserved fragments of continental crust at 4.0–3.8 Ga are chiefly of tonalitic gneisses containing fragments of komatiite and basalt (amphibolite), some of which may be remnants of the early oceanic crust. Model lead ages of the Earth and isotopic ages from meteorites suggest that the earliest terrestrial crust may have formed just after or during the late stages of planetary accretion at about 4.5 Ga. Although the original extent of Early Archean continental fragments is not known, they comprise less than 10 per cent of preserved Archean crust (Figure 5.2). The spar-

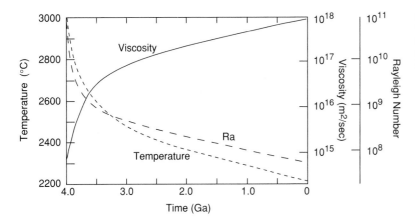

Figure 5.1 Average temperature, viscosity, and Rayleigh number of the mantle with time. Modified after McGovern and Schubert (1989).

sity of rocks older than 3.0 Ga may be related to losses resulting from recycling of early crust into the mantle as discussed later in the chapter. It may be possible to learn more about the first terrestrial crust by comparison with the lunar crust and crusts of other terrestrial planets. The lunar highlands, for instance, are remnants of an early lunar crust (~ 4.5 Ga), which covered most or all of the Moon's surface (Taylor, 1982). Studies of topographic features and cratering histories of Mercury and Mars also suggest the preservation of widespread primitive crusts. If the early history of the Earth was similar to these planets, it also may have had an early crust that covered most of its surface. Because remnants of this crust appear not to be preserved in the continents, if such a crust existed on the Earth soon after planetary accretion, it must have been destroyed by continuing impacts on the surface and/or by recycling into the mantle.

Theories for the origin of the Earth's crust fall into three broad categories:

1 heterogeneous accretion of the Earth
2 impact models
3 terrestrial models.

In the heterogeneous accretion model, the last compounds to condense from the Solar Nebula produce a thin veneer on planetary surfaces rich in alkali and other volatile elements, which may form or evolve into the first crust. A major problem with this model is that many non-volatile elements, such as U, Th, and REE, which are concentrated in the core and lower mantle in a heterogeneously-accreted Earth, are today concentrated in the crust. This necessitates magmatic transfer from within the Earth, thus producing a crust of magmatic origin. Also, as discussed in Chapter 7, heterogeneous accretion of the Earth is faced with other geochemical problems.

Several models have been proposed for crustal origin either directly or indirectly involving impact of accreting objects. All call upon surface impacting that leads to melting in the mantle producing either mafic or felsic magmas that rise to the surface forming a crust. Large impacts may have produced mare-like craters on the

terrestrial surface that were filled with impact-produced magmas (Grieve, 1980). If the magmas or their differentiation products were felsic, continental nuclei may have formed and continued to grow by magmatic additions from within the Earth. Alternatively, if the impact craters were flooded with basalt, they may have become oceanic crust. Although initially attractive, impact models are faced with many difficulties in explaining crustal origin. For instance, most or all of the basalts that flood lunar maria formed later than the impacts and are not related directly to impacting. Also, only relatively small amounts of magma were erupted into lunar mare craters. Perhaps the most significant problem with the lunar mare analogy, is that mare basins formed in still older anorthositic crust.

Models that call upon processes operating within the Earth have been most popular in explaining the origin of the Earth's early crust. The fact that textures and geochemical relationships indicate that the early anorthositic crust on the Moon is a product of magmatic processes favours a similar origin for the Earth's earliest crust. It is likely that sufficient heat was retained in the Earth, after or during the late stages of planetary accretion, to melt the upper mantle partially or entirely. Complete melting of the upper mantle would result in a magma ocean (see Chapter 7) which, upon cooling, should give rise to a widespread crust. Even without a magma ocean, extensive melting in the early upper mantle should produce large quantities of magma, some of which rises to the surface forming an early basaltic crust. Whether plate tectonics was operative at this time is not known. However, some mechanism of plate creation and recycling must have been operative to accommodate the large amounts of heat loss and vigorous convection in the early mantle.

Composition of the primitive crust

Numerous compositions have been suggested for the Earth's earliest crust. Partly responsible for diverging opinions are the different approaches to estimating composition. The most direct approach is to find and describe a relict of the primitive crust. Although some investigators have not given up on this approach, the chances of a remnant of this crust being preserved seem very small. Another approach is to deduce the composition from studies of the preserved Archean crust. However, compositions and field relations of rock types in the oldest preserved Archean terranes may not be representative of the earliest terrestrial crust. Another approach has been to assume that the Earth and Moon have undergone similar early histories and hence to go to the Moon, where the early record is well-preserved, to determine the composition of the Earth's primitive crust. Geochemical models based on crystal-melt equilibria and a falling geothermal gradient with time have also been used to constrain the composition of the early terrestrial crust.

Felsic models

Some models for the production of a primitive felsic or andesitic crust rely on the assumption that low degrees of partial melting in the mantle will be reached before high degrees, and hence felsic magmas should be produced before mafic ones. Other models call upon fractional crystallization of basalt to form andesitic or felsic crust. Shaw (1976) proposed that the mantle cooled and crystallized from the centre outwards, concentrating incompatible elements into a near-surface basaltic magma layer. This layer underwent fractional crystallization, resulting in the accumulation of an anorthositic scum in irregular patches and in residual felsic magmas that crystallized to form the first stable crust by about 4 Ga.

First, the high heat generation in the Early Archean probably produced large degrees of melting of the upper mantle, and it is unlikely therefore that felsic melts could form directly. Although felsic or andesitic crust could be produced by fractional crystallization of basaltic magmas, this requires a large volume of basalt, which probably would have formed the first crust.

Anorthosite models

Studies of lunar samples indicate that the oldest rocks on the lunar surface are gabbroic anorthosites and anorthosites of the lunar highlands, which are remnants of a widespread crust formed between 4.5 and 4.4 Ga (Taylor, 1982). This primitive crust appears to have formed in response to catastrophic heating that led to widespread melting of the lunar interior and production of a voluminous magma ocean. As the magma ocean rapidly cooled and underwent fractional crystallization, pyroxenes and olivine sank and plagioclase (and some pyroxenes) floated, forming a crust of anorthosite and gabbroic anorthosite. Impact disrupted this crust and produced mare craters, and these craters were later filled with basaltic magmas (3.9–2.5 Ga).

As discussed later in the chapter, most Early Archean anorthosites are similar in composition (i.e., high An

content, associated chromite) to lunar anorthosites and not to younger terrestrial anorthosites. It is clear from field relationships, however, that these Archean anorthosites are not remnants of an early terrestrial crust, since they commonly intrude tonalitic gneisses. If, however, the Earth had an early melting history similar to that of the Moon, the first crust may have been composed dominantly of gabbroic anorthosites. In this scenario, preserved Early Archean anorthosites may represent the last stages of anorthosite production, which continued after both mafic and felsic magmas were also being produced.

The increased pressure gradient in the Earth limits the stability range of plagioclase to depths considerably shallower than on the Moon. Experimental data suggest that plagioclase is not a stable phase at depths > 35 km in the Earth. Hence, if such a model is applicable to the Earth, the anorthositic fraction, either as floating crystals or as magmas, must find its way to very shallow depths to be preserved. The most serious problem with the anorthositic model, however, is related to the hydrous nature of the Earth. Plagioclase will readily float in an anhydrous lunar magmatic ocean, but even small amounts of water in the system cause it to sink (Taylor, 1987; 1992). Hence in the terrestrial system, where water was probably abundant in the early mantle, an anorthositic scum on a magma ocean would not form.

Basalt models

In terms of our understanding of the Earth's early thermal history and from the geochemical and experimental data base related to magma production, it seems likely that the Earth's primitive crust was mafic in composition. If a magma ocean existed, cooling would produce a widespread basaltic crust. Without a magma ocean (or after its solidification), basalts also may have composed an important part of the early crust. The importance of basalts in Early Archean greenstone successions attests to its probable importance on the surface of the Earth prior to 4 Ga.

The Earth's oldest rocks

The oldest-preserved rocks occur as small, highly-deformed terranes tectonically incorporated within Archean crustal provinces (Figure 5.2). These terranes are generally < 500 km across, and are separated from surrounding crust by shear zones. Although the oldest-known rocks on Earth are the 4.0-Ga Acasta Gneisses from northwest Canada, the oldest minerals are detrital zircons from the 3-Ga Mt Narryer quartzites in Western Australia. Detrital zircons from these sediments have U–Pb ion probe ages ranging from about 3.5 Ga to 4.3 Ga, although only a small fraction of the zircons are > 4.0 Ga (Froude et al., 1983). Nevertheless, these old zircons are important in that they indicate the presence of felsic sources, some of which contained domains that were 4.3 Ga. These domains may have been remnants of continental crust, although the lateral extent of any given domain may have been much smaller than microcontinents such as Madagascar and the Lord Howe Rise.

The oldest, isotopically-dated rocks on Earth are the Acasta gneisses in north-west Canada (Figure 5.3). These gneisses are a heterogeneous assemblage of highly-deformed tonalites, tectonically interleaved on a centimetre scale with amphibolites, ultramafic rocks, granites and, at a few locations, metasediments (Bowring et al., 1989; 1990). Acasta amphibolites appear to represent basalts and gabbros, many of which are deformed dykes and sills. The metasediments include calc-silicates, quartzites, and biotite-sillimanite schists. The rare occurrence of the assemblage tremolite–serpentine–talc–forsterite in ultramafic rocks indicates that the metamorphic temperature was in the range of 400–650 °C. U–Pb zircon ages from the tonalitic and amphibolite fractions of the gneiss yield ages of 4.03–3.96 Ga with some components, especially the pink granites, having ages as low as 3.6 Ga. Thus, it would appear that this early crustal segment evolved over about 400 My, and developed a full range in composition of igneous rocks from mafic to K-rich felsic types. Because of the severe deformation of the Acasta gneisses, the original field relations between the various lithologies is not well-known. However, the chemical compositions of the Acasta rocks are very much like less deformed Archean greenstone–TTG assemblages, suggesting a similar origin and tectonic setting.

The largest and best-preserved fragment of Early Archean continental crust is the Itsaq Gneiss Complex in Southwest Greenland (Nutman et al., 1993; Nutman et al., 1996). In this area, three terranes have been identified, each with its own tectonic and magmatic history, until their collision at about 2.7 Ga (Figure 5.4) (Friend et al., 1988). The Akulleq terrane is dominated by the Amitsoq TTG complex, most of which formed at 3.9–3.8 Ga and underwent high-grade metamorphism at 3.6 Ga. The Akia terrane in the north comprises 3.2–3.0 Ga tonalitic gneisses that were deformed and metamorphosed at 3.0 Ga, and the Tasiusarsuaq terrane, dominated by 2.9–2.8-Ga rocks, was deformed and metamorphosed when the terranes collided in the Late Archean. Although any single terrane records < 500 My of pre-collisional history, collectively the terranes record over 1 Ga of history before their amalgamation in the Late Archean. Each of the terranes also contains remnants of highly-deformed supracrustal rocks. The most extensively studied is the Isua sequence in the Isukasia area in the northern part of the Akulleq terrane (Figure 5.4). Although highly altered by submarine metasomatism, this succession comprises from bottom to top: basalts and komatiites with intrusive ultramafics interbedded with

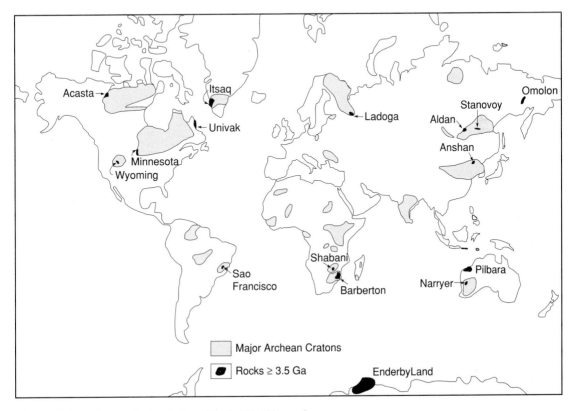

Figure 5.2 Map showing the distribution of the Earth's oldest rocks.

Figure 5.3   The 4.0-Ga Acasta gneisses from the Archean Slave province, north-west of Yellowknife, NWT, Canada. This outcrop, with the founder Sam Bowring, shows interlayered TTG and granite (light bands).

banded iron formation; intrusive sheets of tonalite and granite; basalts and ultramafic rocks; mafic volcanigenic turbidites; and basalts with interbedded banded iron formation (Rosing et al., 1996). The Isua succession is similar to mafic platform successions in Late Archean greenstones, again suggesting that greenstone tectonic settings were around by 3.9 Ga.

The Pilbara craton in Western Australia also comprises a group of accreted terranes, the most widespread of which is the Warrawoona terrane, which formed between 3.5 and 3.2 Ga. Although extensively altered by submarine processes, the Warrawoona sequence is probably the best preserved Early Archean greenstone (Barley, 1993; Krapez, 1993). It rests unconformably on an older greenstone–TTG complex with a U–Pb zircon age of about 3.5 Ga (Buick et al., 1995). This is important because it indicates that not only was the Warrawoona deposited on still older continental crust, but land emerged above sea level by 3.46 Ga in this region. The Warrawoona sequence is a thick succession of submarine volcanics, both tholeiites and calc-alkaline types, and volcaniclastic and chemical (chert, BIF) sediments largely indicative of shallow marine environments, all of which are multiply-deformed and intruded with plutons ranging from tonalite to granodiorite in composition.

From the base upwards it has three groups (Figure 5.5):

1   the Talga Talga Group composed chiefly of gabbro, pillowed basalts, and minor cherty sediments
2   the Duffer Group, a sequence of calc-alkaline andesitic to felsic volcanic and volcaniclastic sediments
3   the Salgash Group, comprising high-Mg basalts interbedded with cherts and volcaniclastic sediments.

Distinctive chert horizons, like the Marble Bar chert, can be traced for large distances in the eastern Pilbara. The Talga Talga Group is a submarine Archean mafic plain succession, although not a submarine plateau because it was erupted on continental crust. The overlying Duffer and Salgash Groups contain lithologic packages similar to those associated with modern continental margin arcs.

The Barberton greenstone in southern Africa is one of the most studied Early Archean greenstones. Together with coeval TTG plutons, the Barberton succession formed at 3.55–3.2 Ga (Kamo and Davis, 1994; Kroner et al., 1996). It includes four tectonically juxtaposed terranes, with similar stratigraphic successions in each terrane (Lowe, 1994). Each succession (known as the Onverwacht Group) begins with submarine basalts and komatiites of the Komati Formation (Figure 5.5), an

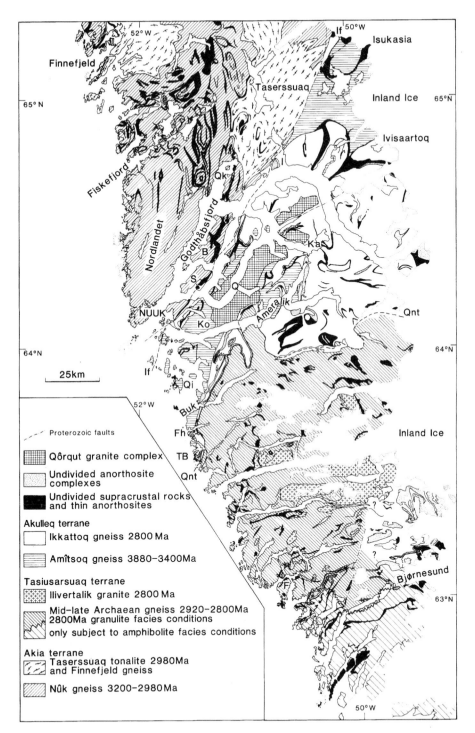

Figure 5.4 Generalized geologic map of the Nuuk region in Southwest Greenland showing three Early Archean terranes. Courtesy of Clark Friend.

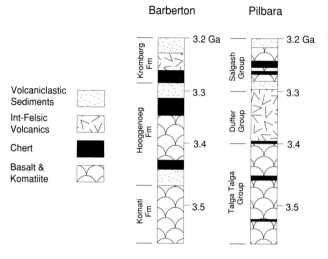

Figure 5.5 Stratigraphic sections of Early Archean greenstones from the Barberton greenstone belt in South Africa and the Warrawoona sequence in the Pilbara craton, Western Australia.

Archean mafic plain succession that could represent a submarine plateau. Overlying the mafic plain succession are the Hooggenoeg and Kromberg Formations, a suite of felsic to basaltic submarine volcanics, fine-grained volcaniclastic sediments, and cherts, possibly representing an oceanic arc. The terminal Moodies Group (not shown in Figure 5.5), which includes orogenic sediments, may have been deposited during amalgamation of the four terranes just after 3.2 Ga.

Unlike most Late Archean greenstones, many of which evolved in less than 50 My, Early Archean greenstones had long histories of > 500 My before colliding and stabilizing as part of a continent (Condie, 1994). In the Barberton greenstone, individual cycles lasted 50–80 My and included rifting and eruption of thick successions of mafic flows, magmatic quiescence with deposition of chemical sediments, and finally crustal thickening caused by intrusion of TTG plutons. Unlike Late Archean terranes, which accreted into cratons almost as they formed, Early Archean terranes appear to have bounced around like dodgem cars for hundreds of millions of years. Why they did not accrete into continents is an important question that still remains unresolved. Perhaps there were too few of these terranes and collisions were infrequent. Alternatively, most of these terranes may have been recycled into the mantle before having a chance to collide and make a continent.

Crustal origin

The probable characteristics of the early oceanic and continental crust are summarized in Table 5.1. Oceanic crust is generated today at ocean ridges by partial melting of the upper mantle, and there is no reason to believe that early oceanic crust did not form in the same way. Just when the first oceanic crust formed is unknown because it was undoubtedly recycled into the mantle, but it is probable that it crystallized from a

Table 5.1 Characteristics of the Earth's early crust

	Oceanic crust	*Continental crust*
First appearance	~ 4.5 Ga	≥ 4.3 Ga
Where formed	Ocean ridges	Submarine plateaux
Composition	Basalt	TTG
Lateral extent	Widespread, rapidly recycled	Local, rapidly recycled
How generated	Partial melting of ultramafic rocks in upper mantle	Partial melting of wet mafic rocks with garnet left in residue

TTG = tonalite–trondhjemite–granodiorite.

magma ocean soon after planetary accretion. Because of the greater amount of heat in the Archean upper mantle, oceanic crust may have been produced 4–6 times faster than at present, and thus would have been considerably thicker than modern oceanic crust. Like modern oceanic crust, however, it was probably widely distributed on the Earth's surface.

The Earth may be the only terrestrial planet with continental crust. If so, what is unique about the Earth which gives rise to continents? Two factors immediately stand out:

1   the Earth is the only planet with significant amounts of water
2   it may be the only planet on which plate tectonics has been operative.

An important constraint on the origin of Archean continents is the composition of Archean TTG (tonalite–trondhjemite–granodiorite). Experimental data favour an origin for Archean tonalites by partial melting of amphibolite or eclogite in the presence of significant amounts of water (Rapp and Watson, 1995). Without the water, magmas of TTG compositions cannot form.

The production of large amounts of Archean continental crust require subduction of large quantities of hydrated basalt and large quantities of water. Hence, with the possible exception of Venus, the absence of continental crust on other terrestrial planets may reflect the small amounts of water and the absence of plate tectonics on these planets.

It is likely that the earliest felsic crust developed from mafic submarine plateaux, either by partial melting of the thickened mafic roots of the plateaux or by melting of slabs subducted around their margins. In either case, the resulting TTG magmas rise and underplate mafic and komatiitic rocks, some of which are preserved today in greenstone belts. True granites do not appear in the geologic record until about 3.2 Ga and do not become important until after 2.6 Ga. Geochemical and experimental data suggest that these granites are produced by partial melting or fractional crystallization of TTG (Condie, 1986), and it is not until TTG is relatively widespread that granites appear in the geologic record. Thus, the story of early continental crust is the story of three rock types: basalt, tonalite, and granite listed in the general order of appearance in the Archean geologic record. Field relations in most Archean granite–greenstone terranes also indicate this order of relative ages. It would appear that Early Archean basalts were hydrated by seafloor alteration and later they partially melted, either in descending slabs or in thickened root zones of submarine plateaux, giving rise to TTG magmas. TTG, in turn, was partially melted or fractionally crystallized to produce granites.

Thus, unlike the first oceanic crust, which probably covered much or all of the Earth's surface, the first continental crust was probably of more local extent associated with subduction zones and submarine plateaux. Now that we have continental crust, the next question is that of how and at what rate did continents grow?

How continents grow

Although most investigators agree that production of post-Archean continental crust is related to subduction, just how continents are produced in arc systems is not well understood. Oceanic terranes such as island arcs and submarine plateaux may be important building blocks for continents as they collide and accrete to continental margins. However, the fact that these terranes are largely mafic (Kay and Kay, 1985; DeBari and Sleep, 1991), yet upper continental crust is felsic in composition, indicates that oceanic terranes must have undergone dramatic changes in composition to become part of the continents. Although details of the mechanisms by which mafic oceanic terranes evolve into continental crust are poorly known, delamination of lower crust during or soon after collision may play a role. Perhaps colliding oceanic terranes partially melt and felsic magmas rise to the upper continental crust leaving a depleted mafic restite in the lower crust. Because there is no apparent seismic

evidence for thick, depleted continental roots beneath recently accreted crust, if this mechanism is important, the depleted root must delaminate and sink into the mantle, perhaps during plate collisions. Supporting the possibility of modern collisional delamination is a major vertical seismic gap in the Western Mediterranean basin. Tomographic images in this region indicate the presence of a high-velocity slab beneath low-velocity mantle, interpreted to be a piece of delaminated continental mantle lithosphere (Seber et al., 1996). As an alternative to delamination, oceanic terranes may evolve into continental crust by the introduction of felsic components coming from mantle wedges above subduction zones.

Various mechanisms have been suggested for the growth of the continents, the most important of which are magma additions by crustal over- and underplating, and by terrane collisions with continental margins (Rudnick, 1995). Magma from the mantle may be added to the crust by underplating, involving the intrusion of sills and plutons, and by overplating of volcanic rocks (Figure 5.6a). Magma additions can occur in a variety of tectonic environments, the most important of which are arcs, continental rifts, and beneath flood basalts. Up to 20 per cent of the crust in the Basin and Range Province in Nevada was added during the Tertiary by juvenile volcanism and plutonism (Johnson, 1993). Large volumes of juvenile magma from the mantle are added to both oceanic and continental margin arcs. Major continental growth by this mechanism can occur during seaward migration of subduction zones, when arc magmatism must keep up with slab migration. Field relationships in exposed lower crustal sections, such as the Ivrea Zone in Italy, suggest that many mafic granulites are intrusive gabbros and that additions of mafic magmas to the lower continental crust may be important. Also, a high-velocity layer at the base of Proterozoic shields has been interpreted as a mafic underplate (Durrheim and Mooney, 1991). This accounts for a difference in average thickness of Archean shields (~ 35 km) and Proterozoic shields (~ 45 km) (Figure 5.7). Just why Archean continents were not also underplated with mafic magma is not understood, but may be related to the thick, depleted lithospheric roots beneath Archean shields that somehow protect the crust from underplating.

Continental growth also occurs when oceanic terranes collide with and become sutured to continents (Figure 5.6b). Collision of submarine plateaux with continental margins may be one of the most important ways in which continents have grown (Abbott and Mooney, 1995; Condie, 1994). Most of the Cordilleran and Appalachian orogens in North America represent collages of oceanic terranes added by collisions, either at convergent plate boundaries or along transform faults. Lithologic associations, chemical compositions, terrane lifespans, and tectonic histories of Cordilleran terranes in northwestern North America are consistent with collisional growth of the continent in this area (Condie and Chomiak, 1996). The Cordilleran crust appears to have formed mostly from oceanic terranes with pre-collisional his-

a. MAGMATIC OVER- AND UNDERPLATING

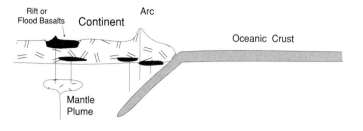

Figure 5.6 Mechanisms of continental growth.

b. TERRANE COLLISIONS

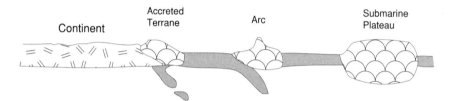

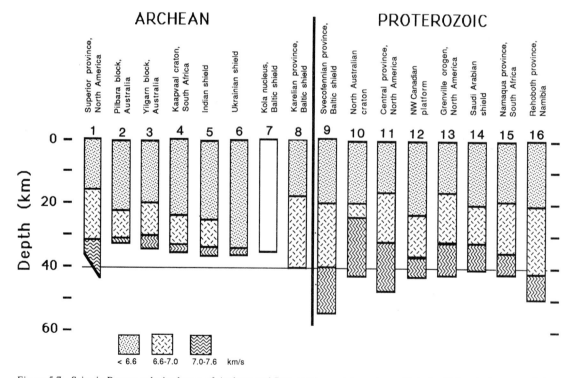

Figure 5.7 Seismic P-wave velocity layers of Archean and Proterozoic crustal provinces. Note the high velocity roots of the Proterozoic provinces. Courtesy of Walter Mooney.

tories of variable complexities and durations. Although some of these terranes began to evolve into immature continental crust before accretion to North America, this evolution most probably began at or not long before the time when they accreted to the continent. This was accomplished largely by incompatible element enrichment resulting from subduction-related processes associated with collisionally-thickened crust. Similar mechanisms have been proposed for Archean continents (Percival and Williams, 1989).

Some orogens, however, such as the 1.7-Ga Yavapai orogen in southwestern North America, appear to have evolved into a continental-margin arc system in a relatively short period of time, perhaps without large lateral displacements of terranes. In fact, data suggest that most Yavapai crust did not pass through an 'oceanic' stage, but evolved directly into mature continental crust (Condie and Chomiak, 1996).

Plate tectonics with time

Petrotectonic assemblages

As data continue to accumulate, it becomes more certain that plate tectonics in some form has been the principal mechanism by which the Earth has cooled for the last 4 Gy. One way of tracking plate tectonics with time is with petrotectonic assemblages as described in Chapter 3. How far back in time are the same petrotectonic assemblages found today, and are their time/space relationships, tectonic histories, and chemical compositions similar to modern assemblages? Except for ophiolites,

the greenstone and TTG assemblages are recognized throughout the geologic record from the oldest known rocks at 4.0–3.6 Ga to the present (Figure 5.8). The oldest well-preserved cratonic/passive margin sediments are in the Moodies Group in South Africa deposited at 3.2 Ga and such sediments are minor, yet widespread in the rock record by 3.0 Ga. Thus, it would appear that cratons, although probably small, were in existence by 3.2–3.0 Ga. Although the oldest isotopically-dated mafic dyke swarm is the Ameralik swarm in southwest Greenland intruded at about 3.25 Ga, deformed remnants of dykes in TTG complexes indicate that there were earlier swarms, perhaps as early as 4 Ga in the Acasta gneisses. The oldest dated anorogenic granite is the Gaborone granite in Botswana emplaced at 2875 Ma. However, clasts of granite with anorogenic characters in conglomerates of the Moodies Group have igneous zircons with U–Pb isotopic ages of 3.6 Ga, indicating that highly-fractionated granites formed in some Early Archean crust. The oldest known continental rift assemblages are in parts of the Dominion and Pongola Supergroups in South Africa, which were deposited on the Kaapvaal craton at about 3 Ga. The oldest accretionary orogens are the Acasta gneisses (4.0 Ga) and the Amitsoq gneisses (3.9 Ga) in northwest Canada and southwest Greenland respectively. Although the oldest well-documented collisional orogens are Early Proterozoic in age (such as the Wopmay orogen in northwest Canada and the Capricorn orogen in Western Australia), it is likely that Late Archean collisional orogens with reworked older crust exist in the granulite terranes of East Antarctica and southern India.

The fact that greenstones, TTG, anorogenic granites, mafic dyke swarms, and accretionary orogens all appear

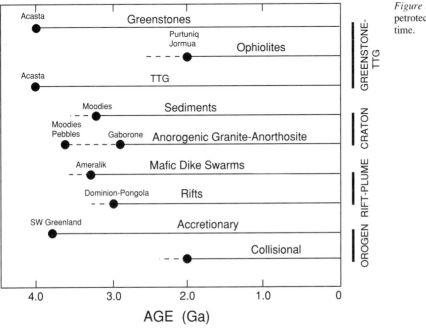

Figure 5.8 Distribution of petrotectonic assemblages with time.

in the very earliest vestiges of our preserved geologic record at 4.0–3.5 Ga, strongly supports some sort of plate tectonics operating on the Earth by this time. By 3.2–3.0 Ga, cratonic/passive margin sediments and continental rifts appeared recording the development of the earliest continental cratons. Although plate tectonics appears to have been with us since at least 4 Ga, there are well-documented differences between some Archean and post-Archean rocks that indicate that Archean tectonic regimes must have differed in some respects from modern ones, and these are reviewed in a later section. These differences have led to the concept of having our cake and eating it too, or in other words, plate tectonics operated in the Archean, but differed in some ways from modern plate tectonics. We are now faced with the question of how and to what degree Archean plate tectonics differed from modern plate tectonics and what these differences mean in terms of the evolution of the Earth.

Seismic reflection profiles

Seismic reflection profiles across Precambrian orogens reveal important constraints on deformational mechanisms. A profile across the Early Proterozoic Svecofennian orogen in the Gulf of Bothnia between Sweden and Finland shows an offset at the Moho of about 10 km and divergent reflectors in the crust, both of which are interpreted as resulting from plate convergence, subduction, and accretion (BABEL, 1990). A similar reflection profile across the Early Proterozoic TransHudson orogen in Canada shows a broadly symmetric structure between the Hearne and Superior Archean cratons (Lucas et al., 1993). Major reflections that dip to the centre of the orogen are associated with arc terranes along both margins. Furthermore, reflection geometries indicate a doubling of crustal thickness beneath the centre of the orogen, a feature consistent with collision of the bounding Archean cratons. These seismic features found in Early Proterozoic orogens are common in Phanerozoic collisional orogens suggesting that collisional orogens have not changed significantly in the last 2 Gy.

A seismic reflection profile across the Archean Superior province in southern Canada shows dipping seismic reflectors that extend 30 km into the mantle and correlate with surface features of Archean age (Figure 5.9) (Calvert et al., 1995). These patterns are very similar to those found in young collisional orogens where terranes are tectonically juxtaposed during collision, and provide convincing fossil seismic reflection evidence of Archean subduction and plate collision.

Archean transcurrent faults

From the studies of Archean transcurrent faults, it is possible to see if large oceanic plates existed in the Archean and if these plates were rigid as modern plates are (Sleep, 1992). The fact that major transcurrent offsets are recorded along Archean faults indicates that Archean plates must have been rigid. Also, two features suggest that oceanic plates comparable in size to modern plates existed in the Archean. First, the length and offset along Archean transcurrent faults in the southern Superior province in Canada are similar to those found on young transcurrent faults in Alaska and elsewhere. Also, it is well-established that the duration of movement with a consistent sense of motion along young transcurrent faults separating oceanic terranes from continents is similar to the length of time which oceanic terranes remain adjacent to continents, which is about 50 My in the northeast Pacific. Thus, a reported duration of consistent transcurrent motion of 30–50 My along Archean faults in the Superior province suggests that Archean plates south of the Superior province had dimensions and velocities similar to modern plates in the northeast Pacific.

If Archean faults from the Superior province are representative of Archean transcurrent faults, it would appear that a hotter Archean mantle did not greatly affect Archean plate sizes or rheological properties.

Episodic age distributions

Introduction

Although still not understood, an episodic distribution of crustal isotopic ages has been well-established since the classic paper of Gastil (1960). Major peaks in granitoid and greenstone ages are recognized at about 2.7, 1.9, and 1.0 Ga. Most investigators interpret these age distributions in terms of episodic growth of the continents (McLennan and Taylor, 1985), although as clearly pointed out by others (Armstrong, 1991), crustal age distributions cannot be equated with crustal growth rates unless there is no recycling of crustal components into the mantle. Although most geoscientists now agree that crustal materials are recycled into the mantle at subduction zones, there is little consensus as to the rate and percentage of recycling and how this rate may have changed with time. This topic will be considered later during discussion of continental growth rates.

The problem of identifying an episodic age distribution is not trivial. First of all, one is faced with resolution of isotopic dating methods. Until recently it was not possible to obtain a resolution better than 20–30 My in the Precambrian, yet Phanerozoic isotopic ages show that some orogenies are separated by time intervals < 30 My. Only one method, the U–Pb zircon method, allows resolution of 5–10 My during the Precambrian, and so only zircon ages should be used to identify episodic age distributions. Still another problem is avoiding geographic bias. Often, many dates are available from a small geographic area and few or none from other areas. Thus, the amplitude of peaks on isotopic age histograms may not be representative of the lateral extent of a given event. Still another problem is the identification of world-wide age gaps. Orogeny may shift from one continent to an-

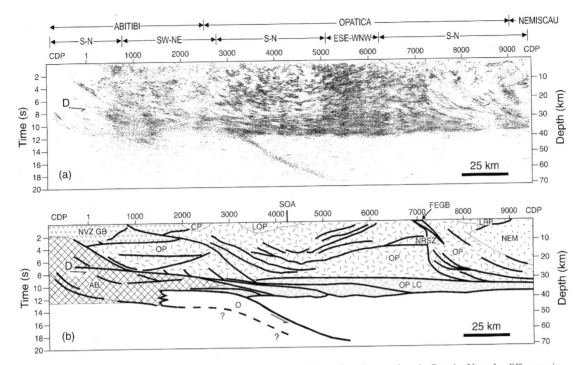

Figure 5.9   (a) Seismic reflection profile across the southeastern Archean Superior province in Ontario. Note the difference in reflective character of the Abitibi and Opatica greenstone belts. D marks the inferred position of lower crustal decollement. (b) Interpretation of the seismic section. OP, Opatica crust (LC, lower crust); AB, Abitibi crust; NVZ GB, FEGB, greenstones; NEM, Nemiscau metasediments; O, inferred oceanic crust; CP, LOP, LRP, plutons; NRSZ, Nottaway River shear zone. Courtesy of A. J. Calvert.

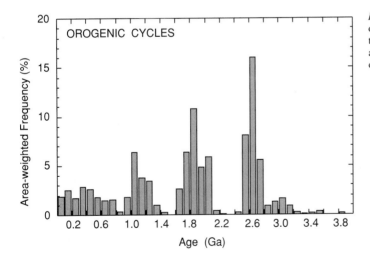

Figure 5.10   Histogram showing distribution of U–Pb zircon ages in continental crust with time. Zircons are from syntectonic granitoids and areas are weighted by areal distribution of crustal provinces.

other, and an age gap on one continent may correspond to an age maximum on another continent.

Granitoid age distributions

Histograms of U–Pb zircon ages from syntectonic granitoids are given in Figure 5.10. In an attempt to overcome geographic bias, the granitoid age distributions in

this figure are weighted according to the areas of crustal provinces which the ages are thought to represent. Anorogenic and post-tectonic granites are not included in the compilation. Several features are apparent from the age distributions as follows:

1   Although episodic ages characterize the continental crust, with three exceptions at 2.7–2.5, 2.0–1.7 and

1.3–1.0 Ga, U–Pb zircon ages do not support correlative world-wide episodicity of orogeny. In addition, age peaks at 2.1–2.0 Ga are recognized in West Africa and the Guiana shield in South America; 2.6–2.5 Ga in India, Antarctica, and North China, and 0.7–0.5 Ga in Africa, southern Eurasia, and South America.

2   When interpreted together with Nd isotopic results, only the 2.7–2.6 and 1.9–1.7 Ga events are times of major world-wide juvenile crust production. Other crustal formation events are recognized in India, Antarctica, and China at 2.6–2.5 Ga, in Africa and South America at 2.1–2.0 and 0.7–0.5 Ga, and in Asia at 3.0–2.9 and 0.4–0.3 Ga.

3   When data from individual geographic areas are examined, it is clear that most orogenies and crustal formation events last for 50–100 My.

4   Orogenic gaps on one continent may be filled in by orogenies on another continent. For instance, a 2.5–2.0-Ga gap in North America is occupied by a 2.1–2.0-Ga event in South America–West Africa and a 2.5-Ga event in India and China. Also, the 1.6–1.3-Ga gap in North America coincides with orogenic events in Africa and Australia. The so-called Lapalian Interval (0.8–0.5 Ga) in North America corresponds to the widespread Pan-African orogeny in Africa and South America.

Greenstones age distributions

The most striking feature of the age distributions in greenstones is the near absence of greenstones with ages in the 2.5–2.2 Ga time window and, with two exceptions, also in the 1.65–1.35 Ga time window (Figure 5.11). Prominent peaks in greenstone eruption age occur at 2.7, 1.9, and 1.3 Ga, similar to the three major peaks in granitoid ages, and support a common explanation for both age distributions. An additional greenstone peak at 2.1 Ga is found in Africa and South America. No age

spikes, however, are apparent in greenstones < 1.3 Ga or > 2.8 Ga.

If data are representative and there are few greenstones preserved within the two 300 My time windows in the Proterozoic, several possible explanations need to be considered (Condie, 1995):

1   plate tectonics stopped during these times and greenstones did not form;

2   platform sediments or glaciers selectively cover greenstones of these ages;

3   greenstones of these ages have been uplifted and removed by erosion;

4   greenstones of these ages have not as yet been sampled and dated; or

5   most greenstones of these ages were recycled into the mantle and are not preserved in the geologic record.

It seems extremely unlikely, in a steadily-cooling mantle in which heat is lost dominantly by plate formation at ocean ridges, that plate tectonics should stop and restart again several times. Supporting this conclusion is the fact that rapid variations in mantle temperature are strongly inhibited by silicate rheology, and thus sharp changes in the cooling history of the mantle are unlikely (Davies, 1992). So the first model is not considered a viable explanation for the absence of greenstones of a given age. It also seems highly implausible that young platform sediments or glaciers should selectively cover greenstones of two ages on all continents. Selective erosion of greenstones of these ages is unlikely in that other supracrustal rocks of these ages are widely preserved. Inadequate regional sampling can certainly lead to isotopic age minima, as evidenced most recently by the discovery of extensive juvenile crust, including greenstones, with an age of about 2.1 Ga in West Africa and in the Guiana shield in South America (Boher et al., 1992). In the last few years, however, our geographic data base of U–Pb zircon ages has increased remarkably,

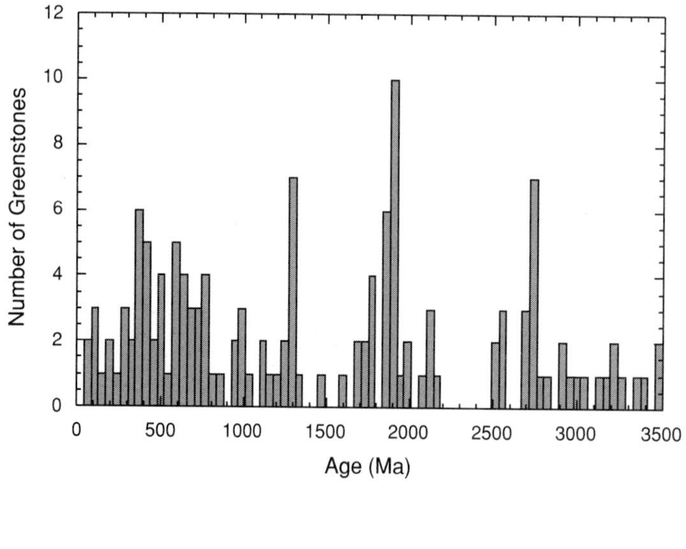

Figure 5.11   Episodic distribution of greenstone U–Pb zircon ages with time. Each age represents one greenstone belt, with no more than one age from any given terrane or stratigraphic succession. From Condie (1995).

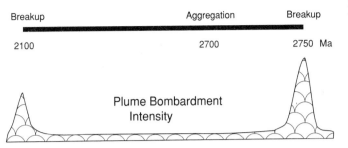

decreasing the probability of completely missing rocks of a given age. This is not to say that more greenstones falling in the two age minima will not be discovered in the future, but that the total volume of greenstones preserved in the continents with these ages is probably very small. We are then left with the final possibility: selective recycling of greenstones of certain ages into the mantle. The fact that there are greenstones in the geologic record is due to their incorporation into relatively buoyant continental crust, which protects them from subduction. They may collide and accrete to continents or they may be underplated with felsic magmas – in either case they are selectively preserved. The question is then, what processes could lead to greenstones being subducted before they collide with continents or before they are underplated with felsic magma?

Episodic ages and supercontinents

Could there have been periods of time during which collisions of greenstones with continents were less frequent, resulting in a larger fraction of greenstones being recycled into the mantle? Perhaps the frequency of encounters between greenstones and continents was less when supercontinents were in existence or breaking up. During these times, subduction zones would be widespread in oceanic areas (over mantle downcurrents), resulting in a greater probability of recycling oceanic terranes (ocean crust, islands, plateaux, arcs) before they had a chance to encounter a continent. Cloos (1993) shows that oceanic terranes should readily subduct if their crustal thickness is less than 20–30 km. It is only the very thick submarine plateaux and arcs that are difficult to subduct, and these should eventually be accreted to continents. The supercontinent model is attractive in attempting to explain the 2.5–2.2 and 1.65–1.35 Ga greenstone minima in that during these times supercontinents appear to have existed (Hoffman, 1989). Periods of supercontinent formation appear to correlate with the three world-wide maxima in age distributions at 2.7–2.6, 1.9–1.7 and 1.3–1.0 Ga (see Figure 5.35).

The only period which shows well-defined episodicity in the ages of greenstones and granitoids is between 2.8–1.3 Ga (Figures 5.10 and 11). Although the number of zircon ages > 2.8 Ga is still relatively small, there is no convincing evidence for greenstone age minima prior to this time. If the supercontinent cycle is necessary for periods of enhanced greenstone preservation and destruction, the absence of pre-2.8-Ga greenstone age minima may mean that there were no supercontinents prior to the Late Archean. However, greenstone age minima are also not recognized in the last 1.3 Gy, when the existence of supercontinents is well-established. Why should the frequency of greenstone preservation by less during times of Early and Middle Proterozoic supercontinents than during times of later supercontinents? Perhaps it is due to a significant overlap between assembly and break up stages of post-1.3-Ga supercontinents. For instance, the Late Proterozoic supercontinent Rodinia was fragmenting as Gondwana was forming at 750–600 Ma (Hoffman, 1991). Not long after the completion of Pangea, marked by the addition of Southeast Asia at 200 Ma, this supercontinent began to break up (about 160 Ma). While still in the supercontinent dispersal phase, terrane collisions began about 100 Ma in western North America and northeastern Asia, and India collided with Tibet about 55 Ma. The widespread greenstone terrane collisions in the Mesozoic and Early Tertiary around the margins of the North Pacific may record a change from the maximum dispersal phase of Pangea to the first assembly phase of a new supercontinent. Also, perhaps contributing to the two Proterozoic age minima is the fact that the margins of supercontinents existing at these times are strongly reworked by later orogenic events, making it more difficult to recognize older greenstones.

A possible scenario for a supercontinent cycle as illustrated for the Late Archean in Figure 5.12 begins with fragmentation of a supercontinent in response to mantle upwelling and plume bombardment from below. This is followed within 50–100 My by assembly of a new supercontinent, which survives 200–300 My before it is fragmented. During the early break up phase (2750 Ma), magmas from plumes should form numerous island arcs, submarine plateaux, and continental flood basalts. During the assembly phase (2700 Ma), characterized by decreasing plume activity, the submarine plateaux and arcs should collide with and be accreted to continents, contributing to increased continental growth. Continents should also grow by mafic underplating from plume-derived magmas.

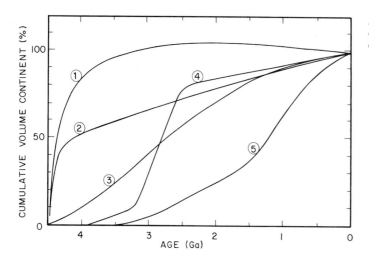

Figure 5.13 Examples of published continental growth rate models. See text for explanation.

Could mantle recycling lead to the near-absence of greenstones in the 2.5–2.2 and 1.65–1.35 Ga intervals? This seems improbable in that not all greenstones should be subducted. It is more likely that there are other greenstones like the Montauban Group (1.45 Ga) in Canada (Nadeau and van Breemen, 1994) and Amal–Horred belt in Sweden (1.6 Ga) (Ahall et al., 1995) that fall in these age intervals, but have not yet been discovered. Surely, some greenstones should survive around the margins of supercontinents. As supercontinents become better defined, we can identify geographic regions in which to concentrate our search for surviving remnants of such greenstones.

A correlation of episodic ages with the supercontinent cycle is testable in that greenstones formed during the two Proterozoic age minima should be chiefly continental-margin arcs, whereas those formed during greenstone age peaks should be a combination or arcs, islands, and submarine plateaux. Our existing geochemical data base seems to support this distribution of greenstones in time (Condie, 1994). In particular, the frequency of greenstones with geochemical features of submarine plateau basalts may be greater at 2.7, 1.9, and 1.3 Ga than at other times. Also, continental flood basalts should be widespread during greenstone age peaks. Although our isotopic age data base for Precambrian flood basalts is still small, what there is tends to support this idea. The oldest isotopically well-dated flood basalts are the Fortescue and Ventersdorp lavas in Western Australia and South Africa, respectively, dated at 2.75–2.7 Ga. These correlate with the first greenstone age maximum. Flood basalts are also recognized in the Wyloo Group in Western Australia and the Jatulian in Scandinavia at about 2.1–2.0 Ga corresponding to the second maximum of greenstone ages. Flood basalts also occur at about 1.8 Ga (in the Belcher Islands) and 1.27 Ga (Coppermine River basalts) in the Canadian shield, both of which lie at or close to peak greenstone activity. Although there are no recognized flood basalts in the Canadian shield at 2.0–1.9 Ga, there are dyke swarms of this age that may have fed now eroded flood basalts.

Continental growth rates

Introduction

Continental growth is the net gain in mass of continental crust per unit of time. Because continental crust is both extracted from and returned to the mantle, continental growth rate can be positive, zero or even negative. Many different models of continental growth rate have been proposed (Figure 5.13). These are based on one or a combination of:

1 Pb, Sr and Nd isotopic data from igneous rocks
2 Sr isotope ratios of marine carbonates
3 the constancy of continental freeboard through time
4 Phanerozoic crustal addition and subtraction rates
5 the areal distribution of isotopic ages
6 estimates of crustal recycling rates into the mantle.

Most investigators have proposed growth models in which the cumulative volume of continental crust has increased with time at the expense of primitive mantle, leaving a complementary depleted mantle behind. Many models tacitly assume that the aerial extent of continental crust of different ages preserved today directly reflects crustal growth through time, an assumption now known to be unrealistic. The earliest models for continental growth were based chiefly on the geographic distribution of isotopic ages on the continents (Hurley and Rand, 1969). These models suggested that continents grew slowly in the Archean and rapidly after 2 Ga (curve 5, Figure 5.13). We now realize that this is not a valid approach to estimating crustal growth rates because many of the Rb–Sr and K–Ar dates used in such studies have been reset during later orogenic events, and the true crustal formation age is older than the reset dates. On the opposite extreme are models that suggest very rapid growth early in the Earth's history, followed by extensive recycling of continental sediment back into the mantle, curves (1) and (2) (Reymer and Schubert, 1984). In the model proposed by Fyfe (1978), curve (1), the volume of continental crust in the Proterozoic actually exceeds that present today. Recycling into the mantle is necessary in the latter two models because we do not see old (> 4 Ga) continental crust on the Earth today. Other growth models fall between these extremes and include approximately linear growth with time, curve (3) and episodic growth, where continents grow rapidly during certain periods of time, such as in the Late Archean (Taylor and McLennan, 1985), curve (4).

The role of recycling

Some general features

As is apparent, the present areal distribution of continental crust shows episodicity in the rates of crustal formation. This distribution is a function of two factors:

1 continental extraction rate from the mantle
2 the rate of recycling of continent back into the mantle.

The difference in these two rates is the net continental growth rate, which can also be considered as the preservation rate of continental crust. The degree and rate at which continental crust is returned to the mantle is a subject of considerable disagreement and radiogenic isotopic data do not clearly constrain the problem (Armstrong, 1991). One of the major lines of evidence used to argue against a large volume of Early Archean continental crust is the small volume of preserved crust ≥ 3.5 Ga, either as crustal blocks or as sediments. Although small in volume, continental crust ≥ 3.5 Ga is remarkably widespread, with one or more examples known on all continents (Figure 5.2), a feature which could indicate that Early Archean continental crust was much more extensive than at present. Recycling of this early crust into the mantle is not surprising, as proposed long ago by Armstrong (1981), in that a hotter mantle in the Early Archean should lead to faster and more vigorous convection, which could result in faster crustal recycling than we see today. The Nd isotopic composition of both detrital and chemical sediments also supports extensive sediment recycling. For instance, on a plot of stratigraphic age versus Nd model age, progressively younger sediments deviate farther from the equal age line (Figure 5.14). This reflects recycling of older continental crust and a progressively greater proportion of older crust entering the sediment record with time. Another process which may have contributed to early crustal recycling is bombardment of the Earth's surface with asteroid-size bodies (McLennan, 1988). It is well-established that planets in the inner Solar System underwent intense bombardment with large impactors until about 3.9 Ga, and such impact on the terrestrial surface

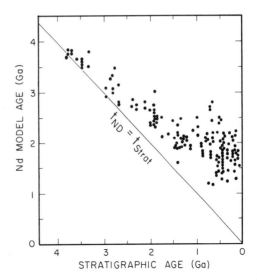

Figure 5.14 Comparison of stratigraphic and Nd model ages of fine-grained terrigenous sediments. Data compiled from many sources.

could have aided in the destruction and recycling of early continental crust.

There are two possible processes by which modern continental crust may be destroyed and returned to the mantle (Reymer and Schubert, 1984; von Huene and Scholl, 1991; Armstrong, 1991). These are sediment subduction and subduction erosion (as discussed in Chapter 3), and delamination and sinking of the lower crust into the mantle during collisional orogeny. Results suggest that about a half of ocean-floor sediment is eventually subducted and does not contribute to accretion in arc margins (von Huene and Scholl, 1991). Veizer and Jansen (1985) have shown that various tectonic settings on the Earth have finite lifetimes in terms of recycling. Contributing to their lifespans are rates of uplift and erosion, as well as subduction and crustal delamination. Active plate settings, such as oceanic crust, arcs, and back-arc basins, are recycled much faster than continental cratons. Most active plate settings have recycling half-lives of < 50 My, whereas collisional orogens and cratons have half-lives of > 350 My, with finite lifetimes (or oblivion ages) of < 100 My and > 1000 My respectively (Figure 5.15). Because there is evidence of small remnants of rocks formed at active plate settings in the geologic record, small amounts of even these tectonic settings may be tectonically trapped in the continents and sheltered from recycling.

Nd isotopes

Perhaps the best way to recognize juvenile continental crust is with Nd isotopes. Positive E_{Nd} values are interpreted to reflect derivation from juvenile sources, whereas negative E_{Nd} values reflect derivation from enriched sources (Arndt and Goldstein, 1987). Ages of crust must

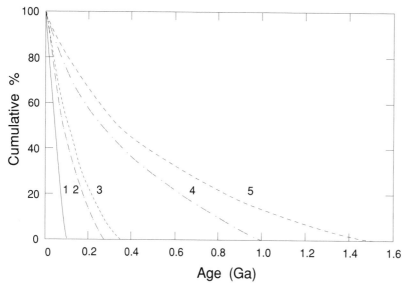

Figure 5.15 Mass-age distributions of tectonic settings. Key to tectonic settings with corresponding half-lives: 1, Oceanic crust (50 My); 2, arc (80 My); 3, passive margin (100 My); 4, collisional orogen (250 My); 5, craton (350 My). Data from Veizer and Jansen (1985).

be determined by some other isotopic system, the most accurate of which is the U–Pb system in zircons. It is now widely recognized that both positive and negative E_{Nd} values are recorded in Precambrian crustal rocks, even in the Earth's oldest known rocks, the Acasta gneisses (Figure 5.16). The vertical arrays of E_{Nd} values on a E_{Nd}–time plot may be explained by a mixing of crustal (negative E_{Nd}) and juvenile mantle melts (positive E_{Nd}). If this is the case, most of the individual data points cannot be used to estimate the rates and amounts of continental growth and corresponding mantle depletion (Bowring and Housh, 1995). Calculated E_{Nd} values as high as +3.5 in 4.0–3.8 Ga rocks indicate the presence of a strongly-depleted mantle reservoir at that time. The isotopic composition of this reservoir changed very little during the Precambrian, and some of it may be the source of modern ocean-ridge basalts. This implies that the isotopic composition of the depleted reservoir was buffered by the addition of either an enriched mantle component or continental crust. Otherwise, the Nd isotopic composition would have evolved along a steep slope, similar to the lunar growth curve in Figure 5.16. It is now recognized that continental crust was available for recycling by 4 Ga, and probably by 4.3 Ga as evidenced by the detrital zircon ages from Western Australia.

The wide range of E_{Nd} values in the Acasta gneisses and other Early Archean rocks requires extreme early fractionation accompanied by efficient recycling to generate these differences by 4 Ga and to prevent unobserved isotopic evolution of the depleted mantle reservoir (Bowring and Housh, 1995). Because of the relative enrichment of Nd in the continents, it easier to buffer mantle evolution with subducted continental sediments than with an enriched mantle reservoir, because of the much smaller volumes of crust required. Armstrong (1981; 1991) proposed a model for continental growth,

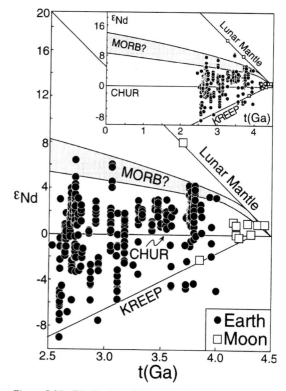

Figure 5.16 Distribution of E_{Nd} values in Archean rocks. Inset shows the evolution of MORB source (depleted mantle), CHUR (chondritic reservoir, similar to primitive mantle), lunar mantle, and KREEP, a high-K component found on the lunar surface. Courtesy of S. A. Bowring.

whereby the amount of recycling decreased with age in parallel with the cooling of the Earth. It is interesting that the calculated growth curve for E_{Nd} for his model falls near the upper limit of E_{Nd} data in Figure 5.16 (bottom of the MORB field). Thus, the paucity of Early Archean rocks in Precambrian shields may reflect the efficiency of this recycling rather than a lack of production of early continental crust. The gradual increase in E_{Nd} values in the last 4 Gy of Earth's history is consistent with a gradual increase in the volume of preserved continental crust with time as a result of cooling of the mantle and consequent decreases in recycling rates. As we shall see, however, there are episodic spikes superimposed on the increasing preservation curve for continental crust.

Freeboard

The freeboard of continents, the mean elevation above sea level, is commonly assumed to be constant with time (Reymer and Schubert, 1984; Armstrong, 1991). If this is the case, and the ocean ridges have diminished in volume with time due to cooling of the mantle, then the continents must have grown at a steady rate to accommodate the decreasing volume of seawater (curve (2), Figure 5.13). However, it is now realized that the constant freeboard assumption may be inaccurate (Galer, 1991). One problem is the thick Archean lithosphere, since a thicker lithosphere tends to offset the effects of shrinking ocean-ridge volume. Another is thicker Archean oceanic crust, which would make the Archean oceanic lithosphere more buoyant, and thus add to the volume discrepancy required by a larger volume of Archean ocean ridges. Also, freeboard is especially sensitive to asthenosphere temperature, which was greater in the Archean. For instance, if the upper mantle temperature in the Early Archean were 1600 °C (about 200 °C hotter than at present), continental crust would have been considerably below sea level and probably would have been subductable (Galer, 1991). In terms of the geologic record, evidence for subaerial weathering and erosion first appears in rocks about 3.8 Ga, and so prior to this time continents may very well have been deeply submerged.

Because of the uncertainties in the freeboard of continents with time, and especially because the magnitudes of all the controlling factors are not yet fully understood, continental freeboard should not be used to constrain continental growth models.

Continental growth in the last 200 My

Reymer and Schubert (1984) were the first to estimate the growth rates of young continental crust by estimating the various crustal addition and subtraction rates. Table 5.2 presents an updated and revised spreadsheet for continental crust produced in the last 200 My. The

Table 5.2 Growth rate of continents in the last 200 My

	Rate (km³/y)
Gains	
Island arcs	1.6
Submarine plateaux	0.8
Oceanic hotspot volcanism	0.2
Continental underplating	0.2
Total	2.8
Losses	
Sediment subduction	0.7
Subduction erosion	0.9
Delamination	0.1 (0–0.2)
Total	1.7
Net growth rate (2.8 km³/y minus 1.7 km³/y)	1.1

In part after Reymer and Schubert (1984) and von Huene and Scholl (1991)

Arcs, submarine plateaux, and hotspot volcanics include also an estimate of those accreted to the continents in the last 200 My

results for arc and submarine plateau crustal addition rates include both arcs and plateaux in ocean basins today and an estimate of the volume of arcs and plateaux accreted to continental margins in the last 200 My. Hence, the arc accretion rate (1.6 km³/y) is considerably greater than the 1.1 km³/y value of Reymer and Schubert based on 'non-accreted' arcs only. The estimate of 0.5 km³/y for submarine plateaux is a minimum value since the volume of plateaux accreted to continents is not well-known. The volume of magma underplated beneath continents is assumed to be equivalent to twice the volume of flood basalts erupted in the last 200 My, and the oceanic hotspot volcanic rate is from Reymer and Schubert (1984). The volume of subducted sediments and material recycled into the mantle by subduction erosion are from von Huene and Scholl (1991). The amount of crust returned to the mantle by delamination is poorly known, with estimates ranging from none to perhaps 0.2 km³/y (a value of 0.1 km³/y has been assumed). Interestingly, despite the major revisions in rates of gains and losses to the continental crust in the last 200 My, the net continental growth rate (gains minus losses) of 1.1 km³/y is similar to the value originally proposed by Reymer and Schubert (1984) (1.06 km³/y).

It is of interest to see if this rate of crustal growth can account for the volume of juvenile continental crust formed in the last 200 My as estimated from precise geochronology and Nd isotope studies. It would appear that about 13 per cent of the continental crust was extracted from the mantle in the last 500 My (see Figure 5.17) and approximately 25 per cent of this formed in the last 200 My. Hence, for a total volume of continental crust of 7.18×10^9 km³ (Cogley, 1984), 270×10^6 km³ formed in the last 200 My. This amount agrees well with the volume calculated using the rate 1.1 km³/y (Table 5.2) of 220×10^6 km³.

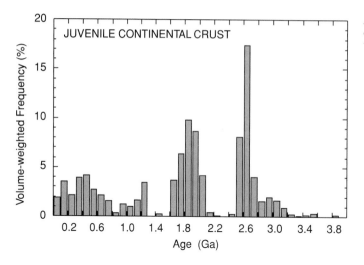

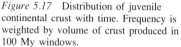

Figure 5.17 Distribution of juvenile continental crust with time. Frequency is weighted by volume of crust produced in 100 My windows.

Towards a continental growth rate model

The approach

Net continental growth rate in the geologic past, which can also be considered as the preservation rate of continental crust, is critically dependent upon two factors (Condie, 1990):

1 the proportion of reworked crust within a given crustal province
2 the growth length intervals assumed.

The volume of juvenile crust, g, extracted from the mantle during a specified time interval, Δt, is given by,

$$g = a - r + m, \qquad (5.1)$$

where a is the volume of crust formed during Δt; r is the volume of reworked crust that must be subtracted; and m is the volume of crust formed in Δt, but now tectonically-trapped as blocks in younger crust. The value of a is determined from the scaled area of crustal provinces/terranes of a given age times the average crustal thickness. The final crustal thickness includes a 10-km thick restoration of crust lost by erosion for a total average thickness of 45 km. Values for r and m are estimated from published Nd isotopic data, U–Pb zircon ages, and detailed geologic maps. Growth is considered in 100 My increments. The amount of reworked crust in a given crustal province is estimated from Nd isotopic data, assuming a mixing of juvenile and evolved end members (Arndt and Goldstein, 1987) and the distribution of detrital and xenocrystic zircon ages. Volumes of reworked crust are redistributed into appropriate earlier growth intervals.

The model

Because of extensive reworking of older crust in some areas, only the most intense and widespread orogenic events are used in delineating the distribution of juvenile crust. Most crustal provinces < 2.5 Ga contain variable amounts of reworked older crust. Early Archean (> 3.5 Ga) crustal provinces, although widely distributed, are very small (chiefly < 500 km across) and may represent remnants of the Earth's early continents. Late Archean provinces (3.0–2.5 Ga) are widespread on all continents and probably underlie much of the platform sediment (or ice) in Canada, Africa, Antarctica, and Siberia. Early Proterozoic provinces (2.0–1.7 Ga) are widespread in North America and in the Baltic shield in Europe, and form less extensive but important orogens in South America, Africa, and Australia. Mid-Proterozoic provinces (chiefly 1.3–1.0 Ga) occur on almost all continents where they form narrow belts along which the Late Proterozoic supercontinent Rodinia was sutured together. Unlike older crustal provinces, little juvenile crust is known of this age. Late Proterozoic provinces (0.8–0.55 Ga) are of importance only in South America, Africa, southern Asia and perhaps in Antarctica.

From the above data base, the volume frequency of juvenile crustal additions to the continents is shown in Figure 5.17. Note the striking similarity to the greenstone age distributions (Figure 5.11). Although as more data is acquired details will probably change on this diagram, the evidence of the episodic nature of crustal growth, which had been suspected for over 30 years, seems to be robust. Most striking are the two periods of rapid crustal growth at about 2.7–2.6 and 2.0–1.8 Ga, and extreme minima at 2.5–2.2 Ga and 1.6–1.3 Ga. By 1.7 Ga, over seventy per cent of the continental crust was in existence, and only about fifteen per cent was added in the last 500 My. Isotopic age data have been compiled and major juvenile crustal provinces are shown on an equal area map projection of the continents in Figure 5.18. Areal distributions indicate that approximately equal volumes of juvenile crust were formed in the Archean and Early Proterozoic (about thirty-five per cent each), and less than half of this amount in each of the Late

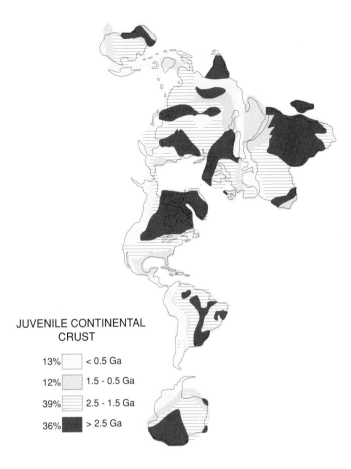

Figure 5.18 Areal distribution of juvenile continental crust shown on an equal-area projection of the continents.

JUVENILE CONTINENTAL
CRUST

13% < 0.5 Ga

12% 1.5 - 0.5 Ga

39% 2.5 - 1.5 Ga

36% > 2.5 Ga

Proterozoic and Phanerozoic. Clearly portrayed on the map is the strikingly irregular distribution of juvenile crust with time. Whereas most of Africa, Antarctica, Australia, and the Americas formed in the Archean and Early Proterozoic, much of Eurasia formed in the Late Proterozoic and Phanerozoic.

Secular changes in the crust–mantle system

Changes in composition of the upper continental crust

Introduction

As a constraint on the evolution of the crust–mantle system, it is important to know if the composition of continental crust has changed with time. Some investigators have used fine-grained terrigenous sediments to monitor changes in upper crustal composition. Justification of this approach relies on the mixing of sediments during erosion and sedimentation, such that a shale, for

instance, reflects the composition of a large geographic region on a continent. Only those elements which are relatively insoluble in natural waters, and thus transferred in bulk to sediments, can be used to estimate crustal composition (such elements as Th, Sc, and the rare earth elements [REE]). Some studies have compared sediments of different ages from different tectonic settings, thus erroneously identifying secular changes in crustal composition which, in fact, reflect the different tectonic settings (Gibbs et al., 1986). The prime example of this is comparison of Archean greenstone sediments (dominantly volcanigenic graywackes) with post-Archean cratonic sediments (dominantly cratonic shales) and interpreting differences in terms of changes at the Archean–Proterozoic boundary (Veizer, 1979; Taylor and McLennan, 1985). Another limitation of sediments in monitoring crustal evolution results from recycling of older sediments, which leads to a buffering effect through which changes in juvenile crustal composition may not be recognizable (Veizer and Jansen, 1985; McLennan, 1988). For sediment geochemical results to be meaningful, sediments should be grouped by average grain size and by lithologic association, which in turn reflects tectonic setting.

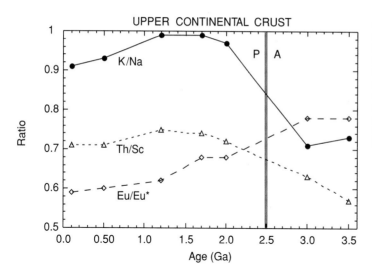

Figure 5.19 Distribution of K/Na, Th/Sc, and Eu/Eu* in upper continental crust with time. Data from Condie (1993). Eu–Eu* is a measure of the Eu anomaly with positve Eu anomalies having values > 1 and negative anomalies with values < 1. Eu/Eu* = $Eu_n/(Sm_n \times Gd_n)^{0.5}$, where n stands for chondrite-normalized values. Estimated errors for ratios range from 1 to 2 times the size of the plotting symbol. A, Archean; P, Proterozoic.

Major elements

There have been numerous studies of secular changes in major element distributions, most of which have concentrated on detrital sediments. Some investigators have suggested that the proportions of sedimentary rocks have changed with time, and are responsible for possible secular changes in major element concentrations (Engel et al., 1974; Schwab, 1978; Ronov et al., 1992). However, all of these studies suffer from the same problem: indiscriminately lumping sediments together from different tectonic settings. Therefore, although significant changes in lithologic proportions and element distributions have been proposed, their existence is questionable.

Only three major element trends are well documented from sampling of Precambrian shields. Data suggest that Ti increases and Mg decreases, and that there is an increase the K/Na ratio near the end of the Archean (Figure 5.19) (Condie, 1993). The changes in Ti and Mg

appear to reflect a decrease in the amount of komatiite and high-Mg basalt in continental sources after the Archean. The increase in K/Na ratio is caused by an increase in K and decrease in Na in average post-Archean granitoids, which dominate in the upper crust.

REE and related elements

At the end of the Archean, average upper continental crust increases in LIL elements, P, Nb, Ta and to some degree in Zr and Hf (Condie, 1993). Also, Rb/Sr and Ba/Sr ratios increase at this time. The Th/Sc ratio shows only a moderate increase at the A–P (Archean–Proterozoic) boundary (Figure 5.19), significantly less than that proposed by Taylor and McLennan (1985) based on comparing unlike sediments. Although the sediment data suggest that Th/U ratio increases with time, there is no evidence for secular variation of this ratio in the shield data (Figure 5.20).

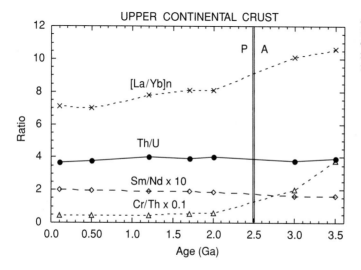

Figure 5.20 Distribution of Cr/Th, Th/U, Sm/Nd and [La/Yb]n in upper continental crust with time. n refers to chondrite-normalized ratio. See Figure 5.19 for other information.

At the end of the Archean, there is also a notable decrease in fractionation of REE, as shown by a decrease in (La/Yb)n ratio and a small increase in Sm/Nd ratio in the upper crust (Figure 5.20). Both of these ratios change in opposite directions to those proposed by Taylor and McLennan (1985) based on sediment data. The Sm/Nd ratio of continental shales has remained about constant with time at 0.18, and agrees well with the average ratio of upper crust today.

Taylor and McLennan (1985) have long maintained that post-Archean sediments differ from Archean sediments by the presence of a negative Eu anomaly. However, it is difficult to test their conclusion because it is based on comparing largely greenstone sediments from the Archean with cratonic sediments in the post-Archean. Precambrian shield results indicate that both Archean shales and Archean upper crust have sizeable negative Eu anomalies (small Eu/Eu* ratios), and that there is only a modest increase at the end of the Archean (Condie, 1993; Gao and Wedepohl, 1995) (Figure 5.19). This increase reflects the importance of Eu anomalies in felsic igneous rocks and sediments derived therefrom, after the end of the Archean. One contributing factor is that unlike Archean TTG, post-Archean TTG generally has significant negative Eu anomalies. Negative Eu anomalies, however, are not limited to post-Archean rocks, and both Archean shales and granites typically show sizeable Eu anomalies.

Thus, although it seems certain that post-Archean upper continental crust has a larger Eu anomaly than its Archean counterpart, it is clear that:

1 both Archean upper crust and Archean sediments have negative Eu anomalies
2 cratonic shales give only a weak suggestion of the increasing Eu anomaly in upper crust formed after the end of the Archean.

Ni, Co and Cr

Decreases in Ni, Co, and Cr as well as in Cr/Th, Co/Th, and Ni/Co ratios are observed at the A–P boundary in both upper continental crust and in fine-grained cratonic sediments (Taylor and McLennan, 1985; Condie, 1993) (Figures 5.20 and 5.21). These changes appear to reflect a decrease in the amount of both high-Mg basalt and komatiite in continental sources after the end of the Archean. A striking example of this decrease occurs in Precambrian cratonic sediments of the Kaapvaal craton in southern Africa (Condie and Wronkiewicz, 1990). A decrease in Cr/Th ratio in Kaapvaal sediments near the A–P boundary appears to reflect a decrease of komatiite-high Mg basalt sources.

Sr isotopes in marine carbonates

The isotopic composition of Sr in seawater is controlled chiefly by the isotopic composition of rivers entering the oceans and the hydrothermal input at ocean ridges

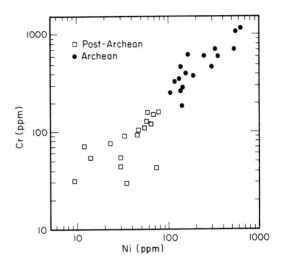

Figure 5.21   Cr–Ni distribution in cratonic fine-grained detrital sediments.

(Veizer, 1989). Today the river influx greatly dominates in controlling seawater $^{87}Sr/^{86}Sr$ ratio. Because of the long residence time of Sr in the oceans (~ 4 My) compared to ocean mixing times (~ 1000 y), the isotopic composition of modern seawater is very uniform (about 0.7099), even in partially landlocked seas like the Black Sea. Because marine carbonates record the seawater Sr-isotopic composition, they have been used to monitor the composition of seawater in the geologic past. However, as many carbonates are altered during diagenesis or metamorphism, it is not clear if the reported $^{87}Sr/^{86}Sr$ ratios for ancient carbonates reflect the actual composition of the seawater from which they were precipitated. Most investigators minimize this problem by using only the lowest $^{87}Sr/^{86}Sr$ ratios in carbonates, since alteration during diagenesis generally increases the Sir-isotopic ratio (Veizer, 1989).

First-order changes

Although several orders of variation in the Sr-isotopic composition of ancient carbonates is recognized, only the first-order changes that occur over times of 500 My or more can be resolved in the Precambrian record. Veizer (1989) and others have interpreted results to indicate that seawater Sr-isotopic composition followed the mantle curve until the Late Archean, when it increased rapidly to about 2 Ga and then increased steadily thereafter (Figure 5.22). The alleged increase in the Late Archean has further been interpreted to reflect a change from mantle-buffered to continent-buffered seawater in response to rapid growth of continents in the Late Archean. Gibbs et al. (1986) were the first to point out, however, that the alleged change at the A–P boundary resulted from comparing Archean greenstone carbonates with post-Archean cratonic carbonates.

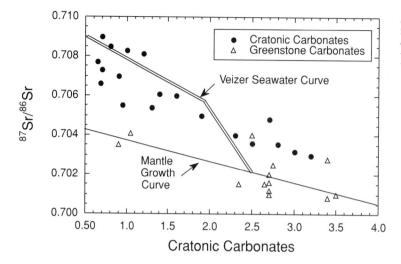

To evaluate this problem further, Condie (1992a) plotted the minimum ^{87}Sr/^{86}Sr ratios of marine carbonates as a function of greenstone or continental (cratonic) tectonic setting (Figure 5.22). Although there is scatter in the data, all but two of the greenstone carbonates fall on or near the mantle growth curve and the cratonic carbonates scatter in a broad band above the mantle growth curve. The clustering of greenstone carbonate data near the mantle growth curve suggests that these carbonates were deposited from seawater buffered by the mantle, perhaps by hydrothermal springs in restricted basins associated with arc systems or submarine plateaux (Veizer et al., 1989).

The significance of continental and greenstone trends on the ^{87}Sr/^{86}Sr–time graph is not yet fully understood. If, as suggested by the modern isotopic composition of seawater, the continental trend monitors the average composition of seawater, then what do the low ^{87}Sr/^{86}Sr ratios in greenstone carbonates tell us? It is well-known that greenstone carbonates are of only local extent and may be deposited by hydrothermal springs on the sea floor. The fact that greenstone carbonate Sr-isotope ratios fall near the mantle evolution growth line may reflect mantle sources for the Sr, consistent with submarine hot springs. If this is the case, why the mantle isotopic signature of the spring waters is not lost by mixing with the much larger volume of seawater with high Sr-isotopic ratios is puzzling. Perhaps the greenstone carbonates were deposited very rapidly near their sources before any substantial mixing with seawater could occur. In any case, sometime before 3 Ga there should be a crossover between continent-buffered and mantle-buffered seawater composition. If continental crust was produced in large volumes in the Early Archean as previously proposed, the crossover may have occurred in the very earliest Archean, some time between 4.5 and 4.3 Ga, and is therefore not recorded in any preserved carbonates.

Changes in the last 1 Gy

During the last 1 Gy, Sr isotopes in marine carbonates show considerable variation (Figure 5.23). Overall, the variation parallels the supercontinent cycle (Richter et al., 1992): the lows occur at the point when supercontinents are fragmenting and thus an increase in the number of ocean ridges between dispersing continental fragments injects more mantle Sr into the oceans, and the ^{87}Sr/^{86}Sr ratio decreases. This is observed when Rodinia began to fragment about 750 Ma. The minimum in the Sr-isotope curve at about 850 Ma may indicate that the break up actually began 50–100 My earlier. Likewise the minima at 500–450 Ma and 160 Ma may reflect with the fragmentation of Laurasia (Laurentia–Baltica) and Pangea, respectively. During the growth of supercontinents, continental relief generally increases leading to a greater continental contribution to the oceans by rivers. The result is an increase in the ^{87}Sr/^{86}Sr ratio in seawater. Examples are the peaks in ^{87}Sr/^{86}Sr that occur at about 550 Ma when Gondwana was forming, and again in the last 60 My, probably in response to the collision of India with Asia, and the uplift of the Himalayas (Harris, 1995). Several other peaks in ^{87}Sr/^{86}Sr ratios in the Paleozoic may reflect continental collisions in the assembly of Pangea, such as the Taconic orogeny in the Ordovician (peak at 420 Ma), the Acadian orogeny in the Devonian (peak at 370 Ma), the Variscan orogeny (peak at about 300 Ma), and the collision that produced the Urals (peak at about 275 Ma).

Rare earth elements and Nd isotopes in siliceous sediments

Introduction

REE and Nd isotopes, like Sr isotopes, monitor the composition of seawater and provide important input in

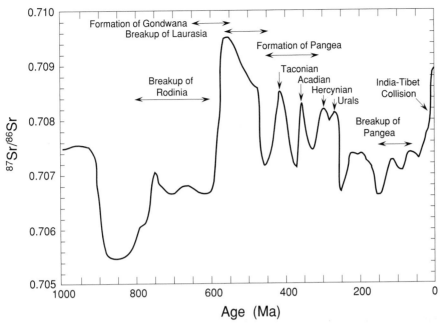

Figure 5.23 The $^{87}Sr/^{86}Sr$ ratio in seawater during the last 1 Gy as inferred from marine carbonates. Data from many sources including those given in Veizer (1989).

studying the relative roles of mantle and crustal sources for seawater (Derry and Jacobsen, 1990; Jacobsen and Pimentel-Klose, 1988). Chemical sediments, such as chert and banded iron formation (BIF), may retain the REE patterns (including Eu anomalies) and the Nd-isotopic compositions of the seawater from which they were deposited, making it possible to track the composition of seawater with time. However, as with Sr isotopes, these distributions can be modified by contamination of chemical sediments with detrital components and by diagenetic/metamorphic effects. These effects can be minimized by comparing rocks of different ages from the same tectonic/diagenetic environment, and by using element and isotopic ratios, which are less likely to change

during diagenesis and metamorphism than absolute concentrations.

Eu anomalies

While Eu anomalies (Eu/Eu* > 1) tend to be positive in Archean cherts and BIF (hereafter referred to collectively as chert), and non-existent or negative in post-Archean cherts (Figure 5.24), it is not clear if this reflects a change in composition in seawater at this time, or if it is an inherent difference between cherts deposited in greenstone and continental tectonic settings. Although many investigators have interpreted the results to mean an increase in continental sources at the end of the

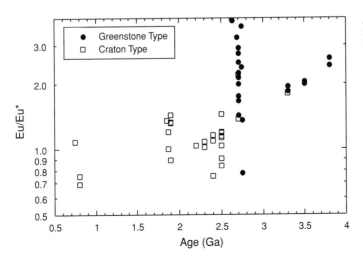

Figure 5.24 Secular distribution of Eu/Eu* in cherts and banded iron formation. See Figure 5.19 for definition of Eu/Eu*.

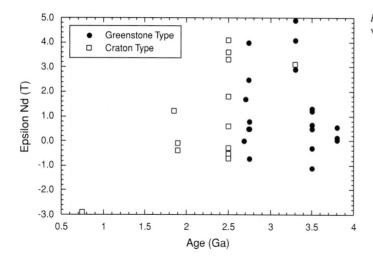

Figure 5.25 Secular distribution of E_{Nd} values in cherts and banded iron formation.

Archean, with two exceptions all the Archean data come from greenstone cherts and all of the post-Archean data from continental cherts. The positive Eu anomalies in greenstone cherts seem to be inherited from mantle-derived hot springs, as evidenced for instance by the positive Eu anomalies in modern submarine hot springs along ocean ridges (Danielson et al., 1992). On the other hand, cherts deposited in continental settings or near continental crust often carry negative Eu anomalies typical of continental crust. As with Sr isotopes in greenstone carbonates, it may be that greenstone cherts record only compositions near hydrothermal vents and not average seawater. Supporting this interpretation is the fact that REE are rapidly removed from modern hot spring waters by deposition of iron hydroxides, and clearly do not reflect compositions of large volumes of seawater (Olivarez and Owen, 1991). Cherts deposited on or near continents, however, appear to reflect continental sources, perhaps from detrital sediment contamination. The oldest continental cherts from the Moodies Group in South Africa (3.2 Ga) support this interpretation. Thus, the apparent drop in Eu/Eu* at 2.6 Ga in Figure 5.24 is probably not real, but an artifact of tectonic setting bias. As with Sr isotopes, if there was a change from mantle- to crustal-controlled seawater composition, it must have occurred very early in the Archean, probably before 4.3 Ga.

Nd isotopes

Because of the short residence time of Nd in seawater (~ 1000 y), large variations in Nd-isotopic composition can occur between ocean basins through time. This makes it very difficult to track average seawater composition with time with Nd isotopes, and indeed contributes to the wide variation in E_{Nd} in cherts and BIF with time (Figure 5.25). There is complete overlap in the E_{Nd} values of continental and greenstone cherts/BIF, and no secular trend is obvious within either population. The present Nd budget of the oceans is dominated by river

waters from continental sources, while ocean ridge hydrothermal sources contribute only about one per cent of the total Nd (Jacobsen and Pimentel-Klose, 1988). Another factor contributing to the wide range in E_{Nd} values in chert/BIF is mixing between surface waters with a continental signature ($E_{Nd} < 0$) and deep waters dominated by hydrothermal sources ($E_{Nd} = +4$). A possible decrease in the E_{Nd} value of seawater with time as suggested by Jacobsen and Pimentel-Klose (1988) is critically dependent on one sample at 750 Ma (Figure 5.25). It would seem that considerably more post-Archean data are needed to justify the existence of such a trend.

Alkaline igneous rocks

Alkaline igneous rocks, such as trachytes, phonolites, basanites, kimberlites, and carbonatites occur on cratons as well as in some continental rifts and oceanic islands. They do not, however, become important in the geologic record until after 200 Ma. Although alkaline igneous rocks are reported in Late Archean greenstones, they are extremely rare and only a relatively small number of occurrences are known of Proterozoic and Paleozoic age. Several factors probably contribute to a decrease in the proportion of alkaline igneous rocks in the Precambrian. In part contributing to this distribution is the fact that continental alkaline igneous centres are small and they are readily removed during uplift and erosion. Veizer et al. (1992) show, for instance, the half-life for the loss of carbonatites by erosion is only about 450 My. Hence, they have a low probability of survival. Alkaline igneous rocks in oceanic islands in accreted terranes of any age are rarely reported, perhaps because they do not survive erosion. Another factor contributing to the rarity of Archean alkaline igneous rocks is the degree of melting in the mantle source. Because very small degrees of melting (< 10 per cent) of the upper-mantle are required to produce these magmas, it is not unexpected that they should be rare in the Archean when mantle temperatures were higher, leading to larger degrees of melting.

Anorthosites

Archean anorthosites often occur in or associated with greenstone belts and typically have megacrystic textures with equidimensional calcic ($> An_{80}$) plagioclase crystals 5–30 cm in diameter in a mafic groundmass (Ashwal and Myers, 1994). Some are associated with cumulate chromite. Field, textural, and geochemical relationships show that most are genetically related to greenstone basalts, where they appear to represent subvolcanic magma chambers that fed submarine eruptions. Parental magmas were Fe- and Ca-rich tholeiites that may have fractionated from komatiitic magmas.

In contrast, most post-Archean anorthosites are associated with anorogenic granites and syenites and contain much less calcic plagioclase (chiefly An_{40-60}) (Wiebe, 1992). They are generally interlayered with gabbros and norites and exhibit cumulus textures and rhythmic layering. Many bodies, which range from $10^2–10^4$ km^2 in surface area, are intruded into older granulite-facies terranes and some are highly fractured. Gravity studies indicate that most bodies are from 2–4 km thick and are sheet-like in shape, suggesting that they represent portions of stratiform igneous intrusions. The close association of Proterozoic anorogenic granites and anorthosites in the Grenville province in eastern Canada suggests a genetic relationship between these rock types. Geochemical and isotopic studies, however, indicate that the anorthosites and granites are not derived from the same parent magma by fractional crystallization or from the same source by partial melting. Data are compatible with an origin for the anorthosites as cumulates from fractional crystallization of high-Al_2O_3 tholeiitic magmas produced in the upper mantle (Emslie, 1978; Wiebe, 1992). The granitic magmas appear to be produced by partial melting of lower crustal rocks, the heat coming from associated basaltic magma that gives rise to the anorthosites by fractional crystallization.

It is still not clear why Archean anorthosites appear to have been produced in oceanic tectonic settings, whereas most post-Archean anorthosites clearly formed in older continental crust.

Ophiolites

As discussed in Chapter 3, a complete ophiolite includes from bottom to top, ultramafic tectonite, layered and non-layered gabbros and ultramafic rocks, sheeted diabase dykes, and pillow basalts. If this definition is strictly adhered to, the oldest known ophiolites are about 2 Ga. Although some Archean ophiolites have been described, they lack one or more of the ophiolite components and so may not be fragments of oceanic crust (Bickle et al., 1994). Harzburgite tectonites are missing in all rock packages described as Archean ophiolites, and only one convincing sheeted dyke complex has been described from an Archean terrane. This dyke complex is in the Kam Group near Yellowknife in northwest Canada (Figure 3.5), in which the dykes have been intruded through older felsic crust and are therefore unlike true ophiolite dyke swarms (Helmstaedt and Scott, 1992).

Why are complete ophiolites apparently missing in the Archean? As discussed in this and other chapters, it seems very likely that plate tectonics was operating in the Archean. This leaves two options:

1 Archean ophiolites have been overlooked or not recognized as such
2 remnants of Archean oceanic crust do not look like complete post-Archean ophiolites.

Considering the detailed geologic mapping in Archean greenstones, it is becoming less and less likely that ophiolites have been overlooked. If the higher heat production in the Archean mantle resulted in production of thicker oceanic crust, because of the greater amount of melting beneath ocean ridges, complete ophiolites may not have been preserved by obduction or underplating mechanisms. Because ophiolites are generally < 7 km thick, a tectonic slice off the top of Archean oceanic crust which was 20 km thick may only include the pillow basalt unit, and pillow basalts are prolific in Archean greenstones. If indeed this were the case, we are immediately faced with the question of how to distinguish slices of oceanic crust from slices of submarine plateaux or arcs in Archean greenstone belts. One has to be very careful in using basalt geochemistry in that plume and depleted mantle components may not have been as well-established in the Archean mantle, and in fact depleted and enriched reservoirs may have been intimately mixed in the upper mantle (Bowring and Housh, 1995). One possible way to distinguish between oceanic crust and submarine plateaux may be the presence of komatiites. If komatiites require plume sources as seems likely, only those greenstone successions without komatiites could be candidates for Archean oceanic crust. Certainly renewed effort seems warranted in our hunt for remnants of Archean oceanic crust.

Greenstones

Although many Archean greenstones differ from younger greenstones by the presence of komatiite, the proportion of basalt also is greater in Archean greenstones (Condie, 1994). In most Archean greenstones the volume of basalt + komatiite exceeds that of intermediate + felsic volcanics (Figure 5.26). Those Archean examples with > 80 per cent basalt + komatiite are the mafic plains such as the Blake River Group in the Abitibi belt and the Talga Talga Group in the Pibara in Western Australia. These may represent submarine plateaux (Kusky and Kidd, 1992), although perhaps some of them are remnants of Archean oceanic crust. The Archean greenstones with 40–80 per cent basalt + komatiite (komatiite is usually absent in these) are more like young oceanic arc systems. On the whole, Proterozoic greenstones for which data are available have smaller proportions of basalt (and usually no komatiite) and range from 20–50 per cent of felsic volcanics + andesite, and in this respect

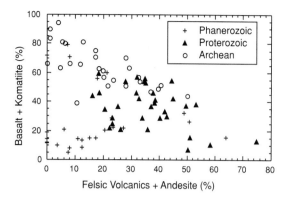

Figure 5.26 Distribution of basalt + komatiite versus felsic volcanics + andesite in greenstones with time From Condie (1994).

are like continental-margin arc systems. Phanerozoic greenstones are more variable than Precambrian greenstones, and appear to represent a greater diversity of oceanic tectonic settings.

Condie (1994) has shown a broad inverse correlation between the abundances of graywacke and basalt + komatiite in greenstones, with most Phanerozoic greenstones having a high proportion of graywacke. In general, the amount of variation increases in the order Archean, Proterozoic, and Phanerozoic. If the sampling is representative in each age group, these results suggest that oceanic tectonic settings were less diverse in the Archean than afterwards. Although there is a great deal of overlap in the proportion of flows to fragmental volcanics in greenstones, the flow/fragmental ratio is generally higher in Archean greenstones than in post-Archean greenstones (Condie, 1994). Also, there appears to be an absence of Archean greenstone assemblages with flow/fragmental ratios < 1. The consistently high flow/fragmental ratio in Archean greenstones suggests they were erupted chiefly as submarine flows in deep water, whereas a greater proportion of post-Archean greenstone volcanics were erupted in shallow water.

Mafic igneous rocks

Greenstone basalts

As mentioned several times in previous sections, to identify secular changes in the chemical compositions of rocks it is important to compare rocks formed in similar tectonic settings to avoid the common pitfall of comparing apples with oranges. As discussed in Chapter 3, lithologic association tracks tectonic setting, and hence to minimize the tectonic setting bias in greenstone basalts a comparison will be made of basalt compositions from similar lithologic associations.

From incompatible element distributions in greenstone basalts, it is clear that basalts with island arc and submarine plateau affinities dominate in the Archean, whereas

basalts with calc-alkaline continental-margin arc affinities dominate in the post-Archean (Condie, 1989). Although Archean basalts show a complete range from MORB-like basalts to basalts with a strong subduction component, as illustrated by the Th/Ta–La/Yb ratios (Condie, 1994), basalts with low Th/Ta ratios dominate in most Archean greenstones (Figure 5.27a). Mafic plain basalts, which share geochemical affinities with submarine plateau or ocean ridge basalts (low Th/Ta and La/Yb ratios, Figure 5.27c), are more common in Archean than in Proterozoic greenstones. In fact, in Early Archean greenstones, like the Barberton in South Africa and the Pilbara in Western Australia, mafic plain basalts greatly dominate in the lower part of the successions (a). In contrast, Proterozoic greenstone basalts are dominantly those with a strong subduction geochemical component (high Th/Ta and La/Yb ratios, b), and mafic plain basalts are rare. Interestingly, none of the Precambrian greenstone basalts seems to record appreciable enriched (EM) or HIMU mantle components in their sources.

It is now well-established that at the same Mg number, which is a measure of magma fractionation, Archean basalts from all lithologic associations are enriched in Fe, Ni, Cr, and Co and depleted in Al compared to most post-Archean basalts (Arndt, 1991). On a Mg number versus Ni plot, modern basalts define one of three trends (Figure 5.28a):

1 the subduction trend in which Ni decreases rapidly with Mg number
2 the MORB trend
3 the submarine plateau trend in which Ni decreases only slightly with Mg number.

The subduction and MORB trends reflect either varying degrees of melting of the source or fractional crystallization. The submarine plateau trend, which is typical also of continental flood basalts, probably reflects a mantle plume source, where the temperature is higher than ambient mantle, and thus the Ni in derivative magmas is greater (Arndt, 1991). Archean basalts, regardless of tectonic setting, show relatively high Ni contents compared with post-Archean basalts at comparable Mg numbers (Figure 5.28b). This probably reflects higher mantle temperatures in the Archean, and correspondingly higher degrees of melting of the mantle. In contrast to the Archean trend, Proterozoic greenstone basalts show both subduction and submarine plateau trends, with the former, however, greatly dominating.

Mafic dykes

Mafic dyke swarms define distinct geochemical populations. Incompatible element ratios in these swarms are relatively insensitive to fractional crystallization and crustal contamination and appear to characterize the mantle sources (Condie, 1996). On a Th/Ta–La/Yb diagram, most swarms record a minimum of three mantle components in their source. Linear distributions showing increasing Th/Ta and La/Yb ratios appear to record

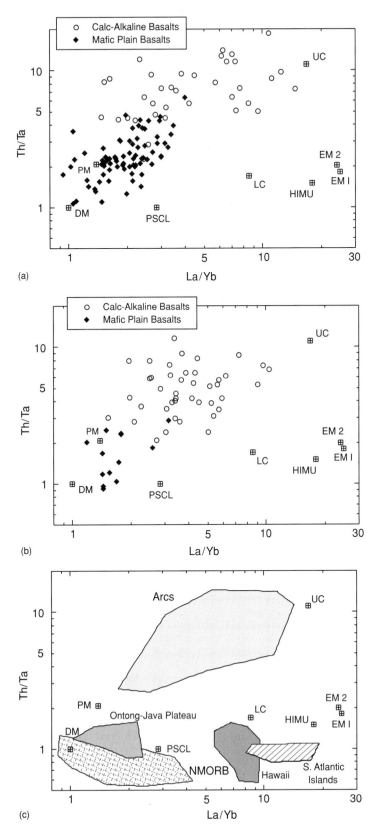

Figure 5.27 Th/Ta–La/Yb graphs showing distribution of basalts from various tectonic settings and ages. (a) Archean greenstone basalts; (b) Proterozoic greenstone basalts; (c) young basalts. DM, depleted mantle; PSCL, post-Archean subcontinental lithosphere; LC, lower continental crust; UC, upper continental crust; HIMU, high U/Pb mantle source; EM1 and EM2, enriched mantle sources. Mantle components defined in Chapter 4.

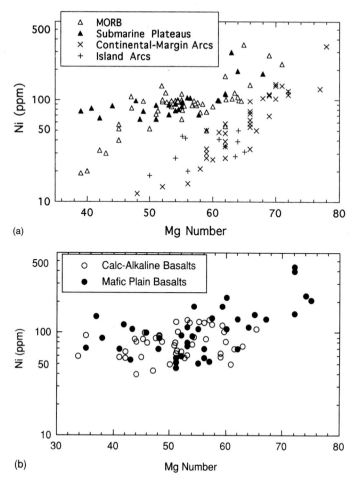

mixing of depleted mantle sources (DM or FOZO; Chapter 4) with Archean subcontinental lithosphere, the latter having a subduction geochemical component. An example is the Early Proterozoic Metachewan swarm in Ontario. Early Proterozoic dyke swarms range from those with a strong contribution of depleted mantle, probably in a plume source, to norite/picrite dyke swarms probably derived from the Archean subcontinental lithosphere. Incompatible element distributions in dykes show a change from the Archean and Early Proterozoic, where depleted and subcontinental lithosphere sources dominated, to the Late Proterozoic where there is also a contribution from enriched (EM) and HIMU mantle sources. The delayed appearance of HIMU–EM sources until the Late Proterozoic may reflect the time it takes to recycle oceanic lithosphere through the lower mantle, beginning in the Late Archean.

Komatiites

Perhaps the most distinctive volcanic rock of the Archean is komatiite. One of the striking differences between the Archean and post-Archean is the relative importance of

komatiites in the Archean (Arndt, 1994). As discussed in Chapter 3, komatiites are ultramafic lava flows (or hypabyssal intrusives) that exhibit a quench texture known as spinifex texture and contain > 18 per cent MgO. Although komatiites are common in some Archean greenstone successions, they are uncommon in the Proterozoic and very rare in the Phanerozoic. Such a secular distribution probably reflects a falling geothermal gradient in the Earth with time.

Although most petrologists agree that the dramatic decrease in abundance of komatiites at the end of the Archean reflects a decrease in mantle temperature, just where and how komatiites are produced in the mantle is not agreed upon. The temperature of the Archean mantle is not well-known and is best constrained by the MgO contents of komatiites, which are directly related to eruptive temperatures. Because of alteration, however, it is difficult to estimate the original MgO content of the magma. Maximum values of MgO range from 29–32 per cent, with the lower value from the least altered komatiites (Nisbet et al., 1993). The eruptive temperature of a komatiite with 30 per cent MgO is about 1600 °C. In comparison, the hottest modern magmas are

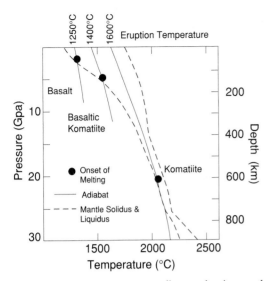

Figure 5.29 Pressure–temperature diagram showing mantle liquidus and solidus and various adiabats for basalts and komatiites.

basaltic komatiites (with 20 per cent MgO) from Gorgona Island (near Columbia) erupted at temperatures of 1400 °C, and typical basalts are not erupted at temperatures over 1300 °C (Figure 5.29). If 1600 °C is taken as the maximum eruption temperature of komatiites in the Archean, following a mantle adiabat suggests that their mantle source would have a temperature near 2000 °C (Figure 5.29). Because this temperature is greater than that predicted by the secular cooling curve of the mantle, it would appear that komatiites come from hot, rising plumes with temperatures 300 °C hotter than surrounding mantle, similar to the temperature difference calculated between modern plumes and surrounding mantle.

Many petrologists agree that komatiites were produced in the hot central tail region of mantle plumes, and associated basalts probably come from the cooler plume head (Campbell and Griffiths, 1992). Each follows a different adiabat (Figure 5.29). If a komatiitic magma with 30 per cent MgO separates from its source and rises to the surface erupting at 1600 °C, its segregation depth should be near 200 km, below which the density of the magma exceeds that of surrounding solids and so cannot separate. As dictated by geochemical models, the degree of melting in the rising magma must reach or exceed 30 per cent, which necessitates that melting begin at depths > 400 km, and perhaps as deep as 600 km as shown in Figure 5.29 (Arndt, 1994). A depletion in Al and heavy REE in some Archean komatiites seems to require fractional crystallization of majorite garnet (see Table 4.2), which requires pressures ≥ 10 Gpa, and perhaps as great as 24 Gpa (Xie et al., 1995). This clearly suggests that the depth of partial melting in Archean plumes was greater than that of modern plumes. In fact, recent experimental melting studies indicate a secular decrease in depth of melting in plumes with time from

< 5 Gpa in Cretaceous plumes to 5–10 Gpa in the Late Archean and 10–20 Gpa in the Early Archean (Herzberg, 1995). This pattern of decreasing depth of plume melting with time is consistent with a cooling Earth.

Blueschists

Blueschists are formed in association with subduction and continental collision and reflect burial to high pressures at relatively low temperatures (Figure 2.14). It has long been recognized that blueschists older than about 1000 Ma are apparently absent in the geologic record (Ernst, 1972). Those with aragonite and jadeitic clinopyroxene, which reflect the highest pressures, are confined to arc terranes < 200 Ma. Three general ideas have been proposed for the absence of pre-1000 Ma blueschists:

1 steeper geotherms beneath pre-1000 Ma arcs prevented rocks from entering the blueschist P–T stability field
2 uplift of blueschists led to recrystallization of lower-pressure mineral assemblages
3 erosion has removed old blueschists.

It may be that all three of these factors contribute to the absence of pre-1000 Ma blueschists. Prior to 2 Ga, steeper subduction geotherms may have prevented blueschist formation. After this time, however, when geotherms were not much steeper than at present, the second two factors may control blueschist preservation. Calculated P–T–time trajectories for blueschists suggest that they may increase in temperature prior to uplift (England and Richardson, 1977), resulting in recrystallization of blueschist-facies assemblages to greenschist- or amphibolite-facies assemblages. After 500 My of uplift and erosion, only the latter two assemblages would be expected to survive at the surface. Uplift and erosion after continental collisions may also remove blueschists. Even in young collisional mountain chains, such as the Himalayas, only a few minor occurrences of blueschist have not been removed by erosion.

Summary of chemical changes at the Archean–Proterozoic (A–P) boundary

The compositional changes in the upper continental crust that occurred at or near the A–P boundary can be related to one or more of four evolutionary changes in Earth history (Condie, 1993). Three of these are direct consequences of cooling of the mantle with time. First is the komatiite effect, which results from relatively large production rates of komatiites and high-Mg basalts in the Archean, and some of these rocks are trapped in the upper continental crust, either tectonically or by underplating with granitoids. Second is the TTG effect, resulting from the voluminous production of TTG in the Archean from partial melting of amphibolite (or eclogite) in which amphibole and/or garnet are left in the residue.

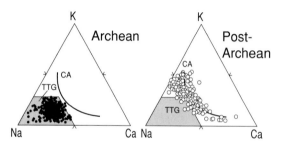

Figure 5.30 K–Na–Ca diagrams showing distribution of Archean and post-Archean granitoids. TTG, tonalite–trondhjemite field; CA, calc-alkaline trend. Courtesy of Herve Martin.

Third is the subduction effect, in which, after the Archean, greenstone basalt sources change from unmetasomatized to metasomatized mantle. At the same time, TTG production shifts from partial melting of wet mafic crust to fractional crystallization of basaltic magmas. Finally, some changes seem to be related to a paleo-weathering affect, in which the intensity of chemical weathering decreased after the Archean. This latter effect will be discussed in Chapter 6.

The komatiite effect

The high Mg, Cr, Co, and Ni in both Archean shales and in Archean upper continental crust argues for the existence of komatiites and high-Mg basalts in Archean crustal sources, but not in post-Archean crustal sources (Taylor and McLennan, 1985; Wronkiewicz and Condie, 1989). This effect also strongly contributes to the high Cr/Th, Co/Th, and Ni/Co ratios and low Th/Sc ratio in Archean upper continental crust and Archean sediments. The sublinear variation in Cr and Ni in post-Archean shales can be explained by mixing felsic and mafic endmembers

(Figure 5.21). The fact that Archean shales do not define a linear array in Ni–Cr space may be caused by selective enrichment of Cr in Archean shales, either by selective adsorption in clays during weathering, or by the presence of fine-grained detrital chromite in Archean sediments.

As previously discussed, the greater amount of komatiite and high-Mg basalt in the Archean reflects higher mantle temperatures at that time.

The TTG effect

Several important geochemical differences exist between Archean and post-Archean TTG (Martin, 1993; 1994). As reflected by their large amounts of Na-plagioclase, Archean TTG are relatively high in Na compared to K and Ca (Figure 5.30). In fact, almost all Archean TTG fall in the tonalite–trondhjemite field and have a limited compositional distribution. On the other hand, post-Archean granitoids are calc-alkaline ranging in composition from diorite to granite, with an abundance of granite-granodiorite. Most striking among the trace elements is the relative depletion in Y and heavy REE in Archean TTG compared to post-Archean TTG. This distinction is particularly well-defined on a La/Yb versus Yb plot (Figure 5.31). The Archean REE patterns are strongly fractionated with low Yb contents. In contrast, post-Archean TTG show only moderate REE fractionation and a broad range of Yb values.

The relatively low heavy REE and Y contents of Archean upper continental crust are well-established and reflect heavy REE-depleted TTG, which dominates in the Archean upper continental crust. Heavy REE depletion in TTG appears to require a mafic source, in which amphibole and/or garnet, both of which concentrate heavy REE, remain in the residue after melt extraction (Martin, 1993; 1994). Low Nb and Ti and low Sm/Nd ratios in Archean upper continental crust also probably

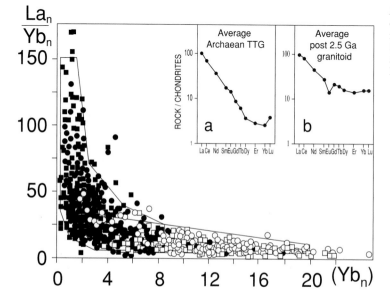

Figure 5.31 La_n/Yb_n ratio versus Yb_n content of TTG. La_n and Yb_n are values normalized to chondritic meteorites. Solid symbols are Archean and open symbols post-Archean. Courtesy of Herve Martin.

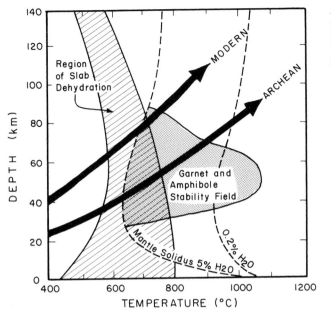

Figure 5.32 Pressure–temperature graph showing possible P–T paths of Archean and modern subduction zones. The 5% H_2O solidus applies to the Archean and the 0.2% H_2O solidus to the present time.

reflect fractionation of REE by amphibole or garnet-rich residues.

These differences in REE and related element behaviour in upper continental crust can be explained if Archean TTG are produced by partial melting of wet, mafic crust with amphibole and/or garnet left in the restite, whereas most post-Archean felsic magmas are produced by fractional crystallization at shallower depths (Martin, 1994). Thus, garnet and amphibole retain Y and heavy REE in the source of the Archean magmas, but do not play a role in the formation of the post-Archean magmas. In the Archean, subducted oceanic crust was relatively young (< 30 My) and warm, and reached melting conditions during slab dehydration due to steeper subduction geotherms (Figure 5.32). These conditions may also have existed in the deep root zones of submarine plateaux. Melting occurs in the stability field of garnet and amphibole and one or both of these phases are left in the restite. In contrast, modern subducted oceanic crust is old and cool and dehydrates before melting. In this case, fluids released into the mantle wedge promote melting leading to the production of basalts that later undergo fractional crystallization to produce felsic magmas. Thus, decreasing geotherms at convergent plate margins may have led to a change in the site of subduction-related magma production from the descending slab and/or deep roots of submarine plateaux in the Archean to the mantle wedge thereafter.

The subduction effect

Taylor and McLennan (1985; 1995) have suggested that LIL element enrichment in post-Archean sediments is inherited from granites produced by intracrustal melting following rapid continental growth at the end of the

Archean. Although the widespread occurrence of cratonic sediments beginning in the Early Proterozoic may reflect the rapid growth of cratons in the Late Archean as previously discussed, the increase in LIL elements in both sediments and average upper continental crust is not necessarily tied to a changing mass of continental crust. It may reflect the formation of an enriched source for granitoids beginning in the Proterozoic, irrespective of continental growth rate. Also, the fact that the TTG/granite ratio in the upper continental crust is roughly constant from the Early Archean onwards (Condie, 1993) does not favour an increasing importance of granite in upper crustal sources after the end of the Archean.

Enrichment in LIL elements in felsic igneous rocks and in upper continental crust at the end of the Archean is accompanied by an increase in the size of the Eu anomaly and small decreases in such ratios as K/Rb, Ba/Rb, and La/Th (Condie, 1993). Similar, but less pronounced changes in LIL element distributions occur in greenstone-related basalts and andesites at the end of the Archean. The increase in LIL elements in arc-related igneous rocks beginning in the Early Proterozoic may be related to an increase in depth of devolatilization of descending slabs (Figure 5.32), which resulted in an increase in the total volume of mantle wedge that was metasomatized by escaping fluids. Because of the greater volume of metasomatized wedge, post-Archean subduction-related magmas may carry a more prominent subduction-zone geochemical signature (LIL element enrichment relative to Nb and Ta) than Archean subduction-related magmas. Another possibility is that Archean TTG was produced by partial melting of unmetasomatized roots of thickened submarine plateaux, and later the source shifted to metasomatized mantle wedges (Condie, 1992b).

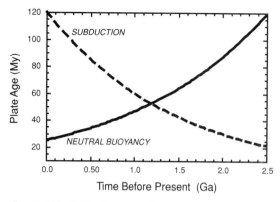

Figure 5.33 Subduction age and neutral buoyancy age versus time. Modified after Davies (1992).

Evolution of the crust–mantle system

Archean plate tectonics

If the Earth has steadily cooled with time as is widely accepted, there is good reason to suspect that plate tectonics may have been different in the Archean. For instance, a hotter mantle in the Archean would have produced more melt at ocean ridges and hence a thicker oceanic crust. Model calculations indicate, in fact, that the Archean oceanic crust should have been about 20 km thick (maybe even thicker in the Early Archean) compared with the present thickness of about 7 km (Sleep and Windley, 1982). Since oceanic crust is less dense than the mantle, the Archean oceanic lithosphere would have been more buoyant, and thus more difficult to subduct. A hotter mantle would also convect faster, probably causing plates to move faster, and thus oceanic lithosphere would have less time to thicken and become gravitationally unstable. Both of these factors would tend to make plates subduct buoyantly in the Archean.

Considering the density and thickness distribution of lithosphere and crust, modern oceanic lithosphere reaches neutral buoyancy in about 20 My (Davies, 1992). After this time, the lithosphere becomes negatively buoyant, but it is not immediately subducted. From the current rate of plate production at ocean ridges ($2.5 \text{ km}^2/\text{y}$) and the total area of sea floor (about $3 \times 10^{14} \text{ m}^2$), the mean age of oceanic crust when it begins to subduct is about 120 My. In a hotter mantle, it takes oceanic plates longer to become neutrally buoyant, and thus, in the Late Archean, neutral buoyancy would not be reached until the plate is ≥ 100 My in age (Figure 5.33). On the other hand, in a hotter mantle plates should move faster in response to faster convection, and thus they should subduct sooner. In the Late Archean they should subduct at about 20 My (Figure 5.33). Hence, while at present the time needed to reach neutral buoyancy is less than the subduction age, as needed for modern-style plate tectonics, in the Archean plates would be ready to subduct

before they reached neutral buoyancy. The crossover in these two ages is about 1 Ga (Figure 5.33).

Although it would have been possible for plates to buoyantly subduct in the Archean, they would not have been moving fast enough to remove the excess Archean heat (Davies, 1992; 1993). How then can this excess Archean heat be lost? It seems that mantle plume activity was probably greater in the Archean than afterwards but plumes cannot be a substitute for heat loss by plate tectonics since they bring heat into the mantle from the core and only marginally increase heat loss from the top of the mantle. In effect, plates cool the mantle while plumes cool the core. What is required is a mechanism in the lithosphere that promotes heat loss at the surface. At least three possibilities merit consideration:

1 A greater total length of the Archean ocean ridge system could help alleviate the problem (Hargraves, 1986). For instance, if Archean heat production were three times that of the present heat production, the total ocean ridge length would have to be twenty-seven times the present to accommodate the additional heat loss.
2 The inversion of basalt to eclogite in buoyantly-subducted Archean oceanic crust may increase the density of the lithosphere sufficiently for it to subduct or delaminate. This mechanism is attractive in that thick Archean oceanic crust provides a large reserve of mafic rocks at depth that could convert to eclogite. Experimental data indicate that conditions are particularly favourable for eclogite formation and delamination at depths of 75–100 km (Rapp and Watson, 1995).
3 The oceanic mantle lithosphere may have been negatively buoyant due to latent heat loss associated with extraction of melt at ocean ridges (Davies, 1993). Two competing factors affect the density of the oceanic mantle lithosphere. The extraction of melt leaves a residue depleted in garnet and thus less dense. On the other hand, the latent heat of melting is carried away by the melt, leaving a residue colder and more dense. If the latent heat effect dominated in the Archean, the oceanic lithosphere could have become negatively buoyant.

Towards an Archean model

Although plate tectonics undoubtedly operated in the Archean, it must have differed in some fundamental aspects from modern plate tectonics for the excess heat from the Earth's upper boundary layer to be effectively removed (Durrheim and Mooney, 1994). As reviewed above, there is an extensive data base to constrain mantle and crustal evolution, both in the Archean and afterwards. Perhaps the single most important factor leading to evolutionary changes is the cooling of the Earth. Most of the changes that occur near the end of the Archean probably reflect cooling of the mantle through threshold temperatures of various mantle processes, which lead to

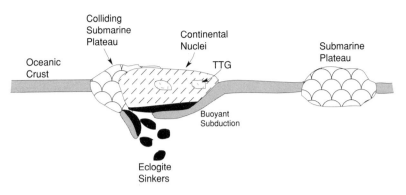

Figure 5.34 Model for the formation of Archean continents.

changes not only in how the mantle evolves, but also in how the crust and atmosphere–ocean system evolve. The following is a consideration of a possible model for the Archean which accommodates the constraints discussed in this chapter.

If fragments of submarine plateaux comprise a significant proportion of Archean greenstones as seems likely, mantle plume activity may have been more widespread than at present. Perhaps the first continental nuclei were submarine plateaux produced from mafic and komatiitic melts extracted from mantle plumes. The plume model provides a means of obtaining komatiites and basalts from the same source: komatiites from the high-temperature plume tail and basalts from the cooler head (Campbell et al., 1989). Spent plume material could be underplated beneath the submarine plateaux causing significant thickening of the lithosphere. This material would be depleted from melt extraction, and thus be relatively low in garnet, which would make it buoyant. Partial melting of the thickened mafic roots of the submarine plateaux could give rise to TTG magmas, producing the first differentiated continental nuclei (Figure 5.34). Buoyant subduction around the perimeters of the plateaux should result in both lateral continental growth and thickening of the lithosphere (Abbott and Mooney, 1995). Thin slabs of oceanic crust may be obducted, remnants of which should be preserved in some Archean greenstones. Because these slabs are thin, they would not include complete ophiolites, and perhaps not even sheeted dykes. Collision of submarine plateaux and island arcs would also contribute to the growing continental nuclei, and both of these should be represented in Archean greenstones. Descending slabs would sink, either because of their negative buoyancy and/or because of eclogite production in buoyantly-subducted oceanic crust (Figure 5.34). TTG could be produced in the mafic descending slabs and/or in the thickened mafic continental roots. In either case, garnet must be left in the restite to explain the heavy REE depletion in Archean TTG. A subduction zone geochemical component would be imparted to the thickening mantle lithosphere by dehydration and partial melting of descending slabs. Some TTG may also be produced in delaminated fragments of eclogite coming from oceanic crust plastered beneath the continental nuclei by buoyant subduction.

Changes at the Archean–Proterozoic boundary

Near the end of the Archean, progressive cooling of the mantle would lead to a change in thermal regime in both the upper and lower mantle. Except during periods of catastrophic overturn as described below, fewer plumes would be produced in the lower mantle, and submarine plateaux would become less frequent at the expense of island arcs. This would explain the apparent decrease in abundance of submarine plateau basalts in post-Archean greenstones. High-Mg komatiites would no longer be produced in mantle plumes, because plume temperatures would not be high enough to generate these magmas. Low-Mg komatiites would continue to be produced in the hot tails of some plumes to the present time. Because of decreasing temperatures at convergent margins, after the Archean descending slabs would no longer melt (except locally), but melting would shift to metasomatized and hydrated mantle wedges. Hence, TTG would no longer show depletions in heavy REE. Arc-related magmas would henceforth come directly or indirectly from metasomatized mantle wedges, and show increased contents of LIL elements. Increased preservation rate, if not increased growth rate, of continents in the Late Archean resulted in the first widespread occurrence of continental crust. Partial melting of this crust during post-Archean collisional orogenies produced granite with negative Eu anomalies, explaining the increase in size of Eu anomalies in post-Archean continental crust. Hence, from the Early Proterozoic onwards, plate tectonics and crustal and mantle processes would operate much like they do today.

Episodic ages and mantle dynamics: a possible connection?

From the age distributions in juvenile continental crust, it would appear that Earth history can be divided into three stages (Condie, 1995) (Figure 5.35):

Stage I   > 2.8 Ga, when greenstones and microcontinents formed and collided continuously, although probably not forming a supercontinent until about 3 Ga

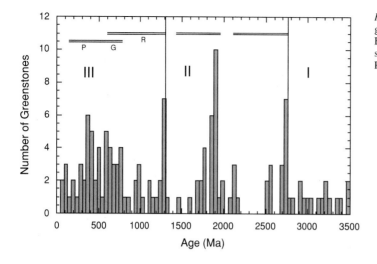

Figure 5.35 Episodic distribution of greenstone ages showing three stages of Earth history. Known or possible supercontinent cycles shown at top. R, Rodinia, G, Gondwana, P, Pangea.

Stage II   > 2.8–1.3 Ga, where a clear episodicity is apparent in ages, and where two greenstone age minima coincide in age with supercontinents

Stage III < 1.3 Ga, when it appears that new greenstones formed and collided continuously with continents.

Why the period of time between 2.8–1.3 Ga was so different from both earlier and later times is unknown. Could it be related to changes in mantle convection patterns?

The peaks in greenstone and continental age distributions at 2.7, 1.9, and 1.3 Ga may reflect catastrophic overturn of the mantle initiated by episodes of sinking of cold lithospheric slabs through the 660-km discontinuity to the D″ layer, where the slabs are heated to generate large plumes that ascend to the base of the lithosphere (Stein and Hofmann, 1994; Condie, 1995;

Davies, 1996). A 2.1-Ga crustal forming episode in West Africa–Guiana may reflect a more localized mantle overturn. The increased plume activity should result in greater production rates of oceanic terranes (especially submarine plateaux), and also in increased continental growth rates as pointed out by Stein and Hofmann (1994). As discussed in Chapter 4, a hotter Archean mantle would result in higher Rayleigh numbers, which in turn would favour layered-mantle convection because the buoyancy effect at the 660-km discontinuity would be enhanced. Hence, descending slabs should be continuously recycled in the upper mantle (Figure 5.36). The rate of oceanic (greenstone) volcanism and continental growth should decrease at a rate similar to the cooling rate of the Earth. Because of recycling of continental crust into the mantle, however, the net growth rate of continental crust would only gradually decrease or remain approximately constant. Thus, during Stage I of Earth history,

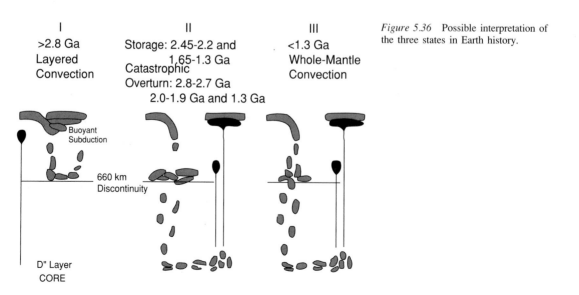

Figure 5.36 Possible interpretation of the three states in Earth history.

there would be gradual but steady changes in magmatic activity both in the oceans and on the continents.

The three periods of rapid increase in continental growth rate in Stage II (Figure 5.35) may be related to one of two factors. Both involve catastrophic plume activity that transports heat into the mantle, which results in a greater magma production rate, and hence an increase in continental growth rate. In the first case, the plumes result from the sudden sinking of slabs through the 660-km discontinuity, whereas in the second they are caused by heat liberated from the core during an episode of rapid growth of the inner core. As the Earth continues to cool, the buoyancy effect of the 660-km phase change should decrease, and slabs begin to penetrate the discontinuity (Davies, 1996). As this process begins, slabs may temporarily accumulate at the 660-km discontinuity, and finally sink as an 'avalanche' of slabs near the end of the Archean. They would sink rapidly to the D'' layer where they would be heated and ascend as mantle plumes. This avalanching could have happened three times during Stage II (Figure 5.36). The first may have occurred about 3 Ga, which led to a plume episode at 2.8 Ga during which time the first supercontinent formed (2.75 Ga, Figure 5.12). Catastrophic slab collapses may also have occurred at 2.1, 1.9 and 1.3 Ga, causing increased greenstone and continent production at these times.

An obvious question is why was there not another catastrophic slab collapse after 1.3 Ga? During Stage III of Earth history there is no evidence for episodic events in the mantle or core. Although it is possible that slab collapse will occur in the future, as discussed in Chapter 4, there is little support in seismic-wave velocity data for accumulation of slabs at the 660-km discontinuity today. Even though some slabs may be locally detained at this discontinuity, seismic tomography results seem more consistent with a steady sinking of slabs into the lower mantle. Thus soon after 1.3 Ga, the Earth may have changed from partially-layered convection to whole-mantle convection. The elimination of catastrophic slab avalanches accompanying this change may explain the cessation of episodicity in the production of juvenile continental crust and of greenstones in the last 1.3 Gy. If this model bears any resemblance to what really happened, just how and why the 660-km discontinuity changed in character at 1.3 Ga is an exciting area for future research.

Summary statements

1 Early in the Earth's history when the mantle was hotter than at present, viscosity was low and convection rapid, perhaps chaotic, as dictated by high Rayleigh numbers. During this period of about 500 My, the Earth cooled rapidly, followed by gradual cooling to the present time.

2 The Earth's earliest crust, which was recycled into the mantle, was probably the product of crystallization of a magma ocean and was probably basaltic in composition.

3 The Earth's oldest rocks at 4.0–3.6 Ga, occur as tectonically-juxtaposed terranes and include tonalitic to granitic gneisses, greenstones (basalts, komatiites), ultramafic rocks, and a variety of volcaniclastic and chemical sediments (chiefly chert/ BIF). They appear to have formed in oceanic tectonic settings such as submarine plateaux and island arcs.

4 The early oceanic crust, although thicker than modern oceanic crust, was widespread and generated by partial melting of ultramafic rocks at ocean ridges. It is likely that the earliest continental crust developed from mafic submarine plateaux, either by partial melting of the thickened mafic roots and/or by melting of oceanic crust in subducted slabs.

5 The two most important ways in which continents grow are magma additions by under- and overplating in arcs, rifts, and flood basalt fields, and by oceanic terrane collisions around continental margins.

6 The occurrence of several modern petrotectonic assemblages at 4.0–3.5 Ga strongly supports some sort of plate tectonics operating on the Earth at this time. By 3.2–3.0 Ga, cratonic/passive margin sediments and continental rifts record the development of continental cratons. Seismic reflection profiles across Archean provinces and studies of motion along Archean transcurrent faults also support the existence of plate tectonics during the Archean.

7 Although episodic ages characterize the continental crust, with three exceptions at 2.7–2.6, 1.9–1.7 and 1.3–1.0 Ga, U–Pb zircon ages do not support correlative worldwide episodicity of orogeny. Only the 2.7–2.6 and 1.9–1.7 Ga events are times of major juvenile crust production.

8 Greenstone age minima at 2.5–2.2 and 1.65–1.35 Ga correlate with supercontinents, and may reflect greater rates of recycling of oceanic terranes into the mantle.

9 Significant amounts of continental crust are destroyed and returned to the mantle by subduction erosion and sediment subduction. An unknown, but probably minor amount of continental crust may be returned to the mantle by delamination of the lower crust during collisional orogeny.

10 The wide range of E_{Nd} values in Early Archean rocks requires extreme early fractionation accompanied by efficient recycling to generate these differences by 4 Ga. The paucity of Early Archean rocks may reflect the efficiency of this recycling rather than a lack of production of early continental crust.

11 The gradual increase in E_{Nd} values in the last 4 Gy is consistent with a gradual increase in the volume of preserved continental crust with time as a result of cooling of the mantle and consequent decreases in recycling rates.

12 Archean upper continental crust is enriched in Mg, Ti, Cr, Ni, and Co, and depleted in LIL elements,

P, Zr, Hf, Nb, Ta, Y, and heavy REE compared with post-Archean upper continental crust. Negative Eu anomalies characterize upper crust of all ages, although they are relatively small in Archean upper crust.

13   Changes in Eu anomalies and in Sr and Nd isotopic composition of seawater at the A–P boundary proposed by earlier investigators are an artifact of comparing data from greenstone and continental tectonic settings. Any transition between mantle- and continent-buffered seawater must have occurred before 4.3 Ga. Changes in the $^{87}Sr/^{86}Sr$ ratio in marine carbonates during the last 1 Gy track the fragmentation and assembly of supercontinents.

14   The apparent absence of Archean ophiolites may be due to thicker oceanic crust in the Archean. Tectonic slices of Archean oceanic crust may only include the upper few kilometres of pillow basalts and not the lower ophiolite units.

15   Mafic plain basalts, which share geochemical affinities with submarine plateau or ocean ridge basalts, are more common in Archean than in Proterozoic greenstones. Proterozoic greenstone basalts are dominantly those with a strong subduction geochemical component.

16   At the same Mg number, Archean basalts from all lithologic associations are enriched in Fe, Ni, Cr, and Co and depleted in Al compared to most post-Archean basalts. This probably reflects higher mantle temperatures in the Archean, and correspondingly higher degrees of melting of the mantle.

17   Geochemical evidence for enriched (EM) and HIMU mantle sources first appears in Late Proterozoic mafic igneous rocks. This delayed appearance may reflect the time it takes to recycle oceanic lithosphere through the lower mantle.

18   A dramatic decrease in the abundance of komatiites at the end of the Archean reflects a decrease in mantle temperatures. Komatiites probably come from hot mantle plumes, and the depth of melting in plumes appears to have decreased from the Archean to the present.

19   Most compositional changes in continental upper crust at the A–P boundary can be explained by: (1) a greater amount of basalt and komatiite in Archean upper continental crust; (2) garnet/amphibole fractionation during production of TTG by partial melting of hydrous basalt; or (3) change in TTG magma source from descending slab or thickened mafic crust to metasomatized mantle. All of these changes re-quire higher temperatures in the Earth during the Archean.

20   Hotter mantle in the Archean may have produced more melt at ocean ridges resulting in a thick oceanic crust, thus reducing the density of the oceanic lithosphere, permitting only buoyant subduction.

21   It is unlikely that modern plate tectonics could remove the excess heat in the Archean mantle. Possible processes that might contribute to cooling the Archean mantle include a greater total length of the ocean ridge system, delamination of eclogite in buoyantly-subducted crust, and subduction of lithosphere with negative buoyancy, imparted by loss of latent heat of melting when ocean ridge basalt is extracted.

22   Earth history can be divided into three stages: (I) > 2.8 Ga, when greenstones and microcontinents formed and collided continuously, although probably not forming a supercontinent until about 3 Ga; (II) 2.8–1.3 Ga, where a clear episodicity is apparent in ages, and where two greenstone age minima coincide in age with supercontinents; and (III) < 1.3 Ga, when it appears that new greenstones formed and collided continuously with continents.

23   The peaks in greenstone and continental age distributions at 2.7, 1.9, and 1.3 Ga may reflect catastrophic overturn of the mantle initiated by episodes of sinking of cold lithospheric slabs through the 660-km discontinuity to the D″ layer, where the slabs were heated to generate plumes that ascended to the base of the lithosphere.

Suggestions for further reading

Bowring, S. A. and Housh, T. (1995). The Earth's early evolution. *Science*, **269**, 1535–1540.

Condie, K. C., editor (1992). *Proterozoic Crustal Evolution*. Amsterdam, Elsevier, 538 pp.

Condie, K. C. (1993). Chemical composition and evolution of the upper continental crust: Contrasting results from surface samples and shales. *Chem. Geol.*, **104**, 1–37.

Condie, K. C., editor (1994). *Archean Crustal Evolution*. Amsterdam, Elsevier, 528 pp.

Rudnick, R. L. (1995). Making continental crust. *Nature*, **378**, 571–577.

Taylor, S. R. and McLennan, S. M. (1985). *The Continental Crust: Its Composition and Evolution*. Oxford, Blackwell Scient., 312 pp.

Chapter 6

The atmosphere, oceans, climates, and life

Introduction

Not only in terms of plate tectonics is the Earth a unique planet in the Solar System, but it is the only planet with oceans and with an oxygen-bearing atmosphere capable of sustaining higher forms of life. How did such an atmosphere–ocean system arise and why only on the Earth? A related question is, once formed how did the atmosphere and oceans evolve with time and, in particular, when and how did free oxygen enter the system? How have climates changed with time and what are the controlling factors, and when and how was life created? What is the role of plate tectonics, mantle plumes, and extraterrestrial impact in the evolution of the atmosphere, oceans, and life? It is these and related questions that will be addressed in this chapter.

General features of the atmosphere

Atmospheres are the gaseous carapaces that surround some planets and, because of gravitational forces, they increase in density towards planetary surfaces. The Earth's atmosphere is divided into six regions as a function of height (Figure 6.1). The magnetosphere, the outermost region, is composed of high-energy nuclear particles trapped in the Earth's magnetic field. This is underlain by the exosphere in which lightweight molecules (such as H_2) occur in extremely low concentrations and escape from the Earth's gravitational field. Temperature decreases rapidly in the ionosphere (to about $-90\,°C$) and then increases again to near $0\,°C$ at the base of the mesosphere. It drops again in the stratosphere and then rises gradually in the troposphere towards the Earth's surface. Since warm air overlies cool air in the stratosphere, this layer is relatively stable and undergoes very little vertical mixing. The temperature maximum at the top of the stratosphere is caused by absorption of ultraviolet (UV) radiation in the ozone layer. The troposphere

is a turbulent region that contains about 80 per cent of the mass of the atmosphere and most of its water vapour. Tropospheric temperature decreases towards the poles which, together with vertical temperature changes, causes continual convective overturn in the troposphere.

The Earth's atmosphere is composed chiefly of nitrogen (78 per cent) and oxygen (21 per cent) with very small amounts of other gases such as argon and carbon dioxide. In this respect, the atmosphere is unique among planetary atmospheres (Table 6.1). Venus and Mars have atmospheres composed largely of CO_2, and the surface pressure on Venus ranges up to 100 times that on Earth while the surface pressure of Mars is $< 10^{-2}$ of that of the Earth. The surface temperatures of Earth, Venus and Mars are also very different (Table 6.1). The outer planets are composed largely of hydrogen and helium and their atmospheres consist chiefly of hydrogen and, in some cases, helium and methane.

The concentrations of minor gases such as CO_2, H_2 and ozone (O_3) in the Earth's atmosphere are controlled primarily by reactions in the stratosphere caused by solar radiation. Solar photons fragment gaseous molecules (such as O_2, H_2, CO_2) in the upper atmosphere producing free radicals (C, H, O), a process called **photolysis**. One important reaction produces free oxygen atoms which are unstable, and which recombine to form ozone. This reaction occurs at heights of 30–60 km, with most ozone collecting in a relatively narrow band at about 25–30 km (Figure 6.1). Ozone, however, is unstable and continually breaks down to form molecular oxygen. The production rate of ozone is approximately equal to the rate of loss, and thus the ozone layer maintains a relatively constant thickness in the stratosphere. Ozone is an important constituent in the atmosphere because it absorbs UV radiation from the Sun, which is lethal to most forms of life. Hence, the ozone layer provides an effective shield that permits a large diversity of living organisms to survive on the Earth. It is for this reason that we must be concerned about the release of synthetic chemicals into the atmosphere which destroy the ozone layer.

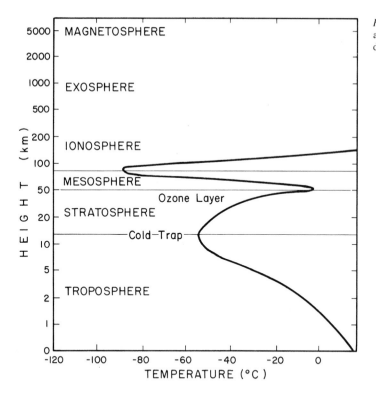

Figure 6.1 Major divisions of the Earth's atmosphere showing average temperature distribution.

Table 6.1 Composition of planetary atmospheres

	Surface temperature (°C)	Surface pressure (bars)	Principal gases
Earth (Early Archean)	~ 85	~ 11	CO_2 (N_2, CO, CH_4)
Earth	−20 to 40	0.1–1	N_2, O_2
Venus	400 to 550	10–100	CO_2 (N_2)
Mars	−130 to 25	≤ 0.01	CO_2 (N_2)
Jupiter	−160 to −90	≤ 2	H_2, He
Saturn	−180 to −120	≤ 2	H_2, He
Uranus	−220 to −120	≤ 5	H_2, CH_4
Neptune	−220 to −120	≤ 10	H_2
Pluto	−220 to −200	≤ 0.005	CH_4

The distributions of N_2, O_2 and CO_2 in the atmosphere are controlled by volcanic eruptions and by interactions between these gases and the solid Earth, the oceans, and living organisms.

The primitive atmosphere

Three possible sources have been considered for the Earth's atmosphere: residual gases remaining after Earth accretion, extraterrestrial sources, and degassing of the Earth by volcanism. Of these, only degassing accommodates a wide variety of geochemical and isotopic con-

straints. One line of evidence supporting a degassing origin for the atmosphere is the large amount of ^{40}Ar in the atmosphere (99.6 per cent) compared with the Sun or a group of primitive meteorites known as carbonaceous chondrites (both of which contain < 0.1 per cent ^{40}Ar). ^{40}Ar is produced by the radioactive decay of ^{40}K in the solid Earth and escapes into the atmosphere chiefly by volcanism. The relatively large amount of this isotope in the terrestrial atmosphere indicates that the Earth is extensively degassed of Ar and, because of a similar behaviour, also of other gases.

Although most investigators agree that the present atmosphere, except for oxygen, is the product of degassing, whether a primitive atmosphere existed and was lost before extensive degassing began is a subject of controversy. One line of evidence supporting the existence of an early atmosphere is the fact that volatile elements should collect around accreting planets during their late stages of accretion. This follows from the very low temperatures at which volatile elements condense from the Solar Nebula (Chapter 7). A significant depletion in rare gases in the Earth compared with carbonaceous chondrites and the Sun indicates that if a primitive atmosphere collected during accretion, it must have been lost. The reason for this is that gases with low atomic weights (CO_2, CH_4, NH_3, H_2 etc.) which probably composed this early atmosphere should be lost even more readily than rare gases with high atomic weights (Ar, Ne, Kr, Xe). Just how such a primitive atmosphere may have been lost is not clear. One possibility is by a T-

Tauri solar wind (see Chapter 7). If the Sun evolved through a T-Tauri stage during or soon after (< 100 My) planetary accretion, this wind of high-energy particles could readily blow volatile elements out of the inner Solar System. Another way in which an early atmosphere could have been lost is by impact with one or more large bodies during the late stages of planetary accretion.

Two models have been proposed for the composition of a primitive atmosphere. The Oparin–Urey model (Oparin, 1953) suggests that the atmosphere was reducing and composed dominantly of CH_4 with smaller amounts of NH_3, H_2, He and H_2O, while the Abelson model (Abelson, 1966) is based on an early atmosphere composed of CO_2, CO, H_2O and N_2. Neither atmosphere allows significant amounts of free oxygen, and experimental studies indicate that reactions may occur in either atmosphere which could produce the first life.

By analogy with the composition of the Sun and the compositions of the atmospheres of the outer planets and of volatile-rich meteorites, an early terrestrial atmosphere may have been rich in such gases as CH_4 and NH_3, and would have been a reducing atmosphere. One of the major problems with an atmosphere in which CH_4 and NH_3 are important is that both of these species are destroyed directly or indirectly by photolysis (Cogley and Henderson-Sellers, 1984). Ammonia is destroyed by UV radiation in as little as 1–10 y. In addition, NH_3 is highly soluble in water and should be removed rapidly from the atmosphere by rain and solution at the ocean surface. Although CH_4 is more stable against photolysis, OH which forms as an intermediary in the methane oxidation chain, is destroyed by photolysis at the Earth's surface in < 50 y. H_2 rapidly escapes from the top of the atmosphere and is also an unlikely major constituent in an early atmosphere. Thus, a reducing primitive atmosphere seems unlikely, and only a non-oxidizing Abelson-type atmosphere seems capable of surviving photolysis.

The degassed atmosphere

Excess volatiles

The present terrestrial atmosphere appears to have formed by the degassing of the mantle and crust. **Degassing** is the liberation of gases from within a planet, and it may occur directly during volcanism, or indirectly by weathering of igneous rocks on a planetary surface. For the Earth, volcanism appears to be most important both in terms of current degassing rates and calculated past rates. The volatiles in the atmosphere, hydrosphere, biosphere, and sediments that cannot be explained by weathering of the crust are known as **excess volatiles** (Rubey, 1951). These include most of the H_2O, CO_2 and N_2 in these near-surface reservoirs. The similarity in the distribution of excess volatiles in volcanic gases to those in near-surface reservoirs (Table 6.2) strongly supports a volcani-

Table 6.2 Excess volatiles in volcanic gases and near-surface terrestrial reservoirs

	Volcanic gases (%)	*Near-surface reservoirs (%)
H_2O	83	87
CO_2	12	12
Cl_1, N_2, S	5	1

*includes atmosphere, hydrosphere, biosphere and sediments

genic origin for these gases, and thus also supports a degassing origin for the atmosphere.

Composition of the early atmosphere

Two models have been proposed for the composition of the early degassed atmosphere depending on whether metallic iron existed in the mantle in the early Archean. If metallic iron was present, equilibrium chemical reactions would liberate large amounts of H_2, CO, and CH_4 and small amounts of CO_2, H_2O, H_2S and N_2 (Holland, 1984). If iron was not present, reactions would liberate mostly CO_2, H_2O and N_2 with minor amounts of H_2, HCl, and SO_2. Because most evidence suggests the core began to form during the late stages of planetary accretion (Chapter 4), it is possible that little if any metallic iron remained in the mantle when degassing occurred. However, if degassing began before the completion of accretion, metallic iron would have been present in the mantle and the first atmosphere would have been a hot, steamy one composed chiefly of H_2, CO_2, H_2O, CO, and CH_4. Because the relative timing of early degassing and core formation are not well-constrained, the composition of the earliest degassed atmosphere is not well-known. Both core formation and most degassing were probably complete in < 100 My after accretion and the composition of the early atmosphere may have changed rapidly during this time interval in response to decreasing amounts of metallic iron in the mantle. It is likely, however, that soon after accretion was complete at 4.45 Ga, H_2 rapidly escaped from the top of the atmosphere and water vapour rained out to form the oceans. This leaves an early atmosphere rich in CO_2, CO, N_2, and CH_4 (Holland et al., 1986; Kasting, 1993). As much as 15 per cent of the carbon now found in the continental crust may have resided in this early atmosphere, which is equivalent to a partial pressure of CO_2 + CO of 10 bars and N_2 of about 1 bar (Table 6.1). The mean surface temperature of such an atmosphere would have been about 85 °C. Even after the main accretionary phase of the Earth had ended, major asteroid and cometary impacts probably continued to about 3.9 Ga, as inferred from the lunar impact record. These impactors could have added more carbon as CO to the atmosphere, and produced NO by shock heating of atmospheric CO_2 and N_2.

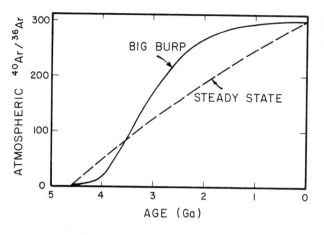

Figure 6.2 Idealized evolution of atmospheric $^{40}Ar/^{39}Ar$ in the steady state and big burp models for terrestrial atmospheric growth.

Growth rate of the atmosphere

Two extreme scenarios are considered for the growth of the atmosphere with time: the **big burp model**, in which the atmosphere grows by rapid degassing during or soon after planetary accretion (Fanale, 1971), and the **steady state model**, where the atmosphere grows slowly over geologic time (Rubey, 1951). One way of distinguishing between these models is to monitor the build-up of ^{40}Ar, ^{129}Xe and ^4He in sedimentary rocks which have equilibrated with the atmosphere–ocean system through time. ^{40}Ar is produced by the radioactive decay of ^{40}K in the Earth, and as it escapes from the mantle it collects in the atmosphere. Since ^{36}Ar is non-radiogenic, the $^{40}Ar/^{36}Ar$ ratio should record distinct evolutionary paths for Earth degassing. The steady state model is characterized by a gradual increase in the $^{40}Ar/^{36}Ar$ ratio with time, while the big burp model should show initial small changes in this ratio followed by rapid increases (Figure 6.2). This is because ^{40}Ar is virtually absent at the time of accretion and hence the $^{40}Ar/^{36}Ar$ ratio is not sensitive to early atmospheric growth. Later in the big burp model, however, as ^{40}Ar begins to be liberated, the $^{40}Ar/^{36}Ar$ ratio grows rapidly, levelling off after about 2 Ga (Sarda et al., 1985). In order to test these two models, it is necessary to determine $^{40}Ar/^{36}Ar$ ratios in rocks that equilibrated with the atmosphere–ocean system in the geologic past. Unfortunately, due to the mobility of argon, reliable samples to study are difficult to find. However, 2-Ga-old cherts, which may effectively trap primitive argon, are reported to have $^{40}Ar/^{36}Ar$ ratios similar to the present atmosphere (295) tending to favour the big burp model. Some argon degassing models suggest that the atmosphere grew very rapidly in the first 100 My during planetary accretion, followed by a more continuous growth to the present (Sarda et al., 1985). These models indicate a mean age for the atmosphere of 4.4 Ga, suggesting very rapid early degassing of the Earth, probably beginning during the late stages of accretion. On the other hand, relatively young K–Ar ages of MORB mantle sources (< 1 Ga) show that the

depleted upper mantle was not completely degassed, and that it decoupled from the atmosphere early in Earth history (Fisher, 1985). These data, together with relatively young U–He and U–Xe ages of depleted mantle, suggest that some degassing has continued up to the present.

Another approach to studying the growth rate of the atmosphere is using rare gases which are trapped in ocean-ridge basalts (MORB). MORB glasses have proved to be particularly useful in that they appear to trap gases in the same proportions as they occur in mantle sources. The most informative gases are those which have isotopes that are products of radioactive decay such as ^4He, ^{40}Ar and various Xe isotopes. The rate at which these isotopes accumulate in the mantle compared with stable isotopes of the same element is related to the degassing rate of the mantle. One potentially important rare gas isotopic system used to study mantle degassing is the ^{129}I–^{129}Xe system (Staudacher and Allegre, 1982). Large concentrations of ^{129}Xe are reported in some MORB glasses. These may be produced by the radioactive decay of ^{129}I, which has a short half-life of only 17 My. If the Earth was largely degassed soon after accretion, much of the Xe, together with other gases, would enter the atmosphere. Iodine, however, which is not a volatile element, should remain within the Earth and ^{129}I should decay to ^{129}Xe producing an enrichment in ^{129}Xe in the mantle, known as an **Xe anomaly** (Figure 6.3). The fact that Xe anomalies are found only in MORB indicates that depleted upper-mantle sources were extensively degassed within 20–30 My after accretion was complete (Staudacher and Allegre, 1982; Allegre et al., 1987). The absence of ^{129}Xe anomalies in island basalts indicates that other parts of the mantle were not extensively degassed during this early event, and that they have degassed more gradually. There are, however, problems with the use of ^{129}Xe in monitoring early degassing, the most important of which is that ^{129}Xe is also produced by the natural fission of ^{238}U in the mantle, and this ^{129}Xe, which is added gradually over geologic time, cannot easily be distinguished from ^{129}Xe formed by the decay of ^{129}I (Ozima et al., 1993).

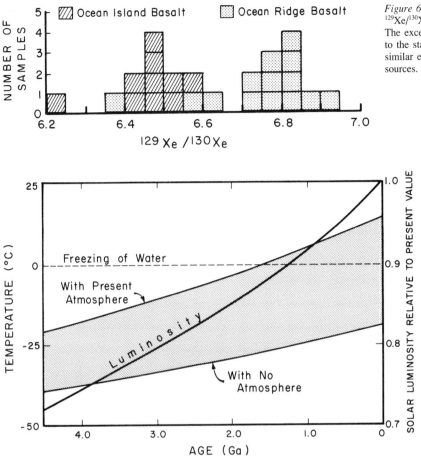

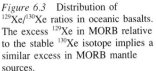

Figure 6.3 Distribution of $^{129}Xe/^{130}Xe$ ratios in oceanic basalts. The excess ^{129}Xe in MORB relative to the stable ^{130}Xe isotope implies a similar excess in MORB mantle sources.

Figure 6.4 The estimated increase in solar luminosity with geologic time and its affect on the Earth's surface temperature.

In conclusion, rare gas studies of terrestrial rocks suggest, but do not prove, that large portions of the mantle underwent extensive degassing within 50 My after the onset of planetary accretion and this was largely complete by 4.45 Ga. However, some degassing of the Earth has continued to the present. The fraction of the atmosphere released during the early degassing event is unknown, but may have been substantial. This idea is consistent with the giant impactor model for the origin of the Moon (Chapter 7), since such an impact should have catastrophically degassed the Earth.

The faint young Sun paradox

Models for the evolution of the Sun indicate that it was less luminous when it entered the main sequence 4.5 Ga. This is because with time the Sun's core becomes denser and therefore hotter as hydrogen is converted to helium. Calculations indicate that the early Sun was 25–30 per cent less luminous than it is today and that its luminosity has increased with time in an approximately linear manner (Kasting, 1987). The paradox associated with this luminosity change is that the Earth's average

surface temperature would have remained below freezing until about 2 Ga for an atmosphere with the same composition at today (Figure 6.4). Yet the presence of sedimentary rocks as old as 3.8 Ga indicates the existence of oceans and running water. A probable solution to the faint young Sun paradox is that the early atmosphere contained a much larger quantity of greenhouse gases than it does today. For instance, CO_2 levels of even a few tenths of a bar could prevent freezing temperatures at the Earth's surface due to an enhanced greenhouse effect. The **greenhouse effect** is caused by gases such as CO_2 and CH_4 that allow sunlight to reach a planetary surface, but absorb infra-red radiation reflected from the surface, which heats both the atmosphere and the planetary surface. An upper bound on the amount of CO_2 in the Early Archean atmosphere is provided by the carbonate–silicate cycle, and appears to be about 1 bar. Although CO_2 was undoubtedly the most important greenhouse gas during the Archean, studies of a 3.5-Ga paleosol (ancient soil horizon) suggest that atmospheric CO_2 levels in the Archean were at least five times lower than required by the faint young Sun paradox (Rye et al., 1995). This constrains the Archean CO_2 levels to

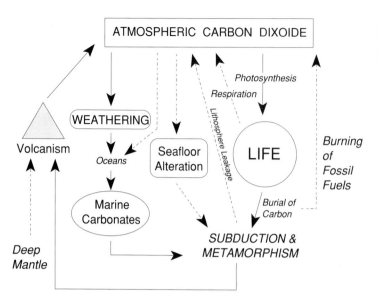

Figure 6.5   The carbonate–silicate cycle. Solid arrows are major controls and dashed arrows minor controls on atmospheric CO$_2$ levels.

about 0.2 bar. Hence, either another greenhouse gas, such as CH$_4$, must have been present in the Archean atmosphere, or some other factor must also have contributed to warming of the atmosphere.

Another factor that may have aided warming the surface of the early Earth is decreased **albedo**, i.e., a decrease in the amount of solar energy reflected by cloud cover. To conserve angular momentum in the Earth–Moon system, the Earth must have rotated faster in the Archean (about 14 hrs/day), which decreases the fraction of global cloud cover by 20 per cent with a corresponding decrease in albedo (Jenkins et al., 1993). However, this effect could be offset by increased cloud cover caused by the near-absence of land areas, at least in the Early Archean when it is likely that the continents were completely submerged beneath seawater (Galer, 1991; Jenkins, 1995).

The carbonate–silicate cycle

One of the important chemical systems controlling the CO$_2$ content of the terrestrial atmosphere is the **carbonate–silicate cycle**. CO$_2$ enters the Earth's atmosphere by volcanic eruptions, the burning of fossil fuels, and by respiration of living organisms (Figure 6.5). Of these, only volcanism appears to have been important in the geologic past, but the burning of fossil fuels is becoming more important today. For instance, records indicate that during the last 100 years the rate of release of CO$_2$ from the burning of fossil fuels has risen 2.5 per cent per year and could rise to three times its present rate in the next 100 years. CO$_2$ returns to the oceans by chemical weathering of silicates, dissolution of atmospheric CO$_2$ in the oceans, and alteration on the sea floor, of which only the first two are significant at present (Figure 6.5). The ultimate sink for CO$_2$ in the oceans is deposition of marine carbonates. Although CO$_2$ is also

removed from the atmosphere by photosynthesis, this is not as important as carbonate deposition. Weathering and deposition reactions can be summarized as follows (Walker, 1990; Brady, 1991):

Weathering,
$$CaSiO_3 + 2CO_2 + 3H_2O \rightarrow Ca^{+2} + 2HCO_3^{-1} + H_4SiO_4$$
Deposition,
$$Ca^{+2} + 2HCO_3^{-1} \rightarrow CaCO_3 + CO_2 + H_2O$$

The cycle is completed when pelagic carbonates are subducted and metamorphosed and CO$_2$ is released and again enters the atmosphere either by volcanism or by leaking through the lithosphere (Figure 6.5). The metamorphic reactions which liberate CO$_2$ can be summarized by the carbonate–silica reaction,

$$CaCO_3 + SiO_2 \rightarrow CaSiO_3 + CO_2.$$

For equilibrium to be maintained in this cycle, increased input of CO$_2$ into the atmosphere results in more weathering and carbonate deposition, thus avoiding build-up of CO$_2$ in the atmosphere. As previously mentioned in Chapter 1, this is known negative feedback. Various negative feedback mechanisms in the carbonate–silicate cycle may have stabilized the Earth's surface temperature in the geologic past (Walker, 1990). As an example, if the solar luminosity were suddenly to drop, the surface temperature would fall causing a decrease in the rate of silicate weathering due to a decrease in evaporation from the oceans (and hence a decrease in precipitation). This results in CO$_2$ accumulation in the atmosphere, which increases the greenhouse effect and restores higher surface temperatures. The converse of this feedback would occur if the surface temperature were suddenly to increase. Although increased CO$_2$ in the Archean atmosphere would result in greenhouse warming of the Earth's surface, it should also cause increased weathering rates resulting in a decrease in CO$_2$.

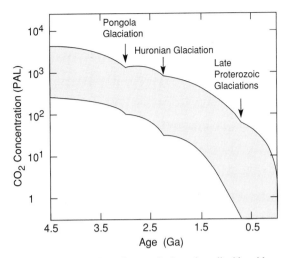

Figure 6.6 Evolution of atmospheric carbon dioxide with time. Shaded area is the range of CO_2 concentrations permitted by geologic indicators. PAL = fraction of present atmospheric level. Modified after Kasting (1993).

However, this would only occur in the Late Archean after a large volume of continental crust had stabilized above sea level to be weathered.

The history of atmospheric CO_2

By 3.8 Ga, it would appear that the atmosphere was composed chiefly of CO_2, N_2, and perhaps CH_4, with small amounts of CO, H_2, H_2O, and reduced sulphur gases (Kasting, 1993; Des Marais, 1994). Based on the paleosol data described above, an average CO_2 level of about 10^3 PAL (PAL = fraction of present atmospheric level) seems reasonable at 4.5 Ga (Figure 6.6). With little if any land area in the Early Archean, removal of CO_2 by seafloor alteration and carbonate deposition should have been more important than today. The fact that Archean carbonates are rare may be due to their being subducted, since there were few if any stable continental shelves on which they could be preserved. As we have seen, although continental crust was being rapidly generated in the Archean, it was also being rapidly recycled into the mantle. Beginning in the Late Archean, however, when continental cratons emerged above sea level and were widely preserved, the volcanic inputs of CO_2 were ultimately balanced by weathering and perhaps to a lesser degree by carbonate deposition. Because of the increasing biomass of algae, photosynthesis also may have contributed to CO_2 drawdown, and rapid chemical weathering was promoted by increased greenhouse warming.

The overall decrease in atmospheric CO_2 level with time appears to be related to changes in the carbonate–silicate cycle, in which CO_2 is removed from the atmosphere by carbonate deposition and photosynthesis, followed by burial of organic matter, faster than it is

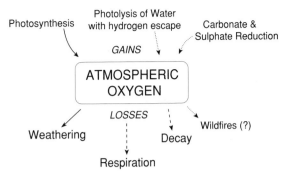

Figure 6.7 Gains and losses of atmospheric O_2. Solid lines, major controls; long-dash lines, intermediate control; short-dash lines, minor controls.

resupplied by volcanism and subduction. Shallow marine carbonates deposited in cratonic basins are effectively removed from the carbonate–silicate cycle and are a major sink for CO_2 from the Mid-Proterozoic onwards. Decreasing solar luminosity and the growth of ozone in the upper atmosphere beginning in the Early Proterozoic also reduced the need of CO_2 as a greenhouse gas. Rapid extraction of CO_2 by depositing marine carbonates may have resulted in sufficient atmospheric cooling to cause glaciations as illustrated in Figure 6.6.

The origin of oxygen

Oxygen controls in the atmosphere

In the modern atmosphere, oxygen is produced almost entirely by **photosynthesis** (Figure 6.7) by the well-known reaction,

$$CO_2 + H_2O \rightarrow CH_2O + O_2$$

A very small amount of O_2 is also produced in the upper atmosphere by photolysis of H_2O molecules. Photolysis is the fragmentation of gaseous molecules by UV radiation from the Sun. For instance, the photolysis of H_2O produces H_2 and O_2 ($H_2O \rightarrow H_2 + 0.5\ O_2$). Oxygen is removed from the atmosphere by respiration and decay, which can be considered as the reverse of the photosynthesis reaction, and by chemical weathering. Virtually all of the O_2 produced by photosynthesis in a given year is lost in < 50 y by oxidation of organic matter. Oxygen is also liberated by the reduction of sulphates and carbonates in marine sediments. Without oxidation of organic matter and sulphide minerals during weathering, the O_2 content of the atmosphere would double in about 10^4 y (Holland et al., 1986). It has also been suggested that wildfires may have contributed in the past to controlling the upper limit of O_2 in the atmosphere. It appears that the O_2 content of the modern atmosphere is maintained at a near constant value by various negative feedback mechanisms involving primarily photosynthesis and decay. If, however, photosynthesis were to stop, respiration and decay would continue until all of the organic matter on Earth was transformed to CO_2 and H_2O. This would occur in about twenty years and involve only a minor decrease in the amount of O_2 in the atmosphere. Weathering would continue to consume O_2 and would take about 4 My to use up the current atmospheric supply.

Prior to the appearance of photosynthetic microorganisms, and probably for a considerable time thereafter, photosynthesis was not an important process in controlling atmospheric oxygen levels. In the primitive atmosphere, O_2 content was controlled by the rate of photolysis of water, hydrogen loss from the top of the atmosphere, and weathering rates at the surface. The rate at which water is supplied by volcanic eruptions is also important because volcanism is the main source of water available for photolysis. If metallic iron was present in the mantle during the earliest stages of degassing of the Earth, H_2 would have been an important component of volcanic gases. Since H_2 rapidly reacts with O_2 to form water, significant amounts of volcanic H_2 would prevent O_2 from accumulating in the early atmosphere. As H_2O instead of H_2 became more important in volcanic gases, in response to removal of iron from the mantle as the core grew, oxygen could begin to accumulate. Also, early photosynthesizing organisms may have contributed to the first oxygen in the atmosphere. As photosynthesis became more widespread, recombination of H_2 and O_2 to form water could not keep pace with O_2 input and the oxygen level in the atmosphere must have increased. This assumes that the rate of weathering did not increase with time, an assumption which is supported by geologic data.

Geologic indicators of ancient atmospheric oxygen levels

Banded iron formation

The distribution of banded iron formation with geologic time provides a constraint on O_2 levels in the oceans and atmosphere. **Banded iron formations** or **BIF** are chemical sediments, typically thin-bedded or laminated with > 15 per cent iron of sedimentary origin (Figure 6.8). They also commonly contain layers of chert and generally have $Fe^{+3}/Fe^{+2} + Fe^{+3}$ ratios in the range of 0.3–0.6 reflecting an abundance of magnetite (Fe_3O_4). BIF are metamorphosed and major minerals include quartz, magnetite, hematite, siderite and various other Fe-rich

Figure 6.8   Early Archean banded iron formation from the Warrawoona Supergroup, Western Australia. Courtesy of Andrew Glikson.

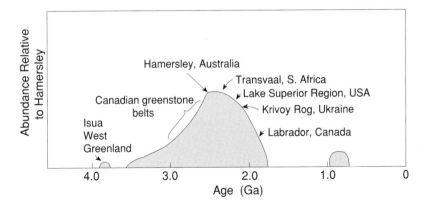

carbonates, amphiboles, and sulphides. Although most abundant in the Late Archean and Early Proterozoic (Figure 6.9), BIF occur in rocks as old as 3.8 Ga (Isua, southwestern Greenland) and as young as 0.8 Ga (Rapitan Group, north-western Canada). The Hamersley basin (2.5 Ga) in Western Australia is the largest known single BIF depository (Klein and Beukes, 1992).

Most investigators consider the large basins of BIF to have formed on cratons or passive margins in shallow marine environments. Many BIF are characterized by thin wave-like laminations that can be correlated over hundreds of kilometres (Trendall, 1983). During the Late Archean and Early Proterozoic, enormous amounts of ferrous iron appear to have entered BIF basins with calculated Fe precipitation rates of 10^{12}–10^{15} gm/y (Holland, 1984). Although some of the Fe^{+2} undoubtedly came from weathering of the continents, most appears to have entered the oceans by submarine volcanic activity (Isley, 1995). Deposition occurred as Fe reacted with dissolved O_2, probably at shallow depths forming flocculent, insoluble ferric and ferro-ferric compounds. Archean BIF are thought to have been deposited in deep water in a stratified ocean (Klein and Beukes, 1992). Beginning at about 2.3 Ga, the stratified ocean began to break down with the deposition of shallow-water oolitic-type BIF, such as those in the Lake Superior area.

Although Precambrian BIF clearly represent a large oxygen sink, the O_2 content of the coexisting atmosphere may have been quite low. For instance, hematite (Fe_2O_3) or $Fe(OH)_3$ can be precipitated over a wide range of P_{O_2} values and Early Archean BIF may have been deposited in reducing marine waters. The large amount of oxygen in Late Archean and especially in Early Proterozoic BIF, however, appears to require the input of photosynthetic oxygen. This agrees with paleontologic data that indicate a rapid increase in numbers of photosynthetic algae during the same period of time. Many BIF contain well-preserved fossil algae remains. Thus, it appears that the increase in abundance of BIF in the Late Archean and Early Proterozoic (Figure 6.9) reflects an increase in oceanic O_2 content in response to increasing numbers of photosynthetic organisms. Only after most of the BIF was deposited and the Fe^{+2} in solution was

exhausted did O_2 begin to escape from the oceans and accumulate in the atmosphere in appreciable quantities (Cloud, 1973). The drop in abundance of BIF at about 1.7 Ga reflects this exhaustion of reduced iron.

Redbeds and sulphates

Redbeds are detrital sedimentary rocks with red ferric oxide cements. They generally form in fluvial or alluvial environments and the red cements are the result of subaerial oxidation (Folk, 1976), thus requiring the presence of O_2 in the atmosphere. The fact that redbeds do not appear in the geologic record until about 2.3 Ga (Eriksson and Cheney, 1992) suggests that oxygen levels were very low in the Archean atmosphere.

Sulphates, primarily gypsum and anhydrite, occur as evaporites. Although evidence of gypsum deposition is found in some of the oldest supracrustal rocks (~ 3.5 Ga), evaporitic sulphates do not become important in the geologic record until after 2 Ga. Since their deposition requires free O_2 in the ocean and atmosphere, their distribution supports rapid growth of oxygen in the atmosphere beginning in the Early Proterozoic. Additional evidence favouring a switch to oxidizing conditions in the Early Proterozoic is provided by sulphur and carbon isotope data (see later discussion). An increase in the range of $\delta^{13}C$ values for Archean organic matter implies relatively large amounts of marine methane, which in turn requires very low atmospheric CO_2 content prior to 2.5 Ga.

Detrital uraninite deposits

Several occurrences of Late Archean to Early Proterozoic detrital uraninite and pyrite are well-documented, the best known of which are those in the Witwatersrand basin in South Africa (~ 2.9 Ga) and those in the Blind River–Elliot Lake area in Canada (~ 2.3 Ga). No significant occurrences, however, are known to be younger than Mid-Proterozoic (Walker et al., 1983). A few very minor occurrences of detrital uraninite are found in young sediments associated with rapidly rising mountain chains like the Himalayas. Both uraninite and pyrite are unsta-

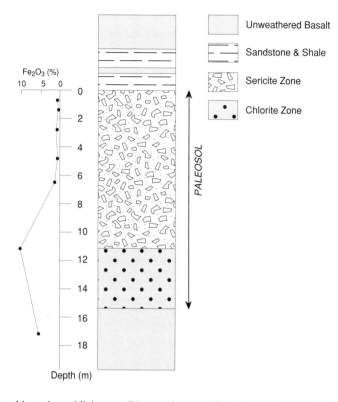

Unweathered Basalt

Sandstone & Shale

Sericite Zone

Chlorite Zone

Figure 6.10 Stratigraphic section of the 2.7-Ga Mount Roe paleosol, Western Australia, showing the concentration of Fe_2O_3 with depth. After MacFarlane et al. (1994).

ble under oxidizing conditions and are rapidly dissolved. The preservation of major Late Archean and Early Proterozoic deposits of detrital uraninite and pyrite in conglomerates and quartzites indicates that weathering did not lead to total oxidation and dissolution of uranium and iron. To preserve uraninite, P_{O_2} must have been 10^{-2} of the present atmospheric level (PAL). The few occurrences of young detrital uraninite are metastable, and the uraninite is preserved only because of extremely rapid sedimentation rates and will not survive for long.

The restriction of major detrital uraninite–pyrite deposits to > 2.3 Ga again favours very low O_2 levels in the atmosphere prior to this time.

Paleosols

Paleosols are preserved ancient weathering profiles or soils that contain information about atmospheric composition (Holland, 1992). Highly-oxidized paleosols retain most, if not all, of the iron in Fe^{+3} and Fe^{+2} compounds, whereas paleosols that formed in non-oxidizing or only slightly oxidizing environments show significant losses of iron in the upper horizons, especially in paleosols developed on mafic parent rocks. This is illustrated for a Late Archean profile from Western Australia in Figure 6.10. A lower atmospheric content of oxygen is necessary for the Fe to be leached from the upper paleosol horizon (MacFarlane et al., 1994). Elements such as Al, which are relatively immobile during weathering, are

enriched in the upper horizons due to the loss of mobile elements such as Fe^{+2}. This indicates that by 2.7 Ga, levels of O_2 in the atmosphere were < 8 per cent of present-day levels.

Beginning in the Early Proterozoic, paleosols do not show this leaching of iron. Results suggest that the O_2 level of the atmosphere rose dramatically from about 1 per cent PAL to > 15 per cent of this level about 2 Ga (Holland and Beukes, 1990).

Biologic indicators

The Precambrian fossil record also provides clues to the growth of atmospheric oxygen. Archean and Early Proterozoic life forms were entirely prokaryotic organisms, the earliest examples of which evolved in anaerobic (oxygen-free) environments. Prokaryotes that produced free oxygen by photosynthesis appear to have evolved by 3.5 Ga. The timing of the transition from an anoxic to oxygeneous atmosphere is not well-constrained by microfossil remains but appears to have begun approximately 2.3 Ga. Certainly by 2.0 Ga when heterocystous cyanobacteria appear, free O_2 was present in the atmosphere in significant amounts, in agreement with the paleosol data. The first appearance of eukaryotes at about 2.4–2.3 Ga indicates that atmospheric oxygen had reached 1 per cent PAL, which is necessary for mitosis to occur. The appearance of simple metazoans at about 2 Ga requires O_2 levels high enough for oxygen to diffuse across membranes (~ 7 per cent PAL).

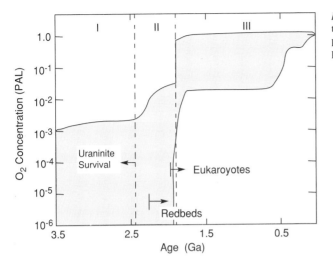

Figure 6.11   Growth in atmospheric oxygen with time. Shaded area is the range of O_2 concentrations permitted by geologic indicators. PAL = fraction of present atmospheric level. After Kasting (1993).

The growth of atmospheric oxygen

The growth of oxygen in the Earth's atmosphere can be divided into three stages (Kasting, 1987; 1991):

Stage I   a *reducing stage* in which free oxygen occurs in neither the oceans nor the atmosphere

Stage II   an *oxidizing stage* in which small amounts of oxygen are present in the atmosphere and surface ocean water, but not in deep ocean waters

Stage III   an *aerobic stage* in which free oxygen pervades the entire ocean–atmosphere system.

These stages are diagramatically illustrated in Figure 6.11 with corresponding geologic indicators.

During Stage I, the ozone shield required to block out lethal UV radiation is absent. If such a shield existed on the primitive Earth, it must have been produced by some gaseous compound other than ozone. Although marine cyanobacteria must have produced O_2 by photosynthesis during this stage, in order for large volumes of BIF to form this oxygen must combine rapidly with Fe^{+2} and be deposited. The atmosphere should have remained in Stage I until the amount of O_2 produced by photosynthesis, followed by organic carbon burial, was enough to overwhelm the input of volcanic gases. The change from Stages I to II occurred during the Early Proterozoic and is marked by the appearance of redbeds at 2.3 Ga, an increase in abundance of evaporitic sulphates, a broad maximum in the deposition rate of BIF at 2.5 Ga, and the end of major detrital uraninite–pyrite deposition. At this time, photosynthetic O_2 entered the atmosphere at a rate sufficient to clean the atmosphere of molecular H_2. Evidence that the lower atmosphere and surface ocean waters during this stage were weakly oxidizing, yet deep marine waters were reducing, is provided by the simultaneous deposition of oxidized surface deposits (redbeds, evaporites) and BIF in a reducing (or at least non-oxidizing) environment. Two models have been proposed for the control of atmospheric O_2 levels during Stage II. One proposes that low O_2 levels are maintained by mantle-derived Fe^{+2} in the oceans and BIF deposition, and the other relies on low photosynthetic productivity in the ocean. Which of these is most important is not as yet clear. In either case, it would appear that Stage II was short-lived (Figure 6.11).

The change from Stages II to III occurred about 2.0 Ga as oxygen rapidly entered the atmosphere (Holland, 1994). The transition is marked by the near disappearance of BIF from the geologic record and the appearance of eukaroytic organisms. The onset of Stage III is defined by the exhaustion of Fe^{+2} from the oceans such that photosynthetic O_2 levels increase and the oceans become oxidizing. Further increases result in liberation of O_2 into the atmosphere. During this stage an effective ozone screen develops, which is probably responsible for the rapid diversification and increase in numbers of micro-organisms in the Mid-Proterozoic. By 540 Ma, atmospheric O_2 must have risen to at least 10 per cent PAL to permit the appearance of carbonate shell-forming organisms.

Phanerozoic atmospheric history

It is possible to track the levels of CO_2 and O_2 in the Earth's atmosphere during the Phanerozoic using the burial and weathering rates of organic carbon, Ca and Mg silicates, and carbonates as deduced from the preserved stratigraphic record. Other factors, such as the affect of changing solar radiation on surface temperature and weathering rates and the use of the $^{87}Sr/^{86}Sr$ seawater curve to estimate the affect of tectonics on weathering and erosion rates, can be used to fine-tune the results (Berner and Canfield, 1989; Berner, 1994). Although CO_2 shows a gradual drop throughout the Phanerozoic, it is very high in the early to middle Paleozoic and shows a pronounced minimum 300 Ma

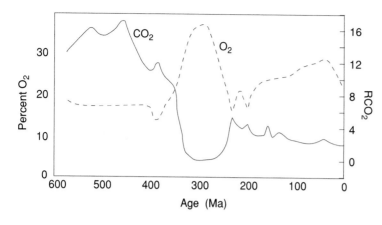

Figure 6.12 Oxygen and CO_2 atmospheric concentrations during the Phanerozoic. RCO_2 = mass ratio of CO_2 at a given time divided by the modern value. After Berner and Canfield (1989) and Berner (1994).

(Figure 6.12). Variations in atmospheric CO_2 are controlled by a combination of tectonic and biologic processes, often with one or the other dominating, and by the increasing luminosity of the Sun. An increase in solar luminosity since the beginning of the Paleozoic is probably responsible for the gradual drop in atmospheric CO_2. This drawdown is caused by enhanced weathering rates owing to the greenhouse warming of the atmosphere. The dramatic minimum in CO_2 about 300 Ma appears to be related to:

1 enhanced silicate weathering
2 the appearance and rapid development of land plants, which through photosynthesis, followed by burial, rapidly removed carbon from the atmosphere–ocean system.

Rapid weathering rates in the early and middle Paleozoic probably reflect a combination of increasing solar luminosity and the fragmentation of Gondwana (early Paleozoic only), during which enhanced ocean-ridge and plume activity pumped large amounts of CO_2 into the atmosphere, resulting in increased weathering rates. The increase in CO_2 in the Mesozoic, which resulted in worldwide warm climates, may be caused by enhanced ridge and plume activity associated with the onset of fragmentation of Pangea. A gradual drop in CO_2 in the Late Cretaceous and Tertiary appears to reflect a combination of decreasing ocean-ridge activity, thus decreasing the CO_2 supply, and increased weathering and erosion rates in response to terrane collisions around the Pacific (exposing more land area for weathering).

Unlike CO_2, the variation in O_2 during the Phanerozoic appears to be controlled chiefly by the burial rate of carbon and sulphur, which in turn is controlled by sedimentation rates. Oxygen levels were relatively low and constant during the early and middle Paleozoic, followed by a high peak about 300 Ma (Figure 6.12). During the Mesozoic, O_2 gradually increased followed by a steep drop in the Tertiary. The peak in O_2 in the late Paleozoic correlates with the minimum in CO_2 and appears to have the same origin: enhanced burial rate of organic carbon accompanying the rise of vascular land plants. This is evidenced today by the widespread coal deposits of this age. The drop in oxygen at the end of the Permian (250 Ma) probably reflects a change in climate to more arid conditions. Depositional basins changed from coal swamps to oxidized terrestrial basins, largely aeolian and fluvial, thus no longer burying large quantities of carbon. Also, a dramatic drop in sea level at the end of the Permian resulted in oxidation and erosion of previously-buried organic material, thus reducing atmospheric O_2 levels. Burial of land plants in coal swamps during the Cretaceous may have contributed to the increased oxygen levels at this time, and again a drop in sea level during the Tertiary exposed these organics to oxidation and erosion, probably causing the fall in atmospheric O_2 during the Tertiary.

The dramatic increase in atmospheric oxygen in the Carboniferous may have been responsible for several important biological changes at this time. The increase should have enhanced diffusion-dependent processes such as respiration, and hence some organisms could attain large body sizes (Graham et al., 1995). Perhaps the most spectacular examples are the insects, where some forms attained gigantic sizes probably in response to increased diffusive permeation of oxygen into the organisms. Also, the increased oxygen levels would result in a denser atmosphere (21 per cent greater than at present), which may have led to the evolution of insect flight by offering greater lift. Increased aquatic oxygen levels during the Carboniferous–Permian would permit greater biomass densities and increased metabolic rates, both of which could lead to increases in radiation of taxa and in organism sizes. Impressive examples of both of these changes occur in the brachiopods, foraminifera, and corals.

The carbon isotope record

Carbon isotopes in carbonates and organic matter offer the most effective way to trace the growth of the crustal reservoir of reduced carbon (Des Marais et al., 1992). The fractionation of ^{13}C and ^{12}C is measured by the $^{13}C/^{12}C$ ratio in samples relative to a standard such that,

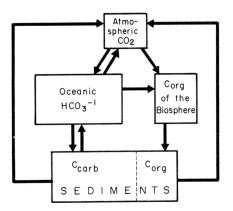

Figure 6.13 Schematic diagram of the carbon cycle. C_{carb} and C_{org} are carbonate and organic carbon respectively.

$$\delta^{13}C \text{ (per mil)} = \{[(^{13}C/^{12}C)_{sample}/(^{13}C/^{12}C)_{std}] - 1\} \times 1000.$$

This expression is used for both carbonate (d_{carb}) and organic (d_{org}) isotopic ratios. The relative abundances of carbon isotopes are controlled chiefly by equilibrium isotopic effects among inorganic carbon species, fractionation associated with the biochemistry of organic matter, and the relative rates of burial of carbonate and organic carbon in sediments. Most of the carbon in the Earth's exogenic systems is stored in sedimentary rocks, with only about 0.1 per cent in living organisms and the atmosphere–hydrosphere. Oxidized carbon occurs primarily as marine carbonates and reduced carbon in organic matter in sediments. In the carbon cycle, CO_2 from the oceans and atmosphere is transferred into sediments as carbonate carbon (C_{carb}) or organic carbon (C_{org}), the former of which monitors the composition of the oceans (Figure 6.13). The cycle is completed by the weathering of sedimentary rocks which returns CO_2 to the atmosphere, most of it from carbonates but with increasing amounts with time from organic carbon.

Because organic matter preferentially incorporates ^{12}C over ^{13}C, there should be an increase in the $^{13}C/^{12}C$ ratio (as measured by $\delta^{13}C$) in buried carbon with time, and indeed this is what is observed (Figure 6.14) (Worsley and Nance, 1989; Des Marais et al., 1992). $\delta^{13}C_{org}$ increases from values < -40 per mil in the Archean to modern values of -20 to -30 per mil. On the other hand, seawater carbon as tracked with $\delta^{13}C_{carb}$, remains roughly constant with time, with $\delta^{13}C_{carb}$ averaging about zero per mil. The percentage of total carbon buried as organic carbon also shows a long-term increase with time with two striking peaks, one at about 2 Ga and another at about 0.9 Ga (Figure 6.15) (Des Marais et al., 1992). The rates of organic burial depend chiefly on two factors:

1 the intensity of biomass production
2 the sedimentation rate.

It would appear that the gradual increase in the amount of buried organic carbon with time reflects the increas-

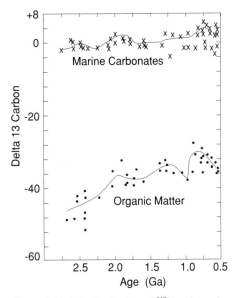

Figure 6.14 The distribution of $\delta^{13}C$ with age in organic matter from sediments and in marine carbonates.

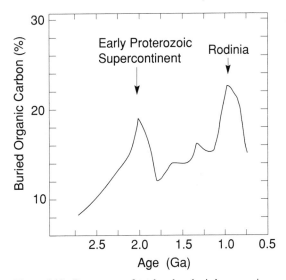

Figure 6.15 Percentages of total carbon buried as organic carbon in the Precambrian. After Des Marais et al. (1992).

ing biomass of micro-organisms, whereas the two peaks reflect increases in sedimentation rates resulting in rapid burial of organics. Sedimentation rate, in turn, is controlled by rates of uplift and erosion. It is interesting that both organic carbon burial peaks correspond roughly to times of major collisional orogeny and supercontinent formation (Chapter 1). The simplest explanation of the increased rates of carbon burial is that they reflect increased erosion and sedimentation associated with collisional uplift during supercontinent amalgamation. Thus, carbon isotopic data may provide yet another piece

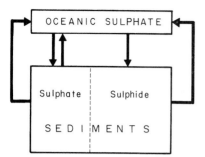

Figure 6.16 Schematic diagram of the sulphur cycle.

of evidence to track the history of supercontinents in the Early and Late Proterozoic.

During the Phanerozoic, there is a major peak in $\delta^{13}C_{carb}$ in the Carboniferous and Permian, and it appears to reflect an increase in burial rate of organic carbon (Frakes et al., 1992). This is because organic matter selectively enriched in ^{12}C depletes seawater in this isotope, thus raising the $\delta^{13}C$ values of seawater. The maximum $\delta^{13}C_{carb}$ corresponds to the rise and spread of vascular land plants in the late Paleozoic, which provided a new source of organic debris for burial (Berner, 1987). Also conducive to the preservation of organic remains at this time were the vast lowlands on Pangea, which appear to have been sites of widespread swamps where bacterial decay of organic matter is minimized. A drop in $\delta^{13}C_{carb}$ at the end of the Permian is not understood. Pangea did not begin to fragment until the Triassic, so the destruction of swampy environments does not seem to be the answer. Perhaps large amounts of photosynthetic O_2 generated by Carboniferous forests led to extensive forest fires that destroyed large numbers of land plants in the Late Permian.

The sulphur isotope record

The geochemical cycle of sulphur resembles that of carbon (Figure 6.16). During this cycle ^{34}S is fractionated from ^{32}S, with the largest fractionation occurring during bacterial reduction of marine sulphate to sulphide. Isotopic fractionation is expressed as $\delta^{34}S$ in a manner similar to that used for carbon isotopes. Sedimentary sulphates appear to record the isotopic composition of sulphur in seawater. Mantle ^{34}S is near zero per mil and bacteria reduction of sulphate strongly prefers ^{32}S, thus reducing $\delta^{34}S$ in organic sulphides to negative values (–180 per mil) and leaving oxidized sulphur species with approximately equivalent positive values (+17 per mil). Hence, the sulphur cycle is largely controlled by the biosphere, and in particular by sulphate-reducing bacteria that inhabit shallow marine waters. Due to the mobility of dissolved sulphate ions and rapid mixing of marine reservoirs, $\delta^{34}S$ values of residual sulphate show limited variability (+17±2 per mil) compared with the spread of values in marine sulphides (–5 to –350 per mil).

The time variation of $\delta^{34}S$ in sulphates for the last 600 My is approximately the mirror image of the $\delta^{13}C$ curve. Thus, a maximum (+32 per mil) is observed for $\delta^{34}S$ in the Cambrian and a minimum (+11 per mil) in the Carboniferous and Permian. This complementary relationship between the two δ values is not well understood, but undoubtedly results from interactions of the carbon and sulphur cycles (Holland et al., 1986). The overall sulphur isotope trends from 3.8 Ga to the present suggest a gradual increase in sulphate $\delta^{34}S$ and a corresponding decrease in sulphide $\delta^{34}S$. Changes in $\delta^{34}S$ with time reflect:

1   changes in the isotopic composition of sulphur entering the oceans from weathering and erosion
2   changes in the relative proportions of sedimentary sulphide and sulphate receiving sulphur from the ocean–atmosphere system
3   the temperature of seawater (Schidlowski et al., 1983; Ohmoto and Felder, 1987).

Because weathering tends to average out sulphide and sulphate input, one or both of the latter two effects probably accounts for oscillations in the sulphate $\delta^{34}S$ curve with time (Figure 6.17). The isotopic composition of evaporitic sulphates should provide a measure of the intensity of bacterial sulphate in the oceans. Hence, the increasing $\delta^{34}S$ in marine sulphates since the Early Archean could be accounted for by increasing sulphide–sulphate ratios in the oceans, reflecting either an increase in the numbers of sulphate-reducing bacteria and/or Archean oceans with much less sulphate than present oceans. The oldest evidence for sulphate-reducing bacteria is $\delta^{34}S$ from pyrites from the 3.5-Ga Barberton greenstone in South Africa (Ohmoto et al., 1993). An alternative explanation for the diverging $\delta^{34}S$ sulphide and sulphate curves with time is that the average surface temperature of the oceans has progressively cooled from 30–50 °C in the Archean to present-day temperatures. This idea is based on results of experiments with living sulphate-reducing bacteria under increased water temperatures (Ohmoto and Felder, 1987). A warmer Archean ocean is in agreement with the probable interpretation of oxygen isotopes in cherts as discussed below.

The oceans

Introduction

The oceans cover 71 per cent of the Earth's surface and contain most of the hydrosphere. The composition of seawater varies geographically and with depth, and is related principally to the composition of input river water and hydrothermal water along ocean ridges. Chemical precipitation, melting of ice, and evaporation rate also affect the salinity of seawater. Dissolved nutrients, such as bicarbonate, nitrate, phosphate and silica, vary considerably in concentration as a result of removal by planktonic micro-organisms such as foraminifera, diatoms, and algae. Seawater temperature is rather constant in the upper

Eh were probably reached rapidly in the early oceans and maintained thereafter by silicate–seawater buffering reactions and recycling at ocean ridges.

Growth rate of the oceans

It is likely that the growth rate of the oceans paralleled that of the atmosphere, perhaps with a slight delay reflecting the time it took for a steam atmosphere to cool and condense. Although continental freeboard has been used to monitor the volume of seawater with time (Wise, 1973), as discussed in Chapter 5, uncertainties in the magnitude of the various factors that control freeboard render any conclusions suspect. In any case, a large fraction of the oceans probably formed during the late stages of planetary accretion, such that by 4.0 Ga most of the current volume of seawater (> 90 per cent) was in place on the Earth's surface.

Sea level

Factors controlling sea level

Changes in sea level leave an imprint in the geologic record by the distribution of sediments and biofacies, the areal extent of shallow seas as manifest by preserved cratonic sediments, and by calculated rates of sedimentation. A widely-used approach to monitor sea level is to map successions of transgressive and regressive facies in marine sediments. During transgression, shallow seas advance on the continents, and during regression they retreat. Transgressions and regressions, however, are not always accompanied by rises or drops in sea level and, in order to estimate sea level changes, it is necessary to correlate facies and unconformities over large geographic regions on different continents and in different tectonic settings using sequence strati-graphy (Steckler, 1984; Vail and Mitchum, 1979). Using this technique, onlap and offlap patterns are identified in seismic reflection profiles and used to construct maps of the landward extent of coastal onlap, from which changes in sea level are estimated.

Although a long-term component (200–400 My) in sea level change during the Phanerozoic seems to be agreed upon by most investigators, short-term components and amplitudes of variation remain uncertain. Major transgressions are recorded in the early Paleozoic and in the Late Cretaceous (Figure 6.18). Estimates of the amplitude of the Cretaceous sea-level rise range from 100–350 m above present sea level, while estimates of the early Paleozoic rise generally fall in the range of tens of metres (Algeo and Seslavinsky, 1995). Cyclicity in sea-level change has been proposed by many investigators with periods in the range of 10–80 My. The Vail sea level curve (Figure 6.18) has been widely accepted although the amplitudes are subject to uncertainty.

Several factors can cause short-term changes in sea level. Among the most important are tectonic controls, chiefly continental collisions, which operate on timescales of 10^6–10^8 y, and glacial controls which operate on

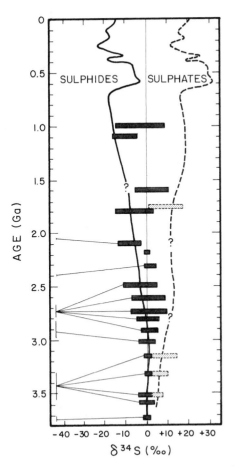

Figure 6.17 The isotopic composition of sulphur in marine sulphates and sulphides with geologic time. After Schidlowski et al. (1983).

300 m (~ 18 °C), then drops dramatically to a depth of about 1 km where it remains relatively constant (3–5 °C) to the ocean bottom. The cold bottom water in the oceans forms in polar regions, where it sinks because of its high density, and travels towards the equator displacing warm surface waters that travel poleward. The geologic history of seawater is closely tied to the history of the atmosphere, and oceanic growth and the concentration of volatile elements in seawater is controlled, in part, by atmospheric growth (Holland et al., 1986). Most paleoclimatologists agree that the oceans formed early in Earth history by condensation of atmospheric water. Some water, however, may have been added to the Earth by early cometary impacts.

Because of the relatively high CO_2 content of the early atmosphere, the first seawater to condense must have had a low pH (< 7) due to dissolved CO_2 and other acidic components (H_2S, HCl etc.). However, this situation was probably short-lived in that submarine volcanic eruptions and seawater recycling through ocean ridges would introduce large amounts of Na, Ca, and Fe. Thus, a pH between 8 and 9 and a neutral or negative

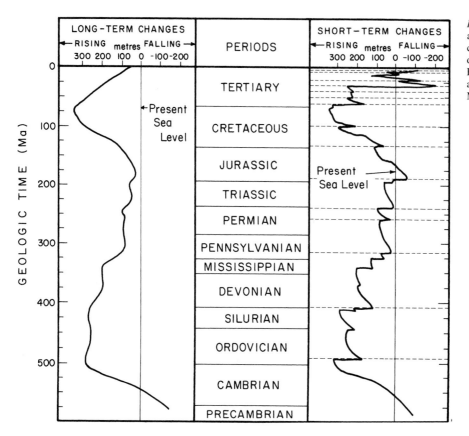

Figure 6.18 Long- and short-term changes in sea level during the Phanerozoic. Modified after Vail and Mitchum (1979).

timescales of 10^4–10^8 y. Tectonic controls seem to produce sea level changes at rates < 1 cm/10^3 y, while glacial controls can produce rate changes up to 10 m/10^3 y. With the present ice cover on the Earth, the potential amplitude for sea level rise if all the ice were to melt is 200 m. The short-term drops in sea level at 500, 400, and 350 Ma (Figure 6.18) probably reflect glaciations at these times. The rise of the Himalayas in the Tertiary produced a drop in sea level of about 50 m.

Long-term changes in sea level are related to:

1 rates of seafloor spreading
2 the characteristics of subduction
3 the motion of continents with respect to geoid highs and lows
4 supercontinent insulation of the mantle (Gurnis, 1993).

One of the most important causes of long-term sea level changes is the volume of ocean-ridge systems. The cross-sectional area of an ocean ridge is dependent upon spreading rate, since its depth is a function of its age. For instance, a ridge that spreads 6 cm/y for 70 My has three times the cross-sectional area of a ridge that spreads at 2 cm/y for the same amount of time. The total ridge volume at any time is dependent upon spreading rate (which determines cross-sectional area) and total ridge length. Times of maximum seafloor spreading correspond

well with the major transgression in the Late Cretaceous, thus supporting the ridge volume model. Calculations indicate that the total ridge volume in the Late Cretaceous could displace enough seawater to raise sea level by about 350 m.

Subduction can also affect sea level. For instance, if additional convergent and divergent plate margins form in an oceanic plate while spreading rate is maintained at a constant value, the average age of oceanic lithosphere decreases and leads to uplift above subduction zones, causing a drop in sea level (Gurnis, 1993). Also, when continental plates move over geoid highs, sea level falls, and conversely, sea level rises when they move over geoid lows. As the mantle warms up beneath a supercontinent, the continental plate rises isostatically, with a corresponding fall in sea level.

Sea level and the supercontinent cycle

Worsley et al. (1984; 1986) have proposed an intriguing model in which Phanerozoic sea level is directly tied to the supercontinent cycle, with a repeat time of about 400 My. The rapid increases in sea level observed at 600–500 Ma and 200–100 Ma correspond to the onset of the break up of Laurasia (Laurentia–Baltica) and Pangea respectively, and the early Paleozoic long-term maximum coincides with the assembly of Pangea (Fig-

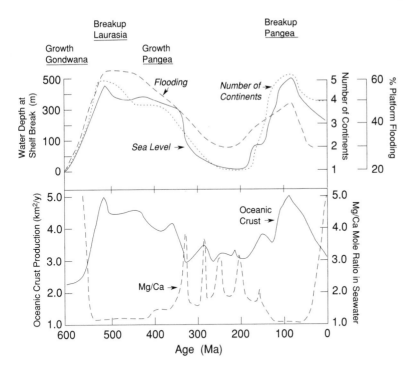

Figure 6.19 Comparison of variation in sea level, number of continents, amount of flooding of shallow shelves by seawater, production rate of oceanic crust, and the Mg/Ca mole ratio in seawater as a function of time. Also indicated are the break up and growth stages of supercontinents. After Worsley et al. (1984), Gaffin (1987), and Hardie (1996).

ure 6.19). The low sea level at about 600 Ma correlates with the assembly of Gondwana and the low at 300–200 Ma correlates with the final assembly of Pangea. The stands of sea level accompanying supercontinents probably reflect the geoid highs developed beneath the supercontinents, whereas the high sea levels accompanying supercontinent dispersal reflect enhanced ocean-ridge activity. Supporting this idea is the fact that the number of continents with time also correlates with sea level (Figure 6.19). Also, as expected, the percentage of flooding of the continents increases during supercontinent dispersal and decreases during supercontinent stability. Because major changes in sea level appear to be related to changes in volume of the ocean-ridge system, times of high sea level should also be times of rapid production of oceanic crust, when ridges are relatively high. Using the Vail sea level curve, calculated rates of production of oceanic crust are also shown in Figure 6.19. Note that ocean crust production rate increases by up to a factor of two during supercontinent break-ups.

The Worsley sea level model is exciting in that it presents a potential method by which older supercontinents can be tracked from sea level data inferred from sequence stratigraphy of pre–1-Ga marine successions.

Changes in the composition of seawater with time

Marine carbonates and evaporites

Although it is commonly assumed that the composition of seawater has not changed appreciably with time, some

observational and theoretical data challenge this assumption. For instance, some Archean marine carbonates contain giant botryoids of aragonite and Mg-calcite beds up to several metres thick that extend for hundreds of kilometres (Grotzinger and Kasting, 1993) (Figure 6.20). By comparison, Early Proterozoic carbonates have much less spectacular occurrences of these minerals, although cement crusts in tidal flat deposits are common. In contrast, all younger marine carbonates lack these features, or in the case of the cements, show a progressive decline in abundance with age. Despite the fact that the modern ocean is over-saturated in carbonate, cement crusts are not deposited today. These secular changes strongly suggest that Precambrian seawater was greatly over-saturated in carbonate and that saturation decreased with time.

Another difference between Precambrian and Phanerozoic sediments is the sequence of evaporite deposition. In Early Proterozoic sediments, halite (NaCl) is often deposited immediately on top of carbonate without intervening sulphates. Although minor occurrences of evaporitic gypsum are reported in the Archean (Buick and Dunlop, 1990), they appear to reflect the local composition of seawater controlled by local source rocks. The first really extensive sulphate deposits appear in the MacArthur basin in Australia at 1.7–1.6 Ga, which essentially coincides with the last examples of deposition of seafloor carbonate cements. The paucity of gypsum evaporites before 1.7 Ga may have been due to low concentrations of sulphate in seawater before this time, or to the possibility that the bicarbonate to Ca ratio was sufficiently high during progressive seawater evapora-

Figure 6.20   Aragonite botryoid pseudomorphs from the Late Archean (2.52 Ga) Transvaal Supergroup in South Africa. Two aragonite fans nucleated on the sediment surface and grew upwards. Width of photograph about 30 cm. Courtesy of John Grotzinger.

tion that Ca was exhausted by carbonate precipitation before the gypsum stability field was reached (Grotzinger and Kasting, 1993).

Evaporite deposition may also have influenced the composition of the oceans during the Phanerozoic. The large mass of evaporites deposited during some time intervals may have been important in lowering the NaCl and $CaSO_4$ contents of world oceans (Holser, 1984). For instance, the vast volumes of salt (chiefly NaCl) deposited in the Late Permian may have caused average ocean salinity to drop significantly during the Triassic and Jurassic (Figure 6.21). Evaporite deposition has been sharply episodic and erosion of older evaporites has been relatively slow in returning salts to the oceans. The distribution of evaporites of all ages, coupled with a slow recycling rate into the oceans, suggests that the mean salinity of seawater may have decreased since the Mid-Proterozoic due to evaporite deposition.

Secular changes in the mineralogy of marine limestones and evaporites during the Phanerozoic may also be related to the supercontinent cycle. Periods of marine aragonite deposition are also times when $MgSO_4$ evaporites are widespread, whereas periods of dominantly calcite deposition correspond to an abundance of KCl

evaporite deposition (Hardie, 1996). This distribution may be controlled by chemical reactions associated with hydrothermal alteration at ocean ridges in which there is a net transfer of Na + Mg + SO_4 from seawater to rock and of Ca + K from rock to seawater. The intensity of these reactions is a function of the volume of seawater circulated through the ocean-ridge system, which in turn is a function of the heat flux associated with ocean crust production. During times of high production of oceanic crust, therefore, the oceans should be depleted in Na–Mg sulphates and enriched in Ca and K ions, as for instance in the early Paleozoic and again in the Late Cretaceous. The Mg/Ca mole ratio in seawater can be used to track these reactions, in that at mole ratios < 2 only calcite is deposited, whereas at higher ratios both high-Mg calcite and aragonite are deposited. Because the production rate of oceanic crust is tied closely to the supercontinent cycle (Figure 6.19), the mineralogy of marine carbonates and evaporites should also reflect this cycle. The Mg/Ca mole ratio of seawater calculated from oceanic crust production rate (which is in turn calculated from sea level) (Figure 6.19) agrees well with the observed secular changes in composition of carbonates and evaporites, supporting this model (Hardie, 1996).

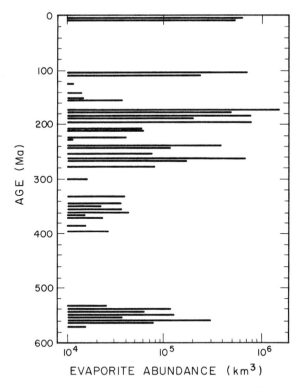

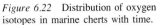

ured by the deviation of $^{18}O/^{16}O$ from standard mean ocean water (SMOW) in terms of $\delta^{18}O$ where,

$$\delta^{18}O \text{ (per mil)} = [\{(^{18}O/^{16}O)_{sample}/(^{18}O/^{16}O)_{SMOW}\} -1] \times 1000.$$

Because marine cherts and banded iron formation (BIF) are resistant to secondary changes in $\delta^{18}O$, they are useful in the monitoring the composition of seawater with geologic time (Knauth and Lowe, 1978; Gregory, 1991). The $\delta^{18}O$ of marine chert is a function of the temperature of deposition, the isotopic composition of seawater, and secondary changes in the chert. A summary of published $\delta^{18}O$ from Precambrian continental and greenstone cherts and BIF clearly shows that Archean greenstone cherts have low $\delta^{18}O$, but they overlap with Proterozoic continental cherts (Figure 6.22). Although few samples are available from Proterozoic continental cherts, there is a clear increase in Proterozoic $\delta^{18}O$ values towards the Phanerozoic values of 25–35 per mil.

Two models have been proposed to explain increasing $\delta^{18}O$ in cherts, and hence in seawater, with geologic time.

1   *The mantle–crust interaction model* (Perry Jr. et al., 1978). If Archean oceans were about the same temperature as modern oceans, they would have a $\delta^{18}O$ of 34 per mil less than cherts precipitated from them, or about 14 per mil. For seawater to evolve to a $\delta^{18}O$ of zero per mil with time, it is necessary for it to react with substances with $\delta^{18}O$ values > 0 per mil. Recycling seawater at ocean ridges and interactions of continent-derived sediments and river waters are processes that can raise seawater $\delta^{18}O$ values. Large volumes of seawater are recycled through the crust at ocean ridges and react with basalt that has a $\delta^{18}O$ of 5.5–6 per mil, and continental sediments have $\delta^{18}O$ values even higher (7–20 per mil). This model is consistent with increasing volumes of continental crust preserved with time.

Figure 6.21 Estimated abundances of marine evaporites ($NaCl+CaSO_4$) during the Phanerozoic. Modified after Holser (1984).

Oxygen isotopes in seawater

Oxygen-18 is fractionated from oxygen-16 in the ocean–atmosphere system during evaporation, with the vapour preferentially enriched in ^{16}O. Fractionation of these isotopes also decreases with increasing temperature and salinity. The fractionation of the two isotopes is meas-

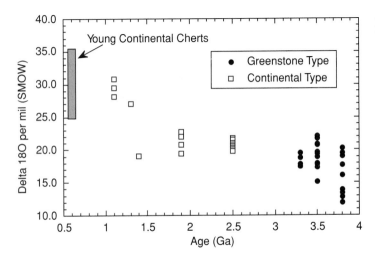

Figure 6.22 Distribution of oxygen isotopes in marine cherts with time.

2 *The seawater-cooling model* (Knauth and Lowe, 1978). In this model, the oxygen isotope trend is explained by a gradual cooling of the Earth's surface temperature from about 80 °C at 3.5 Ga to present temperatures. The $\delta^{18}O$ of seawater may or may not be constant over this time; however, if not constant it must have changed in such a way so as not to have affected the temperature control. Some investigators suggest that the evidence for gypsum deposition in the Archean is an obstacle for this model, since gypsum cannot be deposited at temperatures > 58 °C. However, as discussed above, the only well-documented examples of Archean gypsum deposition appear to reflect local and not world-wide ocean compositions.

Muehlenbachs and Clayton (1976) argued persuasively that recycling of seawater at ocean ridges has caused the oxygen isotopic composition of seawater to remain constant with time. However, it was not until the studies of ophiolite alteration (Holmden and Muehlenbachs, 1993), that it became clear that $\delta^{18}O$ of seawater has probably not changed in the last 2 Gy, since the distribution of $\delta^{18}O$ in low- and high-temperature alteration zones in 2-Ga ophiolites is the same as that found in young ophiolites. Hoffman et al. (1986) furthermore suggest that the stratigraphic distribution of $\delta^{18}O$ in alteration zones of the 3.5-Ga Barberton greenstone belt may also indicate that the oxygen isotopic composition of seawater has not changed appreciably since the Early Archean. Thus, if the secular change in $\delta^{18}O$ in cherts and BIF is real, it would appear that the temperature of Archean seawater must have been greater than at later times, as originally suggested by Knauth and Lowe (1978). As pointed out by some investigators, however, such an interpretation does not easily accommodate the evidence for widespread Early Proterozoic glaciation, which necessitates cool oceans.

The dolomite–limestone problem

It has long been recognized that the ratio of dolomite to calcite in sedimentary carbonates decreases with increasing age for the last 3 Gy. This is also reflected by the MgO/CaO ratio of marine carbonates (Figure 6.23a). Short-term variations in the dolomite/calcite ratio are also resolved in the Phanerozoic carbonate record. Most investigators agree that these trends are primary and are not acquired during later diagenesis, metasomatism, or alteration. The molar MgO/MgO+CaO ratio in igneous rocks (~ 0.2) is close to that observed in pre-Mesozoic marine carbonates (0.2–0.3), which suggests that Mg and Ca were released during weathering in the same proportions as they occur in source rocks prior to 100 Ma (Holland, 1984). The residence times of CO_2, Mg^{+2} and Ca^{+2} in seawater are short and are delicately controlled by input and removal rates of diagenetic silicates and carbonates.

The short-term variations in dolomite/calcite ratio have been explained by changes in the distribution of sedi-

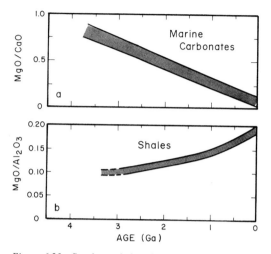

Figure 6.23 Secular variations in, (a) MgO/CaO ratio in marine carbonates, and (b) MgO/Al$_2$O$_3$ ratio in shales. Data in part from Ronov and Migdisov (1971).

mentary environments with time. For instance, dolomite is widespread during times in which large evaporite basins and lagoons were relatively abundant, such as in the Permian (Sun, 1994). Although the origin of the long-term secular change in the dolomite/calcite ratio is not fully understood, one of two explanations has been most widely advocated (Holland, 1984):

1 *Change in Mg depository* Removal of Mg^{+2} from seawater may have changed from a dominantly carbonate depository in the Proterozoic to a dominantly silicate depository in the Late Phanerozoic. There is a tendency for the MgO/Al$_2$O$_3$ in shales to increase over the same time period (Figure 6.23b), providing some support for this mechanism, in that Mg appears to be transferred from carbonate to silicate reservoirs with time.

2 *Decrease in atmospheric CO_2* Silicate-carbonate equilibria are dependent upon the availability of CO_2 in the ocean–atmosphere system. When only small amounts of CO_2 are available in the ocean, silicate–carbonate reactions shift such that Ca^{+2} is precipitated chiefly as calcite rather than in silicates. If the rate of CO_2 input into the ocean exceeds the release rate of Ca^{+2} during weathering, both Mg^{+2} and Ca^{+2} are precipitated as dolomite. Hence, a greater amount of dolomite in the Precambrian may reflect an atmosphere–ocean system with greater amounts of CO_2, which is consistent with the atmosphere model previously discussed.

Archean carbonates

Another attribute of carbonate deposition not well-understood is the sparsity of Archean carbonates. Among the causes proposed for this observation are the following:

1 the pH of Archean seawater was too low for carbonate deposition
2 carbonates were deposited on stable cratonic shelves during the Archean and later eroded away
3 carbonates were deposited in deep ocean basins during the Archean and later destroyed by subduction.

The probable high CO_2 content of the Archean atmosphere has been appealed to by some to explain the sparsity of Archean marine carbonates (Cloud, 1968). The reasoning is that high CO_2 contents in the atmosphere result in more CO_2 dissolved in seawater, thus lowering seawater pH and allowing Ca^{+2} to remain in solution because of its increased solubility. However, the CO_2 content of seawater is not the only factor controlling pH. Mineral stability considerations suggest that although the CO_2–CO_3^{-2} equilibria may have short-term control of pH, silicate equilibria have long-term control. In seawater held at a constant pH by silicate buffering reactions, the solubility of Ca^{+2} actually decreases with increasing CO_2 levels. Hence, the CO_2 mechanism does not seem capable of explaining the sparsity of Archean carbonates.

As for the second explanation, carbonate rocks in Archean continental successions are very rare, and thus removal of such successions by erosion will not solve the missing Archean carbonate problem. The third possibility tentatively holds most promise. Because of the small number of stable cratons in the Archean, carbonate sedimentation was possibly confined to deep ocean basins, controlled perhaps by planktonic algae that played a role similar to that of foraminifera today. If so, these pelagic carbonates would be largely destroyed by subduction, accounting for their near-absence in the geologic record.

Sedimentary phosphates

Phosphorus is a major building block of all forms of life, and so a knowledge of the spatial and temporal distribution of phosphates in sediments should provide insight into patterns of organic productivity (Follmi, 1996). It is well-known that sedimentary phosphates in the Phanerozoic have an episodic distribution, with well-defined periods of major phosphate deposition in the Late Proterozoic–Early Cambrian, Late Cretaceous–Early Tertiary, Miocene, Late Permian, and Late Jurassic (Figure 6.24) (Cook and McElhinny, 1979). Although less well-constrained in terms of age, there also appear to be peaks in phosphate deposition at about 2.0, 1.5, and 1.0 Ga. Most sedimentary phosphates occur at low paleolatitudes ($< 40°$), and appear to have been deposited in arid or semi-arid climates, either in narrow east–west seaways in response to dynamic upwelling of seawater, or in broad north–south seaways in response to upwelling along the east side of the seaway. In both cases, shallow seas are necessary for the upwelling.

The secular distribution of phosphate deposition may be in part related to the supercontinent cycle. For instance, the major peaks of phosphate deposition at 1.5,

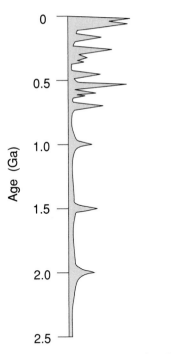

Figure 6.24 Secular variations in sedimentary phosphate deposition. After Cook and McElhinny (1979).

0.8–0.6, and < 0.1 Ga correlate with the break up of supercontinents during which the area of shallow seas increased as continental fragments dispersed. This resulted in a rise in sea level, providing extensive shallow-shelf environments into which there was a transfer of deep phosphorus-rich ocean waters to the shallow photic zone. The rapid increase in abundance of sedimentary phosphates at the base of the Cambrian is particularly intriguing, and appears to be the largest of several phosphate 'events' near the Cambrian–Precambrian boundary (Cook, 1992). The extensive phosphate deposition at this time may have triggered changes in biota, including both an increase in biomass and the appearance of metazoans with hard parts. Some of the Mesozoic–Cenozoic phosphate peaks, such as the Late Cretaceous and Eocene peaks, are related to the formation of narrow east–west seaways accompanying the continuing fragmentation Pangea. The most dramatic example of this is the Tethyan Ocean, an east–west seaway that closed during the last 200 My providing numerous rifted continental fragments with shallow shelves for phosphate deposition.

Paleoclimates

Introduction

Major worldwide changes in climate appear to be related to plate tectonics or to astronomical phenomena. Plate tectonic processes govern the size, shape and dis-

Table 6.3   Paleoclimatic indicators

Indicator	Interpretation
Widespread glacial deposits	Continental glaciation; widespread cold climates
Coal	Abundant vegetation, moist climate, generally mid-latitudes
Aeolian sandstones	Arid or semi-arid; generally but not always hot or warm temperatures
Evaporites	Arid or semi-arid; generally but not always hot or warm temperatures
Redbeds	Arid to subtropical; commonly semi-arid or arid
Laterites and bauxites	Tropical, hot and humid climates
Corals and related invertebrate fauna	Subtropical to tropical, near-shore marine

tribution of continents which, in turn, influence ocean-current patterns. The distribution of mountain ranges, volcanic activity, and changes in sea level are also controlled by plate tectonics, and each of these factors may affect climate. Volcanism influences climate in two ways:

1   volcanic peaks may produce a rainshadow effect similar to other mountain ranges
2   volcanism can introduce large amounts of volcanic dust and CO_2 into the atmosphere, the former of which reduces solar radiation and the latter of which increases the greenhouse effect.

Clearly, the distribution of continents and ocean basins is an important aspect of paleoclimatology and reconstructions of past continental positions are critical to understanding ancient climatic regimes. Astronomical factors, such as changes in the Sun's luminosity, changes in the shape of the Earth's orbit about the Sun, and changes in the angle between the Earth's equator and the ecliptic also may influence climate.

The distribution of ancient climates can be constrained with **paleoclimatic indicators**, which are sediments, fossils or other information that is sensitive to paleoclimate. A summary of major paleoclimatic indicators and their interpretation is given in Table 6.3. Some sediments reflect primarily precipitation regimes rather than temperature regimes. Coal, for instance, requires abundant vegetation and an adequate water supply but can form at various temperature regimes, with the exception of arid hot or arid cold extremes (Robinson, 1973). Aeolian sand dunes may form in cold or hot arid (or semi-arid) environments (as exemplified by modern dunes in Mongolia and the Sahara respectively). Evaporites form in both hot and cold arid environments, although they are far more extensive in the former. Laterites and bauxites seem unambiguously to reflect hot, humid climates. The distribution of various floral and faunal groups may also be useful in reconstructing ancient climatic belts. Modern hermatypic hexacorals, for instance, are limited to warm surface waters (18–25 °C) between latitude 38 ° N and 30 ° S, whereas Late Jurassic hermatypic

corals displaced northwards about 35 degrees. Because paleomagnetic and other data indicate that the continents during the Late Jurassic were at about the same latitude as at present, the coral distributions indicate a warmer climate with tropical conditions extending to high latitudes. It is also possible to estimate paleoclimatic temperatures with oxygen isotopes in marine carbonates and cherts, provided the rocks have not undergone isotopic exchange since their deposition.

Glacial diamictites (poorly-sorted sediments containing coarse- and fine-grained clasts) and related glacial-fluvial and glacial-marine deposits are important in identifying cold climates.

Extreme care must be used in identifying glacial sediments as sediments formed in other environments can possess many of the features characteristic of glaciation. For instance, subaqueous slump, mudflow, and landslide deposits can be mistaken for tillites, and indeed some ancient glaciations have been proposed based on inadequate or ambiguous data (Rampino, 1994). It also has been proposed that fallout from ejecta from large asteroid or comet impacts can produce large debris flows with characteristics similar to glacial tillites, including striated clasts and striated pavements (Oberbeck et al., 1993). At least one Late Proterozoic 'tillite' contains shocked quartz (Rampino, 1994), suggesting that it is an impact deposit, and indeed other alleged tillites in the geologic record need to be examined for shocked minerals. The bottom line is that no single criterion should be accepted in the identification of continental glacial deposits. Only a convergence of evidence from *widespread* locations, such as tillites, glacial pavement, faceted and striated boulders, and glacial dropstones, should be considered satisfactory.

Changes in the surface temperature of the Earth with time are related chiefly to three factors:

1   an increase in solar luminosity
2   variation in albedo
3   variation in the greenhouse effect.

Albedo and the greenhouse effect are party controlled by other factors such as plate tectonics and global volcanism. All three factors play an important role in long-term (10^8–10^9 y) secular climate changes, while plate tectonics and volcanism appear to be important in climate changes occurring on timescales of 10^6–10^8 y (Crowley, 1983).

Glaciation

Direct evidence for major terrestrial glaciations exists for perhaps 5–10 per cent of Earth history. At least nine major periods of continental glaciation are recognized in the geologic record (Table 6.4). The earliest glaciation at about 3 Ga is represented only in southern Africa in the Pongola Supergroup (von Brunn and Gold, 1993) and is probably of only local extent. The first evidence of widespread glaciation is the Huronian glaciation at 2.4–2.3 Ga, evidence for which has been found in

Table 6.4 Summary of major terrestrial glaciations

Age (Ma)	Geographic areas
3000	Pongola Supergroup, southern Africa
2400–2300	Huronian: Laurentia, Baltica, Siberia, South Africa
1000–900	Congo Basin, Africa; Yenisey Uplift, Siberia
750–700	Sturtian: Australia, Laurentia, South Africa
625–580	Vendian: Eurasia, South Africa, Australia, Antarctica
600–500	Sinian: China, Baltica, Laurentia, South America
450–400	Gondwana
350–250	Gondwana
15–0	Antarctica, North America, Eurasia

Laurentia, Baltica, Siberia, and South Africa. At least two major ice caps existed in Laurentia at this time, as recorded by marine glacial deposits around their perimeters (Figure 6.25). Although the evidence for Late Proterozoic glaciations (750–600 Ma) is overwhelming, the geographic distribution and specific ages of individual glaciations are not well-known (Eyles, 1993). Major Late Proterozoic ice sheets were located in central Australia, west-central Africa, the Baltic shield, Greenland and Spitsbergen, western North America and possibly in eastern Asia and South America. At least three glaciations are recognized as given in Table 6.4. Puzzling features of the Late Proterozoic glacial deposits include their association with shallow marine carbonates suggestive of warm climatic regimes. During the

Paleozoic, Gondwana migrated over large latitudinal distances, crossing the South Pole several times. Major glaciations in Gondwana occurred in the Late Ordovician, Late Carboniferous and Early Permian, and are probably responsible for drops in sea level at these times (Figure 6.26). Minor Gondwana glaciations are also recorded in the Cambrian and Devonian. Although minor glaciations may have occurred in the Mesozoic, the next large glaciations were in the Late Cenozoic, culminating with the multiple glaciations in the Quaternary.

As summarized in Table 6.5, most of the Earth's glacial record is preserved in continental rifts, foreland basins, or passive margins (Eyles, 1993). In these areas, uplift was complementary to basinal subsidence, which provided a suitable environment for preservation of both terrestrial and marine glacial deposits. The most common glacial sediments, however, are not terrestrial tillites, but glacial fluvial and debris-flow diamictites that are reworked and deposited in marine basins. Ancient glacial deposits are not common in tectonic settings related to arc systems. This may be due in part to selective destruction of glacial sediments by subduction or incorporation into accretionary prism melanges, where a glacial source may be difficult to recognize. Also, glacial deposits associated with modern arcs are usually related to local alpine glaciers, and thus the initial volume of glacial sediment is small.

Precambrian climates

Using paleoclimatic indicators together with models of atmospheric evolution, it is possible to characterize aver-

Figure 6.25 Distribution of Early Proterozoic (2.4–2.3 Ga) glaciation in Laurentia–Baltica.

Lake Baikal

SIBERIA

BALTICA

Greenland

LAURENTIA

◄⬛ Early Proterozoic Glacial Deposits

▢ Minimum areas of Continental Glaciation

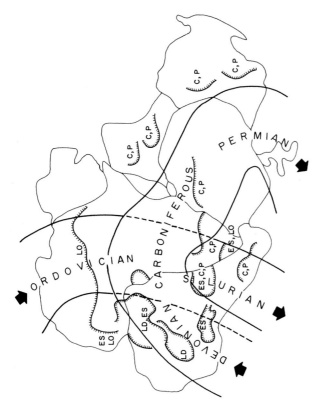

Figure 6.26 Apparent polar wander paths and major glacial centres across western Gondwana during the Paleozoic. Modified after Caputo and Crowell (1985). E, Early; L, Late; O, Ordovician; S, Silurian; D, Devonian; C, Carboniferous; P, Permian.

Table 6.5 Glaciations in terms of tectonic setting

	Forearc	Backarc	Foreland Basin	Intracratonic rift	Passive Margin
Late Cenozoic (< 36 Ma)	Gulf of Alaska	Bransfield Strait, Antarctica		Poland; North Sea rift; Ross Sea rift; Interior Alaska	Eastern Canada; NW Europe
Late Paleozoic (350–250 Ma)	Pacific margin of Gondwana; E Australia; Antarctica		Karoo basin, S. Africa	S. Africa; Arabia; Parana basin, Brazil; Australia	
Ord/Silurian (450–400 Ma)			West Africa; Arabia		
Late Proterozoic (1–0.6 Ga)	Damara orogen, S. Africa; Arabia; Paraguay; North Africa	Newfoundland	West Africa		Margins of Laurentia
Early Proterozoic (2.4–2.3 Ga)				South Africa	Huronian Supergroup, Laurentia; Siberia; Baltica
Archean (3 Ga)				Pongola Supergroup, S. Africa	

In part after Eyles (1993).

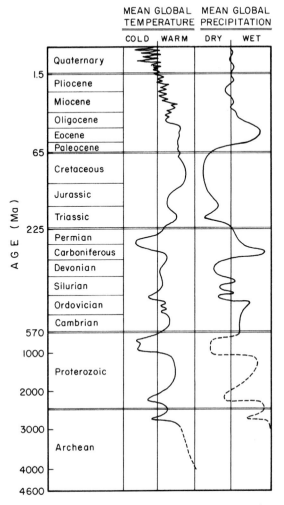

Figure 6.27 Generalized average surface temperature and precipitation history of the Earth. After Frakes (1979).

age surface temperatures and precipitation regimes with time. The abundance of CO_2 in the Archean atmosphere resulted in greenhouse warming to average surface temperatures greater than at present (Figure 6.27). Few climatic indicators are preserved in Archean supracrustal successions. Archean BIF is commonly thought to have been deposited in shallow marine waters, and hence implies low-wind velocities to preserve the remarkable planar stratification. This is consistent with a CO_2-rich atmosphere in which differences in temperature gradient with latitude are small, as are corresponding wind velocities (Frakes et al., 1992). The earliest evaporites and stromatolitic carbonates appear at about 3.5 Ga and also favour generally warm climates. The relative abundance of shallow marine carbonates and the preservation of lateritic paleosols in the Proterozoic suggest a continuation of warm, moist climates after the end of the Archean. The increasing proportion of evaporites and

redbeds after 2 Ga appears to record the increasing importance of semi-arid to arid climatic regimes. Colder climates as recorded by widespread glaciations alternated with warm, moist climates as reflected by marine carbonates in the Late Proterozoic.

Also supporting an unusually warm Archean climate are geochemical data related to weathering rates. Studies of modern weathering show that Ca, Na, and Sr are rapidly lost during chemical weathering and that the amount of these elements lost is proportional to the degree of weathering (Condie, 1993). As shown by the chemical index of weathering (CIW), most post-Archean shales show moderate losses of Ca, Na, and Sr from source weathering, with CIW indices of 80–95 and Sr contents of 75–200 ppm. These values reflect the 'average' intensity of chemical weathering of the shale sources. Most Archean shales, however, show greater losses of all three elements, with typical CIW indices of 90–98 and Sr contents < 100 ppm. This suggests that the average intensity of chemical weathering may have been greater during the Archean than afterwards, a conclusion which is consistent with a CO_2-rich atmosphere in the Archean.

Although data clearly indicate that Archean and Early Proterozoic climates were hotter than at present, at least two glaciations must somehow be accommodated at this time. Calculations indicate that a drop of only 2 °C is enough to precipitate an ice age (Kasting, 1987). Causes of glaciations at 3.0 and 2.4–2.3 Ga remain illusive. The absence of significant amounts of volcanic rock beneath tillites of both ages does not favour volcanic dust injected into the atmosphere (reducing sunlight) as a cause of either glaciation. The 3-Ga glaciation is recorded only in eastern part of the Kaapvaal craton is South Africa, and may have been caused by local conditions. However, to account for the Early Proterozoic glaciations, it would appear that a widespread drawdown mechanism for atmospheric CO_2 is needed. One possibility is that increasing numbers of photosynthetic micro-organisms may have extracted large amounts of CO_2 from the atmosphere faster than it could be returned by the carbonate–silicate cycle, thus reducing the greenhouse effect. However, such a model does not explain how glaciation should end nor why large amounts of marine carbonates produced during the CO_2 withdrawal do not occur beneath Early Precambrian tillites. Another possibility which has not been fully evaluated is the possible effect of rapid continental growth in the Late Archean on the CO_2 content of the atmosphere. Perhaps rapid chemical weathering of the newly-formed cratons reduced CO_2 levels in the atmosphere and led to cooling and glaciation.

The Late Proterozoic glaciations, many of which appear to have occurred at low latitudes, are even more puzzling. Because of inadequate isotopic ages, a given glaciation may not have been world-wide but only of local or continental extent. Paleomagnetic data give conflicting results regarding the latitude of the Late Proterozoic glaciations. Earlier results suggest that all of these glaciations occurred at tropical or subtropical latitudes, whereas recent data suggest that at least some occurred at high latitudes (Meert and van der Voo, 1994). Williams (1975) suggested that the obliquity of the Earth's orbit may have changed with time and that during the Late Proterozoic the equator was tipped at a high angle to the ecliptic resulting in a decrease in mean annual temperature. Such a condition should favour widespread glaciation even at low latitudes. However, this model does not seem capable of explaining the angular momentum distribution in the Earth–Moon system. It is possible that a continued decrease in atmospheric CO_2 level reducing the greenhouse effect was responsible for initiation of the Late Proterozoic glaciations (Brasier, 1992). The widespread shallow marine carbonates associated with Late Proterozoic glacial deposits are consistent with rapid CO_2 removal from the atmosphere. However, renewed ocean ridge activity accompanying the break up of Rodinia during this time should have pumped more CO_2 into the atmosphere, increasing greenhouse warming!

Phanerozoic climates

Paleoclimate summary

Most of the early and middle Paleozoic are characterized by climates that on the whole were warmer than present climates, while late Paleozoic climates show wide variations (Frakes et al., 1992). Europe and North America underwent relatively small changes in climate during the Paleozoic, while Gondwana underwent several major changes. This is due to the fact that the continental fragments comprising Laurasia remained at low latitudes during the Paleozoic, while Gondwana migrated over large latitudinal distances. Major glaciations occurred in Gondwana in Late Ordovician, Late Carboniferous, and Early Permian.

The Mesozoic is characterized by warm, dry climates that extend far north and south of equatorial regions. Paleontologic and paleomagnetic data also support this interpretation. Initial Triassic climates were cool and humid like the late Paleozoic, and they were followed by a warm, drying trend until Late Jurassic. The largest volumes of preserved evaporites are in the Late Triassic and Early Jurassic and reflect widespread arid climates on Pangea. Most of the Cretaceous is characterized by warm, commonly humid climates regardless of latitude, with tropical climates extending from 45° N to perhaps 75° S. Glacial climates are of local extent and only in polar regions. Mean annual surface temperatures during the Mid-Cretaceous, as determined from oxygen isotope data, are 10–15 °C higher than today. The Late Cretaceous is a time of maximum transgression of the continents, and the warm Cretaceous climates appear to reflect a combination of increased atmospheric CO_2 levels and low albedo of the oceans. The Early Tertiary is a time of declining temperatures, significant increased precipitation, and falling sea level. Middle and Late Tertiary are characterized by lower and more variable tempera-

ture and precipitation regimes. These variations continued and became more pronounced in the Quaternary, leading to the alternating glacial and interglacial epochs. Four major Quaternary glaciations are recorded, with the intensity of each glaciation decreasing with time: the Nebraskan (1.46–1.3 Ma), Kansan (0.9–0.7 Ma), Illinoian (550,000–400,000 a), and Wisconsin (80,000–10,000 a). Each glaciation lasted 100–200 thousand years, with interglacial periods of 200–400 thousand years. Smaller glacial cycles with periods of the order of 20–40 thousand years are superimposed on the larger cycles.

Glaciations

Both early and late Paleozoic glaciations appear to be related to plate tectonics. During both of these times, Gondwana was glaciated as it moved over the South Pole (Caputo and Crowell, 1985; Frakes et al., 1992). Major glaciations occurred in central and northern Africa in the Late Ordovician when Africa was centered over the South Pole (Figure 6.26). As Gondwana continued to move over the pole, major ice centres shifted into southern Africa and adjacent parts of South America in the Early Silurian. Gondwana again drifted over the South Pole, beginning with widespread glaciation in the Late Devonian and Early Carboniferous in South America and central Africa, and shifting to India, Antarctica, and Australia in the Late Carboniferous and Permian. Paleozoic glaciations may also be related to prevailing wind and ocean currents associated with the movement of Gondwana over the pole.

The next major glaciations began in the Mid-Tertiary and were episodic with interglacial periods. Between 40–30 Ma, the Drake Passage between Antarctica and South America opened and the Antarctic circumpolar current was established, thus thermally isolating Antarctica (Crowley, 1983). The final separation of Australia and Antarctica at about this time may also have contributed to the cooling of Antarctica by inhibiting poleward transport of warm ocean currents. Glaciation appears to have begun in Antarctica 15–12 Ma and by 3 Ma in the Northern Hemisphere. Initiation of Northern Hemisphere glaciation may have been in response to the formation of Panama which caused the Gulf Stream, which had previously flowed westward about the equator, to flow north along the coasts of North America and Greenland. This shift in current brought warm, moist air into the Arctic, which precipitated and began to form glaciers.

Three factors have been suggested as important in controlling the multiple glaciations that characterize the Pleistocene (Eyles, 1993):

1 episodic variations in volcanic activity
2 cooling and warming of the oceans
3 cyclic changes in the Earth's rotation and orbit.

During the Pleistocene there is no close correlation between volcanic activity and glaciation, and so no support for the first idea. Cyclical changes in temperature of the oceans caused by changes in the balance between evaporation and precipitation may be in part responsible for multiple glaciations. A typical cycle might start as moist, warm air moves into the Arctic region precipitating large amounts of snow that result in glacial growth. As oceanic temperatures drop in response to glacial cooling, evaporation rates from the ocean decrease and the supply of moisture to glaciers also decreases. So glaciers begin to retreat, climates warm and an interglacial stage begins. The warming trend has a negative feedback and eventually increases evaporation rates, and moisture is again supplied to glaciers as the cycle starts again.

Most investigators agree that cyclical changes in the shape of the Earth's orbit about the Sun and in the inclination and wobble of the Earth's rotational axis can bring about cyclical cooling and warming of the Earth. Today, the Earth's equatorial plane is tilted at 23.4 ° to the ecliptic, but this angle varies from 21.5 °–24.5 ° with a period of about 41 000 years. Decreasing the inclination results in cooler summers in the Northern Hemisphere and favours ice accumulation. The Earth's orbit also changes in shape from a perfect circle to a slightly ellipsoidal shape with a period of 92 000 years and the ellipsoidal orbit results in warmer than average winters in the Northern Hemisphere. The Earth also wobbles as it rotates with a period of about 26 000 years and the intensity of seasons varies with the wobble. When variations in oxygen isotopes or fauna are examined in deep-sea cores, all three of these periods, known as the **Milankovitch periods**, can be detected, with the 92 000 year period being prominent. These results suggest that on time scales of 10 000–400 000 years, variations in the Earth's orbit are the fundamental cause of Quaternary glaciations (Imbrie, 1985). Spectral analyses of $\delta^{18}O$ and orbital time series strongly support the model.

Paleoclimates and supercontinents

There are some interesting correlations between paleoclimates and the supercontinent cycle, which strongly suggest that major climatic regimes in the past are at least in part related to supercontinent cycles, which in turn are related to mantle plume activity. As previously noted, the first evidence for a supercontinent is at the end of the Archean. In the earliest Proterozoic, there is a record of the first widespread glaciation, the first redbeds, and perhaps the first phosphate deposition (Worsley and Nance, 1989; Veevers, 1990). As mentioned above, increased weathering associated with this first supercontinent, together with increased biologic productivity and organic burial, may have drawn down CO_2 in the atmosphere, leading to widespread glaciation. Fragmentation of the supercontinent at about 2.2–2.1 Ga should have led to warmer climates as new ridge systems propagated and dispersed the fragments. Increased input of oxygen into the atmosphere at this time as recorded by paleosols, redbeds, and other climatic indicators may have led to the appearance of eukaryotes at about 2 Ga. Although a second supercontinent appears to have formed

between about 2.0–1.9 Ga, there are no widespread glacial deposits of this age, suggesting that climates were not necessarily cold. As this supercontinent fragmented, perhaps between 1.6–1.4 Ga, there is evidence of widespread warm climates, as extensive marine carbonates were deposited on fragmenting cratons. Oxygen levels in the atmosphere continued to increase so that by about 1.7 Ga metazoans appeared, although they did not become widespread until the Late Proterozoic Ediacaran fauna at about 600 Ma. Rodinia formed between 1.3–1.0 Ga and again the climates may have cooled, although the only indications of glaciation of this age are on the Congo craton in Africa and in Siberia (Mid-Riphean), both at about 1 Ga (Table 6.4). The Late Proterozoic glaciations between 750 Ma–550 Ma seem to have occurred as Rodinia was fragmenting which, as previously discussed, is puzzling since the widespread carbonates of this age seem to reflect warm climates. The last glaciations of this period (650–550 Ma), however, also correlate with the formation of Gondwana.

The high stand of sea level in the Cambrian and Ordovician correlates well with continuing dispersal of Rodinian fragments and the formation of Gondwana, and world-wide climates were warm and mild at this time. As Pangea began to form in the Late Ordovician and Silurian, glaciation occurred again and cold climatic regimes returned. After a brief warming in the Devonian, the extensive drawdown of atmospheric CO_2 in the Carboniferous caused by the rapid increase in biomass of vascular plants and the corresponding rapid rate of carbon burial led again to cool climates and widespread glaciation in Gondwana in the latter part of the Paleozoic. As Pangea began to fragment in the early Mesozoic, warm climates returned, followed by cool climates in the Jurassic and Early Cretaceous, in response perhaps to CO_2 drawdown by increased burial rate of carbon. As Pangea continued to fragment, sea level rose and warm climates returned during the Late Cretaceous and Early Tertiary. This may be related to a major superplume event as described below. Increased burial rates of carbon and increased weathering and erosion rates, caused chiefly by the rising Alpine–Himalayan chain, led to CO_2 drawdown and to the cooling mode we have been in for the last 50 My.

Paleoclimates and mantle plumes

The idea that mantle plumes have had an effect on paleoclimates was first proposed by Larson (1991a), who suggested that there was a direct relationship between the frequency and size of mantle plumes, the frequency of magnetic reversals, and the rate of generation of oceanic crust. Calculations of the rate of production of oceanic crust show a 50–75 per cent increase in production rate between 120–80 Ma (Figure 6.28). A similar increase in production rate of submarine plateaux, especially in the Pacific basin, is observed during the same time interval. What is intriguing is the fact that this time of increased production of oceanic crust and submarine

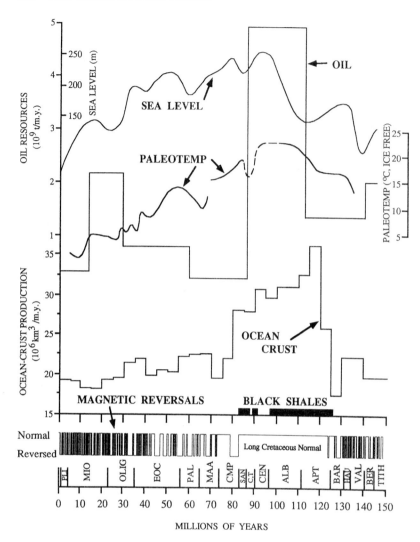

Figure 6.28 Variation in magnetic reversal stratigraphy, oceanic crust production rate, seawater paleotemperature (at high latitude), sea level, black shale deposition, and world oil resources in the last 150 My. Courtesy of Roger Larson.

plateaux coincides with the Cretaceous superchron recording a 40-My period of normal magnetic polarities (Figure 6.28). Larson (1991a) suggested that these features were all caused by a 'superplume event' in the mantle beneath the Pacific basin. About 125 Ma one or more superplumes may have been generated in the D″ layer just above the core–mantle interface, and rose beneath the Pacific basin. A large swell in the mantle beneath Tahiti could be a remnant of the major superplume. How such a superplume may have stopped reversals in the magnetic field for 40 My is unknown, but would appear to be somehow related to a sudden heat loss from the core at 125 Ma. Perhaps the heat loss increased the core cooling rate and convective activity increased in the outer core to restore the heat. This, in turn, decreased the magnetic reversal frequency, and when core convection increased above some critical threshold, reversals stopped altogether, resulting in a superchron.

Of importance to paleoclimates is a possible correlation of the superchron with increased surface temperature of seawater and sea level at this time (Figure 6.28). The paleotemperature high at about 100 Ma appears to require increased CO_2 levels so 6–8 times present-day levels in the atmosphere to enhance the greenhouse effect (Larson, 1991b). Such increases in CO_2 demand a sudden and large input of CO_2 into the atmosphere, far above that easily accommodated by feedback mechanisms in the carbonate–silicate cycle. This CO_2 could have been introduced by increased volcanism both at ocean ridges and in mantle plumes during the possible Cretaceous superplume event (Caldeira and Rampino, 1991). Interestingly, the period of time between 120 and 80 Ma is also a time of widespread deposition of black shales and production of large oil reserves. For instance, 60 per cent of the world's known oil reserves were produced between 112–88 Ma. Cretaceous black shales, such as the Mancos Group in the western United States, have been interpreted as markers of oceanic anoxic events that resulted from increases in organic productivity and partially restricted basins. Both the black shale and oil anomalies in the Cretaceous are caused by increased levels of nutrients and carbon (as CO_2), perhaps supplied by the increase in oceanic-ridge and submarine-plateau volcanic activity in the Cretaceous. There is also a broad peak in gas accumulation for the entire Cretaceous that may be related to the same cause. The high sea level at the same time is caused by the increased volume of ocean ridges. Another superplume event may have occurred in the late Paleozoic when large amounts of coal and gas accumulated on the continents and when sea level was at a high. These events also correlate with the late Paleozoic superchron.

If the ideas of Larson (1991a and b) are correct, episodes of major mantle plume activity, as advocated in Chapter 5 to explain episodic age distributions, may have profound affects on the Earth's paleoclimatic regimes. It will be interesting to see if paleoclimatic indicators in the Precambrian support such a model.

Living systems

General features

Although the distinction between living and non-living matter is obvious for most objects, it is not so easy to draw this line between some unicellular organisms on the one hand, and large non-living molecules, such as amino acids, on the other. It is generally agreed that living matter must be able to reproduce new individuals, it must be capable of growing by using nutrients and energy from its surroundings, and it must respond in some manner to outside stimuli. Another feature of life is its chemical uniformity. Despite the great diversity of living organisms, all life is composed of a few elements (chiefly C, O, H, N, and P) which are grouped into nucleic acids, proteins, carbohydrates, and fats and a few other minor compounds. This suggests that living organisms are related and that they probably had a common origin. Reproduction is accomplished in living matter at the cellular level by two complex nucleic acids, RNA and DNA. Genes are portions of DNA molecules that carry specific hereditary information. Three components are necessary for a living system to self-replicate: RNA and DNA molecules, which provides a list of instructions for replication; proteins that promote replication; and a host organ for the RNA-DNA molecules and proteins. The smallest entities capable of replication are amino acids.

Origin of life

Introduction

Perhaps no other subject in geology has been investigated more than the origin of life (Kvenvolden, 1974;

Oro, 1994). It has been approached from many points of view. Geologists have searched painstakingly for fossil evidences of the earliest life, and biologists and biochemists have provided a variety of evidence from experiments and models that must be incorporated into any model for the origin of life.

Although numerous models have been proposed for the origin of life, two environmental conditions are a prerequisite to all models:

1 the elements and catalysts necessary for the production of organic molecules must be present
2 free oxygen, which would oxidize and destroy organic molecules must not be present.

In the past, the most popular models for the origin of life involve a primordial 'soup' rich in carbonaceous compounds produced by inorganic processes. Reactions in this soup promoted by catalysts such as lightning or UV radiation result in production of organic molecules. Primordial soup models, however, seem unnecessary in terms of rapid degassing of the Earth prior to 4 Ga. Rapid recycling of the early oceans through ocean ridges would not allow concentrated 'soups' to survive, except perhaps locally in evaporite basins after 4 Ga. Because the chances of organic molecules being present in sufficient amounts, in the correct proportions, and in the proper arrangement are very remote, it would seem that the environment in which life formed would have been very widespread in the Early Archean. Possibilities include volcanic environments and hydrothermal vents along ocean ridges.

Simple amino acids have been formed in the laboratory under a variety of conditions. The earliest experiments were those of Miller (1953), who sparked a hydrous mixture of H_2, CH_4 and NH_3 to form a variety of organic molecules including four of the twenty amino acids composing proteins, Similar experiments, using both sparks and UV radiation in gaseous mixtures of H_2O, CO_2, N_2, and CO (a composition more in line with that of the Earth's early degassed atmosphere), also resulted in the production of amino acids, HCN and formaldehydes, the latter of which can combine to form sugars. Heat also may promote similar reactions.

The role of impacts

As indicated by microfossils, life was certainly in existence by 3.5 Ga, and carbon isotope data, although less definitive, suggests that life was present by 3.8 Ga. This being the case, the origin of life must have coincided with the last stage of heavy bombardment of planets in the inner Solar System as indicated by the impact craters on the Moon and other terrestrial planets with ancient surfaces. As an example, the impact record on the Moon shows that crater size, and hence the impact energy, falls exponentially from 4.5 Ga to about 3.0 Ga, decreasing more gradually thereafter (Figure 6.29) (Sleep et al., 1989; Chyba, 1993). Similarity of crater frequency versus diameter relations for Mercury and Mars implies

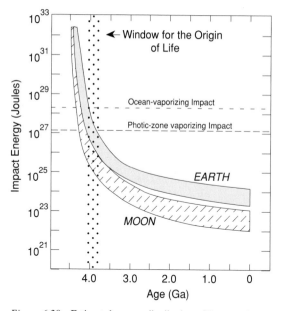

Figure 6.29 Estimated energy distribution of lunar and terrestrial impacts with time. Also shown is the 'window' for the origin of life defined by the termination of ocean-vaporizing impacts (4.1 Ga) and the oldest fossils (3.5 Ga). Modified after Sleep et al. (1989).

that planets in the inner Solar System underwent a similar early bombardment history, although the Earth's history has been destroyed by plate tectonics. A decrease in impact energy with time on Earth is likely to be similar to that on the Moon, except that after 3.0 Ga energies were perhaps an order of magnitude higher on the Earth (Figure 6.29). Because the Earth's gravitational attraction is greater than that of the Moon, it should have been hit with many more large objects than the Moon before 3.5 Ga. Until about 4.1 Ga, the Earth was probably hit by several asteroid-sized bodies with impact energies large enough to vaporize an entire ocean (Figure 6.29). Such atmospheres should persist for several months before recondensing into oceans. Furthermore, such events should lead to greenhouse heating of the atmosphere up to temperatures of about 1700 °C for thousands of years, thus 'sterilizing' the Earth's surface to depths of hundreds of metres. If life was formed during or between these large impacts, it is difficult to see how it could survive, and thus each major impact event should in effect reset the clock for the origin of life (Chyba, 1993). From our knowledge of the timing of these early large impact events it would appear that, for life to survive, it would have at the most several hundred millions of years between the end of ocean-vaporizing impacts at about 4.1 Ga and the oldest known fossils at 3.5 Ga. Because life is relatively advanced by 3.5 Ga, it is likely that the window for the origin of life is between 4.1–3.8 Ga (Figure 6.29).

Even after the large impactor events ceased, it is likely

that the Earth sustained many impacts sufficient to vaporize the photic zone of the oceans (depths of about 200 m) (Sleep et al., 1989). Although adverse surface temperatures would last for only about 300 years after impact in these cases, the outlook is not good for survival of any photosynthetic organisms needing sunlight. This line of reasoning clearly favours an origin for life in the deep oceans well beneath the photic zone, and is consistent with the idea that life formed at hydrothermal vents on the sea floor as discussed below.

Another intriguing aspect of early impact is the possibility that relatively small impactors actually introduced volatile elements and small amounts of organic molecules to the Earth's surface which could be used in the origin of life. The idea that organic substances were brought to the Earth by asteroids or comets is not new, and in fact it was first suggested in the early part of the twentieth century. Lending support to the idea is the recent discovery of in-situ organic-rich grains in comet Halley, and data suggest that up to 25 per cent organic matter may occur in other comets. Carbonaceous chondrites (a type of meteorite), which appear to be fragments of a particular group of asteroids, contain several per cent of organic matter. The problem with such an origin for organic compounds on Earth is how to get these substances to survive impact. Even for small objects (~ 100 m in radius), impact should destroy organic inclusions unless the early atmosphere was very dense (~ 10 bar CO_2), and could sufficiently slow the objects before impact, However, interplanetary dust from colliding comets or asteroids could survive impact and may have introduced significant amounts of organic molecules into the atmosphere or oceans. Whether this possible source of organics was important depends critically on the composition of the early atmosphere. If the atmosphere was rich in CO_2 as suggested earlier, the rate of production of organic molecules was probably quite small, and hence the input of organics by interplanetary dust may have been significant.

The RNA world

Although it seems relatively easy to form amino acids and other simple organic molecules, just how these molecules combined to form the first complex molecules, like RNA, and then evolved into living cells remains largely unknown. Recent studies of RNA suggest that it may have played a major role in the origin of life. RNA molecules have the capability of splitting and producing an enzyme that can act as a catalyst for replication (Zaug and Cech, 1986) (Figure 6.30). Necessary conditions for the production of RNA molecules in the Early Archean include a supply of organic molecules, a mechanism for molecules to react to form RNA, a container mineral to retain detached portions of RNA so that they can aid further replication, a mechanism by which some RNA can escape to colonize other populations, and some means of forming a membrane to surround a proto-cell wall (Nisbet, 1986). During the Archean, hydrothermal sys-

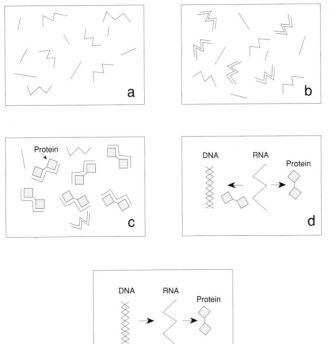

Figure 6.30 Diagrammatic representation of the RNA world. (a) RNA is produced from ribose and other organic compounds; (b) RNA molecules learn to copy themselves; (c) RNA molecules begin to synthesize proteins; (d) the proteins serve as catalysts for RNA replication and the synthesis of more proteins. They also enable RNA to make double-strand molecules that evolve into DNA; (e) DNA takes over and uses RNA to synthesize proteins, which in turn enables DNA to replicate itself and transfer its genetic code to RNA.

tems on the sea floor may have provided these conditions (Gilbert, 1986). In laboratory experiments, RNA splitting occurs at temperatures around 40 °C with a pH varying from 7.5–9 and with Mg present in solution.

The Early Archean 'RNA world' may have existed in clay minerals, zeolites, and in the pore spaces of altered volcanic rocks. The next stage in replication may have been the development of proteins from amino acids that were synthesized from CH_4 and NH_3. Later still, DNA must form and take over as the primary genetic library (Figure 6.30) (Gilbert, 1986). The next stages of development, although poorly understood, seem to involve production of membranes, which allow the managing of energy supply and metabolism, both of which are essential for the development of a living cell.

Hydrothermal vents: a possible site for the origin of life

Hydrothermal vents on the sea floor have been proposed by several investigators as a site for the origin of life (Chang, 1994; Nisbet, 1995). Modern hydrothermal vents have many organisms that live in their own vent ecosystems, including a variety of unicellular types (Tunnicliffe and Fowler, 1996). Vents are attractive in that they supply the gaseous components such as CO_2, CH_4, and nitrogen species from which organic molecules can form, and they also supply nutrients for the metabolism of

organisms such as P, Mn, Fe, Ni, Se, Zn, and Mo (Figure 6.31). Although these elements are present in seawater, it is difficult to imagine how they could have been readily available to primitive life at such low concentrations. Early life would not have had sophisticated mechanisms capable of extracting these trace metals, thus requiring relatively high concentrations that may exist near hydrothermal vents. The chief objection that has been raised to a vent origin for life is the potential problem of both synthesizing and preserving organic molecules necessary for the evolution of cells. The problems is that the temperatures at many or all vents may be too high, and they would destroy, not synthesize, organic molecules (Miller and Bada, 1988). However, many of the requirements for the origin of life seem to be available at submarine hydrothermal vents.

One possible scenario for the origin of life at hydrothermal vents begins with CO_2 and N_2 in vent waters at high temperatures deep in the vent (Shock, 1992). As the vent waters containing these components circulate to shallower levels and lower temperatures, they cool and thermodynamic conditions change such that CH_4 and NH_3 are the dominant gaseous species present. Provided suitable catalysts are available (see below), these components can then react to produce a variety of organic compounds. The next step is more difficult to understand, but somehow simple organic molecules must react with each other to form large molecules such as peptides, lipids, and esters.

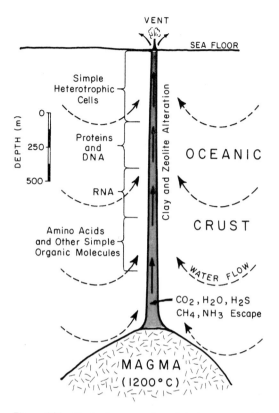

VENT

SEA FLOOR

DEPTH (m)

Simple
Heterotrophic
Cells

Clay and Zeolite Alteration

Proteins
and
DNA

OCEANIC

RNA

CRUST

Amino Acids
and Other Simple
Organic Molecules

WATER FLOW

CO_2, H_2O, H_2S
CH_4, NH_3 Escape

MAGMA
(1200°C)

Figure 6.31 Idealized cross-section of Archean ocean-ridge hydrothermal vent showing possible conditions for the formation of life.

The first life

One of the essential features of life is its ability to reproduce. It is probable that the ability to replicate was acquired long before the first cell appeared on the scene. Cairns-Smith (1982) has suggested that clays may have played an important role in the evolution of organic replication. Organic compounds absorbed in clays may have reacted to form RNA and, through natural selection. RNA molecules eventually disposed of their clay hosts. Because hydrothermal systems appear to have lifetimes of the order of 10^4–10^5 y at any given location, RNA populations must have evolved rapidly into cells, or more likely they were able to colonize new vent systems. Another possible catalyst is zeolites which possess a wide variety of pores of different shapes and sizes that permit small organic molecules to pass through while excluding or trapping larger molecules (Nisbet, 1986). They are also a characteristic secondary mineral around hydrothermal plumbing systems (Figure 6.31). The significance of variable-sized cavities in zeolites is that a split-off RNA molecule may be trapped in such a cavity, where it can aid replication of the parent molecule. Although the probability is small, it is possible that the first polynucleotide chain formed in the plumb-

ing system of an early hydrothermal vent on the sea floor.

The first cells were primitive in that they had poorly-developed metabolic systems and survived by absorbing a variety of nutrients from their surroundings (Kandler, 1994; Pierson, 1994). They must have obtained nutrients and energy from other organic substances by fermentation, which occurs only in anaerobic (oxygen-free) environments. **Fermentation** involves the breakdown of complex organic compounds into simpler compounds that contain less energy, and the energy liberated is used by organisms to grow and reproduce. Cells that obtain their energy and nutrients from their surroundings by fermentation or chemical reactions are known as **heterotrophs**, in contrast to **autotrophs** which are capable of manufacturing their own food. Two types of anaerobic cells evolved from DNA replication. The most primitive group, the archaebacteria, use RNA in the synthesis of proteins, whereas the more advanced group, the eubacteria have advanced replication processes and may have been the first photosynthesizing organisms.

Rapid increases in the numbers of early heterotrophs may have led to severe competition for food supplies. Selection pressures would tend to favour mutations that enabled heterotrophs to manufacture their own food and, thus, to become autotrophs. The first autotrophs had appeared by 3.5 Ga as cyanobacteria. These organisms produced their own food by photosynthesis, perhaps using H_2S rather than water since free oxygen, which is liberated during normal photosynthesis, is lethal to anaerobic cells. Just how photosynthesis evolved is unknown, but perhaps the supply of organic substances and chemical reactions became less plentiful as heterotrophs increased in numbers, and selective pressures increased to develop alternate energy sources. Sunlight would be an obvious source to exploit. H_2S may have been plentiful from hydrothermal vents or decaying organic matter on the sea floor and some cells may have developed the ability to use this gas in manufacturing food. As these cells increased in numbers, the amount of H_2S would not be sufficient to meet their demands and selective pressures would be directed towards alternative substances, of which water is the obvious candidate. Thus, mutant cells able to use water would out-compete forms only able to use H_2S.

The first fossils

Two lines of evidence are available for the recognition of the former existence of living organisms in Early Archean rocks: microfossils and organic geochemical evidence (Schopf, 1994). Many microstructures preserved in rocks can be mistaken for cell-like objects (inclusions, bubbles, microfolds, etc.), and progressive metamorphism can produce structures that look remarkably organic but at the same time can destroy real microfossils. For these reasons caution must be exercised in accepting microstructures as totally biologic.

The oldest well-described assemblage of microfossil-

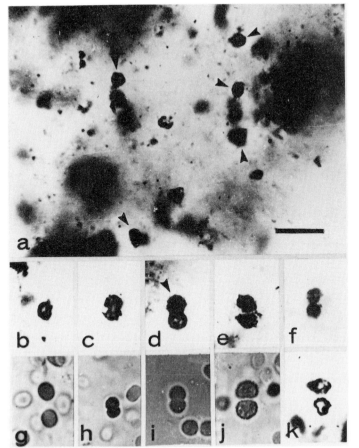

Figure 6.32 Photomicrograph of spheroid microstructures from the Swartkippie Formation, Barberton greenstone belt, South Africa. Arrows note individual cells. Stages in cell division in the Archean samples (b) to (e) are compared to modern prokaryotes in (g) to (j). Scale bar is 10 μm. Courtesy of Andrew Knoll.

like structures comes from cherts in the 3.5-Ga Barberton greenstone belt in South Africa (Figure 6.32), and the oldest unambiguous structures of organic origin are 3.5-Gy-old stromatolites from the Pilbara region of Western Australia (Walter, 1994) (Figure 6.33). Three types of microstructures, ranging in size from < 1 mm to ~ 20 mm, have been reported from the Barberton sequence and from other Archean sediments. These are rod-shaped bodies, filamentous structures, and spheroidal bodies. The spheroidal bodies are similar to alga-like bodies from Proterozoic assemblages and are generally interpreted as such. Of the two known types of cells, prokaryotic and eukaryotic, only prokaryotic types are represented among Archean microfossils. **Prokaryotes** are primitive cells which lack a cell wall around the nucleus and are not capable of cell division: **eukaryotes** possess these features and so are capable of transmitting genetic coding to various cells and to descendants.

Stromatolites

Stromatolites are finely-laminated sediments composed chiefly of carbonate minerals that have formed by the accretion of both detrital and biochemical precipitates on successive layers of micro-organisms (commonly cyanobacteria). They exhibit a variety of domical forms and range in age from about 3.5 Ga to modern. Two parameters are especially important in stromatolite growth: water currents and sunlight. There are serious limitations to interpreting ancient stromatolites in terms of modern ones, however. First of all, modern stromatolites are not well understood and occur in a great variety of aqueous environments (Walter, 1994). The distribution in the past is also controlled by the availability of shallow, stable-shelf environments, the types of organisms producing the stromatolites, the composition of the atmosphere and perhaps the importance of burrowing animals. It is possible to use stromatolites to distinguish deep-water from shallow-water deposition, since reef morphologies are different in these environments.

Although some Early Archean laminated carbonate mats appear to be of inorganic origin, by 3.2 Ga well-preserved organism-built stromatolites are widespread. The oldest relatively unambiguous stromatolite at 3.5 Ga occurs near the town of North Pole in Western Australia, and its age is constrained by U–Pb zircon dates from associated volcanics (Figure 6.33) (Buick et al., 1995). Early Archean stromatolites were probably

Figure 6.33   3.5-Ga-old stromatolites from the Pilbara region in Western Australia. Courtesy of D. R. Lowe.

built by anaerobic photoautotrophs with mucus sheaths (Walter, 1994). These microbes were able to cope with high salinities, desiccation and high sunlight intensities as indicated by their occurrence in evaporitic cherts. Late Archean stromatolites are known from both lagoon and near-shore marine environments and some are very similar to modern stromatolites suggesting that they were constructed by cyanobacteria. Early Proterozoic stromatolites appear to have formed in peritidal and relatively deep subtidal environments, and for the most part appear to be built by cyanobacteria.

Stromatolites increased in numbers and complexity from 2.2 Ga to about 1.2 Ga, after which they decreased rapidly (Figure 6.34) (Grotzinger, 1990; Walter, 1994). Whatever the cause or causes of the decline, it is most apparent initially in quiet subtidal environments and spreads later to the peritidal realm. Numerous causes have been suggested for the decline, of which the two most widely cited are:

1   grazing and burrowing of algal mats by the earliest metazoans
2   decreasing saturation of carbonate in the oceans resulting in decreasing stabilization of algal mats by precipitated carbonate.

Not favouring the grazing idea is the fact that the rapid increase in numbers and diversity of metazoan life forms begins after much of the decline in stromatolites (i.e., at

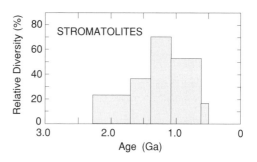

Figure 6.34   The relative diversity of stromatolites through time. After Walter (1994).

about 600 Ma). The extent to which stromatolites can be used to establish a world-wide Proterozoic biostratigraphy is a subject of controversy, which revolves around the role of environment and diagenesis in determining stromatolite shape and the development of an acceptable taxonomy. Because the growth of stromatolites is at least partly controlled by organisms, it should in theory be possible to construct a world-wide biostratigraphic column. Another controversial subject is that of how stromatolite height is related to tidal range. Cloud (1968b) suggests that the height of intertidal stromatolites at maturity reflects the tidal range, whereas Walter (1994) suggests that the situation is much more complex. The distribution of laminations in stromatolites has also been

Figure 6.35 Grypania, the oldest-known fossil of eukaryotic algae from the 2.1-Ga Negaunee Iron Formation, Marquette, Michigan. Scale × 2. Courtesy of Bruce Runnegar.

suggested as a means of studying secular variation of the Earth–Moon system (Chapter 7).

Appearance of eukaryotes

It was not until about 2 Ga, however, that microbes entirely dependent on the use of molecular O_2 appeared in the geologic record (Runnegar, 1994). These are eukaryotes, which are advanced cells with a cell nucleus enclosing DNA and with specialized organs in the cell. Eukaryotes are also able to reproduce sexually. RNA studies of living unicellular eukaryotes suggest that they are derived from archaebacterial prokaryotes some time between about 2.5–2.3 Ga. Although the earliest fossil eukaryotes appear about 2 Ga, they did not become widespread in the geologic record until 1.7–1.5 Ga. The oldest fossil thought to represent a eukaryote is *Grypania* from 2.1-Gy-old sediments in Michigan. *Grypania* is a coiled, cylindrical organism that grew up to about 50 cm in length and 2 mm in diameter (Figure 6.35). Although it has no certain living relatives, it is regarded as a probable eukaryotic alga because of its complexity, structural rigidity and large size.

RNA studies of modern eukaryotes suggest that the earliest forms to evolve were microsploridians, amoebae, and slime moulds. Later branches lead to the metazoans, red algae, fungi, and higher plants, as well as several groups of protistans. A minimum date for this latter radiation is given by well-preserved, multicellular red alga fossils from 1.2–1.0-Ga sediments in Arctic Canada. Again, modern RNA studies suggest that modern algal and plant chloroplasts have a single origin from a free-living cyanobacterium.

The origin of metazoans

Metazoans appear to have evolved from single-celled ancestors that developed a colonial habit. The adaptive value of a multicellular way of life relates chiefly to increases in size and the specialization of cells for different functions. For instance, more suspended food settles on a large organism than on a smaller one. Since all cells do not receive the same input of food, food must be shared among cells and a 'division of labour' develops among cells. Some concentrate on food gathering, others on reproduction, while still others specialize in protection. At some point in time, when intercellular

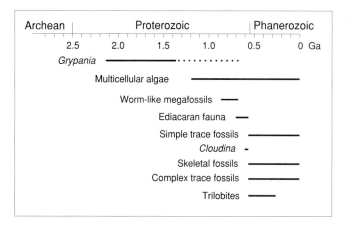

Figure 6.36   Time distribution of various Precambrian fossil groups.

communication was well-developed, cells no longer functioned as a colony of individuals but as an integrated organism.

The trace fossil record suggests that multicellular animals or metazoans were well-established by 1000 Ma (Figure 6.36), and the great diversity of metazoans of this age suggests that more than one evolutionary line led to multicellular development. Recent exciting findings of leaf-shaped fossils in North China suggest that metazoans had evolved by 1.7 Ga (Shixing and Huineng, 1995). On the basis of their size (5–30 mm long), probable development of organs, and possible multicellular structures, these forms are likely benthic multicellular algae (Figure 6.37). Although metazoans appeared by 1.7 Ga, they did not become widespread until after 1 Ga. Because of an inadequate fossil record, we cannot trace these groups of multicellular organisms back to their unicellular ancestors.

Late Proterozoic metazoans

Although most paleontologists regard Ediacaran fossils as metazoans, some have suggested that some or all may represent an extinct line of primitive plant-like organisms similar to algae or fungi (Seilacher, 1994). However, there are more similarities of the Ediacarans to primitive invertebrates than to algae or fungi (Runnegar, 1994; Weiguo, 1994). From the widespread fossil record, some thirty-one Ediacaran species have been described, including forms that may be ancestral to flatworms, coelenterates, annelids, soft-bodied arthropods, and soft-bodied echinoderms. The most convincing evidence for Late Proterozoic animals comes from trace fossils associated with the Ediacaran fauna. Looping and spiralling trails up to several millimetres in width and strings of fecal pellets point to the presence of soft-bodied animals with a well-developed nervous system, asymmetry, and a one-way gut.

Recently reported U-Pb zircon ages from ash beds associated with Ediacaran fossils in Namibia in southwest Africa indicate that this fauna is no older than 550

Figure 6.37   A carbonaceous multicellular fossil (*Antiqufolium clavatum*) from the 1.7-Ga Tuanshanzi Formation, North China. Scale bar 2 mm. Courtesy of Zhu Shixing.

Ma, and that some forms are as young as 543 Ma (Grotzinger et al., 1995). This places their age just before the Cambrian–Precambrian boundary at 540 Ma (Figure 6.36). Prior to these isotopic ages, a large gap was thought to have existed between the Ediacaran fossils and the diverse invertebrate forms that suddenly appear in the Cambrian. It now appears that some of the shelly Cambrian forms overlap with the Ediacaran forms.

These new findings support the idea, but do yet prove, that some Ediacaran forms were ancestral to some Cambrian invertebrates.

The Cambrian explosion and the appearance of skeletons

All of the major invertebrate phyla (except the Proto-zoa) made their appearances in the Cambrian, a feature which is sometimes referred to as the **Cambrian explosion** (Weiguo, 1994; Bengtson, 1994). This increase in numbers and diversity of organisms is matched by a sharp increase in the diversity of trace fossils and the intensity of bioturbation (the churning of subaqueous sediments by burrowing organisms). One of the immediate results of bioturbation was the return of buried organic matter to the carbon cycle and consequently, a decrease in the net release of oxygen due to decay. Two biological inventions permitted organisms to invade sediments (Fischer, 1984). First, the development of exoskeletons allowed organisms (such as trilobites) to dig by using appendages and, second, the appearance of coelomes permitted worm-like organisms to penetrate sediments. Although calcareous and siliceous skeletons did not become widespread until the Cambrian, the oldest known metazoan with a mineralized exoskeleton is *Cloudina*, a tubular fossil of world-wide distribution predating the base of the Cambrian by at least 10 My (appearing about 550 Ma) (Figure 6.36).

The reasons why hard parts were developed in so many different groups at about the same time is a puzzling problem in Earth history. Possibly it was for armour that would give protection against predators. However, one of the earliest groups to develop a hard exoskeleton was the trilobites, which were the major predators of the Cambrian seas (Figure 6.36). Although armour has a role in the development of hard parts in some forms, it is probably not the only or original reason for hard parts. The hard parts in different phyla developed independently, and are made of different materials. More plausible ideas are that hard parts are related to an improvement in feeding behaviour, loco-motion, or support. As an example of improved feeding behaviour, in brachiopods the development of a shell enclosed the filter-feeding 'arms' and permitted the filtration of larger volumes of water, similar in principle to how a vacuum cleaner works. Possibly the appearance of a hard exoskeleton in trilobites permitted a more rapid rate of locomotion by extending the effective length of limbs. Additional structural support may have been the reason why an internal skeleton developed in some echinoderms and in corals.

Recent very precise U–Pb zircon ages that constrain the base of the Cambrian to about 545 Ma have profound implications for the rate of the Cambrian explosion (Bowring et al., 1993). These results show that the onset of rapid diversification of phyla probably began within 10 My of the extinction of the Ediacaran fauna. All of the major groups of marine invertebrate organ-isms reached or approached their Cambrian peaks by 530–525 Ma, and some taxonomic groupings suggest that the number of Cambrian phyla actually exceeded the number known today. Assuming the Cambrian ended at 510–505 Ma, the evolutionary turnover among the trilobites is among the fastest observed in the Phanerozoic record. Using the new age for the base of the Cambrian, the average longevity for Cambrian trilobite genera is only about 1 My, much shorter than was previously thought.

Evolution of Phanerozoic life forms

Cambrian faunas are dominated by trilobites (~ 60 per cent) and brachiopods (~ 30 per cent), and by Late Ordovician most of the common invertebrate classes that occur in modern oceans were well established. Trilobites reached the peak of their development during the Ordovician, with a great variety of shapes, sizes, and shell ornamentation. Bryozoans, which represent the first attached communal organisms, appeared in the Ordovician. Graptolites, cephalopods, crinoids, echino-derms, molluscs, and corals also began to increase in numbers at this time. Vertebrates first appeared during the Ordovician as primitive fish-like forms without jaws. Marine algae and bacteria continued to be the important plant forms during the early Paleozoic.

The late Paleozoic is a time of increasing diversification of plants and vertebrates and of decline in many invertebrate groups. Brachiopods, coelenterates and crinoids all increase in abundance in the late Paleozoic, followed by a rapid decrease in numbers at the end of the Permian. The end of the Paleozoic was also a time of widespread extinction, with trilobites, eurypterids, fusulines and many corals and bryozoa becoming extinct. Insects appeared in the Late Devonian. Fish greatly increased in abundance during the Devonian and Mississippian, amphibians appeared in the Mississippian, and reptiles in the Pennsylvanian. Plants increased in numbers during the late Paleozoic as they moved into terrestrial environments. Psilopsids are most important during the Devonian, with lycopsids, ferns and conifers becoming important thereafter. Perhaps the most important evolutionary event in the Paleozoic was the development of vascular tissue in plants, which made it possible for land plants to survive under extreme climatic conditions. Seed plants also began to become more important relative to spore-bearing plants in the Late Paleozoic and Early Mesozoic. The appearance and rapid evolution of amphibians in the late Paleozoic was closely related to the development of forests, which provided protection for these animals. The appearance for the first time of shell-covered eggs and of scales (in the reptiles) allowed vertebrates to adapt to a greater variety of climatic regimes.

During the Mesozoic, gymnosperms rapidly increased in numbers with cycads, ginkgoes and conifers being most important. During the Early Cretaceous, angio-sperms (flowering plants) made their appearance and rapidly grew in numbers thereafter. The evolutionary success of flowering plants is due to the development of a flower and enclosed seeds. Flowers attract birds and insects that provide pollination, and seeds may develop fleshy fruits which, when eaten by animals, can serve to disperse the seeds. Marine invertebrates, which decrease in numbers at the end of the Permian, make a comeback in the Mesozoic (such as bryozoans, molluscs, echinoderms, and cephalopods). Gastropods, pelecypods, foraminifera and coiled cephalopods are particularly important Mesozoic invertebrate groups. Arthropods in the form of insects, shrimp, crayfish, and crabs also rapidly expand during the Mesozoic. Mesozoic reptiles are represented by a great variety of groups, including dinosaurs. Dinosaurs are of herbiferous and carniverous types, as well as marine and flying forms. Birds and mammals evolve from reptilian ancestors in the Early Jurassic. The development of mammals is a major evolutionary breakthrough, in that their warm-blooded nature allowed them to adapt to a great variety of natural environments (including marine), and their increased brain size allowed them to learn more rapidly than other vertebrates. During the Cenozoic, mammals evolved into large numbers of groups filling numerous ecological niches. Man evolved in the anthropoid group about 4 Ma. The vertebrate groups characteristic of the Mesozoic continued to increase in numbers and angiosperms expanded exponentially.

Biological benchmarks

Many benchmarks have been recognized in the appearance and evolution of new life forms on Earth as summarized in Figure 6.38. The first is that life on Earth originated probably about 4 Ga, but after the cessation of large impact events at the surface. The first autotrophs may have appeared by 3.8 Ga and by 3.5 Ga anaerobic prokaryotes were widespread and stromatolites, probably constructed by cyanobacteria, had appeared. Sulphate-reducing bacteria were also present by 3.5 Ga, although probably not widespread. Some time about 2.4 Ga, oxygen levels in the atmosphere had increased enough for eukaryotes to appear, although the oldest eukaryotic fossils are 2.1 Ga. The oldest metazoan fossils at 1.7 Ga suggest that metazoans may have evolved soon after the appearance of eukaryotic cells. Unicellular eukaryotes became the dominant life forms by 1.7–1.5 Ga and stromatolites, probably constructed chiefly by cyanobacteria, peaked in abundance and diversity about 1.2 Ga. Soft-bodied metazoans increased rapidly in numbers between 1 Ga and 550 Ga, culminating with the Ediacaran fauna at 600–543 Ma. *Cloudina*, the first metazoan with an exoskeleton, appeared about 550 Ma.

Important Phanerozoic benchmarks include the explosion of marine invertebrates between 540 and 530 Ma, the appearance of vertebrates (hemichordates) at about 530 Ma, land plants at 470 Ma, vascular plants at 410 Ma, amphibians at 370 Ma, reptiles and the amniote egg at about 330 Ma, insects at 310 Ma, mammals at 215

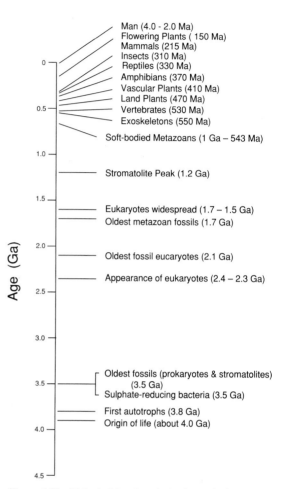

Figure 6.38   Biological benchmarks in the geologic past.

Ma, flowering plants at 150 Ma, and Man at about 4 Ma (*Australopithecus* at 4 Ma, *Homo* at 2 Ma).

Mass extinctions

Introduction

Important **mass extinctions**, which are extinctions of many diverse groups of organisms over short periods of time, occurred at eight different times during the Phanerozoic (McLaren and Goodfellow, 1990). Mass extinction episodes affect a great variety of organisms, marine and terrestrial, stationary and swimming forms, carnivores and herbivores, protozoans and metazoans. Hence the causal processes do not appear to be related to specific ecological, morphological or taxonomic groups.

Causes of mass extinction fall into three groups: extraterrestrial, physical, and biological. Extraterrestrial causes that have been suggested include increased production of cosmic and X-radiation from nearby stars, increased radiation during reversal in the Earth's magnetic field, and climatic changes caused by supernova events or by impact on the Earth's surface. Among the physical environmental changes that have been proposed to explain extinctions are rapid climatic changes, reduction in oceanic salinity caused by widespread evaporite deposition, fluctuations in atmospheric oxygen level and changes in sea level. Collision of continents may also lead to extinction of specialized groups of organisms as discussed in Chapter 1. Rapid changes in environmental factors lead to widespread extinctions, while gradual changes permit organisms to adapt and may lead to diversification. Correlation analysis between times of microfossil extinctions in deep-sea sediments and polarity reversals of the Earth's magnetic field does not support a relationship between the two (Plotnick, 1980). As discussed in the next section, a great deal of evidence seems to support impact on the Earth's surface as a cause for the extinctions at the Cretaceous–Tertiary (K/T) boundary.

Although it is clear that no single cause is responsible for all major extinctions, many mass extinctions share common characteristics suggestive of a catastrophic event. Perhaps no other subject in geology has received more attention or has been more controversial than the question of what catastrophic process or processes is or are responsible for major mass extinctions. In this section, some of the characteristics of mass extinctions are reviewed, concentrating especially on the K/T boundary extinctions, which have attracted the most interest.

Impact extinction mechanisms

It is important to distinguish between the ultimate cause of mass extinctions, such as impact, and the immediate cause(s) such as rapid changes in environment that kill plants and animals in large numbers world-wide. Impact of an asteroid on the Earth's surface clearly will cause changes in the environment, some of which could result in extinctions of various groups of organisms. As an example, the major consequences of collision of 10-km-diameter asteroid on the Earth's surface are summarized as follows.

1 **Darkness**. Fine dust and soot particles would spread world-wide in the upper atmosphere completely cutting out sunlight for a few months after the impact. This would suppress photosynthesis and initiate a collapse in food chains causing death to many groups of organisms by starvation. For instance, major groups that became extinct at the end of the Cretaceous, such as most or all dinosaurs, marine planktonic and nektonic organisms, and benthic filter feeders were in food chains tied directly with living plants. Organisms less affected by extinction, including marine benthic scavengers, deposit feeders, and small insectivorous mammals, are in food chains dependent upon dead-plant material.

2 **Cold**. The dust would produce darkness and be accompanied by extreme cold, especially in continental interiors far from the moderating influence of oceans. Within two or three months, continental surface temperatures would fall to –20 °C (Figure 6.39).

3 **Increased Greenhouse Effect**. If an asteroid collided in the ocean, both dust and water vapour would be spread into the atmosphere. Fallout calculations indicate that after the dust settles, water vapour will remain in the upper atmosphere producing an enhanced greenhouse effect that could raise surface temperatures well in excess of the tolerance limits of many terrestrial organisms.

4 **Acid Rain**. Energy liberated during the impact may cause atmospheric gases to react, producing nitric acid and various nitrogen oxides. Hence, another side effect of an impact bearing on extinctions is the possibility of a nitric acid rain. These acid rains could last as long as a year and would lower the pH of surface water in the upper 100 m sufficiently to kill a large number of planktonic organisms.

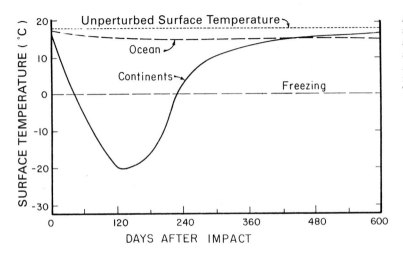

Figure 6.39 Change in terrestrial surface temperature of the oceans and continents following impact with a body about 10 km in diameter. Atmospheric dust density following impact is 1 gm/cm^2. Modified after Toon (1984).

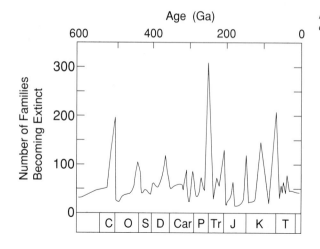

Figure 6.40 Patterns of animal and plant family extinctions during the Phanerozoic. After Benton (1995).

5 *Wildfires*. The soot particles reported in clays at the K/T boundary may be the result of widespread wildfires ignited by infrared radiation from the initial impact. Studies of the soot particles indicate that they come chiefly from the burning of coniferous forests.

6 *Toxic Seawater*. Asteroid collision should also result in the introduction of a wide variety of trace elements, many of which are toxic, into the oceans (such as Hg, Se, Pb, and Cd). Organisms living in surface marine waters would be exposed to prolonged concentrations of these toxic elements, contributing perhaps to their extinctions.

Episodicity of mass extinctions

Eight important mass extinctions are recognized in the Phanerozoic, and the same peaks are found for terrestrial and marine organisms, indicating that the major extinctions affected organisms on land and in the sea at the same time (Sepkoski, 1989; Benton, 1995) (Figure 6.40). The Late Carboniferous, Late Jurassic, and Early Cretaceous extinctions are more prominent for terrestrial than marine organisms. The apparently high extinction rate in the Early Cambrian may not be real, but reflects the low diversity of organisms at that time. Five major mass extinctions are recognized in the data (Figure 6.40): the Late Ordovician, Late Devonian, Late Permian, Late Triassic, and Late Cretaceous. Of these, the Permian extinction rate is highest, with a mean family extinction rate of 61 per cent for all life, 63 per cent for terrestrial organisms, and 49 per cent for marine organisms (Benton, 1995).

Although mass extinctions are clearly episodic, the data in Figure 6.40 do not support the idea of Sepkoski (1989) that they are also periodic with a mean spacing of about 26 My. On the contrary, major events are spaced from 20 to 60 My apart. Hence, major mass extinctions do not seem to favour a periodic cause, like cometary showers.

Extinctions at the K/T boundary

Organisms affected

Although most or all of the dinosaurs did not survive the K/T boundary, numerous terrestrial species, such as lizards, frogs, salamanders, fish, crocodiles, alligators, and turtles show no significant effects across this boundary. At the generic level, the terrestrial K/T extinction was only about 15 per cent. One of the exciting controversies in geology today is that of whether all of the dinosaurs disappeared suddenly at the end of the Cretaceous. Teeth of twelve dinosaur genera have been described above the K/T boundary in Paleocene sediments in eastern Montana (Sloan et al., 1986). If these teeth are in place and have not been reworked from underlying Cretaceous sediments, they clearly indicate that at least some dinosaurs survived the K/T extinction. If, on the other hand, they are reworked, then the oldest *in-place* dinosaur remains are still Late Cretaceous in age. Suggesting these dinosaur teeth are not reworked is the fact that reworked remains of a widespread species of Late Cretaceous mammals common in the underlying Cretaceous sediments are not found in these earliest Paleocene rocks of Montana. Also, the dinosaur teeth do not seem to have eroded edges as they would if they were redeposited. Other possible earliest Tertiary dinosaur remains are known from India, Argentina, and New Mexico. It would appear that debate will continue about the precise age of the final dinosaur extinctions until articulated dinosaur skeletons are found above the K/T boundary layer.

The marine extinctions at the end of the Cretaceous are far more spectacular than the terrestrial extinctions, involving many more species and groups of animals (McLaren and Goodfellow, 1990; Erwin, 1993). At the family level, the marine extinction rate is about 15 per cent, while at the generic and specific levels it is about 70 per cent. Major groups to disappear at the close of the Cretaceous are the ammonites, belemnites,

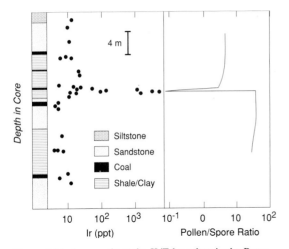

Figure 6.41 Ir anomaly at the K/T boundary in the Raton basin in NE New Mexico. Data are from a drill core. Also shown is the ratio of angiosperm pollen to fern spores in the section. After Orth et al. (1987). ppt = parts per trillion.

inoceramid clams, rudistid pelecypods, mosasaurs, and plesiosaurs. The extinction of benthic forms preferentially takes place in groups that had free-swimming larvae. Among benthic forms such as most molluscs, bryozoans, and echinoids, extinction rates were low, whereas calcareous phytoplankton and planktonic foraminifera show sharp extinctions. The abrupt disappearance of the plankton appears to have stressed marine communities by removing much of the base of the marine food chain. Particularly affected were reef-living bivalves, most oysters, clams, ammonites, many corals, most marine reptiles, and many fishes.

The marine groups usually considered to have become extinct at the K/T boundary show a common pattern, with a gradual decline in variety (some beginning as early as 88 Ma), a sharp decline in the last half million years of the Cretaceous, and an abrupt termination of many species at the K/T boundary. Some of these groups have a few survivors in the earliest Tertiary.

Seeking a cause

Evidence for impact

Iridium anomalies. One of the strongest evidences for impact is the enrichment of iridium (Ir) in a clay layer at the K/T boundary at many locations worldwide (Alvarez et al., 1990). The age of this clay layer has been measured at several places on the continents, and the average age is 66.7 Ma, which is precisely the age of the K/T boundary as dated from deepsea sediments. This enrichment, known as an iridium anomaly (Figure 6.41), cannot be produced from crustal sources because of the exceedingly low iridium content of crustal rocks, but it could come from collision of an asteroid. Following impact iridium, which is very volatile, would have been

injected into the stratosphere and spread over the globe, gradually settling out in dust particles over a few months.

Glass Spherules. Spherules are glassy droplets (a few tenths of a millimetre in diameter) of felsic composition, commonly found in K/T boundary clays (Maurrasse and Sen, 1991). By analogy with tektites, which are impact glasses with diameters up to a few centimetres found on the Earth, the small spherules at the K/T boundary appear to have formed by melting of crustal rocks, followed by rapid chilling as they are thrown into the atmosphere.

Soot. Soot or small carbonaceous particles are also widespread in K/T boundary clays and may be the remains of widespread wildfires which spread through forests following impact (Wolbach et al., 1985).

Shocked Quartz. One of the strongest evidences for impact is the widespread occurrence of shocked quartz in K/T boundary clays (Bohor et al., 1987; Hildebrand et al., 1991). Shock lamellae in quartz are easily identified (Figure 6.42) and are produced by a high-pressure shock wave passing through the rock. Such shocked quartz is common around nuclear weapon test sites and around well-documented impact sites such as Meteor Crater in Arizona.

Stishovite. Stishovite is a high-pressure polymorph of silica formed during impact and has been found in the K/T boundary layer clay (McHone et al., 1989). Like shocked quartz, it has only been reported at known impact and nuclear explosion sites.

Earth-crossing asteroids

Is it possible to assume, from our understanding of asteroid orbits and how they change by collisions in the asteroid belt, that asteroids have collided with the Earth? The answer is yes. Today there are approximately fifty Earth-crossing asteroids with diameters greater than 1 km, and a total population of over 1000. **Earth-crossing asteroids** are asteroids which are capable of colliding with the Earth with only small perturbations of their orbits. To be effective in K/T mass extinctions, the 'killer-asteroid' would have to be at least 10 km in diameter. Today there are about eight Earth-crossing asteroids with diameters ≥ 10 km, and probability calculations suggest that about ten asteroids of this size have collided with the Earth since the end of the Precambrian (McLaren and Goodfellow, 1990).

Comets

Sepkoski (1989) suggested that Mesozoic and Cenozoic mass extinctions of marine invertebrates are periodic with a strong periodicity at 26 My. It has been known for some time that the frequency of comet showers in the inner Solar System has a periodicity of about 30 My. These similar periodicities have suggested to some geologists a cause and effect relationship, in which periodic cometary impacts are responsible for mass extinctions (Hut et al., 1987). The cometary model also has the advantage that cometary showers generally last

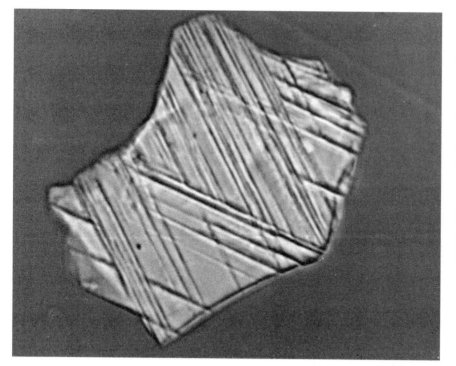

Figure 6.42 SEM photograph of an HF-etched shocked quartz grain from the K/T boundary claystone at Teapot Dome, Wyoming. Two major sets of planar deformation features (shock lamellae) are displayed in this view. Open planar features of the two major sets originally contained a silica glass phase, which has been removed by the acid etching. Grain is 72 μm in maximum dimension. Courtesy of Bruce Bohor.

2–3 My and would spread extinctions over this time, as observed for instance for some organisms at the K/T boundary.

About three long-period and ten short-period (< 20 per year) comets pass inside the Earth's orbit each year. Because long-period comets have velocities greater than asteroids, collisions with the Earth should liberate approximately an order of magnitude more energy. Statistical calculations indicate the Earth-crossing comets make about nine returns to the inner Solar System before being ejected into different orbits, and this corresponds to a mean lifetime of about 500 000 years. Estimates based on frequency of cometary showers suggest that about 50 per cent of the preserved craters on Earth are the products of cometary rather than asteroid or meteorite impact. A major shower involving ~ 10^9 comets of > 3 km in diameter would result in about twenty impacts on the Earth's surface and should occur every 300–500 My, and smaller showers (~ 10^8 comets) involving about two impacts should occur every 30–50 My. The probable ages of known impact structures (and impact glasses) on the Earth's surface suggest episodes of cometary impacting with major peaks at 99, 65, and 35 Ma. Recognized mass extinctions that annihilate 50–95 per cent of ecologically diverse lower taxa are recognized at 93, 66, and 36 Ma. The similarity in timing is consistent with the possibility that cometary showers may be responsible for these extinctions. A major problem with the periodic extinctions model, however, is that periodicity has not been recognized in the Paleozoic and early Mesozoic. It would appear that any explanation of extinctions involving cometary collision must also explain why they only started in the last 100 My or so.

Flood basalts

Evidence has been proposed in recent years to support a volcanic cause for K–T extinctions. The discovery of enriched Ir in atmospheric aerosols erupted from Kilauea in Hawaii indicates that Ir may be concentrated in oceanic plume-fed volcanic eruptions. Although glass spherules can be formed during eruptions of basalt, their distribution is localized around eruptive centres. Large eruptions of flood basalts have been suggested as causes of mass extinctions, and consistent with this possibility is the fact that major flood basalt eruptions show a periodicity of about 30 My, with some peaks coinciding with some major extinction peaks (Courtillot and Cisowski, 1987). Major flood basalt eruptions produce $1-2 \times 10^6$ km^3 of magma and are erupted over short periods of time of less than 1 My (Chapter 3). Isotopic dating of the Deccan traps in India indicate that they were erupted 66 Ma at the K/T boundary and that volcanism occurred in three periods each lasting 50 000–100 000 years. The first of these eruptions occurred before the extinction of dinosaurs, and sauropods and carnosaurs as well as mammals are found between the first two lava eruptions. Large volcanic eruptions are capable of introducing large quantities of sulphate aerosols into the atmosphere, which could cause immense amounts of acid rain, reduce the pH of surface seawater, add both volcanic ash and carbon dioxide to the atmosphere, and

Table 6.6 Comparison of impact and volcanic models for K–T boundary extinctions

Observational evidence	Impact	Flood basalt eruption
Ir anomaly	Yes: asteroid Possibly: comet	Possible, but not likely
Glass spherules	Yes: impact melts	No: of local extent only
Shocked quartz	Yes: common at impact sites	No
Stishovite	Yes: found at some impact sites	No
Soot	Yes: from widespread fires	No: fires only local
Worldwide distribution of evidence	Yes	No
Summary	Acceptable: accounts for all observational evidence	Rejected: cannot explain shocked quartz, stishovite, or soot

In part after Alvarez (1986).

perhaps deplete the ozone layer. Erupted ash could further reinforce the global cooling trend, and the combined effect of these events could result in widespread extinctions, spread over a million years or so.

Conclusions

So where do we stand in terms of understanding the K/T extinctions today? Clearly some extinctions occurred in the 10 My before the K/T boundary and these appear to be due to terrestrial causes, such as a fall in sea level and temperature drops. However, the numerous extinctions which occurred in a short period of time 66 Ma seem to require a catastrophic cause. In Table 6.6, the various evidences for impact are compared with flood basalt eruption to explain the K/T extinctions. Although it would appear that both impact and volcanic causes can explain the Ir anomalies, only impact can readily account for the wide distribution of glass spherules and soot, and the presence of shocked quartz and stishovite.

Chicxulub and the K/T impact site

Another question related to the asteroid impact model is where is the impact crater? Since most of the Earth's surface is covered with oceans, the chances are that it hit the oceans, and any crater formed on the sea floor was probably subducted in the last 65 My. Although the search for a crater of the right age and size (≥ 100 km in diameter) on the continents is still continuing, the current best candidate is the 180-km Chicxulub crater in Yucatan (Figure 6.43) (Hildebrand et al., 1991). Consistent with a location in the Caribbean for a K/T impact site is the common presence of shocked quartz and spherules in Caribbean K/T boundary clays. Both of these features require at least some continental crust and the Caribbean basin contains both oceanic and continental crust. Impact breccia deposits are also widespread in Cuba and Haiti at the K/T boundary, supporting a Caribbean impact site (Figure 6.44). Chicxulub crater is the only known large example of a Caribbean crater. It occurs

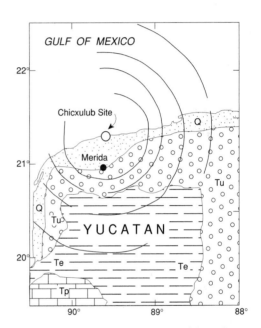

Figure 6.43 Simplified geologic map of the region around Merida, Mexico showing the location of Chicxulub crater buried with Tertiary sediments. Rings are gravity anomalies related to the impact. Q, Quaternary alluvium; Tu, Upper Tertiary sediments; Te, Eocene sediments; Tp, Paleocene sediments.

in Late Cretaceous marine carbonates deposited on the Yucatan platform on continental crust (Figure 6.43). Some of the crater is now filled with impact breccias and volcanic rocks (Figure 6.45), the latter of which have chemical compositions similar to the glass spherules found in K/T boundary clays. Melted crustal rocks from the crater also have high Ir contents. Supporting Chicxulub as the K/T impact site are precise $^{40}Ar/^{39}$ Ar ages of 65–66 Ma from glassy melt rock in the crater which, within the limits of error, is exactly the age of the K/T boundary (Swisher et al., 1992). U–Pb isotopic ages from shocked zircons in K/T boundary-layer clays from widely spaced locations in North America yield

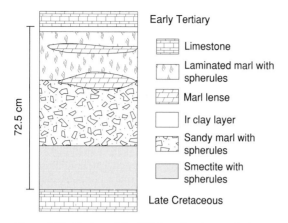

Early Tertiary

Limestone

Laminated marl with
spherules

Marl lense

Ir clay layer

Sandy marl with
spherules

Smectite with
spherules

Late Cretaceous

Figure 6.44 Section of the K/T boundary layer at Beloc, Haiti. The layer is 72.5 cm thick with Ir concentrated in the thin clay layer at the top. After Maurrasse et al. (1991).

identical ages to shocked zircons from Chicxulub crater of 548 Ma, again strongly supporting a Chicxulub source for the fallout (Kamo and Krogh, 1995).

A major environmental consequence of asteroid impact occurring in the oceans is a tsunami. **Tsunamis** are rapidly travelling sea waves caused by catastrophic disturbances such as earthquakes and volcanic eruptions. Calculations indicate that the asteroid which formed Chicxulub crater should have produced successive tsunamis up to 100 m high with periods of less than one hour, and these would have flooded the coastlines around the Gulf of Mexico and the Atlantic within a day of the impact. Tsunamis should cause widespread erosion and deposit poorly-sorted sediments in tidal and beach environments, and such deposits are known at precisely the K/T boundary from the Caribbean, the Gulf of Mexico, and from the Atlantic Coast from points as far away as New Jersey and Denmark.

The Permian/Triassic extinction

Killing about 60 per cent of all organisms on the Earth is remarkably difficult, yet this is precisely what happened at the end of the Permian (McLaren and Goodfellow, 1990; Erwin, 1994). This is the closest metazoans have come to being exterminated during the past 500 My. Also at the end of the Permian, Pangea was nearly complete and sea level dropped significantly, evaporite deposition was widespread, and global warming occurred. The Siberian flood basalts, which have been precisely dated at 250 ± 0.2 Ma, erupted in < 1 My at the P/Tr (Permian/Triassic) boundary (Renne et al., 1995).

The pattern of P/Tr extinctions is complex, with some groups disappearing well below the boundary, others at the boundary, and still others after the boundary. The marine fossil record provides the most complete record of the P/Tr extinction. In South China, where a complete marine succession across the P/Tr boundary is exposed, 91 per cent of all invertebrate species disappear including 98 per cent of ammonoids, 85 per cent of bivalves, and 75 per cent of shallow-water fusulinids. Precise isotopic dating at the P/Tr boundary suggests that many of these extinctions occurred in < 2 My and perhaps < 1 My. Many groups of sessile, filter-feeders disappeared at this time, as did marine invertebrates living in near-shore tropical seas. In terms of terrestrial animals, 80 per cent of the reptile families and six of the nine amphibian families disappeared at the end of the Permian. Among the insects, 8 of the known 27 orders disappeared in Late Permian, four suffered serious declines in diversity, and three became extinct during the Triassic. Land plants show little evidence of extinction at the P/Tr boundary, although pollens underwent significant change.

Unlike the K/T boundary, there is no evidence at the P/Tr boundary for impact fallout. One of the causes considered for the Permian extinctions is the eruption of the Siberian flood basalts, releasing large amounts of

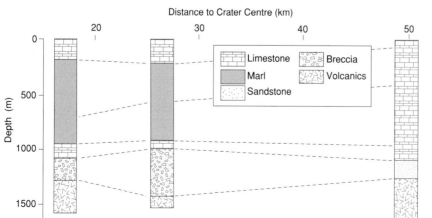

Distance to Crater Centre (km)

Limestone   Breccia
Marl   Volcanics
Sandstone

Depth (m)

Figure 6.45 Correlation of stratigraphic sections across Chicxulub crater from borehole data.

CO_2 causing enhanced greenhouse warming. Extinctions could be related both to changing climatic regimes and to injection of sulphate aerosols into the atmosphere resulting in acid rains (Renne et al., 1995). However, it is difficult to explain all of the extinctions by this mechanism, especially those that occurred long before eruption of the Siberian basalts! Most investigators lean towards multiple causes for the P/Tr extinctions (Erwin, 1994). Important contributing factors may have been the following:

1 loss of shallow marine habitats in response to falling sea level
2 completion of Pangea resulting in increased weathering rates and drawdown of atmospheric CO_2
3 eruption of the Siberian flood basalts causing global warming and acid rains
4 rapid transgression in the Early Triassic destroying near-shore terrestrial habitats, causing shifts in spores and pollen and declines in insects and some vertebrates.

Summary statements

1 The Earth's atmosphere is composed chiefly of N_2 and O_2, while the atmospheres of Mars and Venus are composed largely of CO_2. Atmospheres of the outer planets are chiefly H_2.
2 Depletion of rare gases in the Earth indicates that, if an early primitive atmosphere existed, it was lost during the late stages of planetary accretion (by 4.45 Ga).
3 The terrestrial atmosphere grew by early degassing of the Earth. The most important constituents in the early atmosphere were probably CO_2, N_2, CO and perhaps CH_4. Water would have rained out rapidly from this hot, steamy atmosphere to form the oceans. Isotopic studies of rare gases suggest, but do not prove, extensive degassing of the Earth within 50 My after accretion.
4 In order to avoid freezing surface temperatures on Earth prior to 2 Ga caused by low solar luminosity, an atmosphere rich in CO_2 is likely in order to promote greenhouse warming. To offset the low solar luminosity, additional warming is also required and may have been provided by CH_4 (another greenhouse gas) and/or a reduced albedo caused by faster rotation of the Earth.
5 A decrease in atmospheric CO_2 level with time is related to changes in the carbonate–silicate cycle, in which CO_2 is removed from the atmosphere by carbonate deposition and photosynthesis, followed by burial of organic matter faster than it is resupplied by volcanism and subduction.
6 Atmospheric oxygen is produced by photosynthesis and reduction of marine sulphates and carbonates and is lost principally by weathering, respiration

and decay. Prior to the appearance of photosynthetic organisms, minor amounts O_2 were produced by H_2O photolysis in the upper atmosphere.
7 An increase in abundance of banded iron formation in the Late Archean and Early Proterozoic probably reflects increasing numbers of oxygen-producing microbes. The drop in abundance of banded iron formation after 1.7 Ga reflects exhaustion of ferrous iron in seawater.
8 The growth of O_2 in the atmosphere progressed in three stages: (I) free O_2 occurs in neither the oceans nor atmosphere; (II) small amounts of O_2 occur in the atmosphere and surface ocean water; and (III) O_2 invades the entire atmosphere–ocean system. Stage II began 2.5–2.3 Ga and is marked by the appearance of redbeds, an increase in the abundance of sulphates and near-disappearance of detrital uraninite–pyrite deposits. Stage III began about 2.0 Ga and is characterized by the near-disappearance of banded iron formation and appearance of eukaroytic organisms.
9 During the Phanerozoic, atmospheric CO_2 levels are controlled chiefly by tectonic processes, which affect weathering/erosion rates, the rate of burial of organic carbon, and increasing solar luminosity. O_2 levels are controlled chiefly by the burial rates of organic carbon. A peak in O_2 and a minimum in CO_2 in the late Paleozoic appear to reflect the enhanced burial rate of organic carbon accompanying the rise of vascular land plants.
10 The $^{13}C/^{12}C$ ratio of seawater remains approximately constant with time, while that of buried organic carbon gradually increases reflecting the preferential uptake of ^{12}C by organisms. During the Precambrian, the amount of buried organic carbon gradually increases with time and shows peaks at 2 Ga and 0.9 Ga. It would appear that this gradual increase reflects an increasing biomass, but the peaks are probably caused by increased erosion and sedimentation associated with supercontinent amalgamation.
11 A major peak in $\delta^{13}C$ of marine carbonates in the Carboniferous and Permian appears to reflect an increase in burial rate of organic carbon associated with the rapid spread of vascular plants.
12 The $\delta^{34}S$ of marine sulphates increase with time from 3.8 Ga while $\delta^{34}S$ of marine sulphides decrease. These changes may be caused by an increase in numbers of sulphate-reducing bacteria of by cooling of the oceans with time.
13 Secular decreases in aragonite, high-Mg calcite, and cement crusts in marine carbonates suggest that Precambrian seawater was greatly oversaturated in carbonate and that saturation decreased with time. The paucity of gypsum evaporites before 1.7 Ga may be due to low concentrations of sulphate in seawater before this time, or exhaustion of Ca ion by carbonate deposition before the gypsum stability field was reached.

14   The growth of the Earth's oceans probably paralleled atmospheric growth and reached > 90 per cent of present ocean volume by 4.0 Ga.

15   The ratio of ^{18}O to ^{16}O increases in cherts and banded iron formation (BIF) from the Archean to the present. Because the distribution of $\delta^{18}O$ in alteration zones in 2-Ga ophiolites and 3.5-Ga greenstones is similar to that in young ophiolites, it is unlikely the $\delta^{18}O$ of seawater has changed with time. Hence, the increase in $\delta^{18}O$ of chert/BIF with time probably reflects secular cooling of seawater.

16   Increases in the Ca/Mg and calcite/dolomite ratios in marine carbonates with time may reflect one or some combination of changes in the Mg depository in seawater, a decrease in atmospheric CO_2 level, or losses of Mg due to recycling at ocean ridges.

17   Archean carbonates were probably deposited in the deep ocean basins where they were readily destroyed by subduction, accounting for the sparsity of these rocks in the geologic record.

18   Short-term changes in sea level are caused principally by glaciation and changes in continental elevation due to continental collisions. Long-term changes in sea level are related to rates of seafloor spreading, the characteristics of subduction, the motion of continents with respect to geoid highs and lows, and by supercontinent insulation of the mantle.

19   Important paleoclimatic indicators include glacial deposits, coal, aeolian sandstones, evaporites, redbeds, laterites and bauxites, various floral and faunal populations, and $\delta^{18}O$ values of marine carbonates and cherts. The Earth's surface temperature is related to solar luminosity, albedo, and the greenhouse effect.

20   Nine major periods are recognized in the geologic record (see Table 6.4). Important causes of glaciation include decreases in solar heating due to volcanic eruptions, decreased CO_2 degassing of the Earth, changes in continent–ocean geometry caused by plate tectonics, and changes in the Earth's orbital characteristics.

21   The Precambrian, early Paleozoic, and late Mesozoic are times of widespread warm climates. Cool climates are often associated with supercontinents, whereas warm climates usually occur during supercontinent dispersal.

22   There is a correlation in the Late Cretaceous between increased production rate of oceanic crust and submarine plateaux, lack of magnetic reversals (the Cretaceous superchron), increased surface temperature of seawater, high sea level, and high production rates of black shale and oil. All may be related to a superplume event.

23   Life appears to have formed on Earth about 4 Ga, perhaps originating at a hydrothermal vent on the sea floor. Steps in the formation of a living cell include reactions of gaseous components to produce amino acids, formation of polypeptides including RNA and DNA, and the appearance of membranes. RNA may have played a key role in the development of self-replication.

24   From our knowledge of the timing of early large impact events, it would appear that for life to survive it would have formed between the end of ocean-vaporizing impacts at about 4.1 Ga and about 3.8 Ga.

25   The oldest microfossils and stromatolites are about 3.5 Ga and appear to represent anaerobic prokaryotes. It was not until about 2 Ga that microbes entirely dependent on the use of molecular O_2 appeared in the geologic record Although metazoans had evolved by 1.7 Ga, they did not become widespread until after 1 Ga.

26   The Ediacaran soft-bodied metazoan fauna at 600–543 Ma may have been ancestral to some Cambrian invertebrates. An explosion of multicellular life happened over a very short period of time of 10–20 My near the Precambrian–Cambrian boundary.

27   Processes leading to mass extinctions following cometary or asteroidal impact include darkness, cold temperatures, increased greenhouse effect, acid rain, wildfires, and toxic seawater.

28   Mass extinction at the K/T boundary appears to have resulted from asteroid impact as evidenced by the presence of the following in K/T boundary clay beds: (1) widespread Ir anomalies; (2) glass spherules of impact-melt origin; (3) soot particles remaining from wildfires; (4) shock lamellae in quartz produced by a high-pressure shock waves; and (5) stishovite, a high-pressure form of silica. Most data favour the Chicxulub crater in Yucatan as the K/T boundary impact site.

29   The largest mass extinction of all time occurred at the P/Tr boundary. Contributing to this extinction may have been loss of shallow marine habitats in response to falling sea level; completion of Pangea resulting in drawdown of atmospheric CO_2; eruption of the Siberian flood basalts causing global warming and acid rains; and rapid transgression in the Early Triassic destroying near-shore terrestrial habitats.

Suggestions for further reading

Bengtson, S., editor (1994). *Early Life on Earth*. New York, Columbia Univ. Press, 630 pp.

Berner, R. A. (1994). 3Geocarb II: A revised model of atmospheric CO_2 over Phanerozoic time. *Amer. J. Sci.*, **294**, 56–91.

Dressler, B. O., Grieve, A. F. and Sharpton, V. L. (1992). Large meteorite impacts and planetary evolution. *Geol. Soc. America, Spec. Paper 358*.

Frakes, L. A., Francis, J. E. and Syktus, J. L. (1992). *Climate Modes of the Phanerozoic*. New York, Cambridge Univ. Press, 274 pp.

Hildebrand, A. R. et al. (1991). Chicxulub crater: A possible Cretaceous/Tertiary boundary impact crater on the Yucatan Peninsula, Mexico. *Geology*, **19**, 867–871.

Holland, H. D. (1984). *The Chemical Evolution of the Atmosphere and Oceans*. Princeton, NJ, Princeton Univ. Press, 582 pp.

Kasting, J. F. (1993). Earth's early atmosphere. *Science*, **259**, 920–926.

Larson, R. L. (1991). Geological consequences of superplumes. *Geology*, **19**, 963–966.

Schopf, J. W., editor (1983). *Earth's Earliest Biosphere, Its Origin and Evolution*. Princeton, NJ, Princeton Univ. Press, 543 pp.

Chapter 7

Comparative planetary evolution

Introduction

As members of the Solar System, the Earth and the Moon must be considered within the broader framework of planetary origin. Most data favour an origin for the planets by condensation and accretion of a gaseous Solar Nebula in which the Sun forms at the centre (Boss et al., 1989; Taylor, 1992). Considering that the age of the universe is of the order of 15 Ga, the formation of the Solar System at about 4.6 Ga is a relatively recent event. Since it is not possible as yet to observe planetary formation in other gaseous nebulae, we depend upon a variety of indirect evidence to reconstruct the conditions under which the planets in the Solar System formed. Geophysical and geochemical data provide the most important constraints. Also, because the interiors of planets are not accessible for sampling, we often rely on meteorites to learn more about planetary interiors. Most scientists now agree that the Solar System formed from a gaseous dust cloud, known as the Solar Nebula, which will be discussed later in the chapter.

In this final chapter, we shall look at the Earth as a member of the Solar System by comparing it with other planets. As discussed in earlier chapters, much is known about the structure and history of planet Earth, and it is important to emphasize the uniqueness of Earth in comparison with the other planets. First of all, plate tectonics and continents seem to be unique to the Earth. Why should only one of the terrestrial planets cool in a manner that results in plate tectonics? Although there is no answer to this question yet, there are some possible reasons, especially in comparison with our sister planet Venus. The oceans and the presence of free oxygen in the terrestrial atmosphere are other unique features in the Solar System, and the origin and evolution of the atmosphere–ocean system have been discussed in Chapter 6. And of course, Earth appears to be the only mem-

ber of the Solar System with life, although this may not always have been so. As we review other bodies in our Solar System, and especially as we discuss the origin of the Solar System, let us remember these unique terrestrial features which somehow must be accommodated in any model of planetary origin and evolution.

Impact chronology in the inner Solar System

It is well-known that impacts on planetary surfaces play an important role in planetary evolution (Glikson, 1993). Impact effects are known on varying scales from dust size to planetary size, and have occurred throughout the history of the Solar System. A considerable amount of effort has gone into studying the impact record of the inner Solar System as recorded on the lunar surface, and this record is now used to estimate the ages of craters on other planets. The sequence of events (i.e., volcanism, rifting, erosion, etc.) on a planetary or satellite surface can be deduced from the cross-cutting relationships of craters and, if important events can be dated, this relative history can then be tied to absolute ages (Price et al., 1996). This, of course, assumes that the impact flux rate with time can be estimated. It is well-known that most of the large impacts on the lunar surface were early, terminating about 3850 Ma, with the large Imbrium impact. Although the termination of major impacts in the inner Solar System was probably very early, it may not have been synchronous throughout. However, the striking similarity of age/crater density curves on the Moon, Mercury, Venus, and Mars strongly suggests a similar impact history for the inner part of the Solar System, and so it appears justified to use the lunar time scale throughout this region.

Table 7.1 Properties of the planets

	Mean distance to Sun (AU)	Orbital period (d, days; y, years)	Mean orbital velocity (km/sec)	Mass (Earth = 1)	Equatorial 1 radius (km)	Mean density (g/cm³)	Zero pressure density (g/cm³)	Area/mass (Earth = 1)	Core/mantle (Earth = 1)
Mercury	0.387	88d	48	0.056	2439	5.42	5.3	2.5	12
Venus	0.723	225d	35	0.82	6051	5.24	3.95	1.1	0.9
Earth	1.0	365d	30	1.0	6378	5.52	4.0	1.0	1.0
Moon	1.0	27.3d*	1.0*	0.012	1738	3.34	3.3	6.1	0.12
Mars	1.52	687d	24	0.11	3398	3.94	3.75	2.5	0.8
Jupiter	5.2	11.9y	13	318	71600	1.32			
Saturn	9.2	29.5y	9.7	95.2	60000	0.69			
Uranus	19.1	84y	6.8	14.5	25600	1.27			
Neptune	30.0	164y	5.4	17.2	24760	1.64			

*Period and velocity about the Earth

Members of the Solar System

The planets

Mercury

Mercury, the closest planet to the Sun, is a peculiar planet in several respects. First, the orbital eccentricity and inclination of the orbit to the ecliptic are greater than for any other planet except Pluto (Table 7.1). Also, the high mean density (5.42 g/cm³) of Mercury implies an Fe–Ni core that comprises about 66 per cent of the planet's mass, thus having a core/mantle ratio greater than any other planet. Although Mercury's magnetic field is very weak, like the Earth, it has a dipole field that is probably generated by an active dynamo in the liquid outer core. Results from Mariner 10 show that the surface of Mercury is similar to that of the Moon in terms of crater distributions and probable age. Spectral studies of the mercurian surface are consistent with a crust rich in plagioclase, perhaps much like the lunar highlands crust (Tyler, 1988). The occurrence of a weak sodium cloud around the planet supports this idea. There are two types of plains on Mercury's surface: the early intercrater plains formed prior to 4 Ga and the smooth plains formed at about 3.9 Ga. These plains have been attributed to both debris-sheets from large impacts and to fluid basaltic lavas, and their origin continues to be a subject of controversy (Taylor, 1992). For instance, microwave and infrared radiation reflected from the mercurian surface indicate a very high albedo inconsistent with the presence of basalt (Jeanloz et al., 1995). These data would seem to favour a crust on Mercury, including the plains, composed almost entirely of anorthosite.

Mercury also displays a global network of large lobate scarps up to 1 km in height that may be high-angle reverse faults. They indicate a contraction in planetary radius of the order of 2–4 km, probably due to rapid cooling of the mercurian mantle. The faulting which produced the lobate scarps occurred both before and after major volcanism. Employing the lunar impact time scale, crater distributions indicate that the contraction occurred before massive bombardment of the planet at 3.9 Ga, and probably before 4 Ga.

Two ideas have been suggested to explain the relatively high density of Mercury:

1. high-temperature evaporation of the silicate mantle, thus enhancing the proportion of iron in the planet
2. removal of part of the mantle by collision with another planet during or soon after the accretion of Mercury (Vilas et al., 1988).

The reflectance data suggesting a plagioclase-rich crust and the presence of a sodium cloud around Mercury do not favour the evaporation theory, since volatile elements like sodium should have been removed and lost during the high-temperature evaporation event. The impactor idea has received increased support from stochastic modelling of planetary accretion (discussed later) in the inner Solar System, which indicates that numerous potential impactors up to Mars-size may have existed in the Solar Nebula. To give a core/mantle ratio similar to the Earth and Venus, Mercury would have to lose about half of its original mass during the collision (Figure 7.1). The fate of this fragmented material, mostly silicates and oxides, is unknown, but it may have been swept away by solar radiation, or possibly accreted to Venus or Earth if it crossed their orbits.

Mars

The Mariner and Viking missions indicate that Mars is quite different from the Moon and Mercury (Taylor, 1992; Carr et al., 1993). The martian surface includes major shield volcanoes, fracture zones, and rifts as well as large canyons that appear to have been cut by running water. Also, much of the planet is covered with wind-blown dust, and more than half of the surface is covered by a variably-cratered terrain similar to the surfaces of the Moon and Mercury. Large near-circular basins are similar to lunar mare basins and probably formed at

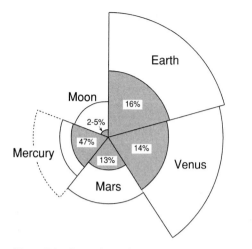

Figure 7.1 Comparison of the core/mantle ratios by volume in the terrestrial planets and the Moon. Dashed line for Mercury is the part of the mantle that may have been lost during collision with another planetary body.

about 3.9 Ga. Much of the northern hemisphere of Mars comprises volcanic plains and large stratovolcanoes. In contrast, the southern hemisphere is covered by the Ancient Cratered Terrain greater than 4 Gy in age. The Tharsis bulge, which was probably formed by a large mantle upwelling, is the dominant structural feature on the Martian surface. It is about 10 km high at the centre and 8000 km across, comprising about 25 per cent of the martian surface. The Tharsis region includes gigantic shield volcanoes and volcanic plains, and the Valles Marineris represents an enormous rift valley that spans one-quarter of the equator. The size of martian volcanoes implies a very thick lithosphere on Mars. The distribution and surface features of lava flows on Mars indicate low viscosities of eruption and a total volume of lavas much greater than on the Earth.

A peculiar feature of Mars in the presence of widespread wind-blown dust, perhaps analogous to terrestrial

loess deposits. The martian dust is the cause of the variations in albedo on the planet's surface, once thought to be canals. In martian bright areas, dust may accumulate to over one metre in thickness. The movement of dust on the surface by gigantic dust storms appears to be related to seasonal changes. On the whole, erosion on the martian surface has been very slow, consistent with the un-weathered chemistry of the rocks analysed by the Viking landers. The redistribution of sediment by wind occurs at a rate much greater than weathering and erosion, such that much of the dusty sediment on Mars was produced early in martian history and has been reworked ever since.

Chemical analyses from the Viking landing sites suggest that the dominant volcanic rocks on Mars are Fe-rich basalts and that weathering of the basalts occurs in a hydrous, oxidizing environment. Surface compositions and SNC meteorites, thought to be derived from the martian surface, indicate a largely mafic or komatiitic crust. The Russian Phobos mission obtained chemical analyses of a large area also consistent with basaltic rocks (Figure 7.2). Remote sensing studies show a strong concentration of Fe^{+3} on the surface indicative of hematite, probably in wind-blown dust. Because Mars has an uncompressed density much less than that of the Earth and Venus (Table 7.1), it must also have a distinctly different composition from these planets. If Mars has a carbonaceous chondrite (C1) composition and is completely differentiated, the core mass is about 21 per cent of total mass and the core radius about 50 per cent of the planetary radius. It would appear that, compared with Venus and Earth, Mars is more volatile-rich by at least a factor of two, and its core probably contains a substantial amount sulphur (Taylor, 1992).

One of the most puzzling aspects of martian geology is the role that water has played on the planetary surface (Carr et al., 1993). Although water is frozen on Mars today, we see evidence for running water in the past as dry valleys and canyons (Figure 7.3). Clearly climatic conditions on Mars must have been warm enough to permit running water some time in the past. The most intriguing features are the large, flat valleys, some over

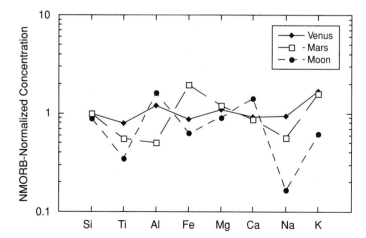

Figure 7.2 NMORB normalized major element distributions in crustal rocks from Venus, Mars, and the Moon. NMORB, normal ocean ridge basalt.

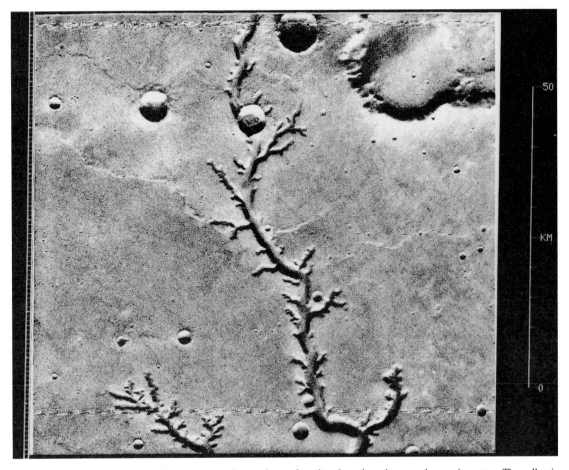

Figure 7.3 Nirgal Vallis, a branching canyon on the martian surface thought to have been cut by running water. The valley is incised into crust probably older than 3.5 Ga. Viking Orbiter frame 466A54. Courtesy of NASA.

1000 km long, that appear to have been cut by running water. These valleys, which generally have few tributaries, may be the products of gigantic floods. What caused the floods remains problematic, and they all may not be of the same origin. One possible cause is volcanism, which could have melted permafrost, suddenly releasing huge volumes of water. Other martian canyons with numerous tributaries look like terrestrial canyons and appear to have been cut more slowly by rivers or spring sapping by ground water (Figure 7.3). Most of these canyons are in the Ancient Cratered Terrain, the oldest part of the martian crust, indicating the presence of transient warm climates on Mars more than 4 Ga.

The geologic history of Mars was probably similar to that of the Moon and Mercury for the first few hundred million years. Core formation was early, probably coinciding with a transient magma ocean. Part of the early crust survived major impacting that terminated at about 3.9 Ga, and occurs as the Ancient Cratered Terrain in the southern hemisphere. This early crust may have dated to 4.5 Ga based on the age of SNC meteorites probably derived from the martian surface. After rapid crystallization of the magma ocean prior to 4.4 Ga, extensive melting in the upper mantle resulted in the formation of a thick basaltic crust. Heat loss was chiefly by mantle plumes. The Tharsis bulge developed at about 3.9–3.8 Ga, perhaps in response to a large mantle plume (Mutch et al., 1976). With a drop in surface temperature permafrost formed, which later locally and perhaps catastrophically melted and caused massive floods which cut the large flat-bottom valleys. The major erosional events probably occurred during or just before the terminal large impact event at about 3.9 Ga. Continued fracturing and volcanism on Mars extended to at least 1000 Ma and perhaps 100 Ma.

Venus

In comparison to Earth

Unlike the other terrestrial planets, Venus is similar to the Earth both in size and mean density (5.24 and

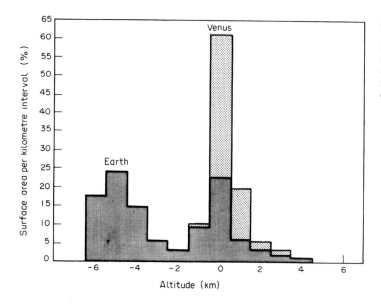

Figure 7.4 Comparison of relief on Venus and Earth. Surface height is plotted in 1 km intervals as a function of surface area. Height is measured from sphere of average planetary radius for Venus and from sea level for Earth. After Pettengill et al. (1980).

5.52 g/cm^3 respectively) (Table 7.1). After correcting for pressure differences, the uncompressed density of Venus is within one per cent of that of the Earth, indicating that both planets are similar in composition, with Venus having a somewhat smaller core/mantle ratio. Although both planets also have similar amounts of N_2 and CO_2, most of the Earth's CO_2 is not in the atmosphere but in carbonates. Venus also differs from the Earth by the near-absence of water, and the high density and temperatures of its atmosphere. As discussed later, Venus may at one time have lost massive amounts of water by the loss of hydrogen from the upper atmosphere. Unlike the Earth, Venus lacks a satellite, has a slow retrograde rotation (225 days/y), and does not have a measurable magnetic field. The absence of a magnetic field in Venus may be due to the absence of a solid inner core since, as discussed in Chapter 5, crystallization of an inner core may be required for a dynamo to operate in the outer core of a planet. Of the total venusian surface, 84 per cent is flat rolling plains, some of which are over 1 km above the average plain elevation. Only 8 per cent of the surface is true highlands, and the remainder (16 per cent) lies below the average radius forming broad shallow basins. This is very much unlike the topographic distribution on the Earth, which is bimodal due to plate tectonics (Figure 7.4). Thus, the unimodal distribution of elevation on Venus does not support the existence of plate tectonics on Venus today.

The spectacular Magellan imagery indicates that unlike Earth, deformation on Venus is distributed over thousands of kilometres rather than occurring in rather narrow orogenic belts (Solomon et al., 1992). There are numerous examples of compressional tectonic features on Venus, such as the western part of Ishtar Terra (Figure 7.5). Ishtar Terra is a highland about 3 km above mean planetary radius surrounded by compressional features that are suggestive of tectonic convergence result-

ing in crustal thickening. Coronae are large circular features (100–1000 km in diameter) that may be the surface expressions of mantle plumes. Venus is the only planet known to have coronae. An approximately inverse correlation between crater and corona density suggests that the volcano–tectonic process that forms coronae may be the same process that destroys craters (Stefanick and Jurdy, 1996). Another unique and peculiar feature of the venusian surface is the closely-packed sets of grooves and ridges known as tesserae, which appear to result from compression.

Perhaps the most important data from the Magellan mission are those related to impact craters (Kaula, 1995). Unlike the Moon, Mars, and Mercury, Venus does not preserve a record of heavy bombardment from the early history of the Solar System (Price and Suppe, 1994). In fact, crater size/age distribution indicates an average age of the venusian surface of only 500–300 My, indicating extensive resurfacing of planetary surface at this time. Most of this resurfacing is with low-viscosity lavas, presumably mostly basalts as inferred from the Venera geochemical data. Crater distribution also indicates a rapid decline in the resurfacing rate within the last few tens of millions of years. However, results suggest that some features, such as some large volcanoes (72 Ma), some basalt flows (128 Ma) and rifts (130 Ma), and many of the coronae (120 Ma), are much younger than the average age of the resurfaced plains, and probably represent ongoing volcanic and tectonic activity (Price et al., 1996).

The differences between Venus and Earth, together with the lower bulk density of Venus, affect the nature and rates of surface processes (weathering, erosion, deposition) as well as tectonic and volcanic processes. Because a planet's thermal and tectonic history is dependent on its size and area/mass ratio as described later, Venus and Earth are expected to have similar histories.

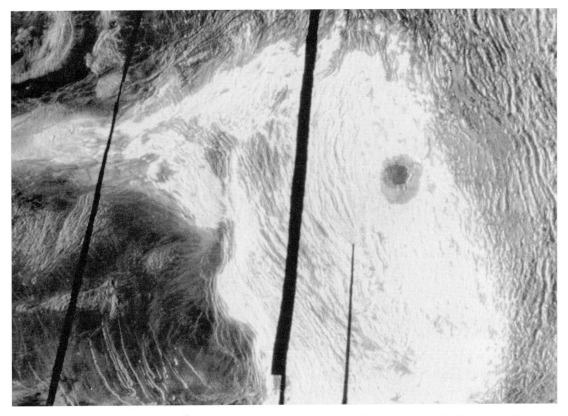

Figure 7.5   Magellan image of Maxwell Montes, the highest mountain range on Venus, which stands 11 km above the average diameter of the planet. The complex pattern of intersecting ridges and valleys reflects intense folding and shearing of the crust. Courtesy of U.S. Geological Survey.

However, the surface features of Venus are quite different from those of the Earth, raising questions about how Venus transfers heat to the surface and whether plate tectonics is active, or has ever been active. The chief differences between Earth and Venus appear to have two underlying causes:

1   small differences in planetary mass leading to different cooling, degassing, and tectonic histories
2   differences in distance from the Sun, resulting in different atmospheric histories.

Surface composition

Much has been learned about the surface of Venus from scientific missions by the United States and Russia. The Russian Venera landings on the venusian surface have provided a large amount of data on the structure and composition of the crust. Results suggest that the majority of the venusian surface is composed of blocky bedrock surfaces and less than one-fourth contains porous, soil-like material (McGill et al., 1983). The Venera landers have also revealed the presence of abundant volcanic features, complex tectonic deformation, and unusual ovoidal features of probable volcanic–tectonic origin. Reflectance studies of the venusian surface suggest that iron oxides may be important components. Partial chemical analyses made by the Venera landers indicate that basalt is the most important rock type. The high K_2O recorded by Venera 8 and 13 is suggestive of alkali basalt, while the results from the other Venera landings clearly indicate tholeiitic basalt, perhaps with geochemical affinities to terrestrial ocean ridge tholeiites (Figure 7.2). A venusian crust composed chiefly of basalt is consistent with the presence of thousands of small shield volcanoes that occur on the volcanic plains, typically 1–10 km in diameter and with slopes of about 5°. The size and distribution of these volcanoes resembles terrestrial oceanic island and seamount volcanoes.

Cooling and tectonics

In order to understand the tectonic and volcanic processes on Venus, it is first necessary to understand how heat is lost from the mantle. Four sources of information are important in this regard: the amount of ^{40}Ar in the venusian atmosphere, lithosphere thickness, topography, and gravity anomalies. The amount of ^{40}Ar in planetary

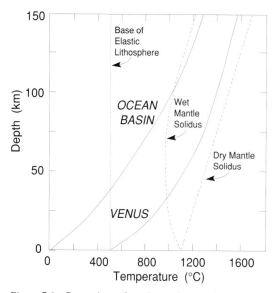

Figure 7.6 Comparison of geotherms from an average terrestrial ocean basin and Venus. Conduction is assumed to be the only mode of lithospheric heat transfer on Venus. Also shown are wet (0.1 per cent H_2O) and dry mantle solidi.

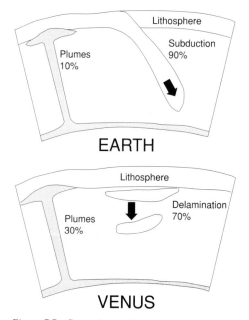

Figure 7.7 Comparison of Earth cooling mechanisms with possible cooling mechanisms on Venus. Estimates of magnitudes given in percentages. Modified after Turcotte (1995).

atmospheres can be used as a rough index of past tectonic and volcanic activity, because it is produced in planetary interiors by radioactive decay and requires tectonic–volcanic processes to escape. Venus has about one-third as much [40]Ar in its atmosphere as does the Earth, which implies less tectonic and volcanic activity for comparable [40]K contents.

In contrast to the Earth, where at least 90 per cent of heat is lost by the production and subduction of oceanic lithosphere, there is no evidence for plate tectonics on Venus. Thus, it would appear that Venus, like the Moon, Mercury, and Mars, must lose its heat through conduction from the lithosphere, perhaps transmitted upwards chiefly by mantle plumes. The base of the thermal lithosphere in terrestrial ocean basins is about 150 km deep, where the average geotherm intersects the wet mantle solidus. On Venus, however, where it is likely that the mantle is dry, an average geotherm does not intersect the dry solidus indicating the absence of a distinct boundary between the lithosphere and mantle (Figure 7.6). The base of the elastic lithosphere in ocean basins is at the 500 °C isotherm or about 50 km deep. Because 500 °C is near the average surface temperature of Venus, there is no elastic lithosphere on Venus. Another important difference between Venus and the Earth is the strong positive correlation between gravity and topography on Venus, implying compensation depths in the venusian mantle of 100–1000 km. This requires strong coupling of the mantle and lithosphere, and hence the absence of an asthenosphere, agreeing with the thermal arguments above. This situation may have arisen from a lack of water in Venus. One of the important

consequences of a stiff mantle is the inability to recycle lithosphere through the mantle, again showing that plate tectonics cannot occur on Venus. The deformed plateaux and the lack of features characteristic of brittle deformation, such as long faults, suggest the venusian lithosphere behaves more like a viscous fluid than a brittle solid, like the Earth's lithosphere. The very steep-sided high elevation plateaux on Venus, however, attest to the high strength of the venusian lithosphere.

Two thermal-tectonic models have been proposed for Venus: the conduction and mantle-plume models (Bindschadler et al., 1992). In the conduction model, Venus loses heat by simple conduction through the lithosphere and tectonics is a result of compression and tension in the lithosphere in response to the changing thermal state of the planet. It is likely that such a model describes the Moon, Mercury and Mars at present. Not favouring a conduction model for Venus, however, is the implication that the topography is young, since it cannot be supported for long with warm, thin lithosphere.

In the mantle-plume model, which is preferred by most investigators, Venus is assumed to lose heat from large mantle plumes coupled with delamination and sinking of the lithosphere (Figure 7.7) (Turcotte, 1995; 1996). Also consistent with plumes are the deep levels of isostatic gravity compensation beneath large topographic features, suggesting the existence of plumes beneath these features. Unlike Earth, most of the topographic and structural features on Venus can be accounted for by mantle plumes, with compressional forces over mantle downwellings responsible for the compressional features on the surface. On the whole, the geophysical observations from Venus support the idea that mantle downwelling is the dominant driving force for deformation of the surface of Venus. The return flow in the mantle would also occur in downwellings and undoubtedly involve delamination and sinking of significant volumes of the lithosphere. This is a striking contrast to the way the Earth cools as shown in Figure 7.7. It is still debated whether the resurfacing of the planet at 500–300 Ma was caused by a catastrophic planet-wide mantle plume event that is in the waning stages today, or is simply ongoing resurfacing of which the mean age is 500–300 Ma (Strom et al., 1994).

The Outer Planets

Jupiter and Saturn, the two largest planets, have densities indicating that they are composed chiefly of hydrogen and helium (Table 7.1). In the outer parts of the planets, these elements occur as ices and gases and at greater depths as fluids. In the deep interior of the outer planets, H_2 and He must occur as metals, and still deeper, the cores of at least Jupiter and Saturn probably include a mixture of H_2–He metal and silicates. Relative to the Sun, all of the outer planets are enriched in elements heavier than He. Magnetic fields of the outer planets vary significantly, both in orientation and magnitude, and the origin of these fields is poorly understood. They are not, however, produced by dynamo action in a liquid Fe core, as is the case in the terrestrial planets. Unlike Jupiter and Saturn, the densities of Uranus and Neptune require a greater silicate fraction in their interiors. Models for Uranus, for instance, suggest a silicate core and icy inner mantle composed chiefly of H_2O, CH_4, and NH_3, and a gaseous and icy outer mantle composed chiefly of H_2 and He. Neptune must have an even greater proportion of silicate and ice. Except for Jupiter, with a 3° inclination to the ecliptic, the outer planets are highly tilted in their orbits (Saturn 26.7°; Uranus 98°; Neptune 29°). Such large tilts probably result from collisions with other giant planets early in the history of the Solar System.

Satellites and planetary rings

General features

Three classes of planetary satellites are recognized as follows:

1 **Regular satellites**, which include most of the larger satellites and many of the smaller, are those which revolve in or near the plane of the planetary equator and revolve in the same direction as the parent planet moves about the Sun.
2 **Irregular satellites** have highly inclined, often retrograde and eccentric orbits, and many are far from the planet. Many of Jupiter's satellites belong to this category as do the outermost satellites of Saturn and

Neptune (Phoebe and Nereid, respectively). Most, if not all, of these satellites were captured by the parent planet.
3 **Collisional shards** are small, often irregular-shaped satellites that appear to have been continually eroded by ongoing collisions with smaller bodies. Many of the satellites of Saturn and Uranus are of this type. Phobos and Deimos, the tiny satellites of Mars, may be captured asteroids.

There are regularities in satellite systems that are important in constraining satellite origin. For instance, the large regular satellites of Jupiter, Saturn, and Uranus have low inclination, prograde orbits indicative of formation from an equatorial disc. Although regular satellites extend to 20–50 planet radii, they do not form a scale model of the Solar System. While the large satellites are mostly rocky or rock-ice mixtures, small satellites tend to be more ice rich, suggesting that some of the larger satellites may have lost ice or accreted rock. Very volatile ices, such as CH_4 and N_2, appear only on satellites distant from both the Sun and the parent planet, reflecting the cold temperatures necessary for their formation. One fact which emerges from an attempt to classify satellites is that no general theory of satellite formation is possible.

Since the Voyager photographs of planetary rings in the outer planets, the origin of planetary rings has taken on new significance. Some investigators have suggested that the rings of Saturn can be used as an analogue for the Solar Nebula from which the Solar System formed. Although Jupiter, Saturn, Uranus, and Neptune are now all known to have ring systems, they are all different, and no common theory can explain all of them. Two major models have attracted most attention. In one, the rings are formed with the planet as remnants of an accretion disc or from planetary spin off. Alternatively, they may be the debris resulting from the disruption of a satellite or a large captured comet. Both theories have problems, although only the satellite/comet break-up theory can overcome the problem of survival of the ring systems for the lifetime of the Solar System. Calculations indicate that planetary rings should survive for only 10^6–10^7 years, and thus it is difficult to see how they could be left over from planetary accretion about 4.6 Ga.

The Moon

As a planetary satellite, there are many unique features about the Moon. Among the more important are the following, all of which must be accommodated by any acceptable model for lunar origin:

1 The orbit of Moon about the Earth is neither in the equatorial plane of the Earth, nor in the ecliptic, but it is inclined at 6.7° to the ecliptic (Figure 7.8).
2 Except for the Pluto–Charon pair, the Moon has the largest mass of any satellite-planetary system.

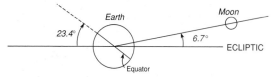

Figure 7.8 Orbital relations of the Earth–Moon system.

3   The Moon has a low density compared to the terrestrial planets, implying a relatively low iron content.
4   The Moon is strongly depleted in volatile elements and enriched in some refractory elements such as Ti, Al, and U.
5   The angular momentum of the Earth–Moon system is anomalously high compared to other planet-satellite systems.
6   The Moon rotates in the same direction as does the Earth.

A great deal has been learned about the geochemistry and geophysics of the Moon from the Apollo landings (Taylor, 1982; 1992). Although average lunar density is much less than average Earth density (Table 7.1), its uncompressed density is about the same as the Earth's mantle, implying that the Moon is composed largely of Fe and Mg silicates. Unlike most other satellites, which are mixtures of silicates and water ice, the Moon must have formed in the inner part of the Solar System.

From seismometers placed on the Moon by astronauts, we can deduce the broad structure of the lunar interior. The Moon has a thick crust (60–100 km) comprising about 12 per cent of the lunar volume, and it appears to have formed very soon after planetary accretion about 4.5 Ga (Taylor, 1992). From analysis of limited sampling by the astronauts, the lunar crust is composed chiefly of anorthosites and gabbroic anorthosites as represented by exposures in the lunar highlands. These rocks typically have cumulus igneous textures, although modified by impact brecciation. Sm–Nd isotopic dating indicates this plagioclase-rich crust formed at about 4.45 Ga. As discussed later, it appears to have formed by crystallization of an extensive magma ocean. Also characteristic of the lunar surface are the **mare basins**, large impact basins formed prior to 3.8 Ga, covering about 17 per cent of the lunar surface (Figure 7.9). These basins are flooded with basalt flows which are only 1–2 km thick, and were probably erupted chiefly from fissures. Isotopically-dated mare basalts range from 3.9–2.5 Ga. The impacts which formed the mare basins did not initiate the melting that produced the basalts, which were erupted up to hundreds of millions of years later and thus represent a secondary crust on the Moon. The youngest basaltic eruptions may be as young as 1 Ga. The lunar crust overlies a mantle comprised of two layers. The upper layer or lithosphere extends to a depth of 400–500 km and is probably composed of cumulate ultramafic rocks. The second layer extends to about 1100 km, where a sharp break in seismic velocities occurs. Although evidence is still not definitive, it appears that the Moon has a small metallic core (300–500 km in diameter), composing 2–5 per cent of the lunar volume (Figure 7.1).

Although the Moon does not presently have a magnetic field, remanent magnetization in lunar rocks suggests a lunarwide magnetic field at least between 3.9–3.6 Ga (Fuller and Cisowski, 1987). The maximum strength of this field was probably only about one-half that of the present Earth's field. It is likely that this field was generated by fluid motions in the lunar core, much like the present Earth's field is produced. A steady decrease in the magnetic field after 3.9 Ga reflects cooling and complete solidification of the lunar core by no later than 3 Ga.

The most popular model for lunar evolution involves the production of an ultramafic magma ocean that covers the entire Moon to depth of 500 km and crystallizes in < 100 My beginning at about 4.45 Ga (Figure 7.10). Plagioclase floats, producing an anorthositic crust, and pyroxenes and olivine largely sink producing an incompatible-element depleted upper mantle. Later partial melting of this mantle produces the mare basalts. Detailed models for crystallization of the magma ocean indicate that the process was complex, involving floating 'rockbergs' and cycles of assimilation, mixing, and trapping of residual liquids. The quenched surface and anorthositic rafts were also continually broken up by impact.

Satellite origin

Regular satellites in the outer Solar System are commonly thought to form in discs around their parent planets. The discs may have formed directly from the Solar Nebula as the planets accreted, and the satellites grew by collision of small bodies within the discs. Alternatively, the discs could form by the break up of planetesimals when they came within the Roche limit (the distance at which tidal forces of the planet fragment a satellite). Still other possible origins for planetary discs include spin-off due to contraction of a planet and outward transfer of angular momentum, and massive collisions between accreting planets. The collisional scenario is particularly interesting in that it forms the basis for the most widely accepted model for the origin of the Moon, as discussed later. Regardless of the way it originates, once a disc is formed, computer modelling indicates that satellites will accrete in very short periods of time of the order of 1 My. The irregular satellites and collisional shards, however, cannot be readily explained by the disc accretion model. This has led to the idea that many satellites have been captured by the gravity field of their parent planet, during a near collision of the two bodies. As dictated by their compositions, some of the rocky satellites in the outer Solar System may have accreted in the inner Solar System, and their orbits were perturbed in such a manner as to take them into the outer Solar System where they were captured by one of the jovian planets.

Figure 7.9 Oblique view of the southern part of Imbrium basin, one of the large mare basins on the Moon. Courtesy of Lunar and Planetary Institute.

Comets

Comets are important probes of the early history of the Solar System, since compared with other bodies in the Solar System, they appear to have been the least affected by thermal and collisional events (Wyckoff, 1991). Comet heads are small (radii of 1–10 km) and have a low density (0.1–1 gm/cm³). Because of their highly elliptical orbits, comets reside in a 'dormant state' most of the time in the outer reaches of the Solar System at temperatures ≤ 180 °C, in what are known as the Oort and Kuiper clouds (30–50 000 AU from the Sun). Perturbations of cometary orbits by passing stars have randomized them, thus making it difficult to determine where comets originally formed in the Solar Nebula. However, the presence of CO_2 and sulphur in comets suggests they formed in the outer cold regions of the nebula. Only for a few months do most comets come close enough to the Sun

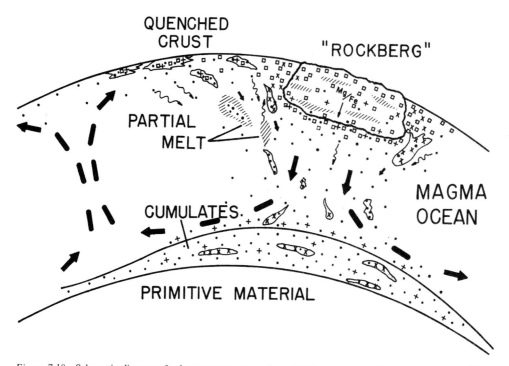

Figure 7.10   Schematic diagram of a lunar magma ocean about 4.5 Ga. Arrows are flow patterns and rockbergs are principally anorthosites. After Longhi (1978).

(0.5–1.5 AU) to form vaporized tails. Short-period comets, which orbit the Sun in less than 200 years, appear to come from just beyond Neptune at 35–50 AU, and they may represent 'leftovers' from the primordial Solar Nebula. Chiron, a dark body with a diameter of only 175 km that revolves about the Sun in a highly elliptical orbit between Saturn and Uranus, may be a comet.

Our knowledge of cometary composition was greatly increased during the Giotto mission in 1986 during a close approach to comet Halley. The nucleus of comet Halley is very irregular in shape and its surface is covered with craters, as well as with a layer of dark dust up to 1 m thick. How much dust (silicates, oxides) resides inside this or other comets is still unknown. Model calculations, however, indicate in general that the dust/ice ratio in comets is the order of 0.5–0.9. Data show that the gaseous component in Halley is composed chiefly of water vapour with only traces of CO_2, CO, CH_4 and NH_3. Since the ratios of these gases are dissimilar from the Sun, it appears that material composing Halley is not primitive, but has been frac-tionated. Also, compared with solar abundances, hydrogen is strongly depleted in Halley.

Asteroids

Asteroids are small planetary bodies, most of which revolve about the Sun in an orbit between Mars and Jupiter (Lebofsky et al., 1989; Taylor, 1992). Of the

4000 or so known asteroids, most occur between 2 and 3 AU from the Sun (Figure 7.11). The total mass of the asteroid belt is only about 5 per cent of that of the Moon. Only a few large asteroids are recognized, the largest of which is Ceres with a diameter of 930 km. Most asteroids are < 100 km in diameter and there is high frequency with diameters of 20–30 km. Three main groups of asteroids are recognized:

1   the near-Earth asteroids (Apollo, Aten, and Amor classes), some of which have orbits that cross that of the Earth
2   the main belt asteroids
3   the Trojans, revolving in the orbit of Jupiter (Figure 7.11).

Most meteorites arriving on Earth are coming from the Apollo asteroids. The orbital gaps where no asteroids occur in the asteroid belt (for instance at 3.8 and 2.1 AU, Figure 7.11), appear to reflect orbital perturbations caused by Jupiter. In terms of spectral studies, asteroids vary significantly in composition (Table 7.2), and as discussed later, some can be matched to meteorite groups. Within the asteroid belt, there is a zonal arrangement that reflects chemical composition. S, C, P, and D asteroid classes occupy successive rings outward in the belt, whereas M-types predominate near the middle, and B- and F-types near the outer edge. The broad pattern is that fractionated asteroids dominate in the inner part of the belt, and the low-albedo primitive types (class C)

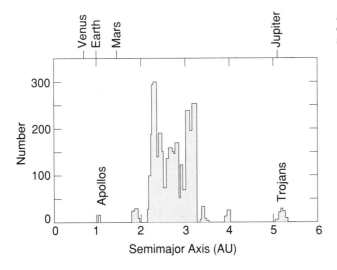

Figure 7.11 Histogram of frequency of asteroid distances from the Sun. Distance from Sun in astronomical units (1 AU = Earth–Sun distance).

Table 7.2 Characteristics of major asteroid classes

Low-Albedo classes

C   Common in the outer part of the asteroid belt at 3 AU; perhaps parental to carbonaceous chondrites

D   Found close to the orbit of Jupiter (5 AU); no meteorite analogues

P   Common near 4 AU; no meteorite analogues

Moderate-albedo classes

M   Common in main asteroid belt (3 AU); similar to Fe meteorites

Q   Apollo and related asteroids that cross the Earth's orbit; parental to major chondrite meteorite groups

S   Common in inner parts of asteroid belt (2 AU); some in Earth-crossing orbits; may be parental to pallasites and some Fe meteorites

V   4 Vesta and related asteroids at about 2 AU; parental to some basaltic achondrites

occur only in the outer portions of the belt. Thus, asteroids inwards of 2 AU are igneous asteroids, and the proportion of igneous to primitive asteroids decreases outwards such that by 3.5 AU there are no igneous types represented. Although asteroids are continually colliding with each other as indicated by the angular and irregular shapes of most, the remarkable compositional zonation in the asteroid belt indicates that mixing and stirring in the belt must be relatively minor.

The existence of the asteroid belt raises some interesting questions about the origin of the Solar System. Why is there such a depletion in mass in this belt in comparison with that predicted by interpolation of planetary masses? Was there ever a single small planet in the asteroid belt and if not, why not? Cooling rate data from iron meteorites (discussed later) which come from the asteroids, as well as an estimate of the tidal forces of Jupiter, indicate that a single planet never existed in the asteroid belt. The tidal forces of Jupiter would fragment the planet before it grew to planetary size. So it appears

that the asteroids accreted directly from the Solar Nebula as small bodies, and have subsequently been broken into even smaller fragments by continuing collisions. The depletion of mass in the asteroid belt may also be due to Jupiter. Because of its large gravitation field, it is likely that Jupiter swept up most of the mass in this part of the Solar System, leaving little for the asteroids. It would appear that asteroid growth stopped in most bodies when they reached about 100 km in diameter, as the belt ran out of material. Furthermore, the preservation of what appears to be basaltic crust on some asteroids suggests that they have survived for 4.55 Ga, when melting occurred in many asteroids as determined by dating fragments of these bodies which arrive on Earth as meteorites.

Meteorites

Introduction

Meteorites are small extraterrestrial bodies that have fallen on the Earth. Most meteorites fall as showers of many fragments, and over 3000 individual meteorites have been described. Meteorites have also been found on the Moon's surface and presumably occur also on other planetary surfaces. One of the best preserved and largest suites of meteorites is found within the ice sheets of Antarctica. In order to avoid contamination, when they are chopped out of the ice they are given special care and documentation similar to the samples collected on the Moon's surface. Trajectories of meteorites entering the Earth's atmosphere have been measured and indicate that they come from the asteroid belt. Although it seems likely that most meteorites come from the asteroid belt, some are fragments ejected off the lunar surface or other planetary surfaces by impact, and some may be remnants of comets. Most meteorites appear to have been produced during collisions between asteroids. Meteorites date to 4.56–4.55 Ga, and thus appear to be fragments of asteroids formed during the early stages

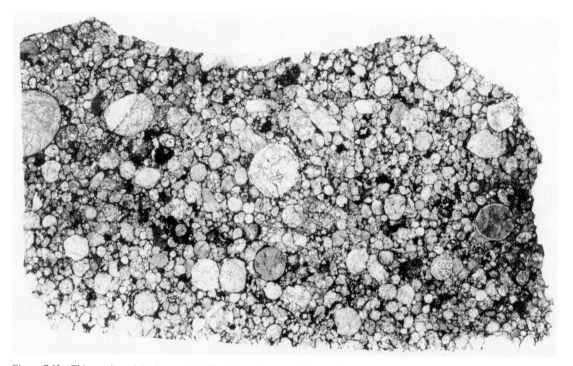

Figure 7.12 Thin section of the Inman chondrite. Photomicrograph shows chondrules (circular bodies), which are about 1 mm in diameter. The chondrules, which formed by rapid cooling of liquid droplets, are composed of olivine and pyroxene crystals surrounded by metallic iron (black). Long dimension about 2.5 cm. Courtesy of Rhian Jones.

of accretion in the Solar Nebula. Many meteorites are breccias in that they are composed of an amalgamation of angular rock fragments tightly welded together. These breccias appear to have formed during collisions on the surfaces of asteroids. In some cases, melting occurred around fragment boundaries as reflected by the presence of glass. Meteorites range in size from dust particles to bodies hundreds of metres across. Those with masses more than 500 gm fall on the Earth at a rate of about one per 10^6 km^2 per year.

Meteorites are classified as *stones*, including chondrites and achondrites, *stony-irons* and *irons* depending on the relative amounts of silicates and Fe–Ni metal phases present. **Chondrites**, the most widespread meteorites, are composed in part of small silicate spheroids known as *chondrules* (Figure 7.12), and have chemical compositions similar to the Sun. **Achondrites**, which lack chondrules, commonly have igneous textures and appear to have crystallized from magmas, and thus preserve the earliest record of magmatism in the Solar System. Some achondrites are breccias that probably formed on asteroid surfaces by impact. Stony-irons and irons have textures and chemical compositions that suggest they formed in asteroid interiors by fractionation and segregation of melts. The metal in meteorites is composed of two phases of Fe–Ni which are intergrown, producing the Widmanstatten structure visible on polished surfaces of iron meteorites.

Chondrites

Chondrites are the most common meteorites and consist of two components: chondrules and matrix (Figure 7.12). Chondrules have a restricted size range of 0.2–4 mm in diameter, with most < 1 mm. Despite considerable diversity in the composition of individual chondrules, mean compositions of chondrules from various groups of meteorites are very similar, suggesting that they were very well-mixed before accreting into parent bodies. One group of chondrites, the carbonaceous chondrites, are of special interest. These meteorites are hydrated, contain carbonaceous matter, and have not been subjected to temperatures $> 200\,°C$ for carbonaceous compounds to survive. They consist of a matrix of hydrated Mg silicates (principally chlorite and serpentine) enclosing chondrules of olivine and pyroxene. An important chemical feature of carbonaceous chondrites is that they contain elements in approximately solar ratios, suggesting they are primitive, and many investigators think one class of carbonaceous chondrites (Cl) are samples of the primitive Solar Nebula from which the Solar System formed. Matrices typically show wide variations in chemical and mineralogical composition that are thought to reflect differences in chemical composition within the Solar Nebula.

Although it is clear that chondrules are the products of rapid cooling of liquid droplets, it is not yet agreed if they formed within the Solar Nebula or later by melting on a planetary surface. Possible mechanisms by which chondrules could form on planetary surfaces include impact melting, collisions between chondrite parent bodies that result in melting, and planetary volcanism. All of these models have serious problems. For instance, all chondrules are ≥ 4.55 Ga, yet impact melting on asteroid surfaces predicts that ages should range to as young as 4.5 Ga. Also, volcanism is not an efficient way to produce liquid ultramafic droplets, nor can it explain the oxygen isotopic compositions of chondrules (Taylor, 1992). Collision of asteroids cannot explain the rather limited size distribution of chondrules nor their wide range in oxygen isotopic compositions. Relative to primitive carbonaceous chondrites, chondrules are enriched in lithophile elements and depleted in siderophile and chalcophile elements. This provides an important boundary condition for chondrule origin, in that the material from which they formed must have undergone earlier melting to fractionate elements prior to chondrule formation. This would seem to eliminate an origin for chondrules by direct condensation from the Solar Nebula. An alternative to nebular condensation is that chondrules formed by melting of pre-existing solids in the Solar Nebula, which is the only mechanism that can explain all of the physical and chemical constraints. It would appear that metal, sulphide, and silicate phases must have been present in the nebula before chondrule formation, and that chondrules formed by rapid melting and cooling of these substances. What caused the melting? Perhaps there were nebular flares, analogous to modern solar flares, that released sudden bursts of energy into the nebular cloud instantly melting local clumps of dust, which chilled forming chondrules, and these chondrules were later accreted into planetary bodies.

SNC meteorites

The SNC meteorites (named after Shergotty, Nakhla, and Chasigny meteorites) have distinct chemical compositions requiring that they come from a rather evolved planet (Marti et al., 1995). SNC meteorites are fine-grained igneous cumulates of mafic or komatiitic composition. The most convincing evidence that they have come from Mars is the presence of a trapped atmospheric component similar to the composition of the martian atmosphere as determined from spectral studies. Also, the major element composition of shergottites is very similar to the compositions measured by the Viking lander on the martian surface. Because of the problems of ejecting material from the martian surface, it is possible that the SNC meteorites were all derived from a single, large impact that occurred about 200 Ma. Sm–Nd ages suggest most of these meteorites crystallized from magmas about 1.3 Ga, at shallow depths. At least one martian meteorite from Antarctica, however, yields an age of about 4.5 Ga, and appears to represent a fragment of the martian Ancient Cratered Terrain, possibly some of the oldest crust preserved in the Solar System.

Refractory inclusions

Meteorite breccias contain a great variety of components, which have been subjected to detailed geochemical studies. Among these are inclusions rich in refractory elements (such as Ca, Al, Ti, Zr) known as CAIs (Ca and Al-rich inclusions), which range in size from dust to a few centimetres across. Stable isotope compositions of these inclusions indicate that they are foreign to our Solar System (Taylor, 1992). Because the sequence of mineral appearance in CAIs does not follow that predicted by condensation in a progressively cooling Solar Nebula, and differs from inclusion to inclusion, it is probable that local rather than widespread heating occurred in their nebular sources. Some appear to be direct condensates from a nebular cloud, while others show evidence favouring an origin as a residue from evaporation. The ages of CAIs indicate that they became incorporated in our Solar Nebula during the early stages of condensation and accretion. Just where they came from and how they became incorporated in the Solar Nebula, however, remains a mystery.

Iron meteorites and parent-body cooling rates

It is likely that most iron meteorites come from the cores of asteroid-size bodies (radii of 300–1000 km). For such cores to form, the parent bodies must have melted soon after or during accretion and molten Fe–Ni settled to the centre of the bodies. It is possible to constrain the cooling rates of iron meteorites from the thickness and Ni content of kamacite (α-iron) bands. This, in turn, provides a means of estimating the size of the parent body, since smaller parent bodies cool faster than larger ones. Calculated cooling rates are generally in the range of 1–10 °C/My, which indicates an upper limit for the radius of parent bodies of 300 km, with most lying between 100 and 200 km. Such results clearly eliminate the possibility that a single planet was the parent body for meteorites and asteroids. This conclusion is consistent with that deduced from estimates of Jupiter's tidal forces, which also indicate that a single planet could not form in the asteroid belt.

Asteroid sources

From a combination of mineralogical and spectral studies of meteorites and from spectral studies of asteroids, it has been possible to assign possible parent bodies of some meteorites to specific asteroid groups (Lebofsky et al., 1989). The most remarkable spectral match is between the visible spectrum of the third largest asteroid, 4 Vesta, and a group of meteorites known as basaltic achondrites (Figure 7.13). Supporting this source is the fact that Vesta occurs in an orbit with a 3:1 resonance, which is an 'escape hatch' for material knocked off Vesta to enter the inner Solar System.

Unfortunately, most groups of meteorites do not seem to have spectral matches among the asteroids, including

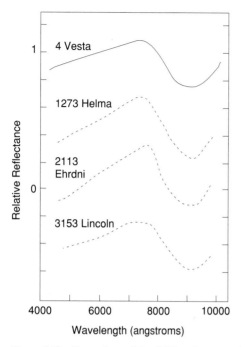

Figure 7.13 Comparison of the visible reflectance of asteroid 4 Vesta with three basaltic achondrite meteorites that may have been derived from Vesta.

the most common meteorites, the chondrites. This may be due to what is generally referred to as 'space weathering' of asteroid surfaces changing their spectral characteristics (Pieters and McFadden, 1994). However, no evidence to support this idea has yet been recovered. Another possibility is that the asteroid parents of most chondrites are smaller than we can resolve with our remote sensing on Earth.

Meteorite chronology

It is possible by using a variety of isotopic dating schemes to learn a great deal about the history of meteorites. Using **extinct radiogenic isotopes** with short half-lives, i.e., isotopes that completely decay in times less than 100 My, it is possible to estimate the **formation interval**, which is the time span between nucleosynthesis (element formation in stars) and meteorite parent-body formation. ^{129}I, which decays to ^{129}Xe with a half-life of 17 My, is useful in estimating the formation interval. If the formation interval were > 100 My (about five half-lives of ^{129}I), essentially all the ^{129}I would have decayed before planetary accretion, whereas if the formation interval was < 100 My, ^{129}I would accrete in the planets before it all decayed. In the latter case, there should be an excess of ^{129}Xe in the planets from the ^{129}I decay, and this is precisely what is observed. An excess of ^{129}Xe in the Earth and in meteorites indicates a formation interval for the Solar System less than 100 My.

Sm–Nd, Rb–Sr, and Re–Os isochron ages record the times of accretion and partial melting in meteorite parent bodies. These ages cluster between 4.56–4.53 Ga, which is currently the best estimate for the age of the Solar System (Figure 7.14). The oldest reliably-dated objects in the Solar System are the CAIs in Allende meteorite at 4566 Ma. The oldest Sm–Nd ages that reflect melting of asteroids are from basaltic achondrites at 4539 ± 4 Ma. Iron meteorite ages range from 4.56–4.46 Ga. K–Ar dates from meteorites reflect cooling ages and generally fall in the range of 4.4–4.0 Ga.

Fragmentation of asteroids by continual collisions exposes new surfaces to bombardment with **cosmic rays**, which are high-energy particles ubiquitous in interplanetary space. Interactions of cosmic rays with elements in the outer metre of meteorites produces radioactive isotopes that can be used to date major times of parent-body break up (Marti and Graf, 1992). Stone meteorites have cosmic-ray exposure ages of 100–5 Ma, as illustrated for instance by L-chondrites (Figure 7.15), while irons are chiefly 1000–200 Ma. The peak at about 40 Ma in the L-chondrite exposure ages is interpreted as a major collisional event between asteroids at this time. The differences in exposure ages between stones and irons reflect chiefly the fact that irons are more resistant to collisional destruction than stones. After a meteorite falls on Earth, it is shielded from cosmic rays and the amount of parent isotope remaining can be used to calculate a so-called **terrestrial age**, which is the time at which the meteorite fell on the Earth's surface and became effectively cut off from a high cosmic-ray flux. Although most terrestrial ages are less than 100 years, some as old as 1.5 Ma have been reported.

Chemical composition of the Earth and Moon

Since it is not possible to sample the interior of the Earth and the Moon, indirect methods must be used to estimate their composition (Hart and Zindler, 1986; McDonough and Sun, 1995). It is generally agreed that the Earth and other bodies in the Solar System formed by condensation and accretion from a Solar Nebula, and that the composition of the Sun roughly reflects the composition of this nebula. Nucleosynthesis models for the origin of the elements also provide limiting conditions on the composition of the planets. As we saw in Chapter 5, it is possible to estimate the composition of the Earth's upper mantle from analysis of mantle xenoliths and basalts, both of which transmit information about the composition of their mantle sources to the surface. Meteorite compositions and high-pressure experimental data also provide important input on the overall composition of the Earth.

Shock-wave experimental results indicate a mean atomic weight for the Earth of about 27 (mantle = 22.4 and core = 47.0) and show that it is composed chiefly of

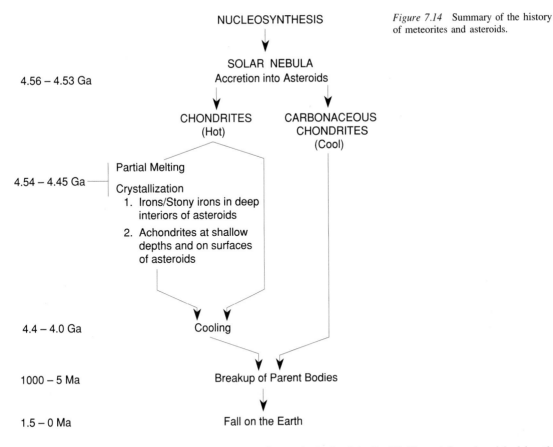

Figure 7.14 Summary of the history of meteorites and asteroids.

iron, silicon, magnesium, and oxygen. When meteorite classes are mixed in such a way as to give the correct core–mantle mass ratio (32/68) and mean atomic weight of the Earth, results indicate that iron and oxygen are the most abundant elements, followed by silicon and magnesium (Table 7.3). Almost 94 per cent of the Earth is composed of these four elements.

From lunar heat-flow results, correlations among refractory elements, and density and moment of inertia considerations, it is possible to estimate bulk lunar composition (Table 7.3). Compared with the Earth, the Moon is depleted in Fe, Ni, Na and S, and enriched in other major elements. The bulk composition of the Moon is commonly likened to the composition of the Earth's mantle because of similar densities. The data indicate, however, that the Moon is enriched in refractory elements like Ti and Al compared with the Earth's mantle.

It is necessary to refer to three geochemical groups of elements as a constraint on the accretion of the Earth and Moon. **Volatile elements** are those elements that can be volatilized from silicate melts under moderately reducing conditions at temperatures below 1400 °C, while

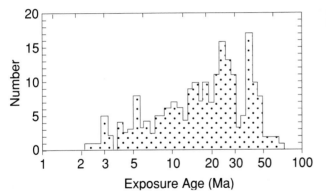

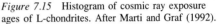

Figure 7.15 Histogram of cosmic ray exposure ages of L-chondrites. After Marti and Graf (1992).

Table 7.3 Major element composition of the Earth and Moon

	Earth		Moon
	1	2	3
Fe	29.9	28.2	8.3
O	30.9	32.4	44.7
Si	17.4	17.2	20.3
Mg	15.9	15.9	19.3
Ca	1.9	1.6	3.2
Al	1.4	1.5	3.2
Ni	1.7	1.6	0.6
Na	0.9	0.25	0.06
K	0.02	0.02	0.01
Ti	0.05	0.07	0.2

Values in weight percentages

1   Nonvolatile portion of Type-I carbonaceous chondrites with FeO/FeO+MgO of 0.12 and sufficient SiO_2 reduced to Si to yield a metal/silicate ratio of 32/68 (Ringwood, 1966)
2   From Allegre et al. (1995b)
3   Based on Ca, Al, Ti = 5 × Type-I carbonaceous chondrites; FeO = 12 per cent to accommodate lunar density; and Si/Mg = chondritic ratio (after Taylor, 1982)

refractory elements are not volatilized under the same conditions. Refractory elements can be further subdivided into **oxyphile** and **siderophile groups**, depending on whether they follow oxygen or iron, respectively, under moderately reducing conditions. Both the Earth and the Moon differ from carbonaceous chondrites in the distribution of elements in these groups (Figure 7.16). Compared with carbonaceous chondrites, both bodies are depleted in volatile and siderophile elements and enriched in oxyphile refractory elements. Any model for the accretion of the Earth and the Moon from the Solar Nebula must explain these peculiar element distribution patterns.

Vertical zoning of elements has occurred in both the Earth and the Moon. This zoning is the result of element **fractionation**, which is the segregation of elements with similar geochemical properties. Fractionation results from physical and chemical processes such as condensation, melting, and fractional crystallization. Large-ion lithophile elements, such as K, Rb, Th, and U, have been strongly enriched in the Earth's upper mantle and even more so in the crust in relation to the mantle and core. In contrast, siderophile refractory elements (such as Mn, Fe, Ni, and Co) are concentrated chiefly in the mantle or core, and oxyphile refractory elements (such as Ti, Zr, and La) in the mantle.

Age of the Earth and Solar System

Because the Earth is mostly inaccessible to sampling and is a continuously-evolving system, it is difficult to date. Also contributing to this difficulty is the fact that the Earth accreted over some time interval and hence, if an age is obtained, what event in this time interval does it date? The first isotopic ages of the Earth were model Pb ages of about 4.55 Ga obtained from sediments and oceanic basalts (Patterson et al., 1955). Until recently precise ages were obtained from meteorites, and other attempts to date the Earth have yielded ages in the range of 4.55–4.45 Ga. Reconsideration of model Pb ages indicates that the first major differentiation event in the Earth occurred at about 4.45 Ga. The ^{129}I–^{129}Xe isotopic system, previously discussed, can also be used to calculate an age for the Earth and results yield a value of 4.46 ± 0.02 Ga (Allegre et al., 1995). It would thus seem that the Earth is the order 100 My younger than the most primitive meteorites.

As previously mentioned, refractory inclusions (CAIs) in Allende meteorite are the oldest dated objects in the Solar System at 4566 Ma. The accretion of most chondrites began a maximum of about 3 My later and lasted for no more than 8 My. The model Pb ages from

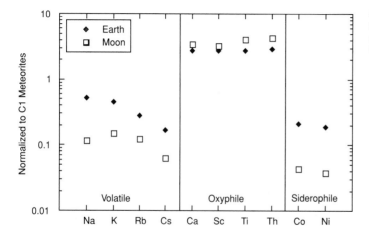

Figure 7.16 Volatile, oxyphile, and siderophile element distributions in the Earth and Moon normalized to C1 meteorites. C1 = Type-I carbonaceous chondrites.

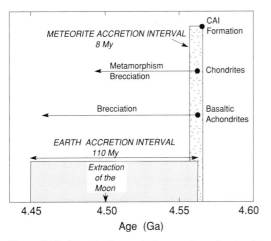

Figure 7.17 Isotopic time scale for accretion of meteorites, Earth, and Moon. Based on Pb–Pb, Sm–Nd, and $^{40}Ar/^{39}Ar$ isotopic systems. After Allegre et al. (1995b).

the Earth probably represent the mean age of core formation, when iron was separated from the mantle and U and Pb were fractionated from each other. On the other hand, the I–Xe age appears to reflect the mean age of degassing of the atmosphere from the mantle. The fact that both of these ages are strikingly similar at 4460–4450 Ma suggests that both processes went on simultaneously during the terminal stages of planetary accretion (Figure 7.17). This being the case, the age of the Earth we obtain is the age of early differentiation, and just how this age is related to the onset of planetary accretion is not well-known. Although the oldest ages of rocks dated from the Moon are about 4450 Ma, model ages

suggest that the Moon accreted 4500–4480 Ma. If the Moon was formed by accretion of material left over after a Mars-sized body hit the Earth, as discussed later, it would appear that this impact occurred at 4.5–4.48 Ga. If the Earth began to accrete at the same time as the CAIs accreted at 4566 Ma, then the total duration of accretion up to the major melting event is in the order of 110 My (Figure 7.17). Hence, if the Moon formed at about 4500 Ma, it would appear that it formed while the Earth was still accreting and the core was not completely formed.

Comparative evolution of the atmospheres of Earth, Venus and Mars

The only three terrestrial planets which have retained atmospheres are Earth, Venus and Mars, yet the compositions and densities of their atmospheres differ significantly from each other (Table 6.1). The greenhouse effect, the distance from the Sun, and planetary mass all may have played a role in giving rise to such different atmospheres (Prinn and Fegley, 1987; Hunten, 1993). Planetary surface temperature is controlled largely by the greenhouse effect and so only when significant amounts of CO_2, H_2O, or CH_4 accumulate in the atmosphere will the surface temperature begin to rise. This can be illustrated by the effect of progressively increasing water pressure in a planetary atmosphere (Walker, 1977). Beginning with estimated surface temperatures as dictated by the distance from the Sun, water vapour will increase in each atmosphere in response to planetary degassing until intersecting the water vapour saturation curve (Figure 7.18). For the Earth, this occurs at about 0.01 atm where

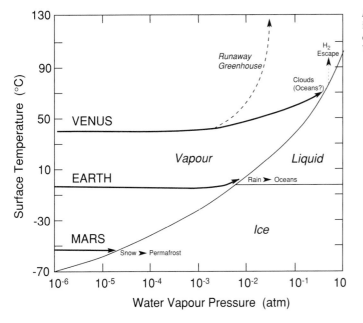

Figure 7.18 Idealized trajectories for increasing atmospheric water vapour pressure on Venus, Earth and Mars. Modified after Walker (1977).

Table 7.4 Comparison of the present atmosphere of Venus and Earth (volume percentages)

	Earth		Venus
	Atmosphere	*Near-surface reservoirs	
N_2	78	2	2
O_2	21	–	tr
CO_2	0.05	98	98
N_2O	≤ 1	≤ 1	≤ 1
Pressure (bars)	1	0.1–1	10–100

* Includes atmosphere, hydrosphere, biosphere and sediments

H_2O begins to condense and precipitate on the surface. Further degassing does not increase the H_2O content of the atmosphere or the surface temperature, but rains out and begins to form the oceans, a process which occurred on Earth probably between 4.5–4.4 Ga. The growing oceans also stabilize the CO_2 content of the atmosphere via the carbonate–silicate cycle as discussed in Chapter 6. Because Venus is closer to the Sun it had a higher initial surface temperature than Earth. This results in the greenhouse effect raising the surface temperature before the vapour saturation curve is intersected. As degassing continues, H_2O and CO_2 may rapidly accumulate in the atmosphere producing a **runaway greenhouse** in which atmospheric temperature continues to rise indefinitely (Figure 7.18). The reason why water is virtually absent in the venusian atmosphere today is due to photolysis in the upper atmosphere producing H_2, which readily escapes, and O_2 which may be incorporated in weathered rocks on the surface and perhaps returned to the venusian mantle by tectonic processes. It is noteworthy that the amount of CO_2 in the venusian atmosphere is approximately equivalent to the amount in near-surface reservoirs on Earth (most of it in marine carbonates) (Table 7.4). Hence, if it were not for the oceans that provide a vital link in the carbonate–silicate cycle, Earth would probably have an uninhabitable atmosphere like Venus.

Mars is farther from the Sun than Earth and has a colder initial surface temperature. Hence, as water increases in the atmosphere and intersects the vapour saturation curve, snow rather than liquid water precipitates (Figure 7.18). As with Earth, neither the surface temperature nor atmospheric water vapour pressure increases with further degassing. Upon further cooling (due to lack of a greenhouse effect), CO_2 would also freeze forming the Martian polar ice caps. The sparsity of light gases like H_2 and N_2 on Mars probably reflects the relatively small mass of the planet and its inability to retain light gases which escape from its gravitational field.

Although this model accounts for the general features of the evolution of the atmospheres of Earth, Venus, and Mars, some aspects require modification. For instance, the model assumes that Mars was cold and frozen from the beginning, an assumption that seems unlikely in terms

of the large canyons on the surface which were almost certainly cut by running water before 4 Ga (Carr, 1987). These canyons and their tributary systems are remarkably similar to terrestrial canyons (Figure 7.3) and seem to require surface temperatures above the freezing point of water during the time they formed. For this reason, it is probable that an early CO_2 greenhouse effect existed on Mars and this necessitates some kind of tectonic recycling of CO_2. Because Mars is too small to retain oceans, and it is unlikely that plate tectonics ever operated, it is also unlikely that an Earth-like carbonate–silicate cycle controlled CO_2 levels in the early atmosphere. Although the nature of the martian early recycling system is unknown, the planet must have cooled more rapidly than the Earth, perhaps trapping most of the CO_2 in its interior. Due to insufficient greenhouse heating, the surface cooled and most of the remaining CO_2 condensed as ice on the surface. It is probable that today most of the degassed water and some of the CO_2 on Mars occur as permafrost near the surface.

The runaway greenhouse model for Venus assumes that Venus never had oceans, another assumption that is questionable in terms of the high D/H (deuterium/hydrogen) ratio in trace amounts of water in the Venusian atmosphere. This ratio, which is 100 times higher than that in the Earth's water, may have developed during preferential loss of H_2 from the venusian atmosphere enriching the residual water in the heavier deuterium isotope, which has a much smaller escape probability than hydrogen (Gurwell, 1995). Although this observation does not prove the existence of early oceans on Venus, the planet at one time must have had at least 10^3 times more water than at present, either in oceans or as water vapour in the atmosphere. An alternative and preferred evolutionary path for Venus leads to condensation of water at 4.5–4.4 Ga, formation of water vapour clouds, and possibly of oceans (Figure 7.18). On the Earth, water is blocked from entering the stratosphere by a cold trap (a temperature minimum) at 10–15 km above the surface (Figure 6.1), and most water vapour condenses when it reaches the cold trap. However, if water vapour in the lower atmosphere exceeds 20 per cent, as it must have on Venus, it produces greenhouse warming. This results in movement of the cold trap to a high altitude (~ 100 km) where it is no longer effective in preventing water vapour from rising into the upper atmosphere, where it can undergo photolysis and hydrogen escape. Thus, on Venus hydrogen escape could have eliminated an oceanic volume of water in < 30 My. Following this early loss of water, the venusian atmosphere would rapidly evolve into a runaway greenhouse caused by the remaining CO_2, and this is the situation that exists today on Venus.

The continuously habitable zone

Living organisms require a very narrow range in surface temperature and hence a narrow range in the content of greenhouse gases. From a knowledge of temperature and gas distributions in the inner Solar System, the width of this zone, known as the **continuously habitable zone**, can be estimated. If the Earth, for instance, had accreted in an orbit 5 per cent closer to the Sun, the atmosphere would continue to rise in temperature and the oceans would evaporate leading to a runaway greenhouse as described above. The outer extreme of the continuously habitable zone is less certain, but appears to extend somewhat beyond the orbit of Mars. Thus, the continuously habitable zone in the Solar System would appear to extend from about 0.95 to 1.5 AU (1 AU = Earth–Sun distance).

This observation is important because the continuously habitable zone is wide enough to allow for habitable planets existing in other planetary systems.

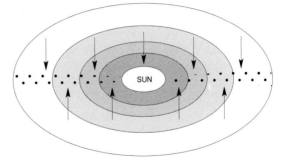

Figure 7.19   Schematic cross-section of the primitive Solar Nebula. Temperatures in the nebula decrease outwards from the Sun from hottest (dark grey) to coolest (white). Arrows indicate direction of gravitational sinking of solid particles into the ecliptic.

Condensation and accretion of the planets

The Solar Nebula model

Although many models have been proposed for the origin of the Solar System, only those that begin with a gaseous Solar Nebula appear to be consistent with data from both astrophysics and cosmochemistry. Stars form by contraction of gaseous nebulae and planets and other bodies form by condensation and accretion as nebulae cool (Hartmann, 1983). Just how the Solar System formed can be considered in terms of three questions:

1 How the Sun acquired the gaseous material from which the planets formed.
2 The history of condensation of the gaseous material.
3 The processes and history of planetary accretion.

Regarding the first question, one viewpoint is that the Sun, already in existence, attracted material into a gaseous nebula about itself. Another proposes that the pre-existing Sun captured a Solar Nebula of appropriate mass and angular momentum to form the Solar System. Most theories, however, call upon condensation and accretion of the Sun and planets from the same cloud at approximately the same time. All models have in common a gaseous nebula from which the planets form. The minimum mass of such a nebula is about 1 per cent of a solar mass. The mechanisms by which the nebula becomes concentrated into a disc with the Sun at the centre are not well-understood. One possibility is the transferring of angular momentum from the Sun to the nebula, caused either by hydromagnetic coupling during rotational instability of the Sun or by turbulent convection in the nebula. Another possibility is a nearby supernova, which may have triggered the collapse of the Solar Nebula. In either case, condensing matter rapidly collapses into a disc about the Sun (Boss et al., 1989; Wetherill, 1994) or into a series of Saturn-like rings, which condense and accrete into the planets (Figure 7.19). Small planetesimals

form within the cloud and spiral toward the ecliptic plane, where they begin to collide with one another growing into 'planetary embryos', which range in size from that of the Moon to Mercury. Finally, these embryos collide and grow into the terrestrial planets and the cores of the giant outer planets. Regardless of the specific mechanism, the collapse of the nebula results in a cold cloud (< 0 °C) except in the region near the protoSun within the orbit of Mercury.

Two processes are important in planet formation. **Condensation** is the production of solid dust grains as the gases in the Solar Nebula cool. **Accretion** is the collision of the dust grains to form clumps and progressively larger bodies, some of which grow into planets (Figure 7.20). Condensation begins during collapse of the nebula into a disc, and leads to the production of silicate and oxide particles as well as to other compounds, all of which are composed chiefly of Mg, Al, Na, O, Al, Si, Fe, Ca, and Ni. In the cooler parts of the nebula, ices of C, N, O, Ne, S, Ar and halogens form. The remaining gaseous mixture is composed chiefly of H_2 and He. The ices are especially abundant in the outer part of the disc where the giant planets form, and silicate–oxide particles are concentrated in the inner part and give rise to the terrestrial planets. Thermodynamic considerations indicate that iron should be present in the cloud initially only in an oxidized state. Type-I carbonaceous chondrites (C1) may represent a sample of this primitive nebula. These meteorites contain only oxidized iron and large amounts of volatile components of the ices listed above (including organic compounds), and they have not been heated to more than 200 °C to preserve their volatile constituents.

Although it is commonly assumed that the Solar Nebula was well mixed, differences in isotopic abundances and ages of various meteorites suggest that it was not (Wasson, 1985; Taylor, 1992). Variations in the abundance of rare gases and isotopic variations of such elements as Mg, O, Si, Ca, and Ba indicate large mass-dependent fractionation effects. The isotopic anomalies

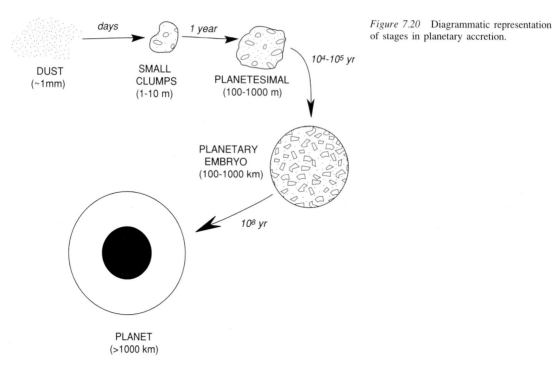

Figure 7.20 Diagrammatic representation of stages in planetary accretion.

appear to require a sudden injection of neutrons into the early Solar Nebula. The source of such a neutron burst may have been one or more supernova, which also may have triggered the collapse of the Solar Nebula. Planetary accretion models fall into two general categories. **Homogeneous accretion** involves condensation of a nebula followed by accretion of planets and other bodies as the nebula continues to cool. In contrast, **heterogeneous accretion** is when condensation and accretion occur at the same time. The specific mechanisms by which particulate matter evolves into small metre-sized bodies and these bodies, in turn, grow into planetesimals (small planets) and planets are becoming better understood. Most experimental data suggest that grains stick together during the early stages of accretion by weak electrostatic forces (Van der Waal forces) that are important in producing centimetre-sized objects. Theoretical studies indicate that a large fraction of the gaseous cloud will condense and accrete rapidly into several hundred asteroid-sized bodies rather than into a swarm of small particles.

Computer models suggest that the first generation of planetesimals will increase in radius to the order of 100–1000 m in times of about one year after condensation (Figure 7.20). Wetherill (1986; 1994) has simulated further planetary growth by stochastic modelling using a high-speed computer. His results indicate that the planetesimals grow into planets in two distinct stages. During the first stage (10^4–10^5 y), planetesimals collide at low velocities and rapidly grow into planetary embryos (Figure 7.20). These embryos are commonly referred to as runaway embryos because in each local zone of the nebula, a single body grows more rapidly than its

neighbours within the zone, and before it stops growing it cannibalistically consumes its neighbours. In the second stage, which lasts about 10^8 years, these embryos collide with each other to form the terrestrial planets and cores of the outer giant planets. During this stage, there is widespread mixing of accreting materials in the inner Solar System as well as 'giant impacts', when one embryo collides with another. It is during this stage that a Mars-sized body may have collided with Earth to form the Moon (see below) and another body collided with Mercury, stripping it of much of its mantle. The computer models indicate that the terrestrial planets should grow to within 50 km of their final radius in the first 25 My, and accretion should be 98 per cent complete by 10^8 years.

During the time of supercollisions, the Earth was probably struck at high velocity many times by bodies ranging up to Mercury in size and at least once by a Mars-sized body. Much of the kinetic energy associated with these collisions would have been expended in heating the interior of the planet, and calculations show that the Earth should be largely melted long before it completes its growth. Thus a widespread magma ocean would exist on the planet. It is during this stage that the Earth's core formed and the planet was extensively degassed to form the atmosphere. Core formation was the first step in producing a zoned planet. The giant planets, Jupiter and Saturn, which consist chiefly of H_2 and He, accreted onto rocky cores that also grew by planetesimal collisions. Once a planet reaches 10–50 Earth masses, it is able to attract H_2 and He from the nebula and to grow into a giant planet. Two models have been proposed for

Table 7.5 Condensation sequence from Solar Nebula at 10^{-4} atm

Phases	Temperature (°C)
Ca, Al, Ti and related refractory oxides	1250–1600
Metallic Fe–Ni	1030–1200
Forsterite (Mg_2SiO_4) and enstatite ($MgSiO_3$)	1030–1170
Ca-plagioclase	900–1100
Na–K feldspars	~ 730
Troilite (FeS)	430
Fe–Mg pyroxenes and olivine	300
Fe oxides	100–200
Carbonaceous compounds	100–200
Hydrated Mg-silicates	0–100
Ices	< 0

After Grossman (1972)

the origin of the outer planets (Pollack and Bodenheimer, 1989). In the first model, the cores of these planets form by aggregation of refractory components, with volatile elements accreting to make the mantles. One problem with this model is that the elements in the core should be highly soluble at the high temperatures and pressures in the planetary interiors, and so a core should not form. The model consistent with most data involves accretion of silicates and ices to form planetary cores, followed by accretion of gaseous components to about fifteen Earth masses, and then by very rapid accretion of H_2 and He.

Heterogeneous accretion models

Heterogeneous accretion models call upon planetary growth by simultaneous condensation and accretion of various compounds as the temperature falls in an originally hot Solar Nebula. The end product is a zoned planet (Grossman, 1972). Cameron (1973) has suggested a model whereby the planets develop in a rotating, disc-shaped Solar Nebula as it cools. Cooling of the gas as dissipation proceeds results in condensation over a wide range of temperature. The sequence of compounds condensed from the nebula at a pressure of 10^{-4} atm is summarized in Table 7.5. Heterogeneous accretion should produce a planet with high-temperature refractory components in the centre, overlain sequentially by metal, various silicates, compounds with oxidized iron, and hydrated silicates/oxides and ices.

Although appealing in that we can produce a zoned planet directly, heterogeneous accretion models are faced with many obstacles. For instance, there is a large degree of overlap in the condensation temperatures given in Table 7.5, and hence a zoned planet is not clearly predicted. Also, the model predicts an Earth with a small core of Ca–Al–Ti oxides, which is not consistent with geophysical data. The problem is not alleviated if iron melts and sinks to the centre displacing the oxide layer upwards, since there is no seismic evidence for a Ca–Al–Ti layer above the core. These, together with several

other significant geochemical and isotopic problems, seem to present insurmountable stumbling blocks for heterogeneous accretion models.

Homogeneous accretion models

In homogeneous accretion models condensation is essentially complete before accretion begins. Compositional zonation in the Solar Nebula is caused by decreasing temperature outward from the Sun (Figure 7.19) and this is reflected in planetary compositions. Refractory oxides, metals, and Mg-silicates are enriched in the inner part of the cloud where Mercury accretes; Mg–Fe silicates and metal in the region from Venus to the asteroids; and mixed silicates and ices in the outer part of the nebula where the giant planets accrete. Homogeneous accretion models produce an amazingly good match between predicted and observed planetary compositions (Figure 7.21). If, for instance, volatiles were blown outwards when Mercury was accreting refractory oxides, metal, and Mg-silicates, Venus and Earth should be enriched in volatile components compared with Mercury. The lower mean density of Venus and Earth is consistent with this prediction. Mars has a still lower density, and it is probably enriched in oxidized iron compounds relative to the Earth and Venus. Likewise, the giant outer planets are composed of mixed silicates and ices, reducing their densities dramatically.

If homogeneously accreted, how do planets become zoned? It would appear that zoning is caused by melting, which results in segregation by such processes as fractional crystallization and sinking of molten iron to the core. Where does the heat come from to melt the planets and asteroids? Some of the major known heat sources are as follows:

1 **Accretional energy**. This energy is dependent upon impact velocities of accreting bodies and the amount of input energy retained by the growing planet. Accretional energy alone appears to have been sufficient, if entirely retained in the planet, to largely melt the terrestrial planets while they were accreting.

2 **Gravitational collapse**. As a planet grows, the interior is subjected to higher pressures and minerals undergo phase changes to phases with more densely-packed structures Most of these changes are exothermic as discussed in Chapter 4, and large amounts of energy are liberated into planetary interiors.

3 **Radiogenic heat sources**. Radioactive isotopes liberate significant amounts of heat during decay. Short-lived radioactive isotopes, such as ^{26}Al and ^{244}Pu, may have contributed significant quantities of heat to planets during accretion. Long-lived isotopes, principally ^{40}K, ^{235}U, ^{238}U and ^{232}Th, are important heat producers throughout planetary history.

4 **Core formation**. Core formation is a strongly exothermic process and appears to have occurred over a relatively short period of time (≤ 100 Ma) beginning during the late stages of planetary accretion.

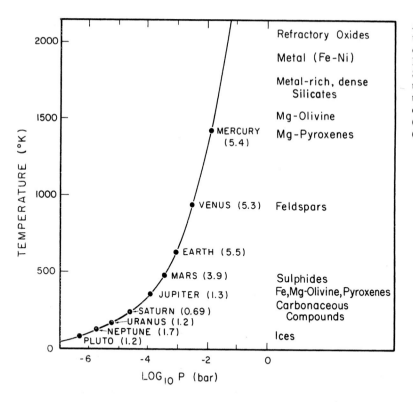

Figure 7.21 Calculated temperature and pressure distribution in the Solar Nebula late in planetary accretion history. Shown also is the distribution of the planets and their densities and the predicted composition from condensation experimental data (Table 7.5). After Hartmann (1983).

Order of magnitude estimates indicate that there was enough heat available during accretion of the terrestrial planets and the Moon, if retained, to melt these bodies completely. Because volatiles are retained in the terrestrial planets, however, it is unlikely that they were ever completely molten. So it is necessary to remove the early heat very rapidly and convection seems the only process capable of bringing heat from planetary interiors to the surface in a short enough time interval (≤ 100 Ma) to escape complete melting. The temperature dependence of mantle viscosity appears to be the most important factor controlling planetary thermal history (Schubert, 1979; Carlson, 1994). Initially, when a planet is hot and viscosity is low, chaotic mantle convection rapidly cools the planet and crystallizes magma oceans. As mantle viscosity increases, beginning 100 My after accretion, convection should cool planetary bodies at reduced rates.

Accretion of the Earth

Together with astrophysical models, the chemical composition of planets provides an important constraint on planetary accretion (Greenberg, 1989; Wetherill, 1990; Taylor, 1992). Any model for accretion of the Earth must account for a depletion in volatile elements in the silicate portion of the Earth (V to K, Figure 7.22) relative to the Solar Nebula, the composition of which is assumed to be that of Type-I carbonaceous chondrites. This increasing depletion with increasing condensation temperature may reflect depletions in the nebular material in the region from which the Earth accreted. Alternatively, the trend could result from mixing between two components in the nebula, one a 'normal' component relatively enriched in refractory elements (Al to La, Figure 7.22), and one strongly depleted in volatiles. In this case, the earliest stage of Earth accretion would be dominated by refractory elements, followed by a second stage in which mixed refractory and volatile components accrete as the temperature of the nebula continues to decrease (partly heterogeneous accretion). Depletions in volatile elements in portions of the inner Solar Nebula may have been caused by an intense solar wind, known as a T-Tauri wind, emitted from the Sun during its early history. This wind could have blown volatile constituents from the inner to the outer part of the Solar System.

What about the depletions in siderophile elements in the silicate fraction of the Earth (Figure 7.16)? These can best be explained by core formation during which these elements (such as Co and Ni), with their strong affinities for iron, are purged from the mantle as molten iron sinks to the Earth's centre to form the core. Thus, it would appear that the volatile element depletions in the Earth occurred during the early stages of planetary accretion and the siderophile element depletions during the late stages.

Origin of the Moon

Scientific results from the Apollo missions to the Moon have provided a voluminous amount of data on the struc-

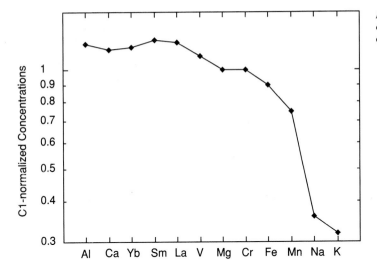

Figure 7.22 C1-normalized element distributions in the Earth. C1 = Type-I carbonaceous chondrites.

ture, composition, and history of the Moon (Taylor, 1982; 1992). Seismic data are generally interpreted in terms of four zones within the Moon, which are, from the surface inwards: a plagioclase-rich crust (dominantly gabbroic anorthosite) 60–100 km thick; an upper mantle, composed chiefly of pyroxenes and olivine, 300–400 km thick; a lower mantle of the same composition (with olivine dominating), about 500 km thick; and 200–400 km of iron core.

Among the more important constraints which any model for lunar origin must satisfy are the following:

1 The Moon does not revolve in the equatorial plane of the Earth, nor in the ecliptic. The lunar orbit is inclined at 6.7° to the ecliptic, whereas the Earth's equatorial plane is inclined at 23.4° (Figure 7.8).
2 Tidal dissipation calculations indicate that the Moon is retreating from the Earth, resulting in an increase in the length of the day of 15 sec/y. Orbital calculations and the Roche limit indicate that the Moon has not been closer to the Earth than about 240 000 km.
3 The Moon is enriched in refractory oxyphile elements and depleted in refractory siderophile and volatile element relative to the Earth (Figure 7.16). Particularly important is the low density of the Moon (3.33 gm/cm^3) compared with other terrestrial planets, which indicates that the Moon is significantly lower in iron than these planets.
4 The Earth–Moon system has an anomalously large amount of angular momentum (3.45×10^{41} gm/cm^2/sec) compared to the other planets.
5 The oxygen isotopic composition of lunar igneous rocks collected during the Apollo missions is the same as that of mantle-derived rocks from the Earth. Because oxygen isotope composition seems to vary with position in the Solar System, the similarity of oxygen isotopes in lunar and terrestrial igneous rocks suggests that both bodies formed in the same part of

the Solar System at approximately the same distance from the Sun.
6 Isotopic ages from igneous rocks on the lunar surface range from about 4.46–3.1 Ga. Model ages indicate that the anorthositic rocks of the lunar highlands crust formed at about 4.46–4.45 Ga.

Models for the origin of the Moon generally fall into one of four categories: (1) fission from the Earth; (2) the double planet scenario in which the Moon accretes from a sediment ring around the Earth; (3) capture by the Earth; and (4) impact on the Earth's surface by a Mars-sized body (Figure 7.23). Any acceptable model must account for the constraints listed above, and so far none of these models is completely acceptable. Each of the hypotheses will be discussed briefly, and a summary of just how well each model complies with major constraints is given in Table 7.6.

Fission models

Fission models involve the separation of the Moon from the Earth during an early stage of rapid spinning when tidal forces overcome gravitational forces. One version of the fission hypothesis (Wise, 1963) suggests that formation of the Earth's core reduced the amount of inertia, increasing the rotational rate and spinning off material to form the Moon. Such a model is attractive in that it accounts for the similarity in density between the Earth's mantle and the Moon and for the absence of a large metallic lunar core. The hypothesis is also consistent with the fact that the Moon revolves in the same direction as the Earth rotates, the circular shape of the lunar orbit, the existence of a lunar bulge facing the Earth, and similar oxygen isotope ratios of the two bodies. The model also explains the iron-poor character of the Moon if fission occurred after core formation in the Earth, since the Moon would be formed largely from the Earth's mantle.

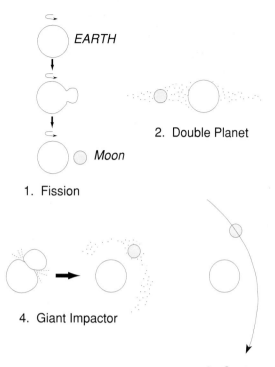

2. Double Planet

1. Fission

4. Giant Impactor

3. Capture

Figure 7.23   Models for the origin of the Moon.

Double planet models

Double planet models involve an accreting Earth with simultaneous accretion of the Moon from orbiting solid particles (Harris and Kaula, 1975). A major advantage of these models, also known as co-accretion or precipitation models, is that they do not invoke special, low-probability events. The models assume that as the Earth accreted, solid particles accumulated in orbit about the Earth and accreted to form the Moon. The general scenario is as follows: the Earth accretes first and its core forms during accretion; as the Earth heats up, material is vaporized from the surface forming a ring around the Earth from which the Moon accretes. Because core formation extracts siderophile elements from the mantle, the material vaporized from the Earth is depleted in these elements, and hence the Moon, which accretes from this material, is also depleted in these elements. Because volatile elements are largely lost by intense solar radiation from a T-Tauri wind after the Earth accretes (but before the Moon accretes), the material from which the Moon accretes is depleted in volatile elements relative to the Earth. This leaves material from which the Moon accretes relatively enriched in oxyphile refractory elements.

The most serious problems with the double planet models are that they do not seem capable of explaining the large amount of angular momentum in the Earth–Moon system (Wood, 1986) and they do not readily explain the inclined lunar orbit.

Capture models

Capture models propose that the Moon and the Earth formed in different parts of the Solar Nebula and that early in the history of the Solar System the Moon or its predecessor approached the Earth and was captured (Taylor, 1992). Both catastrophic and non-catastrophic models of lunar capture have been described, involving retrograde and prograde orbits for the Moon prior to capture.

Capture models fall into two categories: **intact capture**, where a fully-accreted Moon is captured by the Earth, and **disintegrative capture**, where a planetesimal comes within the Earth's Roche limit, is fragmented by tidal forces with most of the debris captured in orbit

Fission models, however, face several major obstacles. For instance, they do not explain the inclination of the lunar orbit, and they require more than four times the total angular momentum than is available in the present Earth–Moon system! If the Earth–Moon system ever had this excess angular momentum, no acceptable mechanism has been proposed to lose it (Wood, 1986). Also, lunar igneous rocks are more depleted in siderophile and volatile elements than terrestrial igneous rocks, indicating that the lunar interior is not similar in composition to the Earth's mantle. Although some investigators still favour a fission model (Binden, 1986), the above problems seem to render the model highly implausible (Table 7.6).

Table 7.6   Summary of major constraints on models of lunar origin

		Fission	Capture	Double-planet	Giant impactor
1	Angular momentum	I	C (I)	I	C
2	Lunar rotation	C	C (C)	C	C
3	Orbital characteristics	I	C (C)	I	C
4	Chemical composition	I	I (I)	C	C

Values in parenthesis for disintegrative capture
C, consistent; I, inconsistent

about the Earth, and the debris re-accretes to form the Moon (Wood and Mitler, 1974). Although intact capture models may explain the high angular momentum and inclined lunar orbit, they cannot readily account for geochemical differences between the two bodies. The similar oxygen isotopic ratios between lunar and terrestrial igneous rocks suggests that both bodies formed in the same part of the Solar System, yet the capture model does not offer a ready explanation for the depletion of siderophile and volatile elements in the Moon. Also, intact lunar capture is improbable because it requires a very specific approach velocity and trajectory. Disintegrative capture models cannot account for the high angular momentum in the Earth–Moon system.

Giant impactor model

The giant impactor model involves a glancing collision of the Earth with a Mars-sized body during which debris from both the Earth and the colliding planet collect and accrete in orbit about the Earth to form the Moon (Hartmann, 1986; Cameron, 1986). Such a model has the potential of eliminating the angular momentum and non-equatorial lunar orbit problems as well as providing a means of explaining chemical differences between the Earth and the Moon. One of the major factors that led to the giant impactor model is the stochastic models of Wetherill (1985), which indicate that numerous large bodies formed in the inner Solar System during the early stages of planetary accretion. Results suggest the existence of at least ten bodies larger in size than Mercury and several equal to or larger than Mars. Wetherill (1986) estimates that about one-third of these objects collided with and accreted to the Earth, providing 50–75 per cent of the Earth's mass. The obliquities of planets and the slow retrograde motion of Venus are also most reasonably explained by late-stage impact of large planetesimals.

In the giant impactor model, a small planet (0.1–0.2 Earth masses) collides with the Earth during the late stages of accretion (Taylor, 1993) (Figure 7.24). Such an impact can easily account for the anomalously high angular momentum in the Earth–Moon system (Newsom and Taylor, 1989). The Moon is derived from the mantle of the impactor, not that of the Earth, thus accounting for the geochemical differences between the two bodies. The high energy of the impact results not only in complete disruption of the impacting planet, but in widespread melting of the Earth producing a magma ocean. Computer simulations of the impact event indicate that the material from which the Moon accretes comes primarily from the mantle of the impactor (Figure 7.24). Because iron had already segregated into the Earth's core, a low iron (and other siderophile element) content of the Moon is predicted by the model. An additional attribute of the model is that the material which escapes from the system after impact is mostly liquid or vapour phases, thus leaving dust enriched in refractory elements from which the Moon forms and so accounting for the

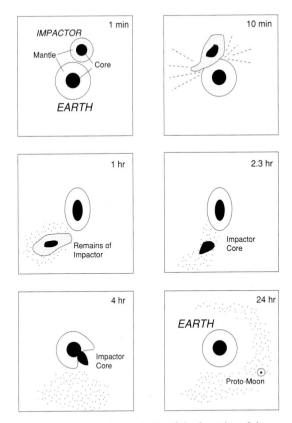

Figure 7.24 Computer simulation of the formation of the Moon by a giant impactor. Model is for an oblique collision with the Earth of impactor of 0.14 Earth masses at a velocity of 5 km/sec. Time elapsed since impact is given in each box. After Newsom and Taylor (1989).

extreme depletion in volatile elements in the Moon. Whether the remaining material immediately accretes into the Moon or forms several small lumps that coalesce to form the Moon on timescales of the order of 100 years is unknown. In either case, the Moon heats up rapidly forming a widespread magma ocean.

Although the giant impactor model needs to be more fully evaluated, it appears to be capable of accommodating more of the constraints related to lunar origin than any of the competing models (Table 7.6).

The Earth's rotational history

Integration of equations of motion of the Moon indicate that there has been a minimum in the Earth–Moon distance in the geologic past, that the inclination of the lunar orbit has decreased with time, and that the eccentricity of the lunar orbit has increased with time as the Earth–Moon distance has increased (Lambeck, 1980). It has long been known that angular momentum is being transferred from the Earth's spin to lunar orbital motion,

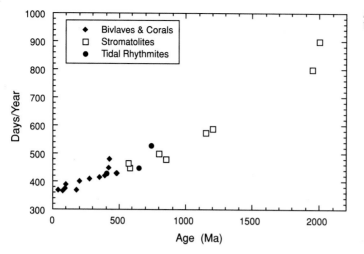

which results in the Moon moving away from the Earth. The current rate of this retreat is about 5 cm/y. The corresponding rotational rate of the Earth has been decreasing at a rate of about 5×10^{-22} rad/sec. Tidal torques cause the inclination of the Moon's orbit to vary slowly with time as a function of the Earth–Moon distance. When this distance is greater than ~ 10 Earth radii, the lunar orbital plane moves towards the ecliptic, and when it is < 10 Earth radii it moves towards the Earth's equatorial plane. For an initially eccentric orbit, the transfer of angular momentum is greater at perigee than apogee, and hence the degree of eccentricity increases with time. These evolutionary changes in angular momentum of the Earth–Moon system result in changes in the length of the terrestrial day and month.

Many groups of organisms secrete sequential layers that are known to be related to cyclical astronomical phenomena. Such organisms, which are known as **biological rhythmites** (Pannella, 1972), provide a means of independently evaluating the calculated orbital retreat of the Moon. The most important groups are corals and bivalves. Daily increments in these organisms are controlled by successive alternation of daylight and darkness. Seasonal increments reflect changes in the length of sunlight per day, seasonal changes in food supply, and in some instances, tidal changes. Actual growth patterns are complex, and local environmental factors may cause difficulties in identifying periodic growth patterns. Results from Phanerozoic bivalves and corals, however, are consistent and suggest a decreasing rotational rate of the Earth of about 5.3×10^{-22} rad/sec (Figure 7.25), which is in good agreement with astronomical values. This corresponds to 420–430 days/y at 500–400 Ma. Stromatolites also deposit regular bands, although a quantitative relationship of banding to astronomical rhythm has not yet been well established (Cao, 1991). The interpretation of stromatolite bands from the Biwabik Iron Formation in Minnesota is that at 2000 Ma there were 800–900 days/y (Figure 7.25) (Mohr, 1975). Another type of geochronometer has been recognized in

laminated fine-grained sediments, known as **sedimentological rhythmites**. Cyclically laminated Late Proterozoic (750–650 Ma) tidal sediments from Australia have been interpreted to record paleotidal periods as well as paleorotation of the Earth (Williams, 1989). Results suggest that 650 Ma, there were 13 months/y and 450 days/y. When both biological and sedimentary rhythmite data are plotted with time, results suggest a sublinear relationship with the number of days per year decreasing at about 0.2 d/My, in good agreement with the extrapolation of astronomical calculations.

If the Moon came within the Roche limit of the Earth (~ 2.9 Earth radii) after 4.5 Ga, a record of such a close encounter should be preserved. Even if the Moon survived this encounter without disintegrating, the energy dissipated in the two bodies would largely melt both bodies and completely disrupt and recycle any crust. The preservation of 4.46-Ga crust on the Moon strongly indicates that the Moon has not been within the Earth's Roche limit since it formed about 4.5 Ga, and thus the relationship in Figure 7.25 cannot be extrapolated much before 2 Ga.

Comparative planetary evolution

From this survey of the Solar System, it seems clear that no two bodies in the Solar System have identical histories. Although the terrestrial planets have many features in common, as do the outer giant planets, it would appear that each planet has its own unique history. At least from our perspective, Earth seems to be the most peculiar. Not only is it the only planet with oceans, an oxygen-bearing atmosphere, and living organisms, but it is the only planet in which plate-tectonic processes are known to be active. As a final topic in this chapter, let us compare the evolutionary histories of the terrestrial planets, and in doing so identify some of the important variables that control and direct the paths of planetary evolution.

As we have seen, the terrestrial planets and the Moon have similar densities (Table 7.1) and thus, on the whole, similar bulk compositions. Each of them is evolving towards a stage of thermal and tectonic stability and quiescence as they cool. The rate at which a planet approaches this final stage is dependent upon a variety of factors, which directly or indirectly control the loss of heat (Schubert, 1979; Carlson, 1994). First of all, the position of a planet in the Solar System is important because, as we have seen, it reflects the condensation sequence of elements from the cooling Solar Nebula. Also important are the abundances of radiogenic isotopes that contribute to heating planetary bodies. The Moon, for instance, contains considerably smaller amounts of U, Th and K than the Earth, and as such will not produce as much radiogenic heat. Analyses of fine-grained materials from the Viking landing sites suggest that Mars is also depleted in radiogenic isotopes compared with the Earth. Planetary mass is important in that the amount of accretional and gravitational energy is directly dependent upon mass. Planetary size is also important in that greater area/mass ratios result in more rapid heat loss from planetary surfaces. For instance, the Moon, Mars, and Mercury should cool much faster than Venus and Earth because of their higher area/mass ratios (Table 7.1). Also important is the size of the iron core, in that much of initial planetary heat is produced during core formation. Except for Mercury, Earth has the highest core/mantle ratio, followed by slightly lower values for Venus and Mars. As discussed previously, the very high core/mantle ratio for Mercury is probably a result of loss of some of the mercurian mantle by an early giant impact, and thus is not indicative of a large contribution of heat from core formation. The volatile contents and especially the water content of planetary mantles and the rate of volatile release are important in controlling atmosphere development, the amount of melting, fractional crystallization trends, and the viscosity of planetary interiors which, in turn, affects the rate of cooling. Convection and/or mantle plume activity appear to be the primary mechanisms by which heat is lost from the terrestrial planets. Only the Earth, however, requires mantle convection and supports plate tectonics.

Planetary crusts are of three types. Primary crust forms during or immediately after planetary accretion by cooling at the surface. Secondary crusts arise later from partial melting of recycled primary crust, or from partial melting of planetary interiors. By analogy with the preserved primary crust on the Moon and probably on Mercury, primary crusts on the terrestrial planets may have been anorthositic in composition, produced by flotation of plagioclase during rapid crystallization of magma oceans. Secondary crusts, however, are typically basaltic in composition, and form only after crystallization of magma oceans. They are produced by partial melting of ultramafic rocks in planetary mantles and in mantle plumes. Examples include the lunar maria basalts, the Earth's oceanic crust, and perhaps most of the crust preserved on Mars and Venus. Tertiary crust is formed by partial melting and further processing of secondary crust, and the Earth's continents may be the only example of tertiary crust in the Solar System.

Although every planet has its own unique history, the primary differences in planetary thermal history are controlled chiefly by heat productivities, volatile-element contents, and cooling rates. Distance from the Sun is also an important variable, especially in terms of planetary composition. A qualitative portrayal of planetary thermal histories is illustrated in Figure 7.26. The temperature scale is schematic. All terrestrial planets underwent rapid heating during late stages of accretion, reaching maximum temperatures at about 4.5 Ga, at

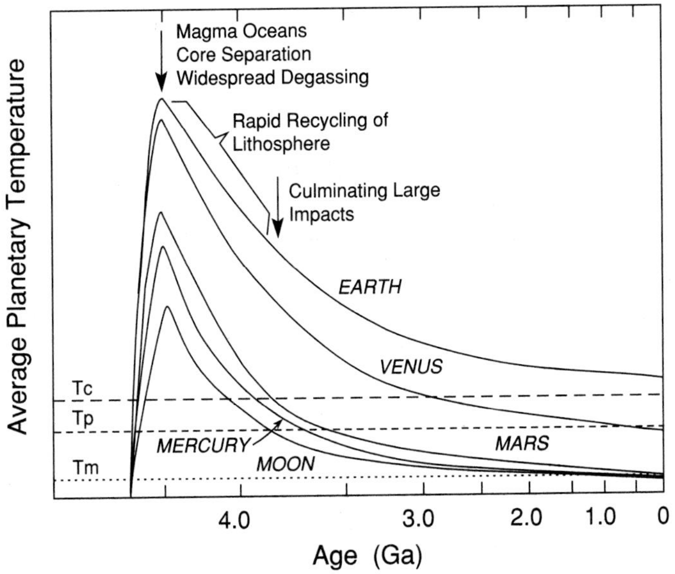

Figure 7.26 Schematic thermal evolution of the terrestrial planets and the Moon. Lower threshold temperatures for planetary processes: Tc, convection; Tp, mantle plume generation at the core–mantle boundary; Tm, magmatism.

which time widespread magma oceans were produced. Also, at this time molten iron descended to planetary centres forming metal cores, and planetary mantles rapidly degassed. Rapid, chaotic convection in the magma oceans resulted in rapid cooling and crystallization producing a transient primary crust, probably of anorthositic composition. Partial melting of recycled crust and/or mantle ultramafic rocks gave rise to widespread secondary basaltic crusts on all of the terrestrial planets, a process that is still occurring on the Earth as ocean ridge basalts and submarine plateaux are produced. At least on Earth and Venus, where substantial heat was available, early crust and lithosphere were rapidly recycled into the mantle, aided by intense giant impacts, which continued to about 3.85 Ga (Figure 7.26). Although continental crust is not preserved on Earth until 4 Ga, it may have been produced and rapidly recycled into the mantle before this time (Chapter 5). Rapid cooling of the smaller planets, including Mercury and probably Mars, led to rapidly-thickened and strong lithospheres by 4.45 Ga. These planets, as well as the Moon, are known as **one-plate planets**, because their thick, strong planetwide lithosphere is basically all one plate. Most of the magmatic activity on these planets resulted from mantle plume activity and occurred before 3.9 Ga. The Moon formed by accretion of material blasted from the Earth by a giant impact at 4.5 Ga, and it rapidly heated and melted forming a magma ocean, which crystallized by 4.45 Ga producing the anorthositic lunar highlands crust. The youngest volcanism on the Moon appears to have been about 2.5 Ga, perhaps 1.5 Ga on Mercury, and possibly as recent as 200 Ma on Mars.

Because of their greater initial heat, Venus and Earth cooled more slowly than the other terrestrial planets. Venus may have passed through the minimum temperature for convection some 3 Ga, while Earth has not yet reached that point in its cooling history (Figure 7.26). If the resurfacing of Venus at 500–300 Ma was caused by a catastrophic mantle plume event, it may have been the last such event, as suggested by the intersection of the Venus cooling curve with Tp, the lower temperature limit for plume production at the core–mantle interface. Volcanism may still be active on Venus, however. The most intriguing question, i.e., why Venus and Earth followed such different cooling paths while both planets are similar in mass and density, remains unresolved. One possibility is that while Earth cooled chiefly by convection and plate tectonics, Venus cooled by mantle plumes and conduction through the lithosphere. If this is the case, we are still left with the question of why different cooling mechanisms dominated in each planet. Some investigators suggest that it has to do with the absence of water on Venus. They argue that in a dry planet the lithosphere is thick and strong as it is on Venus, and there is no melting to produce an LVZ, and thus plates cannot move about or subduct (see Figure 7.6). Taking this one step further, we might ask why Venus is dry, yet the Earth is wet? As suggested previously, Venus may have rapidly lost its water as hydrogen escaped

from the atmosphere soon after or during planetary accretion, because the surface of the planet was too hot for oceans to survive. And why was the surface too hot? Perhaps because Venus is closer to the Sun than the Earth. If this line of reasoning is correct, it may be that the position where a planet accretes in a gaseous nebula is one of the most important variables controlling its evolution. The reason Mars did not sustain plate tectonics or accumulate an ocean may be due to its small mass and rapid cooling resulting in a lithosphere too thick and strong to subduct. If there was enough water to form martian oceans, it either escaped by hydrogen loss or was not effectively degassed from the planet because of the rapid cooling rate.

If the above scenario bears any resemblance to what really happened, it would appear that two important features led to a unique history for the Earth: (1) its position in the Solar System, and (2) its relatively large mass. Without both of these, the Earth may have evolved into quite a different planet from the one we live on.

Summary statements

1 Infrared reflectance indicates that the crust of Mercury is probably composed largely of anorthositic rocks, with little if any basalt. An anomalously high core/mantle ratio required by the high mean density of the planet may reflect removal of part of the mantle during collision with another planet.

2 The martian surface includes large shield volcanoes, fracture zones, and rifts as well as large canyons that appear to have been cut by running water. Chemical analyses suggest the dominant volcanic rocks on Mars are Fe-rich basalts and that weathering of the basalts occurs in a hydrous, oxidizing environment. Most volcanism, deformation, and erosion on Mars occurred at or before 4 Ga.

3 Most differences between Earth and Venus can be related to small differences in planetary mass leading to different cooling, degassing, and tectonic histories, or to differences in distance from the Sun, resulting in different atmospheric histories. Strong positive correlation between gravity and topography on Venus implies a strong coupling of the mantle and lithosphere, and the absence of plate tectonics. Crater distributions on Venus indicate extensive resurfacing with volcanics 500–300 Ma, perhaps in response to a massive mantle plume event. In contrast to the Earth, Venus may cool by mantle plumes and lithosphere delamination, and mantle downwellings appear to be the dominant force responsible for deformation on the surface.

4 Satellites fall into three categories: (1) regular satellites, which include most of the larger satellites and many of the smaller satellites, are those that revolve in or near the plane of the planetary equator and revolve in the same direction as the parent

planet moves about the Sun; (2) irregular satellites have highly inclined, often retrograde and eccentric orbits, and appear to have been captured by the parent planet; and (3) collisional shards are small, often irregular-shaped satellites that are continually eroded by ongoing collisions with smaller bodies.

5   The Moon has a low density compared with the terrestrial planets, implying a relatively low iron content. Also, the angular momentum of the Earth–Moon system is anomalously high compared with other planet–satellite systems. The lunar crust is composed chiefly of anorthosites and gabbroic anorthosites and formed at about 4.45 Ga from crystallization of a magma ocean. Large mare basins on the Moon are flooded with basalts that erupted between 3.9–2.5 Ga.

6   Within the asteroid belt, there is a zonal arrangement of asteroids that reflects chemical composition. Fractionated asteroids dominate in the inner part of the belt, and primitive, unfractionated types occur only in the outer portions of the belt. Because of the tidal forces of Jupiter, it is unlikely that a single planet ever existed in the asteroid belt.

7   Most meteorites come from the asteroid belt, but some are fragments ejected off the Moon or Mars. Chondrules (small silicate spheroids in some meteorites) appear to have formed by melting of pre-existing solids in the Solar Nebula. Decay products of extinct radionuclides in meteorites indicate a time interval between nucleosynthesis and planetary accretion of < 100 My.

8   Both the Earth and Moon are composed chiefly of only four elements: iron, silicon, magnesium, and oxygen. Compared with the primitive Solar Nebula (as represented by C1 meteorites), both bodies are depleted in volatile and siderophile elements and enriched in oxyphile refractory elements. Both bodies are also chemically zoned: large-ion lithophile elements are enriched in the upper mantle and even more so in the crust in relation to the mantle and core. In contrast, siderophile refractory elements are concentrated chiefly in the mantle or core, and oxyphile refractory elements in the mantle.

9   Accretion of the Solar System appears to have begun about 4566 Ma and most meteorite parent bodies formed between 4560–4550 Ma. The Earth accreted over a time interval of about 110 My (4560–4450 Ma), and the Moon formed at about 4500 Ma.

10   The Solar System probably condensed and accreted from a hot gaseous nebula as it cooled. Accretion occurred in two stages: first, small planetesimals form and spiral towards the ecliptic plane where they collide with one another, growing into planetary embryos; second, these embryos collide and grow into the terrestrial planets and the cores of the giant outer planets.

11   During accretion, Earth was probably struck at high velocity many times by bodies ranging up to Mercury in size and at least once by a Mars-sized body. Much of the kinetic energy associated with these collisions melted the interior of the planet, probably producing a widespread magma ocean.

12   Compositional zonation in the Solar Nebula is caused by decreasing temperature outwards from the Sun and this is reflected in planetary compositions. Refractory oxides, metals, and Mg-silicates are enriched in the inner part of the cloud where Mercury accreted; Mg-Fe silicates and metal in the region from Venus to the asteroids; and mixed silicates and ices in the outer part of the nebula where the giant planets accreted. Depletions in volatile elements in the inner Solar Nebula may have been caused by an intense solar wind, and depletions in siderophile elements in the silicate fraction of terrestrial planets by core formation.

13   Differences in the atmospheres of Earth, Venus and Mars reflect distance from the Sun, which controlled initial surface temperatures, planetary mass, and the role of greenhouse gases. Light gases (N_2, H_2) were lost from Mars and rapid planetary cooling resulted in trapping most of its CO_2 in the interior. Venus may have had early oceans that evaporated. H_2O moved into the upper venusian atmosphere where photolysis produced H_2, which escaped, and O_2 which may have oxidized surface rocks, leaving a dense CO_2 atmosphere.

14   Models for lunar origin must account for, (1) the large amount of angular momentum in the Earth–Moon system; (2) the fact that the lunar orbit is not coincident with the Earth's equatorial plane; (3) the similarity in oxygen isotopic composition of the Earth and Moon; and (4) the depletion in volatile and siderophile refractory elements and enrichment in oxyphile refractory elements in the Moon relative to the Earth. Of the models for lunar origin, only the giant impactor model appears capable of accommodating all of these constraints.

15   Astronomical and paleontological data indicate that the Moon has been retreating from the Earth since 2.0 Ga. The preservation of 4.46-Ga crust on the Moon, however, indicates that the Moon has not been within the Earth's Roche limit since it formed at 4.5 Ga.

16   Primary crusts form on magma oceans and are probably anorthositic in composition. Secondary crusts, which are most abundant on the terrestrial planets, are generally basaltic and arise from partial melting of recycled primary crust or from partial melting of planetary interiors. Tertiary crust is formed by partial melting and further processing of secondary crust, and the Earth's continents may be the only example of tertiary crust in the Solar System.

17   Differences in evolution of the terrestrial planets is controlled by, (1) cooling rate, which varies with planetary mass, area/mass ratio, and core/mantle ratio; (2) planetary composition and especially volatile content; and (3) by distance from the Sun. Of

these, planetary mass and distance from the Sun probably account for the unique evolutionary history of Earth compared with Venus and Mars.

Suggestions for further reading

Allegre, C. J., Manhes, G. and Gopel, C. (1995). The age of the Earth. *Geochim. Cosmochim. Acta*, **59**, 1445–1456.

Atreya, S. K., Pollack, J. B. and Matthews, M. S., editors (1989). *Origin and Evolution of Planetary and Satellite Atmospheres*. Tucson, AZ, Univ. Arizona Press, 881 pp.

Cattermole, P. (1995). *Earth and Other Planets*. Oxford, Oxford University Press, 160 pp.

Hartmann, W. K. (1983). *Moons and Planets*. Belmont, CA, Wadsworth Publ. Co., 509 pp.

Kaula, W. M. (1995). Venus reconsidered. *Science*, **270**, 1460–1464.

Lewis, J. S. (1995). *Physics and Chemistry of the Solar System*. San Diego, CA, Academic Press, 556 pp.

Rosenberg, G. D. and Runcorn, S. K., editors (1975). *Growth Rhythms and the History of the Earth's Rotation*. New York, J. Wiley, 559 pp.

Taylor, S. R. (1992). *Solar System Evolution: A New Perspective*. Cambridge, Cambridge Univ. Press, 307 pp.

Wasson, J. T. (1985). *Meteorites*. New York, W. H. Freeman, 267 pp.

References

Abbott, D. H. and Hoffman, S. E. (1984). Archean plate tectonics revisited 1. *Tectonics*, **3**, 429–448.

Abbott, D. H. and Mooney, W. (1995). The structural and geochemical evolution of the continental crust: Support for the oceanic plateau model of continental growth. *Revs. Geophys. Space Phys., Suppplement*, 231–242.

Abelson, P. H. (1966). Chemical events on the primitive Earth. *Proc. Nat. Acad. Sci.*, **55**, 1365–1372.

Ahall, K-I., Persson, P-O. and Skiold, T. (1995). Westward accretion of the Baltic Shield: implications from the 1.6 Ga Amal–Horred belt, SW Sweden. *Precamb. Res.*, **70**, 235–251.

Alabaster, T., Pearce, J. A. and Malpas, J. (1982). The volcanic stratigraphy and petrogenesis of the Oman ophiolitic complex. *Contrib. Mineral. Petrol.*, **81**, 168–183.

Algeo, T. J. and Seslavinsky, K. B. (1995). The Paleozoic world: Continental flooding, hypsometry, and sea level. *Amer. J. Sci.*, **295**, 787–822.

Allegre, C. J. (1982). Chemical geodynamics. *Tectonics*, **81**, 109–132.

Allegre, C. J. and Minster, J. F. (1978). Quantitative models of trace element behavior in magmatic processes. *Earth Planet. Sci. Lett.*, **38**, 1–25.

Allegre, C. J., Brevart, O., Dupre, B. and Minster, J. F. (1980). Isotopic and chemical effects produced in a continuously differentiating convecting earth mantle. *Phil. Trans. Roy. Soc. London*, **297A**, 447–477.

Allegre, C. J., Hamelin, B., Provost, A. and Dupre, B. (1987). Geology in isotopic multispace and origin of mantle chemical heterogeneities. *Earth Planet. Sci. Lett.*, **81**, 319–337.

Allegre, C. J., Moreira, M. and Staudacher, T. (1995a). $^4He/^3He$ dispersion and mantle convection. *Geophys. Res. Lett.*, **22**, 2325–2328.

Allegre, C. J., Poirier, J-P., Humler, E. and Hofmann, A. W. (1995b). The chemical composition of the Earth. *Earth Planet. Sci. Lett.*, **134**, 515–526.

Allen, P. A. and Homewood, P., editors (1986). *Foreland Basins*. Oxford, Blackwell Scient., 453 pp.

Alvarez, W. L. (1986). Toward a theory of impact crises. *EOS*, Sept. 1986, 649–655.

Alvarez, W., Asaro, F., and Montanari, A. (1990). Iridium profile for 10 million years across the Cretaceous–Tertiary boundary at Gubbio (Italy). *Science*, **250**, 1700–1702.

Anderson, D. L. (1989). *Theory of the Earth*. Brookline Village, Maine, Blackwell, 366 pp.

Anderson, D. L., Sammis, C. and Jordan, T. (1971). Composition and evolution of the mantle and core. *Science*, **171**, 1103–1112.

Anderson, J. L. (1983). Proterozoic anorogenic plutonism of North America. *Geol. Soc. America Bull.*, **161**, 133–154.

Anderson, J. L. and Morrison, J. (1992). The role of anorogenic granites in the Proterozoic crustal development of North America. In K. C. Condie (editor), *Proterozoic Crustal Evolution*. Amsterdam, Elsevier, pp. 263–300.

Arculus, R. J. and Johnson, R. W. (1978). Criticism of generalized models for the magmatic evolution of arc–trench systems. *Earth Planet. Sci. Lett.*, **39**, 118–126.

Argus, D. F. and Heflin, M. B. (1995). Plate motion and crustal deformation estimated with geodetic data from the GPS. *Geophys. Res. Lett.*, **22**, 1973–1976.

Armstrong, R. L. (1981). Radiogenic isotopes: The case for crustal recycling on a near-steady-state no-continental-growth Earth. *Phil. Trans. Roy. Soc. London*, **A301**, 443–472.

Armstrong, R. L. (1991). The persistent myth of crust growth. *Austral. J. Earth Sci.*, **38**, 613–630.

Arndt, N. T. (1983). Role of a thin, komatiite-rich oceanic crust in the Archean plate-tectonic process. *Geology*, **11**, 372–375.

Arndt, N. T. (1991). High Ni in Archean tholeiites. *Tectonophys.*, **187**, 411–420.

Arndt, N. T. (1994). Archean komatiites. In K. C. Condie (editor), *Archean Crustal Evolution*. Amsterdam, Elsevier, pp. 11–44.

Arndt, N. T. and Goldstein, S. L. (1987). Use and abuse of crust formation ages. *Geology*, **15**, 893–895.

Ashwal, L. D. and Myers, J. S. (1994). Archean anorthosites. In K. C. Condie (editor), *Archean Crustal Evolution*. Amsterdam, Elsevier, pp. 315–355.

Ave'Lallemant, H. G. and Carter, N. L. (1970). Syntectonic recrystallization of olivine and modes of flow in the upper mantle. *Geol. Soc. America Bull.*, **81**, 2203–2220.

Ayers, L. D. and Thurston, P. C. (1985). Archean supracrustal sequences in the Canadian Shield: an overview. *Geol. Assoc. Canada, Spec. Paper* **28**, 343–380.

BABEL Working Group (1990). Evidence for Early Proterozoic plate tectonics from seismic reflection profiles in the Baltic shield. *Nature*, **348**, 34–38.

Baragar, W. R. A., Ernst, R. E., Hulbert, L. and Peterson, T. (1996). Longitudinal petrochemical variation in the Mackenzie dyke swarm, NW Canadian shield. *J. Petrol.*, **37**, 317–359.

Barley, M. E. (1993). Volcanic, sedimentary and tectono-stratigraphic environments of the 3.46 Ga Warrawoona Megasequence: A review. *Precambrian Res.*, **60**, 47–67.

Barton, J. M., Jr. et al. (1994). Discrete metamorphic events in the Limpopo belt, southern Africa: Implications for the application of P–T paths in complex metamorphic terrains. *Geology*, **22**, 1035–1038.

Bednarz, U. and Schmincke, H. (1994). Petrological and chemical evolution of the NE Troodos extrusive series, Cyprus. *J. Petrol.*, **35**, 489–523.

Bengtson, S. (1994). The advent of animal skeletons. In S. Bengtson (editor), *Early Life on Earth*. New York, Columbia Univ. Press, pp. 412–425.

Benton, M. J. (1995). Diversification and extinction in the history of life. *Science*, **268**, 52–58.

Benz, W. and Cameron, A. G. W. (1990). Terrestrial effects of the giant impact. In H. E. Newsom and J. H. Jones (editors), *Origin of the Earth*. New York, Oxford Univ. Press, pp. 61–67.

Berger, W. H., et al. (1992). The record of Ontong Java Plateau: main results of ODP Leg 130. *Geol. Soc. America Bull.*, **104**, 954–972.

Berner, R. A. (1987). Models for carbon and sulfur cycles and atmospheric oxygen: application to Paleozoic history. *Amer. J. Sci.*, **287**, 177–196.

Berner, R. A. (1994). 3Geocarb II: A revised model of atmospheric CO_2 over Phanerozoic time. *Amer. J. Sci.*, **294**, 56–91.

Berner, R. A. and Canfield, D. E. (1989). A new model for atmospheric oxygen over Phanerozoic time. *Amer. J. Sci.*, **289**, 333–361.

Bickle, M. J., Nisbet, E. G. and Martin, A. (1994). Archean greenstone belts are not oceanic crust. *J. Geol.*, **102**, 121–137.

Binden, A. B. (1986). The binary fission origin of the moon. In W. K. Hartmann, R. J. Phillips and G. J. Taylor (editors), *Origin of the Moon*. Houston, TX, Lunar & Planet. Institute, pp. 519–550.

Bindschadler, D. L., Schubert, G. and Kaula, W. M. (1992). Coldspots and hotspots: Global tectonics and mantle dynamics of Venus. *J. Geophys. Res.*, **97**, 13 495–532.

Blackwell, D. D. (1971). The thermal structure of the continental crust. *Amer. Geophys. Union Mon. No. 14*, 169–191.

Bloxham, J. (1992). The steady part of the secular variation of Earth's magnetic field. *J. Geophys. Res.*, **97**, 19 565–579.

Bodri, L. and Bodri, B. (1985). On the correlation between heat flow and crustal thickness. *Tectonophys.*, **120**, 69–81.

Bogue, S. W. and Merrill, R. T. (1992). The character of the field during geomagnetic reversals. *Ann. Rev. Earth Planet. Sci.*, **20**, 181–219.

Boher, M., Abouchami, W., Michard, A., Albarede, F. and Arndt, N. T. (1992). Crustal growth in West Africa a 2.1 Ga, *J. Geophys. Res.*, **97**, 345–369.

Bohlen, S. R. (1991). On the formation of granulites. *J. Met. Geol.*, **9**, 223–229.

Bohor, B. F., Modreski, P. J. and Foord, E. E. (1987). Shocked quartz in the K/T boundary clays: Evidence for a global distribution. *Science*, **236**, 705–709.

Bonati, E. (1967). Mechanisms of deepsea volcanism in the South Pacific. In P. H. Abelson (editor), *Researches in Geochemistry*. New York, J. Wiley, pp. 453–491.

Boryta, M. and Condie, K. C. (1990). Geochemistry and origin of the Archean Beit Bridge complex, Limpopo belt, South Africa. *J. Geol. Soc. London*, **147**, 229–239.

Boss, A. P., Morfill, G. E. and Tscharnuter, W. M. (1989). Models of the formation and evolution of the solar nebula. In S. K. Atreya, J. B. Pollack and M. S. Matthews (editors), *Origin and Evolution of Planetary and Satellite Atmospheres*. Tucson, Univ. Ariz. Press, pp. 35–77.

Bott, M. H. P. (1979). Subsidence mechanisms at passive continental margins. *Amer. Assoc. Petrol. Geol., Mem.*, **29**, 3–9.

Bottinga, Y. and Allegre, C. J. (1973). Thermal aspects of sea-floor spreading and the nature of the oceanic crust. *Tectonophys.*, **18**, 1–17.

Bowring, S. A. (1990). The Acasta gneisses: Remnant of Earth's early crust. In H. E. Newsom and J. H. Jones (editors), *Origin of the Earth*. Oxford, Oxford Univ. Press, pp. 319–343.

Bowring, S. A. and Housh, T. (1995). The Earth's early evolution. *Science*, **269**, 1535–1540.

Bowring, S. A., Williams, I. S. and Compston, W. (1989). 3.96 Ga gneisses from the Slave province, NW territories, Canada. *Geology*, **17**, 971–975.

Bowring, S. A. et al. (1993). Calibrating rates of Early Cambrian Evolution. *Science*, **261**, 1293–1298.

Boyd, F. R. (1973). A pyroxene geotherm. *Geochim. Cosmochim. Acta*, **37**, 2533–2546.

Boyd, F. R. (1989). Compositional distinction between oceanic and cratonic lithosphere. *Earth Planet. Sci. Lett.*, **96**, 15–26.

Bradshaw, T. K., Hawkesworth, C. J. and Gallagher, K. (1993). Basaltic volcanism in the southern Basin and

Range: No role for a mantle plume. *Earth Planet. Sci. Lett.*, **116**, 45–62.

Brady, P. V. (1991). The effect of silicate weathering on global temperature and atmorspheric CO_2. *J. Geophys. Res.*, **96**, 18 101–106.

Brasier, M. D. (1992). Global ocean–atmosphere change across the Precambrian–Cambrian transition. *Geol. Mag.*, **129**, 161–168.

Breuer, D. and Spohn, T. (1995). Possible flush instability in mantle convection at the Archean–Proterozoic transition. *Nature*, **378**, 608–610.

Brooks, C., Hart, S. R., Hofman, A. and Jarnes, D. E. (1976). Rb–Sr mantle isochrons from oceanic areas. *Earth Planet. Sci. Lett.*, **32**, 51–61.

Brown, M. (1993). P–T–t evolution of orogenic belts and the causes of regional metamorphism. *J. Geol. Soc. London*, **150**, 227–241.

Brown, M., Rushmer, T. and Sawyer, E. W. (1995). Introduction to special section: Mechanisms and consequences of melt segregation from crustal protoliths. *J. Geophys. Res.*, **100**, 15 551–563.

Bryan, P. and Gordon, R. G. (1986). Errors in minimum plate velocity determined from paleomagnetic data. *J. Geophys. Res.*, **91**, 462–470.

Buffett, B. A. et al. (1996). On the thermal evolution of the Earth's core. *J. Geophys. Res.*, **101**, 7989–8006.

Buick, R. and Dunlop, J. S. R. (1990). Evaporitic sediments of Early Archean age from the Warrawoona Group, Western Australia. *Sediment.*, **37**, 247–278.

Buick, R., Groves, D. I. and Dunlop, J. S. R. (1995). Geological origin of described stromatolites older the 3.2 Ga: Comment and reply. *Geology*, **23**, 191–192.

Buick, R. et al. (1995). Record of emergent continental crust 3.5 Ga in the Pilbara craton of Australia. *Nature*, **375**, 574–577.

Burke, K., Dewey, J. F. and Kidd, W. S. F. (1976). Precambrian paleomagnetic results compatible with contemporary operation of the Wilson cycle. *Tectonophys.*, **33**, 287–299.

Cairns-Smith, A. G. (1982). *Genetic Takeover and Mineral Origins of Life*. Cambridge, Cambridge Univ. Press, 477 pp.

Caldeira, K. and Rampino, M. R. (1991). The Mid-Cretaceous superplume, carbon dioxide, and global warming. *Geophys. Res. Lett.*, **18**, 987–990.

Calvert, A. J., Sawyer, E. W., Davis, W. J. and Ludden, J. N. (1995). Archean subduction inferred from seismic images of a mantle suture in the Superior province. *Nature*, **375**, 670–674.

Cameron, A. G. W. (1973). Accumulation processes in the primitive solar nebula. *Icarus*, **18**, 407–450.

Cameron, A. G. W. (1986). The impact theory for origin of the Moon. In W. K. Hartmann, R. J. Phillips and G. J. Taylor (editors), *Origin of the Moon*. Houston, TX, Lunar & Planet. Institute, pp. 609–616.

Camfield, P. A. and Gough, D. I. (1977). A possible Proterozoic plate boundary in North America. *Can. Jour. Earth Sci.*, **14**, 1229–1238.

Campbell, I. H. and Griffiths, R. W. (1992). The changing nature of mantle hotspots through time: Implications for the chemical evolution of the mantle. *J. Geol.*, **92**, 497–523.

Campbell, I. H. and Griffiths, R. W. (1992). The evolution of the mantle's chemical structure. *Lithos*, **30**, 389–399.

Campbell, I. H., Griffiths, R. W. and Hill, R. I. (1989). Melting in an Archean mantle plume: Heads it's basalts, tails it's komatiites. *Nature*, **339**, 697–699.

Cande, S. C., Leslie, R. B., Parra, J. C. and Hobart, M. (1987). Interaction between the Chile ridge and the Chile trench: Geophysical and geothermal evidence. *J. Geophys. Res.*, **92**, 495–520.

Cao, R. (1991). Origin and order of cyclic growth patterns in mat-ministromatolite bioherms from the Proterozoic Wumishan formation, North China. *Precamb. Res.*, **52**, 167–178.

Caputo, M. V. and Crowell, J. C. (1985). Migration of glacial centers across Gondwana during Paleozoic era. *Geol. Soc. America Bull.*, **96**, 1020–1036.

Card, K. D. and Ciesielski, A. (1986). Subdivisions of the Superior Province of the Canadian Shield. *Geosci. Canada*, **13**, 5–13.

Carey, S. and Sigurdsson, H. A. (1984). A model of volcanigenic sedimentation in marginal basins. *Geol. Soc. London Spec. Publ.*, **16**, 37–58.

Carlson, R. L. (1995). A plate cooling model relating rates of plate motion to the age of the lithosphere at trenches. *Geophys. Res. Lett.*, **22**, 1977–1980.

Carlson, R. W. (1991). Physical and chemical evidence on the cause and source characteristics of flood basalt volcanism. *Austral. Jour. Earth Sci.*, **38**, 525–544.

Carlson, R. W. (1994). Mechanisms of Earth differentiation: Consequences for the chemical structure of the mantle. *Revs. Geophys. Space Phys.*, **32**, 337–361.

Carlson, R. W. (1995). Physical and chemical evidence on the cause and source characteristics of flood basalt volcanism. *Austral. J. Earth Sci.*, **38**, 525–544.

Carlson, R. W., Shirey, S. B., Pearson, D. G. and Boyd, F. R. (1994). The mantle beneath continents. Carnegie Inst. Washington, *Year Book 93*, 109–117.

Carr, M. H. (1987). Water on Mars. *Nature*, **326**, 30–35.

Carr, M. H., Kuzmin, R. O. and Masson, P. L. (1993). Geology of Mars. *Episodes*, **16**, 307–315.

Cawood, P. A. and Suhr, G. (1992). Generation and obduction of ophiolites: Constraints from the Bay of Islands Complex, Western Newfoundland. *Tectonics*, **11**, 884–897.

Chandler, F. W. (1988). Quartz arenites: Review and interpretation. *Sed. Geol.*, **58**, 105–126.

Chang, S. (1994). The planetary setting of prebiotic evolution. In S. Bengtson (editor), *Early Life on Earth*. New York, Columbia Univ. Press, pp. 10–23.

Chapman, D. S. (1986). Thermal gradients in the continental crust. In J. B. Dawson, D. A. Carswell, J. Hall and K. H. Wedepohl (editors), *The Nature of the Lower Continental Crust*. London, *Geol. Soc. London Spec. Publ. No. 24*, 63–70.

Chapman, D. S. and Furlong, K. P. (1992). Thermal state of the continental lower crust. In D. M. Fountain, R. Arculus and R. W. Kay (editors), *Continental Lower Crust*. Amsterdam, Elsevier, pp. 179–199.

Christensen, N. I. (1966). Elasticity of ultrabasic rocks. *J. Geophys. Res.*, **71**, 5921–5931.

Christensen, N. I. and Mooney, W. D. (1995). Seismic velocity structure and composition of the continental crust: A global view. *J. Geophys. Res.*, **100**, 9761–9788.

Christensen, U. R. (1995). Effects of phase transitions on mantle convection. *Ann. Rev. Earth Planet. Sci.*, **23**, 65–87.

Christensen, U. R. and Yuen, D. A. (1985). Layered convection induced by phase transitions. *J. Geophys. Res.*, **90**, 10 291–300.

Chyba, C. F. (1993). The violent environment of the origin of life: Progress and uncertainties. *Geochim. Cosmochim. Acta*, **57**, 3351–3358.

Cloos, M. (1993). Lithospheric buoyancy and collisional orogenesis: subduction of oceanic plateaus, continental margins, island arcs, spreading ridges, and seamounts. *Geol. Soc. America Bull.*, **105**, 715–737.

Cloud, P. (1968). Atmospheric and hydrospheric evolution on the primitive Earth. *Science*, **160**, 729–736.

Cloud, P. (1973). Paleoecological significance of the banded iron formation. *Econ. Geol.*, **68**, 1135–1143.

Coffin, M. F. and Eldholm, O. (1994). Large igneous provinces: crustal structure, dimensions, and external consequences. *Revs. Geophys. Space Phys.*, **32**, 1–36.

Cogley, J. G. (1984). Continental margins and the extent and number of continents. *Revs. Geophys. Space Phys.*, **22**, 101–122.

Cogley, J. G. and Henderson-Sellers, A. (1984). The origin and earliest state of the Earth's hydrosphere. *Revs. Geophys. Space Phys.*, **22**, 131–175.

Coleman, R. G. (1977). *Ophiolites*. Berlin, Springer-Verlag, 229 pp.

Condie, K. C. (1981). *Archean Greenstone Belts*. Amsterdam, Elsevier, 434 pp.

Condie, K. C. (1986). Origin and early growth rate of continents. *Precamb. Res.*, **32**, 261–278.

Condie, K. C. (1989). Geochemical changes in basalts and andesites across the Archean–Proterozoic boundary: Identification and significance. *Lithos*, **23**, 1–18.

Condie, K. C. (1990). Geochemical characteristics of Precambrian basaltic greenstones. In R. P. Hall and D. J. Hughes (editors), *Early Precambrian Basic Magmatism*. Glasgow, Blackie & Son, pp. 40–55.

Condie, K. C. (1990). Growth and accretion of continental crust: Inferences based on Laurentia. *Chem. Geol.*, **83**, 183–194.

Condie, K. C. (1992a). Evolutionary changes at the Archean–Proterozoic boundary. In J. E. Glover and S. E. Ho (editors), *The Archean: Terrains, Processes and Metallogeny. Proceeding Volume of the Third International Archean Symposium 1990*. Perth, Australia: *Univ. Western Australia Publ. No. 22*, pp. 177–189.

Condie, K. C. (1992b). Proterozoic terranes and continental accretion in SW North America. In K. C. Condie (editor), *Proterozoic Crustal Evolution*. Amsterdam, Elsevier, pp. 447–480.

Condie, K. C. (1993). Chemical composition and evolution of the upper continental crust: Contrasting results from surface samples and shales. *Chem. Geol.*, **104**, 1–37.

Condie, K. C. (1994). Greenstones through time. In K. C. Condie (editor), *Archean Crustal Evolution*. Amsterdam, Elsevier, pp. 85–120.

Condie, K. C. (1995). Episodic ages of greenstones: A key to mantle dynamics. *Geophys. Res. Lett.*, **22**, 2215–2218.

Condie, K. C. (1996). Sources of Proterozoic mafic dyke swarms: Constraints from Th/Ta-La/Yb ratios. *Precamb. Res.*

Condie, K. C. and Chomiak, B. (1996). Continental accretion: Contrasting Mesozoic and Early Proterozoic tectonic regimes in the North America. *Tectonophys.*

Condie, K. C. and Rosen, O. M. (1994). Laurentia–Siberia connection revisited. *Geology*, **22**, 168–170.

Condie, K. C. and Wronkiewicz, D. J. (1990). The Cr/Th ratio in Precambrian pelites from the Kaapvaal craton as an index of craton evolution. *Earth Planet. Sci. Lett.*, **97**, 256–267.

Cook, F. A., Brown, L. D., Kaufman, S., Oliver, J. E. and Petersen, J. (1981). COCORP seismic profiling of the Appalachian orogen beneath the coastal plain of Georgia. *Geol. Soc. America Bull.*, **92**, 738–748.

Cook, P. J. (1992). Phosphogenesis around the Proterozoic–Phanerozoic transition. *J. Geol. Soc. London*, **149**, 615–620.

Cook, P. J. and McElhinny, M. W. (1979). A reevaluation of the spatial and temporal distribution of sedimentary phosphate deposits in the light of plate tectonics. *Econ. Geol.*, **74**, 315–330.

Coulton, A. J., Harper, G. D. and O'Hanley, D. S. (1995). Oceanic versus emplacement age serpentinization in the Josephine ophiolite: Implications for the nature of the Moho at intermediate and slow spreading ridges. *J. Geophys. Res.*, **100,** 22 245–260.

Courtillot, V. E. and Cisowski, S. (1987). The K/T boundary events: External or internal causes? *EOS*, **68**, 193.

Cowan, D. S. (1986). The origin of some common types of melange in the western Cordillera of North America. In N. Nasu, K. Kobayashi, S. Nyeda, I. Kushiro and H. Kagami (editors), *Formation of Active Ocean Margins*. Dordrecht, D. Reidel Publ., pp. 257–272.

Coward, M. P., Jan, M. Q., Rex, O., Tarney, J., Thirwall, M. and Windley, B. F. (1982). Geotectonic framework of the Himalaya of N. Pakistan. *J. Geol. Soc. London*, **139**, 299–308.

Coward, M. P., Broughton, R. D., Luff, I. W., Peterson, M. G., Pudsey, C. J., Rex, D. C. and Asifkhan, M. (1986). Collision tectonics in the NW Himalayas. *Geol. Soc. London, Spec. Publ.*, **19**, 203–220.

Cox, A. (1969). Geomagnetic reversals. *Science*, **163**, 237–245.

Cronin, V. S. (1987). Cycloid kinematics of relative plate motion. *Geology*, **15**, 1006–1009.

Cronin, V. S. (1991). The cycloid relative-motion model and the kinematics of transform faulting. *Tectonophys.*, **187**, 215–249.

Crough, S. T. (1983). Hotspot swells. *Ann. Rev. Earth Planet. Sci.*, **11**, 165–193.

Crough, S. T., Morgan, W. J. and Hargraves, R. B. (1980). Kimberlites: Their relation to mantle hotspots. *Earth Planet. Sci. Lett.*, **50**, 260–274.

Crowley, K. D. and Kuhlman, S. L. (1988). Apatite thermchronometry of Western Canadian shield: Implications for origin of the Williston basin. *Geophys. Res. Lett.*, **15**, 221–224.

Crowley, T. J. (1983). The geologic record of climatic change. *Revs. Geophys. Space Phys.*, **21**, 828–877.

Culotta, R. C., Pratt, T. and Oliver, J. (1990). A tale of two sutures: COCORP's deep seismic surveys of the Grenville province in the eastern U. S. midcontinent. *Geology*, **18**, 646–649.

Daggett, P. H., Keller, G. R., Morgan, P. and Wen, C. L. (1986). Structure of the southern Rio Grande rift from gravity interpretation. *J. Geophys. Res.*, **91**, 6157–6167.

Dahlen, F. A. and Suppe, J. (1988). Mechanics, growth and erosion of mountain belts. *Geol. Soc. America Spec. Paper, 218*, 161–178.

Dallmeyer, R. D. and Brown, M. (1992). Rapid Variscan exhumation of Eo-Variscan metamorphic rocks from S. Brittany, France. *Geol. Soc. Am. Progs. with Abstracts*, **24**, A236.

Dalziel, E. W. D., Dalla Salda, L. H. and Gahagan, L. M. (1994). Paleozoic Laurentia–Gondwana interaction and the origin of the Appalachian–Andean mountain system. *Geol. Soc. America Bull.*, **106**, 243–252.

Danielson, A., Moller, P. and Dulski, P. (1992). The Eu anomalies in BIFs and the thermal history of the oceanic crust. *Chem. Geol.*, **97**, 89–100.

Davies, G. F. (1980). Review of oceanic and global heat flow estimates. *Revs. Geophys. Space Phys.*, **18**, 718–722.

Davies, G. F. (1992). On the emergence of plate tectonics. *Geology*, **20**, 963–966.

Davies, G. F. (1993). Conjectures on the thermal and tectonic evolution of the Earth. *Lithos*, **30**, 281–289.

Davies, G. F. (1995). Penetration of plates and plumes trough the mantle transition zone. *Earth Planet. Sci. Lett.*, **133**, 507–516.

Davies, G. F. (1995). Punctuated tectonic evolution of the earth. *Earth Planet. Sci. Lettr.*, **136**, 363–379.

Davies, G. F. and Richards, M. A. (1992). Mantle convection. *J. Geol.*, **100**, 151–206.

DeBari S. M. and Sleep, N. H. (1991). Low-Al bulk composition of the Talkeetna island arc, Alaska: Implications for primary magmas and the nature of the arc crust. *Geol. Soc. America Bull.*, **103**, 37–47.

DeLong, S. E. and Fox, P. J. (1977). Geological consequence of ridge subduction. In M. Talwani and W. C. Pittman III (editors), *Island Arcs, Deep-sea trenches, and Back arc basins.* Washington DC, *Amer. Geophys. Union, Maurice Ewing Series 1*, pp. 221–228.

DeMets, C., Gordon, R. G., Stein, S. and Argus, D. F. (1990). Current plate motions. *Geophys. Res. Lett.*, **14**, 911–914.

Derry, L. A. and Jacobsen, S. R. (1990). The chemical evolution of Precambrian seawater: Evidence from REEs in BIF. *Geochim. Cosmochim. Acta*, **54**, 2965–2977.

Des Marais, D. J. (1994). The Archean atmosphere: Its composition and fate. In K. C. Condie (editor), *Archean Crustal Evolution*, Amsterdam, Elsevier, pp. 505–523.

Des Marais, D. J., Strauss, H., Summons, R. E., and Hayes, J. M. (1992). Carbon isotope evidence for the stepwise oxidation of the Proterozoic environment. *Nature*, **359**, 605–609.

Desrochers, J. P., Hubert C., Ludden, J. N. and Pilote, P. (1993). Accretion of Archean oceanic plateau fragments in the Abitibi greenstone belt, Canada. *Geology*, **21**, 451–454.

Dewey, J. F. (1988). Extensional collapse of orogens. *Tectonics*, **7**, 1123–1139.

Dewey, J. F. and Bird, J. M. (1970). Mountain belts and new global tectonics. *Jour. Geophys. Res.*, **75**, 2625–2647.

Dewey, J. F. and Kidd, W. S. F. (1977). Geometry of plate accretion. *Geol. Soc. America Bull.*, **88**, 960–968.

Dewey, J. F., Hempton, M. R., Kidd, W. S. F., Saroglu, F. and Sengor, A. M. C. (1986). Shortening of continental lithosphere: The neotectonics of eastern Anatolia–a young collision zone. *Geol. Soc. London Spec. Publ., 19*, 3–36.

deWit, M. J., Hart, R. A. and Hart, R. J. (1987). The Jamestown ophiolite complex, Barberton mountain belt: A section through 3.5 Ga oceanic crust. *J. African Earth Sci.*, **6**, 681–730.

de Wit, M. J. et al. (1992). Formation of an Archean continent. *Nature*, **357**, 553–562.

Dickinson, W. R. (1970). Relations of andesites, granites, and derivative sandstones to arc–trench tectonics. *Revs. Geophys. Space Phys.*, **8**, 813–860.

Dickinson, W. R. (1973). Widths of modern arc–trench gaps proportional to past duration of igneous activity in associated magmatic arcs. *J. Geophys. Res.*, **78**, 3376–3389.

Dickinson, W. R. and Seely, D. R. (1986). Structure and stratigraphy of forearc regions. *Amer. Assoc. Petrol. Geol. Bull.*, **63**, 2–31.

Dickinson, W. R. and Suczek, C. A. (1979). Plate tectonics and sandstone compositions. *Amer. Assoc. Petrol. Geol. Bull.*, **63**, 2164–2182.

Dilek, Y. and Eddy, E. A. (1992). The Troodos and Kizaldag ophiolites as structural models for slow-spreading ridge segments. *J. Geol.*, **100**, 305–322.

Duba, A. (1992). Earth's core not so hot. *Nature*, **359**, 197–198.

Duffy, T. S., Zha, C., Downs, R. T., Mao, H. and Hemley, R. J. (1995). Elasticity of forsterite to 16 GPa and the composition of the upper mantle. *Nature*, **378**, 170–173.

Duncan, R. A. (1991). Ocean drilling and the volcanic record of hotspots. *GSA Today*, **1**, No. 10, October, 1991.

Duncan, R. A. and Richards, M. A. (1991). Hotspots, mantle plumes, flood basalts, and true polar wander. *Revs. Geophys. Space Phys.*, **29**, 31–50.

Dunlop, D. J. (1995). Magnetism in rocks. *J. Geophys. Res.*, **100**, 2161–2174.

Durrheim, R. J. and Mooney, W. D. (1991). Archean and Proterozoic crustal evolution: Evidence from crustal seismology. *Geology*, **19**, 606–609.

Durrheim, R. J. and Mooney, W. D. (1994). Evolution of the Precambrian lithosphere: Seismological and geochemical constraints. *J. Geophys. Res.*, **99**, 15 359–374.

DuToit, A. (1937). *Our Wandering Continents*. London, Oliver and Boyd, 366 pp.

Elthon, D. and Scarfe, C. M. (1984). High pressure phase equilibria of a high-MgO basalt and the genesis of primary oceanic basalts. *Amer. Mineral.*, **69**, 1–15.

Emslie, R. F. (1978). Anorthosite massifs, rapakivi granites, and Late Proterozoic rifting of North America. *Precamb. Res.*, **7**, 61–98.

Engel, A. E. J., Itson, S. P., Engel, C. G., Stickney, D. M., and Cray, Jr., D. J. (1974). Crustal evolution and global tectonics: a petrogenetic view. *Geol. Soc. America Bull.*, **85**, 843–858.

England, P. C. and Richardson, S. W. (1977). The influence of erosion upon the mineral facies of rocks from different metamorphic environments. *J. Geol. Soc. London*, **134**, 201–213.

Epp, D. (1984). Possible perturbations to hotspot traces and implications for the origin and structure of the Line Islands. *J. Geophys. Res.*, **89**, 11, 273–286.

Eriksson, K. A. and Fedo, C. M. (1994). Archean synrift and stable-shelf sedimentary successions. In K. C. Condie (editor), *Archean Crustal Evolution*. Amsterdam, Elsevier, pp. 171–203.

Eriksson, P. G. and Cheney, E. S. (1992). Evidence for the transition to an oxygen-rich atmosphere during the evolution of redbeds in the Lower Proterozoic sequences of southern Africa. *Precamb. Res.*, **54**, 257–269.

Ernst, W. G. (1972). Occurrence and mineralogic evolution of bluschist belts with time. *Amer. J. Sci.*, **272**, 657–668.

Ernst, R. E., Head, J. W., Parfitt, E., Grosfils, E. and Wilson, L. (1995). Giant radiating dyke swarms on Earth and Venus. *Earth-Sci. Rev.*, **39**, 1–58.

Erwin, D. H. (1993). *The Great Paleozoic Crisis*. NY, Columbia Univ. Press.

Erwin, D. H. (1994). The Permo–Triassic extinction. *Nature*, **367**, 231–235.

Estey, L. H. and Douglas, B. J. (1986). Upper mantle anisotropy: A preliminary model. *J. Geophys. Res.*, **91**, 11 393–406.

Eyles, N. (1993). Earth's glacial record and its tectonic setting. *Earth-Sci. Rev.*, **35**, 1–248.

Fahrig, W. F. (1987). The tectonic settings of continental mafic dyke swarms: Failed arm and early passive margin. *Geol. Assoc. Canada, Spec. Paper 34*, 331–348.

Fanale, F. P. (1971). A case for catastrophic early degassing of the Earth. *Chem. Geol.*, **8**, 79–105.

Fischer, A. G. (1984). Biological innovations and the sedimentary record. In H. D. Holland and A. F. Trendall (editors), *Patterns of Change in Earth Evolution*. Berlin, Springer-Verlag, pp. 145–157.

Fisher, D. E. (1985). Radiogenic rare gases and the evolutionary history of the depleted mantle. *J. Geophys. Res.*, **90**, 1801–1807.

Fisher, R. V. (1984). Submarine volcaniclastic rocks. *Geol. Soc. London, Spec. Publ., 16*, 5–28.

Folk, R. L. (1976). Reddening of desert sands: Simpson Desert, N.T., Australia. *Jour. Sed. Petrol.*, **46**, 604–615.

Follmi, K. B. (1996). The phosphorus cycle, phosphogenesis and marine phosphate-rich deposits. *Earth-Sci. Rev.*, **40**, 55–124.

Forsyth, D. W. (1996). Partial melting beneath a Mid-Atlantic segment detected by teleseismic PKP delays. *Geophys. Res. Lett.*, **23**, 463–466.

Fountain, D. M. and Salisbury, M. H. (1981). Exposed crustal-sections through the continental crust: Implications for crustal structure, petrology and evolution. *Earth Planet. Sci. Lett.*, **56**, 263–277.

Fountain, D. M., Furlong, K. P. and Salisbury, M. H. (1987). A heat production model of a shield area and its implications for the heat flow–heat production relationship. *Geophys. Res. Lett.*, **14**, 283–286.

Frakes, L. A. (1979). *Climates Throughout Geologic Time*. Amsterdam, Elsevier, 310 pp.

Frakes, L. A., Francis, J. E. and Syktus, J. L. (1992). *Climate Modes of the Phanerozoic*. New York, Cambridge Univ. Press, 274 pp.

Friend, C. R. L., Nutman, A. P. and McGregor, V. R. (1988). Late Archean terrane accretion in the Godthab region, southern West Greenland. *Nature*, **335**, 535–538.

Froude, D. O., Ireland, T. R., Kinny, P. O., Williams, I. S. and Compston, W. (1983). Ion microprobe identification of 4100–4200 Ma-old terrestrial zircons. *Nature*, **304**, 616–618.

Fryer, P. (1996). Evolution of the Mariana convergent plate margin system. *Revs. Geophys. Space Phys.*, **34**, 89–125.

Fuller, M. and Cisowski, S. M. (1987). Review of lunar magnetism. In J. A. Jacobs (editor), *Geomagnetism*. NY, Academic Press.

Fyfe, W. S. (1978). The evolution of the Earth's crust: Modern plate tectonics to ancient hot spot tectonics? *Chem. Geol.*, **23**, 89–114.

Gaffin, S. (1987). Ridge volume dependence on seafloor generation rate and inversion using long term sea level change. *Amer. J. Sci.*, **287**, 596–611.

Gaherty, J. B. and Jordan, T. H. (1995). Lehmann discontinuity as the base of an anisotropic layer beneath continents. *Science*, **268**, 1468–1471.

Galer, S. J. G. (1991). Interrelationships between continental freeboard, tectonics and mantle temperature. *Earth Planet. Sci. Lett.*, **105**, 214–228.

Gans, P. B. (1987). An open-system two layer crustal stretching model for the eastern Great Basin. *Tectonics*, **6**, 1–12.

Gao, S. and Wedepohl, K. H. (1995). The negative Eu anomaly in Archean sedimentary rocks: Implications for decomposition, age and importance of their granitic sources. *Earth Planet. Sci. Lett.*, **133**, 81–94.

Garfunkel, Z. (1986). Review of oceanic transform activity and development. *J. Geol. Soc. London*, **143**, 775–784.

Garfunkel, Z., Anderson, C. A. and Schubert, G. (1986). Mantle circulation and the lateral migration of subducted slabs. *J. Geophys. Res.*, **91**, 7205–7223.

Gastil, G. (1960). The distribution of mineral dates in time and space. *Amer. J. Sci.*, **258**, 1–35.

Gibbs, A. K., Montgomery, C. W., O'Day, P. A. and Erslev, E. A. (1986). The Archean/Proterozoic transition: Evidence from the geochemistry of metasedimetnary rocks of Guyana and Montana. *Geochim. Cosmochim. Acta*, **50**, 2125–2141.

Gilbert, W. (1986). The RNA world. *Nature*, **319**, 618.

Gill, J. B. (1987). Early geochemical evolution of an oceanic island arc and backarc: Fiji and the south Fiji basin. *J. Geol.*, **95**, 589–615.

Glikson, A. Y. (1993). Asteroids and early Precambrian crustal evolution. *Earth-Sci. Rev.*, **35**, 285–319.

Goff, J. A., Holliger, K. and Levander, A. (1994). Modal fields: A new method for characterization of random velocity heterogeneity. *Geophys. Res. Lett.*, **21**, 493–496.

Goldstein, S. J. and Jacobsen, S. B. (1988). Nd and Sr isotopic systematics of river water suspended material: Implications for crustal evolution: *Earth Planet. Sci. Lett.*, **87**, 249–265.

Gordon, R. G. and Stein, S. (1992). Global tectonics and space geodesy. *Science*, **256**, 333–342.

Goto, K., Hamaguchi, H. and Suzuki, Z. (1985). Earthquake generating stresses in a descending slab. *Tectonophys.*, **112**, 111–128.

Graham, J. B., Dudley, R., Aguilar, N. M. and Gans, C. (1995). Implications of the late Paleozoic oxygen pulse for physiology and evolution. *Nature*, **375**, 117–120.

Graham, S. A. et al. (1986). Provenance modeling as a technique for analyzing source terrane evolution and controls on foreland sedimentation. In A. Allen and P. Homewood (editors) *Foreland Basins*. Oxford, Blackwell Scient., pp. 425–436.

Grand, S. P. (1987). Tomographic inversion for shear velocity beneath the North American plate. *J. Geophys. Res.*, **92**, 14 065–90.

Grand, S. P. and Helmbuger, D. V. (1984). Upper mantle shear structure beneath the northwest Atlantic Ocean. *J. Geophys. Res.*, **89**, 11 465–475.

Greenberg, R. (1989). Planetary accretion. In S. K. Atreya, J. B. Pollack and M. S. Matthews (editors), *Origin and Evolution of Planetary and Satellite Atmospheres*. Tucson, Univ. Ariz. Press, pp. 137–165.

Gregory, R. T. (1991). Oxygen isotope history of seawater revisited: Time scales for boundary event changes in the oxygen isotope composition of seawater. In H. P. Taylor, Jr. et al. (editors), *Stable Isotope Geochemistry, Geochem. Society Spec. Publ. No. 3*: 65–76.

Grieve, R. A. F. (1980). Impact bombardment and its role in protocontinental growth on the early Earth. *Precamb. Res.*, **10**, 217–247.

Griffin, W. L. and O'Reilly, S. Y. (1987). Is the continental Moho the crust–mantle boundary? *Geology*, **15**, 241–244.

Grossman, L. (1972). Condensation in the primitive solar nebula. *Geochim. Cosmochim. Acta*, **36**, 597–619.

Grotzinger, J. P. (1990). Geochemical model for Proterozoic stromatolite decline. *Amer. J. Sci.*, **290-A**, 80–103.

Grotzinger, J. P. and Kasting, J. F. (1993). New constraints on Precambrian ocean composition. *J. Geol.*, **101**, 235–243.

Grotzinger, J. P., Bowring, S. A., Saylor, B. Z. and Kaufman, A. J. (1995). Biostratigraphic and geochronologic constraints on early animal evolution. *Science*, **270**, 589–604.

Groves, D. I. and Barley, M. E. (1994). Archean Mineralization. In K. C. Condie (editor), *Archean Crustal Evolution*, Amsterdam, Elsevier, pp. 461–504.

Guillou, L. and Jaupart, C. (1995). On the effect of continents on mantle convection. *J. Geophys. Res.*, **100**, 24 217–238.

Gurnis, M. (1988). Large-scale mantle convection and the aggregation and dispersal of supercontinents. *Nature*, **332**, 695–699.

Gurnis, M. (1993). Phanerozoic marine inundation of continents driven by dynamic topography above subducting slabs. *Nature*, **364**, 589–593.

Gurnis, M. and Davies, G. F. (1986). Apparent episodic crustal growth arising from a smoothly evolving mantle. *Geology*, **14**, 396–399.

Gurwell, M. A., 1995. Evolution of deuterium on Venus. *Nature*, **378**, 22–23.

Haapala, I. and Ramo, T. (1990). Petrogenesis of the Proterozoic rapakivi granites of Finland. *Geol. Soc. America Spec. Paper 246*, 275–286.

Hager, B. H. and Clayton, R. W. (1989). Constraints on the structure of mantle convection using seismic observations, flow models, and the geoid. In W. R. Peltier (editor), *Mantle Convection*. New York, Gordon & Breach, pp. 657–763.

Hager, B. H., Clayton, R. W., Richards, M. A., Comer, R. P. and Dziewonski, A. M. (1985). Lower mantle heterogeneity, dynamic topography, and the geoid. *Nature*, **313**, 541–545.

Haggerty, S. E. and Sautter, V. (1990). Ultradeep ultramafic upper mantle xenoliths. *Science*, **248**, 993–996.

Hale, C. J. (1987). Paleomagnetic data suggest link between the Archean/Proterozoic boundary and inner core nucleation. *Nature*, **329**, 233–236.

Hallam, A. (1973). Provinciality, diversity, and extinction of Mesozoic marine invertebrates in relation to plate movements. In D. H. Tarling and S. K. Runcorn (editors), *Implications of Continental Drift to the Earth Sciences, Vol. 1*. London, Academic Press, pp. 287–294.

Hallam, A. (1974). Changing patterns of provinciality and diversity of fossil animals in relation to plate tectonics. *J. Biogeo.*, **1**, 213–225.

Hamilton, W. B. (1988). Plate tectonics and island arcs. *Geol. Soc. America Bull.*, **100**, 1503–1527.

Hardie, L. A. (1996). Secular variation in seawater chemistry: An explanation for the coupled secular variation in the mineralogies of marine limestones and potash evaporites over the past 600 My. *Geology*, **24**, 279–283.

Hargraves, R. B. (1986). Faster spreading or greater ridge length in the Archean? *Geology*, **14**, 750–752.

Harley, S. L. and Black, L. P. (1987). The Archean geological evolution of EnderbyLand, Antarctica. *Geol. Soc. London Spec. Publ. 27*, 285–296.

Harper, C. L. and Jacobsen, S. B. (1996). Evidence for 182Hf in the early Solar System and constraints on the timescale of terrestrial accretion and core formation. *Geochim. Cosmochim. Acta*, **60**, 1131–1153.

Harris, A. W. and Kaula, W. M. (1975). A co-accretional model of satellite formation. *Icarus*, **24**, 516–524.

Harris, N. B. W. (1995). Significance of weathering Himalayan metasedimentary rocks and leucogranites for the Sr isotope evolution of seawater during the Early Miocene. *Geology*, **23**, 795–798.

Harris, N. B. W., Pearce, J. A. and Tindle, A. G. (1986). Geochemical characteristics of collision-zone magmatism. In M. P. Coward and A. C. Ries (editors), *Collision Tectonics. Geol. Soc. London Spec. Publ. 19*, 67–81.

Harrison, C. G. A. (1987). Marine magnetic anomalies—the origin of the stripes. *Ann. Rev. Earth Planet. Sci.*, **15**, 505–534.

Harrison, T. M., Copeland, P., Kidd, W. S. F. and Yin, A. (1992). Raising Tibet. *Science*, **255**, 1663–1670.

Hart, S. R. (1984). A large-scale anomaly in the Southern Hemisphere mantle. *Nature*, **309**, 753–757.

Hart, S. R. (1986). In search of a bulk-Earth composition. *Chem. Geol.*, **57**, 247–267.

Hart, S. R. (1988). Heterogeneous mantle domains: Signatures, genesis and mixing chronologies. *Earth Planet. Sci. Lett.*, **90**, 273–296.

Hart, S. R., Hauri, E. H., Oschmann, L. A. and Whitehead, J. A. (1992). Mantle plumes and entrainment: Isotopic evidence. *Science*, **256**, 517–519.

Hartmann, W. K. (1983). *Moons and Planets*. Belmont, CA, Wadsworth Publ. Co., 509 pp.

Hartmann, W. K. (1986). Moon origin: the impact-trigger hypothesis. In W. K. Hartmann, R. J. Phillips, G. J. Taylor (editors), *Origin of the Moon*. Houston, TX, Lunar & Planet. Institute, pp. 579–608.

Hartnady, C. J. H. (1991). About turn for supercontinents. *Nature*, **352**, 476–478.

Hartnady, C. J. H. and leRoex, A. P. (1985). Southern ocean hotspot tracks and the Cenozoic absolute motion of the African, Antarctic and South American plates. *Earth Planet. Sci. Lett.*, **75**, 245–257.

Hauri., E. H. and Hart, S. R. (1993). Re–Os isotope systematics of HIMU and EMII oceanic island basalts from the south Pacific Ocean. *Earth Planet. Sci. Lett.*, **114**, 353–371.

Hauri, E. H., Whitehead, J. A. and Hart, S. R. (1994). Fluid dynamic and geochemical aspects of entrainment in mantle plumes. *J. Geophys. Res.*, **99**, 24,275–300.

Hawkesworth, C. J., Erlank, A. J., Kempton, P. D. and Waters, F. G. (1990). Mantle metasomatism: Isotope and trace-element trends in xenoliths from Kimberley, South Africa. *Chem. Geol.*, **85**, 19–34.

Hawkesworth, C. J., Gallagher, K., Hergt, J. M. and McDermott, F. (1994). Destructive plate margin magmatism: Geochemistry and melt generation. *Lithos*, **33**, 169–188.

Hedges, S. B., Parker, P. H., Sibley, C. G. and Kumar, S. (1996). Continental breakup and the ordinal diversification of birds and mammals. *Nature*, **381**, 226–229.

Heirtzler, J. R. et al. (1968). Marine magnetic anomalies, geomagnetic field reversals, and motions of the ocean floor and continents. *J. Geophys. Res.*, **73**, 2119–2135.

Heizler, M. T. (1993). Thermal history of the continental lithosphere using multiple diffusion domain thermochronometry. *UCLA, PhD thesis*, p. 295.

Helmstaedt, H. H. and Scott, D. J. (1992). The Proterozoic ophiolite problem. In K. C. Condie (editor), *Proterozoic Crustal Evolution*. Amsterdam, Elsevier, pp. 55–95.

Helmstaedt, H., Padgham, W. A. and Brophy, J. A. (1986). Multiple dykes in lower Kam Group, Yellowknife greenstone belt: Evidence for Archean seafloor spreading? *Geology*, **14**, 562–566.

Herzberg, C. (1995). Generation of plume magmas through time: An experimental perspective. *Chem. Geol.*, **126**, 1–16.

Hey, R., Duennebier, F. K. and Morgan, W. J. (1980). Propagating rifts on mid-ocean ridges. *J. Geophys. Res.*, **85**, 3647–3658.

Hildebrand, A. R. et al. (1991). Chicxulub crater: A possible Cretaceous/Tertiary boundary impact crater on the Yucatan Peninsula, Mexico. *Geology*, **19**, 867–871.

Hirn, A., Nercessian, A., Sapin, M., Jobert, G. and Xu, Z. X. (1984). Lhasa block and bordering sutures – a continuation of the 500-km Moho traverse through Tibet. *Nature*, **307**, 25–27.

Hoak, V. and Hutton, R. (1986). Electrical resistivity in continental lower crust. In J. B. Dawson, D. A. Carswell, J. Hall and K. H. Wedephol (editors), *The Nature of the Lower Continental Crust. London, Geol. Soc London Spec. Publ. No. 24*, 35–50.

Hoffman, P. F. (1988). United plates of America, the birth of a craton. *Ann. Rev. Earth Planet. Sci.*, **16**, 543–603.

Hoffman, P. F. (1989). Speculations on Laurentia's first gigayear. *Geology*, **17**, 135–138.

Hoffman, P. F. (1990). Geological constraints on the origin of the mantle root beneath the Canadian shield. *Phil. Trans. Roy. Soc. London*, **331A**, 523–532.

Hoffman, P. F. (1991). Did the breakout of Laurentia turn Gondwanaland inside-out? *Science*, **252**, 1409–1412.

Hoffman, S. E., Wilson, M. and Stakes, D. S. (1986). Inferred oxygen isotope profile of Archean oceanic crust, Onverwacht Group, South Africa. *Nature*, **321**, 55–58.

Hofmann, A. W. (1988). Chemical differentiation of the Earth: The relationship between mantle, continental crust, and oceanic crust. *Earth Planet. Sci. Lett.*, **90**, 297–314.

Holbrook, W. S., Mooney, W. D. and Christensen, N. J. (1992). The seismic velocity structure of the deep continental crust. In D. M. Fountain, R. Arculus and R. W. Kay (editors), *Continental Lower Crust*. Amsterdam, Elsevier, pp. 1–43.

Holland, H. D. (1984). *The Chemical Evolution of the Atmosphere and Oceans*. Princeton, NJ, Princeton Univ. Press, 582 pp.

Holland, H. D. (1992). Distribution and paleoenvironmental interpretation of Proterozoic paleosols. In J. W. Schopf and C. Klein (editors), *The Proterozoic Bisophere: A Multidisciplinary Study*. New York, Cambridge Univ. Press, pp. 153–155.

Holland, H. D. (1994). Early Proterozoic atmospheric change. In S. Bergtson (editor), *Early Life on Earth*. New York, Columbia Univ. Press, pp. 237–244.

Holland, H. D. and Beukes, N. J. (1990). A paleoweathering profile from Griqualand West, South Africa: Evidence for a dramatic rise in atmospheric oxygen between 2.2 and 1.9 Ga. *Amer. J. Sci.*, **290-A**, 1–34.

Holland, H. D., Lazar, B. and McGaffrey, M. (1986). Evolution of the atmosphere and oceans. *Nature*, **320**, 27–33.

Hollerbach, R. and Jones, C. A. (1993). Influence of the Earth's inner core on geomagnetic fluctuations and reversals. *Nature*, **365**, 541–543.

Holliger, K., Levander, A. R. and Goff, J. A. (1993). Stochastic modeling of the reflective lower crust: Petrophysical and geological evidence from the Ivrea zone. *J. Geophys. Res.*, **98**, 11 967–980.

Holmden, C. and Muehlenbachs, K. (1993). The $^{18}O/^{16}O$ ratio of 2-Ga seawater inferred from ancient oceanic crust. *Science*, **259**, 1733–1736.

Holser, W. T. (1984). Gradual and abrupt shifts in ocean chemistry during Phanerozoic time. In H. D. Holland and A. F. Trendall (editors), *Patterns of Change in Earth Evolution*. Berlin, Springer-Verlag, pp. 123–143.

Homewood, P., Allen, P. A. and Williams, G. D. (1986). Dynamics of the Molasse basin of western Switzerland. In A. Allen and P. Homewood (editors), *Foreland Basins*. Oxford, Blackwell Scient., pp. 199–218.

Honkura, Y. (1978). Electrical conductivity anomalies in the Earth. *Geophys. Surveys*, **3**, 225–253.

Humphris, S. E. et al. (1995). The internal structure of an active sea floor massive sulphide deposit. *Nature*, **377**, 713–716.

Hunten, D. M. (1993). Atmospheric evolution of the terrestrial planets. *Science*, **259**, 915–920.

Hunter, D. R. (1975). *The regional geological setting of the Bushveld Complex*. Johannesburg, Univ. Witwatersrand, Econ. Geol. Research Unit, 18 pp.

Hurley, P. M. and Rand, J. R. (1969). Predrift continental nuclei. *Science*, **164**, 1229–1242.

Hut, P. et al. (1987). Comet showers as a cause of mass extinctions. *Nature*, **329**, 118–125.

Imbrie, J. (1985). A theoretical framework for the Pleistocene ice ages. *Quart. Jour. Geol. London*, **142**, 417–432.

Ingersoll, R. V. (1988). Tectonics of sedimentary basins. *Geol. Soc. America Bull.*, **100**, 1704–1719.

Irving, E. and Puelaiah, G. (1976). Reversals of the geomagnetic field, magnetostratigraphy and relative magnitude of paleosecular variation in the Phanerozoic. *Earth-Sci. Revs.*, **12**, 35–64.

Isley, A. E. (1995). Hydrothermal plumes and the delivery of iron to BIF. *J. Geol.*, **103**, 169–185.

Ita, J. and Stixrude, L. (1992). Petrology, elasticity, and composition of the mantle transition zone. *J. Geophys. Res.*, **97**, 6849–66.

Ito, E., Morooka, K., Ujike, O. and Katsura, T. (1995). Reactions between molten iron and silicate melts at high pressure: Implications for the chemical evolution of Earth's core. *J. Geophys. Res.*, **100**, 5901–5910.

Jacobs, J. A. (1992). *Deep Interior of the Earth*. London, Chapman & Hall, 167 pp.

Jacobs, J. A. (1995). The Earth's inner core and the geodynamo: Determining their roles in the Earth's history. *EOS, 76, No. 25*, June, 1995.

Jacobsen, S. B. and Pimentel-Klose, M. (1988). Nd isotopic variations in Precambrian BIF. *Geophys. Res. Lett.*, **15**, 393–396.

Jarchow, C. M. and Thompson, G. A. (1989). The nature of the Mohorovicic discontinuity. *Ann. Rev. Earth Planet. Sci.*, **17**, 475–506.

Jeanloz, R. (1990). The nature of the Earth's core. *Ann. Rev. Earth Planet. Sci.*, **18**, 357–386.

Jeanloz, R., Mitchell, D. L., Sprague, A. L. and de Pater, I. (1995). Evidence for a basalt-free surface on Mercury and implications for internal heat. *Science*, **268**, 1455–1457.

Jenkins, G. S. (1995). Early Earth's climate: Cloud feedback from reduced land fraction and ozone concentrations. *Geophys. Res. Lett.*, **22**, 1513–1516.

Jenkins, G. S., Marshall, H. G. and Kuhn, W. R. (1993). Precambrian climate: The effects of land area and Earth's rotation rate. *J. Geophys. Res.*, **98**, 8785–8791.

Jephcoat, A. and Olson, P. (1987). Is the inner core of the Earth pure iron? *Nature*, **325**, 332.

Jessberger, E. K. and Kissel, J. (1989). The composition of comets. In S. K. Atreya, J. B. Pollack and M. S. Matthews (editors), *Origins and Evolution of Planetary and Satellite Atmospheres*. Tucson, Univ. Ariz. Press, pp. 167–191.

Jessop, A. M. and Lewis, T. (1978). Heat flow and heat generation in the Superior province of the Canadian shield. *Tectonophys*, **50**, 55–77.

Johnson, D. M. (1993). Mesozoic and Cenozoic contributions to crustal growth in the SW United States. *Earth Planet. Sci. Lett.*, **118**, 75–89.

Jolivet, L., Huchon, P. and Ranguin, C. (1989). Tectonic setting of Western Pacific marginal basins. *Tectonophys.*, **160**, 23–47.

Jones, A. G. (1992). Electrical conductivity of the continental lower crust. In D. M. Fountain, R. Arculus and R. W. Kay (editors), *Continental Lower Crust*. Amsterdam, Elsevier, pp. 81–143.

Jurdy, D. M., Stefanick, M. and Scotese, C. R. (1995). Paleozoic plate dynamics. *J. Geophys. Res.*, **100**, 17 965–975.

Kamo, S. L. and Davis, D. W. (1994). Reassessment of Archean crust development in the Barberton Mountain Land, South Africa, based on U–Pb dating. *Tectonics*, **13**, 167–192.

Kamo, S. L. and Krogh, T. E. (1995). Chicxulub crater source for shocked zircon crystals from the Cretaceous/Tertiary boundary layer, Saskatchewan: Evidence from new U–Pb data. *Geology*, **23**, 281–284.

Kandler, O. (1994). The early diversification of life. In S. Bengtson (editor), *Early Life on Earth*. New York, Columbia Univ. Press, pp. 152–160.

Kasting, J. F. (1987). Theoretical constraints on oxygen and carbon dioxide concentrations in the Precambrian atmosphere. *Precamb. Res.*, **34**, 205–229.

Kasting, J. F. (1991). Box models for the evolution of atmospheric oxygen: An update. *Paleogeog., Paleoclimat., Paleoecol.*, **97**, 125–131.

Kasting, J. F. (1993). Earth's early atmosphere. *Science*, **259**, 920–926.

Kato, T. and Ringwood, A. E. (1989). Melting relationships in the system Fe–FeO at high pressures: Implications for the composition and formation of the Earth's core. *Phys. Chem. Miner.*, **16**, 524–238.

Kaula, W. M. (1995). Venus reconsidered. *Science*, **270**, 1460–1464.

Kay, R. W. and Kay, S. M. (1981). The nature of the lower continental crust: Inferences from geophysics, surface geology and crustal xenoliths. *Revs. Geophys. Space Phys.*, **19**, 271–297.

Kay, S. M. and Kay, R. W. (1985). Role of crystal cumulates and the oceanic crust in the formation of the lower crust of the Aleutian arc. *Geology*, **13**, 461–464.

Kay, S. M., Ramos, V. A., Mpodozis, C. and Sruoga, P. (1989). Late Paleozoic to Jurassic silicic magmatism at the Gondwana margin: Analogy to the Middle Proterozoic in North America? *Geology*, **17**, 324–328.

Kellogg, L. H. (1992). Mixing in the mantle. *Ann. Rev. Earth Planet. Sci.*, **20**, 365–388.

Kendall, J. M. and Silver, P. G. (1996). Constraints from seismic anisotropy on the nature of the lowermost mantle. *Nature*, **381**, 409–412.

Keppler, H. (1996). Constraints from partitioning experiments of the composition of subduction-zone fluids. *Nature*, **380**, 237–240.

Kingma, K. J., Cohen, R. E., Hemley, R. J. and Mao, H. K. (1995). Transformation of stishovite to a denser phase at lower mantle pressures. *Nature*, **374**, 243–245.

Klein, C. and Beukes, N. J. (1992). Proterozoic Iron-formations. In K. C. Condie (editor), *Proterozoic Crustal Evolution*. Amsterdam, Elsevier, pp. 383–417.

Klein, G. deVries (1982). Probale sequential arrangement of depositional systems on cratons. *Geology*, **10**, 17–22.

Klein, G. deVries (1986). Sedimentation patterns in relation to rifting, arc volcanism and tectonic uplift in back-arc basins of the western Pacific Ocean. In N. Nasu, K. Kobayashi, S. Uyeda, I. Kushiro and H. Kagami (editors), *Formation of Active Continental Margins*. Dordrecht, D. Reidel, pp. 517–550.

Klemperer, S. L. (1987). A relation between continental heat flow and the seismic reflectivity of the lower crust. *J. Geophys.*, **61**, 1–11.

Knauth, L. P. and Lowe, D. R. (1978). Oxygen isotope geochemistry of cherts from the Onverwacht Group, South Africa. *Earth Planet. Sci. Lett.*, **41**, 209–222.

Knittle, E. and Jeanloz, R. (1991). Earth's core-mantle boundary: results of experiments at high pressures and temperatures. *Science*, **251**, 1438–1443.

Kokelaar, P. (1986). Magma–water interactions in subaqueous and emergent basaltic volcanism. *Bull. Volcanol.*, **48**, 275–289.

Kominz, M. A. and Bond, G. C. (1991). Unusually large subsidence and sea level events during middle Paleozoic time: New evidence supporting mantle convection models for supercontinent assembly. *Geology*, **19**, 56–60.

Kontak, D. J. and Reynolds, P. H. (1994). $^{40}Ar/^{39}Ar$ dating of metamorphic and igneous rocks of the Liscomb Complex, Meguma terrane, S Nova Scotia, Canada. *Can. Jour. Earth Sci.*, **31**, 1643–1653.

Krapez, B. (1993). Sequence stratigraphy of the Archean supracrustal belts of the Pilbara block, Western Australia. *Precambrian Res.*, **60**, 1–45.

Kroner, A. and Layer, P. W. (1992). Crust formation and plate motion in the Early Archean. *Science*, **256**, 1405–1411.

Kroner, A., Compston, W. and Williams, I. S. (1989). Growth of Early Archean crust in the Ancient Gneiss Complex of Swaziland as revealed by single zircon dating. *Tectonophys.*, **161**, 271–298.

Kroner, A., Hegner, E., Wendt, J. I. and Byerly, G. R. (1996). The oldest part of the Barberton granitoid–greenstone terrain, South Africa: Evidence for crust formation between 3.5 and 3.7 Ga. *Precamb. Res.*, **78**, 105–124.

Kumazawa, M., Helstaedt, H. and Masaki, K. (1971). Elastic properties of eclogite xenoliths form diatremes of the east Colorado Plateau and their implication to the upper mantle structure. *J. Geophys. Res.*, **76**, 1231–1247.

Kusky, T. M. and Kidd, W. S. F. (1992). Remnants of an Archean oceanic plateau, Belingwe greenstone belt, Zimbabwe. *Geology*, **20**, 43–46.

Kvenvolden, K. A. (1974). Natural evidence for chemical and early biologic evolution. *Origins of Life*, **5**, 71–86.

Lambeck, K. (1980). *The Earth's Rotation: Geophysical Cause and Consequences*. Cambridge, Cambridge Univ. Press, 449 pp.

Larson, R. L. (1991a). Latest pulse of Earth: Evidence for a Mid-Cretaceous superplume. *Geology*, **19**, 547–550.

Larson, R. L. (1991b). Geological consequences of superplumes. *Geology*, **19**, 963–966.

Lawton, T. F. (1986). Compositional trends within a clastic wedge adjacent to a fold-thrust belt: Indianola Group, central Utah, Utah, USA. In A. Allen and P. Homewood (editors), *Foreland Basins*. Oxford, Blackwell Scient., pp. 411–424.

Lebofsky, L. A., Jones, T. D. and Herbert, F. (1989). Asteroid volatile inventories. In S. K. Atreya, J. B. Pollack and M. S. Matthews (editors), *Origin and Evolution of Planetary and Satellite Atmospheres*. Tucson, Univ. Ariz. Press, pp. 192–229.

Lee, D. C. and Halliday, A. N. (1995). Hf-W chonometry and the timing of terrestrial core formation. *Nature*, **378**, 771–774.

Leitch, E. C. (1984). Marginal basins of the SW Pacific and the preservation and recognition of their ancient analogues: A review. *Geol. Soc. London Spec. Publ. 16*, 97–108.

Lister, G. S. and Davis, G. A. (1989). The origin of metamorphic core complexes and detachment faults formed during Tertiary continental extension in the northern Colorado River region. *J. Struct. Geol.*, **11**, 65–94.

Litak, R. K. and Brown, L. D. (1989). A modern perspective on the Conrad discontinuity. *Amer. Geophys. Union, EOS, 70, No. 29*.

Lithgow-Bertelloni, C. and Richards, M. A. (1995). Cenozoic plate driving forces. *Geophys. Res. Lett.*, **22**, 1317–1320.

Liu, M. (1994). Asymmetric phase effects and mantle convection patterns. *Science*, **264**, 1904–1907.

Longhi, J. (1978). Pyroxene stability and composition of the lunar magma ocean. *Proc. Lunar Planet. Sci. Conf. 9th*, pp. 285–306.

Loper, D. E. and Lay, T. (1995). The core–mantle boundary region. *J. Geophys. Res.*, **100**, 6397–6420.

Lowe, D. R. (1994a). Archean greenstone-related sedimentary rocks. In K. C. Condie (editor), *Archean Crustal Evolution*. Amsterdam, Elsevier, pp. 121–170.

Lowe, D. R. (1994b). Accretionary history of the Archean Barberton greenstone belt, southern Africa. *Geology*, **22**, 1099–1102.

Lucas, S. B. et al. (1993). Deep seismic profile across a Proterozoic collision zone: Surprises at depth. *Nature*, **363**, 339–342.

Lund, S. P. (1996). A comparison of Holocene paleomagnetic secular variation records from North America. *J. Geophys. Res.*, **101**, 8007–8024.

MacDonald, K. C. (1982). Mid-ocean ridges: Fine scale tectonic, volcanic and hydrothermal processes within the plate boundary zone. *Ann. Rev. Earth Planet. Sci.*, **10**, 155–190.

MacFarlane, A. W., Danielson, A. and Holland, H. D. (1994). Geology and major and trace element chemistry of Late Archean weathering profiles in the Fortescue Group, Western Australia: Implications for atmospheric PO_2. *Precamb. Res.*, **65**, 297–317.

MacGregor, I. D. (1974). The system $MgO–Al_2O_3–SiO_2$: Solubility of alumina in enstatite for spinel and garnet peridotite compositions. *Amer. Mineral.*, **59**, 110–119.

Marko, G. M. (1980). Velocity and attenuation in partially molten rocks. *J. Geophys. Res.*, **85**, 5173–5189.

Marti, K. and Graf, T. (1992). Cosmic-ray exposure history of ordinary chondrites. *Ann. Rev. Earth Planet. Sci.*, **20**, 221–243.

Marti, K., Kim, J. S., Thakur, A. N., McCoy, T. J. and Keil, K. (1995). Signatures of the martian atmosphere in glass of the Zagami meteorite. *Science*, **267**, 1981–1984.

Martignole, J. (1992). Exhumation of high-grade terranes – a review. *Can. Jour. Earth Sci.*, **29**, 737–745.

Martin, H. (1993). The mechanisms of petrogenesis of the Archean continental crust – comparison with modern processes. *Lithos*, **30**: 373–388.

Martin, H. (1994). Archean grey gneisses and the genesis of continental crust. In K. C. Condie (editor), *Archean Crustal Evolution*. Amsterdam, Elsevier, pp. 205–260.

Massey, N. W. D. (1986). Metchosin igneous complex, southern Vancouver Island: Ophiolite stratigraphy developed in an emergent island setting. *Geology*, **14**, 602–605.

Maurrasse, F. J. M. and Sen, G. (1991). Impacts, tsunamis, and the Haitian K/T boundary layer. *Science*, **252**, 1690–1693.

McBirney, A. R. (1976). Some geologic constraints on models for magma generation in orogenic environments. *Canad. Mineral.*, **14**, 245–254.

McCabe, R. (1984). Implications of paleomagnetic data on the collision related bending of island arcs. *Tectonics*, **3**, 409–428.

McClain, J. S. and Atallah, C. A. (1986). Thickening of the oceanic crust with age. *Geology*, **14**, 574–576.

McCulloch, M. T. and Gamble, J. A. (1991). Geochemical and geodynamical constraints on subduction zone magmatism. *Earth Planet. Sci. Lett.*, **102**, 358–374.

McDonough, W. F. (1990). Constraints on the composition of the continental lithospheric mantle. *Earth Planet. Sci. Lett.*, **101**, 1–18.

McDonough, W. F. and Sun, S. S. (1995). The composition of the Earth. *Chem. Geol.*, **120**, 223–253.

McDougall, I. and Harrison, T. M. (1988). Geo-chronology and Thermochronology by the $^{40}Ar/^{39}Ar$ method. New York, Oxford Univ. Press, 212 pp.

McGeary, S., Nur, A. and Ben-Avraham, Z. (1985). Spatial gaps in arc volcanism: The effect of collision or subduction of oceanic plateaus. *Tectonophys.*, **119**, 195–221.

McGill, G. E., Warner, J. L., Malin, M. C., Arvidson, R. E., Eliason, E., Nozette, S. and Reasenberg, R. D. (1983). Topography, surface properties and tectonic evolution of Venus. In D. M. Hunter, L. Colin, T. M. Donahue and V. I. Moroz (editors), *Venus*. Tucson, Univ. Ariz. Press, pp. 69–130.

McGovern, P. J. and Schubert, G. (1989). Thermal evolution of the Earth: Effects of volatile exchange between atmosphere and interior. *Earth Planet. Sci. Lett.*, **96**, 27–37.

McHone, J. F., Niema, R. A., Lewis, C. F. and Yates, A. M. (1989). Stishovite at the K/T boundary, Raton, New Mexico. *Science*, **243**, 1182–1184.

McKenzie, D. P. (1978). Some remarks on the development of sedimentary basins. *Earth Planet. Sci. Lett.*, **40**, 25–32.

McKenzie, D. P. and Morgan, W. J. (1969). Evolution of triple junctions. *Nature*, **224**, 125–133.

McLaren, D. J. and Goodfellow, W. D. (1990). Geological and biological consequences of giant impacts. *Ann. Rev. Earth Planet. Sci.*, **18**, 123–171.

McLennan, S. M. (1988). Recycling of the continental crust. *Pure Appl. Geophys.*, **128**, 683–898.

McLennan, S. M. and Hemming, S. (1992). Sm/Nd elemental and isotopic systematics in sedimentary rocks. *Geochim. Cosmochim. Acta*, **56**, 887–898.

McLennan, S. M. and Taylor, S. R. (1985). *The Continental Crust: Its Composition and Evolution*, Oxford, Blackwell, 312 pp.

Meert, J. G. and van der Voo, R. (1994). The Neoproterozoic glacial intervals: No more snowball Earth? *Earth Planet. Sci. Lett.*, **123**, 1–13.

Melville, R. (1973). Continental drift and plant distribution. In D. H. Tarling and S. K. Runcorn (editors), *Implications of Continental Drift to the Earth Sciences*, Vol. 1. London, Academic Press, pp. 439–446.

Menard, H. W. (1967). Transitional types of crust under small ocean basins. *J. Geophys. Res.*, **72**, 3061–3073.

Menzies, M. A. (1990). Archean, Proterozoic, and Phanerozoic lithospheres. In M. A. Menzies (editor), *Continental Mantle*. Oxford, Clarendon Press, pp. 67–85.

Mezger, K., Rawnsley, C. M., Bohlen, S. R. and Hanson, G. N. (1991). U–Pb garnet, sphene, monazite, and rutile ages: Implications for the duration of high-grade metamorphism and cooling histories, Adirondack Mts, NY. *J. Geol.*, **99**, 415–428.

Miller, S. L. (1953). A production of amino acids under possible primitive Earth conditions. *Science*, **117**, 528–529.

Miller, S. L. and Bada, J. L. (1988). Submarine hot springs and the origin of life. *Nature*, **334**, 609–611.

Mitchell, B. and Landisman, M. (1971). Electrical and seismic properties of the Earth's crust in the SW Great Plains. *Geophys.*, **36**, 141–159.

Mohr, P. (1982). Musings on continental rifts. *American Geophys. Union – Geol. Society of America, Geodynam. Series*, **8**, 293–309.

Mohr, R. E. (1975). Measured periodicities of the Biwabik stromatolites and their geophysical significance. In G. D. Rosenberg and S. K. Runcorn (editors), *Growth Rhythms and the History of the Earth's Rotation*. New York, J. Wiley, pp. 43–55.

Molnar, P. and Stock, J. (1987). Relative motions of hotspots in the Pacific, Atlantic and Indian Oceans since late Cretaceous time. *Nature*, **327**, 587–591.

Mooney, W. D. and Meissner, R. (1992). Multi-genetic origin of crustal reflectivity: A review of seismic reflection profiling of the continental lower crust and Moho. In D. M. Fountain, R. Arculus and R. W. Kay (editors), *Continental Lower Crust*. Amsterdam, Elsevier, pp. 45–79.

Moore, G. F., Curray, J. R. and Emmel, F. J. (1982). Sedimentation in the Sunda trends and forearc region. *Geol. Soc. London Spec. Publ.* **10**, 245–258.

Moores, E. M. (1982). Origin and emplacement of ophiolites. *Revs. Geophys. Space Phys.*, **20**, 735–760.

Moores, E. M. (1991). SW US–East Antarctic (SWEAT) connection: A hypothesis. *Geology*, **19**, 425–428.

Morgan, J. P., Parmentier, E. M. and Lin, J. (1987). Mechanisms for the origin of mid-ocean ridge axial topography: Implications for the thermal and mechanical structure of accreting plate boundaries. *J. Geophys. Res.*, **92**, 12 823–836.

Morgan, P. (1985). Crustal radiogenic heat production and the selective survival of ancient continental crust. *J. Geophys. Res.*, **90**, C561–C570.

Morley, L. W., MacLaren, A. S. and Charbonneau, B. W. (1967). Magnetic Anomaly Map of Canada. *Geol. Surv. Canada, Map No. 1255A*.

Moyes, A. B., Barton Jr., J. M. and Groenewald, P. B. (1993). Late Proterozoic to Early Paleozoic tectonism in Dronning Maud Land, Antarctica: Supercontinental fragmentation and amalgamation. *J. Geol. Soc. London*, **150**, 833–842.

Muehlenbachs, K. and Clayton, R. N. (1976). Oxygen isotope composition of oceanic crust and its bearing on seawater. *J. Geophys. Res.*, **81**, 4365–69.

Mueller, S. and Phillips, R. J. (1991). On the initiation of subduction. *J. Geophys. Res.*, **96**, 651–665.

Mueller, W. (1991). Volcanism and related slope to shallow-marine volcaniclastic sedimentation: An Archean example near Chibougamau, Quebec, Canada. *Precamb. Res.*, **49**, 1–22.

Mutch, T., Arndson, R. E., Head, J. W., Jones, K. L. and Saunders, R. S. (1976). A summary of Martian geologic history. In T. Mutch et al. (editors), *The Geology of Mars*. Princeton, Princeton Univ. Press, pp. 316–319.

Murthy, V. R. (1991). Early differentiation of the Earth and the problem of mantle siderophile elements: A new approach. *Science*, **253**, 303–306.

Nadeau, L. and van Breemen, O. (1994). Do the 1.45–1.39 Ga Montauban Group and the La Bostonnais Complex constitute a Grenvillian accreted terrane? *Geol. Assoc. Canada, Program with Abstracts*, **19**, A81.

Nance, R. D., Worsley, T. R. and Moody, J. B. (1986). Post Archean biogeochemical cycles and long-term episodicity in tectonic processes. *Geology*, **14**, 524–518.

Nelson, B. K. and DePaolo, D. J. (1985). Rapid production of continental crust 1.7 to 1.9 by ago: Nd isotopic evidence from the basement of the North American mid-continent. *Geol. Soc. America Bull.*, **96**, 746–754.

Nelson, K. D. (1991). A unified view of craton evolution motivated by recent deep seismic reflection and refraction results. *Geophys. J. Int.*, **105**, 25–35.

Newsom, H. E. and Sims, K. W. W. (1991). Core formation during early accretion of the Earth. *Science*, **252**, 926–933.

Newsom, H. E. and Taylor, S. R. (1989). Geochemical implications of the formation of the Moon by a single giant impact. *Nature*, **338**, 29–34.

Nicholas, A. (1986). Structure and petrology of peridotites: clues to their geodynamic environment. *Revs. Geophys. Space Phys.*, **24**, 875–895.

Nisbet, E. G. (1986). RNA, hydrothermal systems, zeolites and the origin of life. *Episodes*, **9**, 83–90.

Nisbet, E. G. (1995). Archean ecology: A review of evidence for the early development of bacterail biomes, and speculations on the development of a global-scale biosphere. In M. P. Coward and A. C. Ries (editors), *Early Precambrian Processes. Geol. Soc. London Spec. Publ. 95*, 27–51.

Nisbet, E. G., Cheadle, M. J., Arndt, N. T. and Bickle, M. J. (1993). Constraining the potential temperature of the Archean mantle: A review of the evidence from komatiites. *Lithos*, **30**, 291–307.

Nixon, P. H., Rogers, N. W., Gibson, I. L. and Grey, A. (1981). Depleted and fertile mantle xenoliths from southern African kimberlites. *Ann. Rev. Earth Planet. Sci.*, **9**, 285–309.

Norton, I. O. (1995). Plate motions in the North Pacific: The 43 Ma nonevent. *Tectonics*, **14**, 1080–1094.

Nutman, A. P., Friend, C. R., Kinny, P. D. and McGregor, V. R. (1993). Anatomy of an Early Archean gneiss complex: 3900 to 3600 Ma crustal evolution in southern West Greenland. *Geology*, **21**, 415–418.

Nutman, A. P., McGregor, V. R., Friend, C. R. L., Bennett, V. C. and Kinny, P. D. (1996). The Itsaq gneiss complex of southern West Greenland; The world's most extensive record of early crustal evolution. *Precamb. Res.*, **78**, 1–39.

Nyblade, A. A. and Pollack, H. N. (1993). A global analysis of heat flow from Precambrian terrains: Implications for the thermal structure of Archean and Proterozoic lithosphere. *J. Geophys. Res.*, **98**, 12 207–218.

Oberbeck, V. R., Marshall, J. R. and Aggarwal, H. (1993). Impacts, tillites, and the breakup of Gondwanaland. *J. Geol.*, **101**, 1–19.

O'Connor, J. M. and Duncan, R. A. (1990). Evolution of the Walvis ridge and Rio Grande rise hotspot system: Implications for African and South American plate motions over plumes. *J. Geophys. Res.*, **95**, 17 475–502.

Ohmoto, H. and Felder, R. P. (1987). Bacterial activity in the warmer, sulfate-bearing, Archean oceans. *Nature*, **328**, 244–246.

Ohmoto, H., Kadegawa, T. and Lowe, D. R. (1993). 3.4-Ga biogenic pyrites from Barberton, South Africa: Sulfur isotope evidence. *Science*, **262**, 555–557.

Ohtani, E., Shibata, T., Kubo, T. and Katao, T. (1995). Stability of hydrous phases in the transition zone and the upper most part of the lower mantle. *Geophys. Res. Lett.*, **22**, 2553–2556.

Olivarez, A. M. and Owen, R. M. (1991). The Eu anomaly of seawater: Implications for fluvial versus hydrothermal REE inputs to the oceans. *Chem. Geol.*, **92**, 317–328.

Olson, P., Schuber, G. and Anderson, C. (1987). Plume formation in the D" layer and the roughness of the core-mantle boundary. *Nature*, **327**, 409–413.

Omar, G. I. and Steckler, M. S. (1995). Fission track evidence on the initial rifting of the Red Sea: Two pulses, no propagation. *Science*, **270**, 1341–1344.

O'Nions, R. K. (1987). Relationships between chemical and convective layering in the Earth. *J. Geol. Soc. London*, **144**, 259–274.

Oparin, A. I. (1953). *The Origin of Life*. New York, Dover, 157 pp.

Oro, J. (1994). Early chemical stages in the origin of life. In S. Bengtson (editor), *Early Life on Earth*. New York, Columbia Univ. Press, pp. 48–59.

Orth, C. J., Gilmore, J. S. and Knight, J. D. (1987). Ir anomaly at the K/T boundary in the Raton basin. *New Mexico Geol. Soc. Guidebook, 38th Field Conf.*, pp. 265–269.

Ozima, M., Azuma, S., Zashu, S. and Hiyagon, H. (1993). ^{244}Pu fission Xe in the mantle and mantle degassing chronology. In H. Oya (editor), *Primitive Solar Nebula and Origin of Planets*. Tokyo, Terra Scient. Publ., pp. 503–517.

Pannella, G. (1972). Paleontologic evidence of the Earth's rotational history since early Precambrian. *Astrophys. Space Sci.*, **16**, 212–237.

Pasteris, J. D. (1984). Kimberlites: Complex mantle melts. *Ann. Rev. Earth Planet. Sci.*, **12**, 133–153.

Patchett, J. P. and Arndt, N. T. (1986). Nd isotopes and tectonics of 1.9–1.7 Ga crustal genesis. *Earth Planet. Sci. Lett.*, **78**, 329–338.

Patterson, C. C., Tilton, G. R. and Inghram, M. G. (1955). Age of the Earth. *Science*, **121**, 69–75.

Pearce, J. A. and Peate, D. W. (1995). Tectonic implications of the composition of volcanic arc magmas. *Ann. Rev. Earth Planet. Sci.*, **23**, 251–285.

Pearce, J. A., Harris, N. B. W. and Tindle, A. G. (1984). Trace element discrimination diagrams for the tectonic interpretation of granitic rocks. *J. Petrol.*, **25**, 956–983.

Pearson, D. G. et al. (1995). Re–Os, Sm–Nd and Rb–Sr isotope evidence for thick Archean lithospheric mantle beneath the Siberian craton modified by multistage metasomatism. *Geochim. Cosmochim. Acta*, **59**, 959–977.

Percival, J. A. and West, G. F. (1994). The Kapuskasing uplift: A geological and geophysical synthesis. *Can. Jour. Earth Sci.*, **31**, 1256–1286.

Percival, J. A. and Williams, H. R. (1989). Late Archean Quetico accretionary complex, Superior province, Canada. *Geology*, **17**, 23–25.

Percival, J. A., Fountain, D. M. and Salisbury, M. H. (1992). Exposed crustal cross sections as windows on the lower crust. In D. M. Fountain, R. Arculus, and R. W. Kay (editors), *Continental Lower Crust*. Amsterdam, Elsevier, pp. 317–362.

Percival, J. A., Green, A. G., Milkereit, B., Cook, F. A., Geis, W. and West, G. F. (1989). Seismic reflection profiles across deep continental crust exposed in the Kapuskasing uplift structure. *Nature*, **342**, 416–419.

Perry Jr., E. C., Ahmad, S. N. and Swulius, T. M. (1978). The oxygen isotope composition of 3800-Ma metamorphosed chert and iron formation from Isukasia, W Greenland. *J. Geol.*, **86**, 223–239.

Pettengill, G. H., Campbell, D. B. and Masursky, H. (1980). The surface of Venus. *Scient. Amer.*, **243**, 54–65.

Pfiffner, A. (1992). Alpine orogeny. In D. Blundell, R. Freeman and S. Mueller (editors), *A Continent Revealed: The European Geotraverse*. Cambridge, Cambridge Univ. Press, pp. 180–190.

Pierson, B. K. (1994). The emergence, diversification, and role of photosynthetic eubacteria. In S. Bengtson (editor), *Early Life on Earth*. New York, Columbia Univ. Press, pp. 161–180.

Pieters, C. M. and McFadden, L. A. (1994). Meteorite and asteroid reflectance spectroscopy: Clues to early Solar System processes. *Ann. Rev. Earth Planet. Sci.*, **22**, 457–497.

Piper, J. D. A. (1987). *Paleomagnetism and the Continental Crust*. New York, J. Wiley, 434 pp.

Pitman, W. C. and Talwani, M. (1972). Seafloor spreading in the North Atlantic. *Geol. Soc. America Bull.*, **83**, 619–646.

Plotnick, R. E. (1980). Relationship between biological extinctions and geomagnetic reversals. *Geology*, **8**, 578–581.

Polet, J. and Anderson, D. L. (1995). Depth extent of cratons as inferred from tomographic studies. *Geology*, **23**, 205–208.

Poli, S. and Schmidt, M. W. (1995). H_2O transport and release in subduction zones: Experimental constraints on basaltic and andesitic systems. *J. Geophys. Res.*, **100**, 22 299–314.

Pollack, J. B. and Bodenheimer, P. (1989). Theories of the origin and evolution of the giant planets. In S. K. Atreya, J. B. Pollack and M. S. Matthews (editors), *Origin and Evolution of Planetary and Satellite Atmospheres*. Tucson, Univ. Ariz. Press, pp. 564–602.

Powell, C. McA. et al. (1993). Paleomagnetic constraints on timing of the Neoproterozoic breakup of Rodinia and the Cambrian formation of Gondwana. *Geology*, **21**, 889–892.

Pretorius, D. A. (1976). The nature of the Witwatersrand gold–uranium deposits. In K. H. Wolfe (editor), *Handbook of Stratabound Ore Deposits*. Amsterdam, Elsevier, pp. 29–88.

Prevot, M., Mankineu, E. A., Gromme, C. S. and Coe, R. S. (1985). How geomagnetic field vector reverses polarity. *Nature*, **316**, 230–234.

Price, M. H. and Suppe, J. (1994). Mean age of rifting and volcanism on Venus deduced from impact crater densities. *Nature*, **372**, 756–759.

Price, M. H., Watson, G., Suppe, J. and Brankman, C. (1996). Dating volcanism and rifting on Venus using impact crater densities. *J. Geophys. Res., E2*, **101**, 4657–4671.

Prinn, R. G. and Fegley Jr., B. (1987). The atmosphere of Venus, Earth and Mars: A critical comparison. *Ann. Rev. Earth Planet. Sci.*, **15**, 171–212.

Raff, A. D. and Mason, R. G. (1961). Magnetic survey off the west coast of North America, 40° N to $51\frac{1}{2}$° N. *Geol. Soc. America Bull.*, **72**, 1259–1265.

Raitt, R. W., Shor, G. G., Francis, T. J. and Morris, G. B. (1969). Anisotropy of the Pacific upper mantle. *J. Geophys. Res.*, **74**, 3074–3109.

Rampino, M. R. (1994). Tillites, diamictites, and ballistic ejecta of large impacts. *J. Geol.*, **102**, 439–456.

Ranalli, G. (1991). Regional variations in lithosphere rheology from heat flow observations. In V. Cermak and L. Rybach (editors), *Exploration of the Deep Continental Crust*. Berlin, Springer-Verlag, pp. 1–22.

Rapp, R. P. and Watson, E. B. (1995). Dehydration melting of metabasalt at 8–32kb: Implications for continental growth and crust–mantle recycling. *J. Petrol.*, **36**, 891–931.

Rehrig, W. A. (1986). Processes of regional extension in the western Cordillera: Insights from the metamorphic core complexes. *Geol. Soc. America Spec. Paper 208*, 97–122.

Renne, P. R. et al. (1995). Syncrony and causal relations between Permian–Triassic boundary crises and Siberian volcanism. *Science*, **269**, 1413–1416.

Reymer, A. and Schubert, G., (1984). Phanerozoic addition rates to the continental crust and crustal growth. *Tectonics*, **3**, 63–77.

Richards, M. A. and Engebretson, D. C. (1992). The history of subduction, and large-scale mantle convection. *Nature*, **355**, 437–440.

Richardson, S. H. (1990). Age and early evolution of the continental mantel. In M. A. Menzies (editor), *Continental Mantle*. Oxford, Clarendon Press, pp. 55–65.

Richardson, S. H., Harris, J. W. and Gurney, J. J. (1993). Three generations of diamonds from old continental mantle. *Nature*, **366**, 256–258.

Richter, F. M. (1988). A major change in the thermal state of the Earth at the Archean–Proterozoic boundary: Consequences for the nature and preservation of continental lithosphere. *J. Petrol., Spec. Lithosphere issue*, pp. 39–52.

Richter, F. M., Rowley, D. B. and DePaolo, D. J. (1992). Sr isotope evolution of seawater: The role of tectonics. *Earth Planet. Sci. Lett.*, **109**, 11–23.

Rikitake, T. (1973). Global electrical conductivity of the Earth. *Phys. Earth Planet. Interiors*, **7**, 245–250.

Ringwood, A. E. (1966). The chemical composition and origin of the earth. In P. M. Hurley (editor), *Advances in Earth Sciences*. Cambridge, Massachusetts Institute of Technology Press, pp. 287–326.

Roberts, R. G. (1986). The deep electrical structre of the Earth. *Geophys. Jour. Royal Astro. Soc.*, **85**, 583–600.

Robertson, D. S. (1991). Geophysical applications of very-long-baseline interferometry. *Rev. Modern Phys.*, **63**, 899–918.

Robinson, P. I. (1973). Paleoclimatology and continental drift. In D. H. Tarling and S. K. Runcorn (editors), *Implications of Continental Drift to the Earth Sciences, Vol. 1*. London, Academic Press, pp. 451–476.

Roering, C. et al. (1992). Tectonic model for the evolution of the Limpopo belt. *Precamb. Res.*, **55**, 539–552.

Rona, P. A. (1977). Plate tectonics, energy and mineral resources: Basic research leading to payoff. *EOS*, **58**, 629–639.

Rona, P. A., Klinkhammer, G., Nelson, T. A., Trefry, H. and Elderfield, H. (1986). Black smokers, massive sulfides and vent biota at the Mid-Atlantic Ridge. *Nature*, **321**, 33–37.

Ronov, A. B. and Migdisov, A. A. (1971). Geochemical history of the crystalline basement and the sedimentary cover of the Russian and North America platforms. *Sediment.*, **16**, 137–185.

Ronov, A. B. and Yaroshevsky, A. A. (1969). Chemical composition of the Earth's crust. *Amer. Geophys. Union Mon.*, **13**, 37–57.

Ronov, A. B., Bredanova, N. V. and Migdisov, A. A. (1992). General trends in the evolution of the chemical composition of sedimentary and magmatic rocks of the continental Earth crust. *Sov. Sci. Riev. G. Geol.*, **1**, 1–37.

Rosing, M. T., Rose, N. R., Bridgwater, D. and Thomsen, H. S. (1996). Earliest part of Earth's stratigraphic record: A reappraisal of the > 3.7 Ga Isua supracrust sequence. *Geology*, **24**, 43–46.

Ross, M. I. and Scotese, C. R. (1988). A hierarchical tectonic model of the Gulf of Mexico and Caribbean region. *Tectonophys.*, **155**, 139–168.

Rubey, W. W. (1951). Geologic history of sea water. *Geol. Soc. America Bull.*, **62**, 1111–1148.

Rudnick, R. L. (1992). Xenoliths – samples of the lower continental crust. In D. M. Fountain, R. Arculus, and R. W. Kay (editors), *Continental Lower Crust*. Amsterdam, Elsevier, pp. 269–316.

Rudnick, R. L. (1995). Making continental crust. *Nature*, **378**, 571–578.

Rudnick, R. L. and Fountain, D. M. (1995). Nature and composition of the continental crust: A lower crustal

perspective. *Revs. Geophys. Space Phys.*, **33**, 267–309.

Rudnick, R. L. and Taylor, S. R. (1987). The composition and petrogenesis of the lower crust: A xenolith study. *J. Geophys. Res.*, **92**, 13 981–14 005.

Runnegar, B. (1994). Proterozoic eukaryotes: Evidence form biology and geology. In S. Bengtson (editor), *Early Life on Earth*. New York, Columbia Univ. Press, pp. 287–297.

Ruppel, C. (1995). Extensional processes in continental lithosphere. *J. Geophys. Res.*, **100**, 24 187–215.

Rutter, E. H. and Brodie, K. H. (1992). Rheology of the lower crust. In D. M. Fountain, R. Arculus, and R. W. Kay (editors), *Continental Lower Crust*. Amsterdam, Elsevier, pp. 201–267.

Rutter, E. H. and Neumann, D. (1995). Experimental deformation of partially molten Westerly granite under fluid-absent condition, with implications for the extraction of granitic magmas. *J. Geophys. Res.*, **100**, 15 679–716.

Rye, R., Kuo, P. H. and Holland, H. D. (1995). Atmospheric CO_2 concentrations before 2.2 Ga. *Nature*, **378**, 603–605.

Sacks, I. S. (1983). The subduction of young lithosphere. *J. Geophys. Res.*, **88**, 3355–3366.

Sakuyama, M. and Nesbitt, R. W. (1986). Geochemistry of the Quaternary volcanic rocks of the northeast Japan arc. *J. Volcanol. Geotherm. Res.*, **29**, 413–450.

Sample, J. C. and Fisher, D. M. (1986). Duplex accretion and underthrusting in an ancient accretionary complex, Kodiak Islands, Alaska. *Geology*, **14**, 160–163.

Sanford, A. R. and Einarsson, P. (1982). Magma chambers in rifts. *American Geophys. Union – Geol. Society of America, Geodynam. Series*, **8**, 147–168.

Sarda, P., Staudacher, T. and Allegre, C. J. (1985). ^{40}Ar/^{39}Ar in MORB glasses: Constraints on atmosphere and mantle evolution. *Earth Planet. Sci. Lett.*, **72**, 357–375.

Saunders, A. D., Tarney, J. and Weaver, S. D. (1980). Transverse geochemical variations across the Antarctic Peninsula: Implications for the genesis of calc-alkaline magmas. *Earth Planet. Sci. Lett.*, **46**, 344–360.

Sawkins, F. J. (1990). *Metal Deposits in Relation to Plate Tectonics*. Berlin, Springer-Verlag.

Schermer, E. R., Howell, D. G. and Jones, D. L. (1984). The origin of allochthonous terranes. *Ann. Rev. Earth Planet. Sci.*, **12**, 107–131.

Schidlowski, M., Hayes, J. M. and Kaplan, I. R. (1983). Isotopic inferences of ancient biochemistries: Carbon, sulfur, hydrogen and nitrogen. In J. W. Schopf (editor), *The Earth's Earliest Biosphere: Its Origin and Evolution*. Princeton, NJ, Princeton Univ. Press, pp. 149–186.

Scholl, D. W., von Huene, R., Vallier, T. L. and Howell, D. G. (1980). Sedimentary masses and concepts about tectonic processes at underthrust ocean margins. *Geology*, **8**, 564–568.

Scholz, C. H. and Campos, J. (1995). On the mechanism of seismic decoupling and back arc spreading at subduction zones. *J. Geophys. Res.*, **100**, 22 103–115.

Schopf, J. W. (1994). The oldest known records of life: Early Archean stromatolites, microfossils, and organic matter. In S. Bengtson (editor), *Early Life on Earth*. New York, Columbia Univ. Press, pp. 193–206.

Schreyer, W. (1995). Ultradeep metamorphic rocks: The retrospective viewpoint. *J. Geophys. Res.*, **100**, 8353–8366.

Schubert, G. (1979). Subsolidus convection in the mantles of terrestrial planets. *Ann. Rev. Earth Planet Sci.*, **7**, 287–343.

Schwab, F. L. (1978). Secular trends in the composition of sedimentary rocks assemblages – Archean through Pherozoic time. *Geology*, **6**, 532–536.

Sclater, J. G., Jaupart, C. and Galson, D. (1980). The heat flow through oceanic and continental crust and the heat loss of the Earth. *Rev. Geophys. Space Phys.*, **18**, 269–311.

Scotese, C. R. (1984). Paleomagnetism and the assembly of Pangaea. *American Geophys. Union – Geol. Society of America, Geodynam. Series*, **12**, 1–9.

Scotese, C. R. (1991). Jurassic and Cretaceous plate tectonic reconstructions. *Paleogeog., Paleoclimat., Paleoecol.*, **87**, 493–501.

Scott, D. J., Helmstaedt, H. and Bickle, M. J. (1992). Purtuniq ophiolite, Cape Smith belt, northern Quebec, Canada: A reconstructed section of Early Proterozoic oceanic crust. *Geology*, **20**, 173–176.

Searle, M. P. et al. (1987). The closing of Tethys and the tectonics of the Himalaya. *Geol. Soc. America Bull.*, **98**, 678–701.

Seber, D., Barazangi, M., Ibenbrahim, A. and Demnati, A. (1996). Geophsical evidence for llithosphereic delamination beneath the Alboran Sea and Rit-Betic Mountains. *Nature*, **379**, 785–790.

Seilacher, A. (1994). Early multicellular life: Late Proterozoic fossils and the Cambrian explosion. In S. Bengtson (editor), *Early Life on Earth*. New York, Columbia Univ. Press, pp. 389–400.

Sengor, A. M. C. and Burke, K. (1978). Relative timing of rifting and volcanism on Earth and its tectonic implications. *Geophys. Res. Lett.*, **5**, 419–421.

Sengor, A. M. C., Natal'in, B. A. and Burtman, V. S. (1993). Evolution of the Altaid tectonic collage and Paleozoic crustal growth in Eurasia. *Nature*, **364**, 299–307.

Sepkoski Jr., J. J. (1989). Periodicity in extinction and the problem of catastrophism in the history of life. *J. Geol. Soc. London*, **146**, 7–19.

Shaw, D. M. (1976). Development of the early continental crust, Part 2. In B. F. Windley (editor), *The Early History of the Earth*. New York, J. Wiley, pp. 33–54.

Shaw, D. M., Cramer, J. J., Higgins, M. D. and Truscott, M. G. (1986). Composition of the Canadian Precambrian shield and the continental crust of the Earth. *Geol. Soc. London Spec. Publ. No. 24*, 275–282.

Sherman, D. M. (1995). Stability of possible Fe–FeS and Fe–FeO alloy phases at high pressure and the composition of the Earth's core. *Earth Planet. Sci. Lett.*, **132**, 87–98.

Shimamoto, T. (1985). The origin of large or great thrust-type earthquakes along subducting plate boundaries. *Tectonophys.*, **119**, 37–65.

Shive, P. N., Blakely, R. J., Frost, B. R. and Fountain, D. M. (1992). Magnetic properties of the lower continental crust. In D. M. Fountain, R. Arculus and R. W. Kay (editor), *Continental Lower Crust*. Amsterdam, Elsevier, pp. 145–177.

Shixing, Z. and Huineng, C. (1995). Megascopic multicellular organisms form the 1700-My-old Tuanshanzi Formation in the Jixian area, North China. *Science*, **270**, 620–622.

Shock, E. L. (1992). Chemical environment of submarine hydrothermal systems. *Origin Life Evol. Biosphere*, **22**, 67–108.

Sillitoe, R. H. (1976). Andean mineralization: A model for the metallogeny of convergent plate margins. *Geol. Assoc. Canada Spec. Paper, 14*, 59–100.

Silver, P. G. and Chan, W. W. (1991). Shear wave splitting and subcontinental mantle deformation. *J. Geophys. Res.*, **96**, 16 429–454.

Sims, K. W. W., Newsom, H. E. and Gladney, E. S. (1990). Chemical fractionation during formation of the Earth's core and continental crust: Clues from As, Sb, W, and Mo. In H. E. Newsom and J. H. Jones (editors), *Origin of the Earth*. New York, Oxford Univ. Press, pp. 291–317.

Sinigoi, S. et al. (1994). Chemical evolution of a large mafic intrusion in the lower crust, Ivrea-Verbano zone, N Italy. *J. Geophys. Res.*, **99**, 21 575–590.

Sleep, N. H. (1990). Hotspots and mantle plumes: Some phenomenology. *J. Geophys. Res.*, **95**, 6715–6736.

Sleep, N. H. (1992). Archean plate tectonics: What can be learned from continental geology? *Can. Jour. Earth Sci.*, **29**, 2066–2071.

Sleep, N. H. and Windley, B. F. (1982). Archean plate tectonics: Constraints and inferences. *J. Geol.*, **90**, 363–379.

Sleep, N. H., Nunn, J. A. and Chou, L. (1980). Platform basins. *Ann. Rev. Earth Planet. Sci.*, **8**, 17–34.

Sleep, N. H., Zahnle, K. J., Kasting, J. F. and Morowitz, H. J. (1989). Annihilation of ecosystems by large asteroid impacts on the early Earth. *Nature*, **342**, 139–142.

Sloan, R. E., Rigby Jr., J. K., Van Valen, L. M. and Gabriel, D. (1986). Gradual dinosaur extinction and simultaneous ungulate radiation in the Hell Creek Formation. *Science*, **232**, 629–633.

Small, C. (1995). Observations of ridge-hotspot interactions in the southern Ocean. *J. Geophys. Res.*, **100**, 17 931–946.

Smith, A. G. and Woodcock, N. H. (1982). Tectonic synthesis of the Alpine–Mediterranean region: A review. *American Geophys. Union – Geol. Society of America, Geodynam. Series*, **7**, 15–38.

Smith, D. E. et al. (1994). Contemporary global horizontal crustal motion. *Geophys. J. Inter.*, **119**, 511–520.

Smith, R. B. and Braile, L. W. (1994). The Yellowstone hotspot. *J. Volcan. Geotherm. Res.*, **61**, 121–187.

Solomon, S. C. and Toomey, D. R. (1992). The structure of mid-ocean ridges. *Ann. Rev. Earth Planet. Sci.*, **20**, 329–364.

Solomon, S. C. et al. (1992). Venus tectonics: An overview of Magellan observations. *J. Geophys. Res.*, **97**, 13,199–255.

Spence, W. (1987). Slab pull and the seismotectonics of subducting lithosphere. *Revs. Geophys. Space Phys.*, **25**, 55–69.

Spray, J. G. (1984). Possible causes and consequences of upper mantle decoupling and ophiolite displacement. *Geol. Soc. London Spec. Publ. 13*, 255–267.

Staudacher, T. and Allegre, C. J. (1982). Terrestrial xenology. *Earth Planet. Sci. Lett.*, **60**, 389–406.

Steckler, M. (1984). Changes in sea level. In H. D. Holland and A. F. Trendall (editors), *Patterns of Change in Earth Evolution*. Berlin, Springer-Verlag, pp. 103–121.

Stefanick, M. and Jurdy, D. M. (1984). The distribution of hotspots. *J. Geophys. Res.*, **89**, 9919–9925.

Stefanick, M. and Jurdy, D. M. (1996). Venus coronae, craters, and chasmata. *Jour. Geophys. Res.*, **101**, 4637–4643.

Stephen, R. A. (1985). Seismic anisotrophy in the upper oceanic crust. *J. Geophys. Res.*, **90**, 11, 383–396.

Stein, M. and Hofmann, A. W. (1994). Mantle plumes and episodic crustal growth. *Nature*, **372**, 63–68.

Stockwell, C. H. (1965). Structural trends in the Canadian shield. *Amer. Assoc. Petrol. Geol. Bull.*, **49**, 887–893.

Stockwell, G. S., Beaumont, C. and Boutilier, R. (1986). Geodynamic models of convergent margin tectonics: The transition from rifted margin to overthrust belt and the consequences for foreland basin development. *Amer. Assoc. Petrol. Geol. Bull.*, **70**, 181–190.

Stoddard, P. R. and Abbott, D. (1996). Influence of the tectosphere upon plate motion. *J. Geophys. Res.*, **101**, 5425–5433.

Storey, B. C. (1995). The role of mantle plumes in continental breakup: Case histories from Gondwanaland. *Nature*, **377**, 301–308.

Strom, R. G., Schaber, G. G. and Dawson, D. D. (1994). The global resurfacing of Venus. *J. Geophys. Res.*, **99**, 10 899–926.

Sun, S. Q. (1994). A reappraisal of dolomite abundance and occurrence in the Phanerozoic. *J. Sed. Res.*, **A64**, 396–404.

Sun, S. and McDonough, W. F. (1989). Chemical and isotopic systematics of oceanic basalts: Implications for mantle compostion and processes. In A. D. Saunders and J. J. Norry (editors), *Magmatism in the Ocean Basins. Geol. Soc. London Spec. Publ. 42*, 313–345.

Swisher, C. C. et al. (1992). Coeval $^{40}Ar/^{39}Ar$ ages of 65 Ma from Chicxulub crater melt rock and K/T boundary tektites. *Science*, **257**, 954–958.

Sylvester, P. J. (1994). Archean granite plutons. In K. C. Condie (editor), *Archean Crustal Evolution*. Amsterdam, Elsevier, pp. 261–314.

Tackley, P. J., Stevenson, D. J., Glatzmaier, G. A. and Schubert, G. (1994). Effects of multiple phase transitions in a three-dimensional spherical model of convection in the Earth's mantle. *J. Geophys. Res.*, **99**, 15 877–901.

Talwani, J., LePichon, X. and Ewing, M. (1965). Crustal structure of the mid-ocean ridges, Part 2. *J. Geophys. Res.*, **70**, 341–352.

Tapponnier, P., Peltzer, G. and Armijo, R. (1986). On the mechanics of the collision between India and Asia. *Geol. Soc. London, Spec. Publ. 19*, 115–158.

Tarduno, J. A. and Gee, J. (1995). Large-scale motion between Pacific and Atlantic hotspots. *Nature*, **378**, 477–479.

Tarits, P. (1986). Conductivity and fluids in the upper oceanic upper mantle. *Phys. Earth Planet. Int.*, **42**, 215–226.

Tarney, J. (1992). Geochemistry and significance of mafic dyke swarms in the Proterozoic. In K. C. Condie (editor), *Proterozoic Crustal Evolution*, Amsterdam, Elsevier, pp. 151–179.

Tarney, J., Pickering, K. T., Knipe, R. J. and Dewey, J. F. (editors) (1991). The Behaviour and influence of fluids in subduction zones. *Roy. Soc. London, Spec. Publ. A335*.

Tarney, J., Wood, D. A., Varet, J., Saunders, A. D. and Cann, J. R. (1979). Nature of mantle heterogeneity in the North Atlantic: evidence from Leg 49 basalts. Maurce Ewing Series 2, *American Geophys. Union*, pp. 285–301.

Taylor, S. R. (1982). *Planetary Science: A Lunar Perspective*. Houston TX, Lunar & Planet. Institute, 481 pp.

Taylor, S. R. (1987). The unique lunar composition and its bearing on the origin of the Moon. *Geochim. Cosmochim. Acta*, **51**, 1297–1309.

Taylor, S. R. (1992). *Solar System Evolution: A New Perspective*. Cambridge, Cambridge Univ. Press, 307 pp.

Taylor, S. R. (1993). Early accretional history of the Earth and the Moon-forming event. *Lithos*, **30**, 207–221.

Taylor, S. R. and McLennan, S. M. (1985). *The Continental Crust: Its Composition and Evolution*. Oxford, Blackwell Scient., 312 pp.

Taylor, S. R. and McLennan, S. M. (1995). The geochemical evolution of the continental crust. *Revs. Geophys. Space Phys.*, **33**, 241–265.

Thompson, A. B. and Ridley, J. R. (1987). P–T–t histories of orogenic belts. *Phil. Trans. Roy. Soc. London*, **A321**, 27–45.

Thompson, R. N. and Gibson, S. A. (1994). Magmatic expression of lithospheric thinning across continental rifts. *Tectonophys.*, **233**, 41–68.

Thurston, P. C. (1994). Archean volcanic patterns. In K. C. Condie (editor), *Archean Crustal Evolution*. Amsterdam, Elsevier, pp. 45–84.

Thurston, P. C. and Chivers, K. M. (1990). Secular variation in greenstone sequence development emphasizing Superior province, Canada. *Precamb. Res.*, **46**, 21–58.

Toksoz, N. M., Minear, J. W. and Julian, B. R. (1971). Temperature field and geophysical effects of the downgoing slab. *J. Geophys. Res.*, **76**, 1113–1137.

Toon, D. B. (1984). Sudden changes in atmospheric compostion and climate. In H. D. Holland and A. F. Trendall (editors), *Patterns of Change in Earth Evolution*. Berlin, Springer-Verlag, pp. 41–61.

Trendall, A. F. (1983). Precambrian iron formation of Australia. *Econ. Geol.*, **68**, 1023–1034.

Tunnicliffe, V. and Fowler, M. R. (1996). Influence of seafloor spreading on the global hydrothermal vent fauna. *Nature*, **379**, 531–533.

Turcotte, D. L. (1995). How does Venus lose heat? *J. Geophys. Res.*, **100**, 16 931–940.

Turcotte, D. L. (1996). Magellan and comparative planetology. *J. Geophys. Res.*, **101**, 4765–4773.

Tyler, A. L., Kozlowski, R. W. H. and Lebofsky, L. A. (1988). Determination of rock type on Mercury and the Moon through remote sensing in the thermal infrared. *Geophys. Res. Lett.*, **15**, 808–811.

Unrug, R. (1993). The supercontinent cycle and Gondwanaland assembly: Composition of cratons and the timing of suturing events. *J. Geodynam.*, **16**, 215–240.

Uyeda, S. (1983). Comparative subductology. *Episodes*, **2**, 19–24.

Uyeda, S. and Miyashiro, A. (1974). Plate tectonics and the Japanese islands: A synthesis. *Geol. Soc. America Bull.*, **85**, 1159–1170.

Vail, P. R. and Mitchum Jr., R. M. (1979). Global cycles of relative changes of sea level from seismic stratigraphy. *Amer. Assoc. Petrol. Geol. Mem.*, **29**, 469–472.

Valet, J. P. and Meynadier, L. (1993). Geomagnetic field intensity and reversals during the past 4 My. *Nature*, **336**, 234–238.

Valet, J. P., Laj, C. and Tucholka, P. (1986). High resolution sedimentary record of a geomagnetic reversal. *Nature*, **322**, 27–32.

van der Hilst, R. (1995). Complex morphology of subducted lithosphere in the mantle beneath the Tonga trench. *Nature*, **374**, 154–157.

van der Hilst, R., Engdahl, R., Spakman, W. and Nolet, G. (1991). Tomographic imaging of subducted lithosphere below NW Pacific island arcs. *Nature*, **353**, 37–43.

Van Fossen, M. C. and Kent, D. V. (1992). Paleomagnetism of 122 Ma plutons in New England and the Mid-Cretaceous paleomagnetic field in North America: True polar wander or large-scale differential mantle motion? *J. Geophys. Res.*, **97**, 19 651–661.

Veevers, J. J. (1990). Tectonic-climatic supercycle in the billion-year plate-tectonic eon: Permian Pangean icehouse alternates with Cretaceous dispersed-continents greenhouse *Sed. Geol.*, **68**, 1–16.

Veizer, J. (1979). Secular variations in chemical composition of sediments: A review. *Phys. Chem. Earth*, **11**, 269–278.

Veizer, J. (1989). Strontium isotopes in seawater through time. *Ann. Rev. Earth Planet. Sci.*, **17**, 141–187.

Veizer, J. and Jansen, S. L. (1985). Basement and sedimentary recycling -2: Time dimension to global tectonics. *J. Geol.*, **93**, 625–643.

Veizer, J., Bell, K. and Jansen, S. L. (1992). Temporal distribution of carbonatites. *Geology*, **20**, 1147–1149.

Veizer, J., Hoefs, J., Lowe, D. R. and Thurston, P. C. (1989). Geochemistry of Precambrian carbonates: II. Archean greenstone belts and Archean seawater. *Geochim. Cosmochim. Acta*, **53**, 859–871.

Vidal, Ph., Bernard-Griffiths, J., Cocheric, A., LeFort, P., Peucat, J. J. and Sheppard, S. M. (1984). Geochemical comparison between Himalayan and Hercynian leucogranites. *Phys. Earth Planet. Interiors*, **35**, 179–190.

Vidale, J. E., Ding, X. Y. and Grand, S. P. (1995). The 410-km-depth discontinuity: A sharpness estimate from near-critical reflections. *Geophys. Res. Lett.*, **22**, 2557–2560.

Vielzeuf, D., Clemens, J. D., Pin, C. and Moinet, E. (1990). Granites, granulites, and crustal differentiation. In D. Vielzeuf and P. H. Vidal (editors), *Granulites and Crustal Evolution*. Norwell, MA, Kluwer Acad. Press, pp. 59–85.

Vigny, C., Ricard, Y. and Froidevaux, C. (1991). The driving mechanism of plate tectonics. *Tectonophys.*, **187**, 345–360.

Vilas, F. et al. (1988). *Mercury*. Tucson, Univ. Arizona Press.

Vine, F. (1966). Spreading of the ocean floor: Evidence. *Science*, **154**, 1405–1415.

Vine, F. and Matthews, D. H. (1963). Magnetic anomalies over oceanic ridges. *Nature*, **199**, 947–949.

Vinnik, L. P., Green, R. W. E., and Nicolaysen, L. O. (1995). Recent deformations of the deep continental root beneath southern Africa. *Nature*, **375**, 50–52.

Vitorello, I. and Pollack, H. N. (1980). On the variation of continental heat flow with age and the thermal evolution of continents. *J. Geophys. Res.*, **85**, 983–995.

von Brunn, V. and Gold, D. J. D. (1993). Diamictite in the Archean Pongola sequence of southern Africa. *J. Afric. Earth Sci.*, **16**, 367–374.

Von Damm, K. L. (1990). Seafloor hydrothermal activity: Black smoker chemistry and chimneys. *Ann. Rev. Earth Planet. Sci.*, **18**, 173–204.

von Huene, R. and Scholl, D. W. (1991). Observations at convergent margins concerning sediment subduction, subduction erosion, and the growth of continental crust. *Revs. Geophys. Space Phys.*, **29**, 279–316.

Walck, M. C. (1985). The upper mantle beneath the northeast Pacific rim: A comparison with the Gulf of California. *Geophys. Jour. Roy. Astr. Society*, **81**, 243–276.

Walker, J. C. G. (1977). *Evolution of the Atmosphere*. New York, Macmillan, 318 pp.

Walker, J. C. G. (1990). Precambrian evolution of the climate system. *Paleogeog., Paleoclimat., Paleoecol.*, **82**, 261–289.

Walker, J. C. G., Klein, C., Schidlowski, M., Schopf, J. W., Stevenson, D. J. and Walter, M. R. (1983). Environmental evolution of the Archean–Early Proterozoic

Earth. In J. W. Schopf (editor), *The Earth's Earliest Biosphere: Its Origin and Evolution*. Princeton, NJ, Princeton Univ. Press, pp. 260–290.

Walker, R. J., Morgan, J. W. and Horan, M. F. (1995). Osmium-187 enrichment in some plumes: Evidence for core–mantle interaction? *Science*, **269**, 819–821.

Walter, M. R. (1994). Stromatolites: The main geological source of information on the evolution of the early benthos. In S. Bengtson (editor), *Early Life on Earth*. New York, Columbia Univ. Press, pp. 270–286.

Wang, Y., Weidner, D. J., Liebermann, R. C. and Zhao, Y. (1994). P–V–T equation of state of perovskite: Constraints on composition of the lower mantle. *Phys. Earth Planet. Interiors*, **83**, 13–40.

Warner, M. et al. (1996). Seismic reflections from the mantle represent relict subduction zones within the continental lithosphere. *Geology*, **24**, 39–42.

Wasson, J. T. (1985). *Meteorites*. New York, W. H. Freeman, 276 pp.

Waters, F. G. and Erlank, A. J. (1988). Assessment of the vertical extent and distribution of mantle metasomatism below Kimberley, South Africa. *J. Petrol., Spec. Lithosphere Issue*, pp. 185–204.

Wedepohl, K. H. (1995). The composition of the continental crust. *Geochim. Cosmochim. Acta*, **59**, 1217–1232.

Wegener, A. (1912). Die Entstehung der Kontinente. *Geol. Rund.*, **3**, 276–292.

Weiguo, S. (1994). Early multicellular fossils. In S. Bengtson (editor), *Early Life on Earth*. New York, Columbia Univ. Press, pp. 358–369.

Wendlandt, E., DePaolo, D. J. and Baldridge, W. S. (1993). Nd and Sr isotope chronostratigraphy of Colorado Plateau lithosphere: Implications for magmatic and tectonic underplating of the continental crust. *Earth Planet. Sci. Lett.*, **116**, 23–43.

Wernicke, B. P. (1985). Uniform-sense normal simple shear of the continental lithosphere. *Can. Jour. Earth Sci.*, **22**, 108–125.

Wetherill, G. W. (1985). Occurrence of giant impacts during the growth of the terrestrial planets. *Science*, **221**, 877–879.

Wetherill, G. W. (1986). Accumulation of the terrestrial planets and implications concerning lunar origin. In W. K. Hartmann, R. J. Phillips, G. J. Taylor (editors), *Origin of the Moon*. Houston, TX, Lunar & Planet. Institute, pp. 519–550.

Wetherill, G. W. (1990). Formation of the Earth. *Ann. Rev. Earth Planet. Sci.*, **18**, 205–256.

Wetherill, G. W. (1994). Provenance of the terrestrial planets. *Geochim. Cosmochim. Acta*, **58**, 4513–4520.

White, R. S. (1988). The Earth's crust and lithosphere. *J. Petrol., Spec. Lithosphere Issue*, pp. 1–10.

White, R. S. and McKenzie, D. (1995). Mantle plumes and flood basalts. *J. Geophys. Res.*, **100**, 17 543–585.

Wickham, S. M. (1992). Fluids in the deep crust – petrological and isotopic evidence. In D. M. Fountain, R. Arculus and R. W. Kay (editors), *Continental Lower Crust*. Amsterdam, Elsevier, pp. 391–421.

Wiebe, R. A. (1992). Proterozoic anorthosite complexes. In K. C. Condie (editor), *Proterozoic Crustal Evolution*. Amsterdam, Elsevier, pp. 215–261.

Wiens, D. A. and Stein, S. (1985). Implications of oceanic intraplate seismicity for plate stresses, driving forces and rheology. *Tectonophys.*, **116**, 143–162.

Williams, G. E. (1975). Late Precambrian glacial climate and the Earth's obliquity. *Geol. Mag.*, **112**, 441–465.

Williams, G. E. (1989). Precambrian tidal sedimentary cycles and Earth's paleorotation. *EOS*, **70**, 40–41.

Williams, H. R. et al. (1992). Tectonic evolution of Ontario: Summary and synthesis. *Ontario Geol. Surv., Spec. Vol. No. 4*, 1255–1324.

Williams, L. A. J. (1982). Physical aspects of magmatism in continental rifts. *American Geophys. Union–Geol. Society America, Geodynam. Series*, **8**, 193–219.

Wilson, J. T. (1963). Evidence from islands on the spreading of the ocean floor. *Nature*, **197**, 536–538.

Wilson, J. T. (1965). Transform faults, oceanic ridges and magnetic anomalies SW of Vancouver Island. *Science*, **150**, 482–485.

Wilson, M. (1993). Magmatism and the geodynamics of basin formation. *Sed. Geol.*, **86**, 5–29.

Windley, B. F. (1992). Proterozoic collision and accretionary orogens. In K. C. Condie (editor), *Proterozoic Crustal Evolution*. Amsterdam, Elsevier, pp. 419–446.

Windley, B. F. (1993). Proterozoic anorogenic magmatism and its orogenic connections. *J. Geol. Soc. London*, **150**, 39–50.

Windley, B. F. (1995). *The Evolving Continents* 3rd Edn. New York, J. Wiley, 526 pp.

Wise, D. U. (1963). An origin of the Moon by rotational fission during formation of the Earth's core. *J. Geophys. Res.*, **68**, 1547–1554.

Wise, D. U. (1973). Freeboard of continents through time. *Geol. Soc. America Mem.*, **132**, 87–100.

Wolbach, W. S., Lewis, R. S. and Anders, E. (1985). Cretaceous extinctions: Evidence for wildfires and search for meteoric material. *Science*, **230**, 167–170.

Wood, B. J. (1995). The effect of water on the 410-km seismic discontinuity. *Science*, **268**, 74–76.

Wood, D. A. (1979). A variably veined suboceanic upper mantle – genetic significance for mid-ocean ridge basalts from geochemical evidence. *Geology*, **7**, 499–503.

Wood, J. A. (1986). Moon over Mauna Loa: a review of hypotheses of formation of the Earth's moon. In W. K. Hartmann, R. J. Phillips and G. J. Taylor (editors), *Origin of the Moon*, Houston, TX, Lunar & Planet. Institute, pp. 17–56.

Wood, J. A. and Mitler, H. E. (1974). Origin of the Moon by a modified capture mechanism, or half a loaf is better than a whole one. *Lunar Science*, **5**, 851–853.

Woodhouse, J. H. and Dziewonski, A. M. (1984). Mapping the upper mantle: Three-dimensional modeling of Earth structure by inversion of seismic wave forms. *J. Geophys. Res.*, **89**, 5953–5986.

Woollard, G. P. (1968). The interrelationship of the crust, upper mantle, and isostatic gravity anomalies in the United States. *Amer. Geophys. Union Mon., 12*, 312–341.

Worsley, T. R. and Nance, R. D. (1989). Carbon redox and climate control through Earth history: A speculative reconstruction. *Paleogeog., Paleoclimat., Paleoecol., 75*, 259–282.

Worsley, T. R., Nance, R. D. and Moody, J. B. (1984). Global tectonics and eustasy for the past 2 billion years. *Marine Geol., 58*, 373–400.

Worsley, T. R., Nance, R. D. and Moody, J. B. (1986). Tectonic cycles and the history of the Earth's biogeochemical and paleoceanographic record. *Paleoceanog., 1*, 233–263.

Wronkiewicz, D. J. and Condie, K. C. (1989). Geochemistry and provenance of sediments from the Pongola Supergroup South Africa: Evidence for a 3.0-Ga-old continental craton. *Geochim. Cosmochim. Acta, 53*, 1537–1549.

Wyckoff, S. (1991). Comets: Clues to the early history of the Solar System. *Earth-Sci. Rev., 30*, 125–174.

Wyllie, P. J. (1971). The role of water in magma generation and initiation of diapiric uprise in the mantle. *J. Geophys. Res., 76*, 1328–1338.

Xie, Q., McCuaig, T. C. and Kerrich, R. (1995). Secular trends in the melting depths of mantle plumes: Evidence from HFSE/REE systematics of Archean high-Mg lavas and modern oceanic basalts. *Chem. Geol., 126*, 29–42.

Young, C. J. and Lay, T. (1987). The core–mantle boundary. *Ann. Rev. Earth Planet. Sci., 15*, 25–46.

Zandt, G. and Ammon, C. J. (1995). Continental crust composition constrained by measurements of crustal Poisson's ratio. *Nature, 374*, 152–154.

Zaug, A. J. and Cech, T. R. (1986). The interviewing sequence RNA of Tetrahymena is an enzyme. *Science, 231*, 47–475.

Zhang, Y. S. and Tanimoto, T. (1992). Ridges, hotspots and their interaction as observed in seismic velocity maps. *Nature, 355*, 45–49.

Zhang, Y. S. and Tanimoto, T. (1993). High-resolution global upper mantle structure and plate tectonics. *J. Geophys. Res., 98*, 9793–9823.

Zhao, D., Akira, H. and Horiuchi, S. (1992). Tomographic imaging of P and S wave velocity structure beneath NE Japan. *J. Geophys. Res., 97*, 19 909–928.

Zhong, S. and Gurnis, M. (1994). Role of plates and temperature-dependent viscosity in phase change dynamics. *J. Geophys. Res., 99*, 903–917.

Zhong, S. and Gurnis, M. (1995). Mantle convection with plates and mobile, faulted plate margins. *Science, 267*, 838–843.

Zindler, A., and Hart, S. R. (1986). Chemical geodynamics. *Ann. Rev. Earth Planet. Sci., 14*, 493–571.

Zoback, M. L. and Zoback, M. (1980). State of stress in the coterminous United States. *J. Geophys. Res., 85*, 6113–6145.

Index

Abyssal sediments, 72
Acasta gneisses, 147
accretion, planetary, 247–50
accretionary orogen, 92–8
accretionary prism, 12, 83–5, 87
achondrite, 240
active continental margin, 6
active rift, 80–2
albedo, 186
alkaline igneous rocks, 168
allochthon, 92
anorogenic granite, 98–9
anorthosites, 99, 169
apparent polar wander path, 21–3
arc, 38, 42, 83–92, 106
    compositional polarity, 90–1
    igneous rocks, 89–91
arc processes, 87–9
arc–trench gap, 87
Archean:
    carbonates, 200–1
    geotherm, 175
    plate tectonics, 176–7
    sediments, 102–3
    transcurrent faults, 154
Archean–Proterozoic boundary, 173–7
aseismic ridge, 38, 74–7
ash flow, 85, 89
asteroid, 238–9
    sources, 241–2
asthenosphere, 4
atmosphere:
    composition, 183
    degassed, 183–7
    general features, 181–2
    growth rate, 184
    history, 191–2
    origin, 245–7
    oxygen, 187–91
    planetary, 245–7
    primitive, 182–3
    reducing, 183
aulacogen, 38
autotroph, 212

back-arc basin, 38, 41, 85
banded iron formation, 106–7, 188–9
Barberton greenstone, 148–9
basalts, 170
Benioff zone, 8
bimodal volcanics, 80
biogeography, 27–8
biological benchmarks, 218
blueschists, 89, 173
Bouguer gravity anomaly, 37, 44–5
brittle–ductile transition, 53
buoyant subduction, 10–11, 176–7
Bushveld complex, 80

Cambrian life, 217
carbon, burial of, 193–4
carbon dioxide, 186
carbon isotopes, 192–4
carbonates, 197–8, 200–1
    Sr isotopes, 165–6
carbonate–silicate cycle, 186–7
characteristic depth, 48–9
chemical remanent magnetization, 17, 21–2
chert, 103, 167, 168, 199
Chicxulub impact site, 223–4
chondrite, 240–1
chondrule, 240–1
chron, 18
condensation, 247–50
Conrad discontinuity, 40
continental crust, 39, 59–61
continuous habitable zone, 246–7
collisional:
    orogen, 39, 44, 51–2, 92–8, 106, 154, 166–7
    plate boundary, 11–12
comets, 221–2, 237–8
continental:
    growth, 151–3, 161–3
    growth rate, 158–63
    margin arc, 38, 42
    rift, 38, 42, 52, 79–83, 106
continents:
    chemical composition, 164–8
    size, 39

convection, 134–7
core, 4, 137–41
  age, 138–9
  composition, 138
  fluid motions, 139
  inner, 138
  temperature, 137–8
core complex, 82
core–mantle interface, 113–14
cosmic rays, 242
craton, 39, 43, 52–3, 77–9, 106–7
cratonic basin, 77–9
cratonization, 50–3
crust, 1, 4, 36–68
  composition, 54–62, 146–7
  continental crust, 39, 59–61
  electrical conductivity, 45
  exhumed, 50–3, 58
  fluids, 54
  gravity, 44–5
  juvenile, 64
  lower, 57–8
  melting of, 54
  oceanic, 36–8, 40–2, 61–2
  origin, 144–6, 150–1
  primitive, 144–7
  seismic structure, 40–4, 55–7
  thickness, 49
  transitional, 38, 42–3
crustal province, 62–7
crustal type, 36–44
crustal xenolith, 59
Curie temperature, 17, 21

debris flow, 84–6
D″ layer, 4, 126–7
degassing, 183
depleted mantle, 130–4
detachment fault, 82
detrital quartz, 78–9
detrital remanent magnetization, 17, 21–2
diamictite, 202
dinosaurs, 220
divergent plate boundary, 7
dolomite, 200
Dupal anomaly, 131–2
duplex accretion, 88–9
dyke swarms, 77

Earth:
  accretion of, 250
  age, 244–5
  composition, 242–4
  crossing asteroid, 221
  rotational history, 253–4
  structure, 4–5
  systems, 33–4, 207–9
  thermal history, 144
earthquakes, 6–8, 9, 10
eclogite, 40, 56, 57
Ediacaran fauna, 216
electrical conductivity, 113–14
energy deposits, 107
enriched mantle, 130–4
episodic ages, 154–8, 177–9
Eu anomaly, 164–5, 167–8

eukaryote, 190, 213, 215–16
evaporites, 189, 197–8
excess volatiles, 183
extinct radiogenic isotopes, 242
extinctions, 27–8

faint young Sun paradox, 185–6
fermentation, 212
first motion studies, 7, 9
flake, 92
flood basalts, 76, 222–3
forearc basin, 85
foreland basin, 86, 96
formation interval, 242
fossils, 212–14
410-km discontinuity, 123
FOZO, 132–4
fractionation, 244
fracture zones, 8
freeboard, 161

geochemical tracer, 130
geodesy, 16–17
geodynamics, 139, 140
geoid, 113
geoid anomaly, 113
Geomagnetic Time Scale, 14–15, 18–19
geomagnetism, 17–23
geotherm, 48, 117
geothermal energy, 107
glacial deposits, 202–3
glaciation, 202–3, 205–7
Gondwana, 29–32
granitoid age distribution, 155–6
granitoids, 71, 98–9, 104, 155–6
gravity, 44–5
greenhouse effect, 185, 208, 219, 246
greenstone, 99–104, 169–70
  age distribution, 156–7
  Archean, 99–104
  basalts, 170
  sediments, 102–3

hard parts, evolution of, 217
heat flow, 47–50
heat production, 48–9
heterogeneous accretion, 248–9
heterotroph, 212
Himalayas, 96–7
HIMU, 130–4
homogeneous accretion, 247–50
hotspot track, 23, 24–6
hotspots, 23–6
hydrocarbons, 107
hydrothermal vent, 105–6, 211

igneous rocks, 89–91
impacts, 209–10, 221–4, 219
incompatible elements, 60–2, 70, 72, 76, 91, 103, 119
inland-sea basin, 38–9, 43
inner core, 134
Ir anomaly, 221
iron formation, 106–7
iron meteorites, 241
island arc, 38, 42
isostatic gravity anomaly, 45

Japan trench, 10–11
juvenile crust, 64

kimberlite, 79
komatiites, 101–3, 172–4
K/T boundary, 220–4
K/T impact site, 223–4

Laurentia, 29
layered convection, 134–6
layered igneous intrusion, 80, 106
lead isotopes, 130–1
Lehmann discontinuity, 123
lherzolite, 117
life, 209–14
    early, 209–14
    evolution of, 217–18
    origin of, 209–12
LIL elements, 69, 175, 244
lithosphere, 1, 4, 119–23
    age, 122
    elastic, 119
    origin, 123
    rheology, 53–4
    subcontinental, 122–3
    thermal, 119
    thermal structure, 121–2
    thickness, 120
lower crust, 57–8
lower mantle, 111–13, 126–7
low-velocity zone, 4–5, 123
lunar crust, 145

mafic dykes, 77, 170–2
mafic plain, 100, 103
magma, 54
magma ocean, 140–1, 248
magnetic anomaly, 5, 14–15, 19, 45
magnetic field, 20, 45, 139–40
magnetic reversals, 18–21, 140
mantle:
    composition, 116–19
    convection, 134–7
    depleted, 130–4
    electrical conductivity, 113–14
    enriched, 130–4
    geochemical component, 130–4, 170–2
    heterogeneities, 133
    isochron, 130
    lithosphere, 119–23
    lower, 111–13, 126–7
    metasomatism, 119
    primitive, 119
    seismic anisotropy, 120–1
    solidus, 115
    structure, 4–5, 110–13
    temperature, 5, 113–15
    transition zone, 123–6
    upper, 4, 110
    upwellings, 113, 129
    wedge, 10, 90–1
    xenoliths, 117
mantle plumes, 23–6, 74–9, 127–9, 173, 178–9, 207–9
Mars, 229–31
mass extinctions, 27–8, 218–25
melange, 83–4

melting, 54
Mercury, 229
mesosphere, 4
metamorphic facies, 51
metasomatism, 119
metazoan, 215–17
meteorites, 239
    chronology, 242
Milankovitch periods, 207
mineral deposits, 104–7
Moho, 4, 40, 56–7
Moon, 235–6, 250–3
    capture origin, 252–3
    composition, 242–4
    double planet origin, 252
    fission origin, 251
    giant impactor origin, 253
MORB, 69, 131

Nd isotopes, 130–1, 159–61, 168
negative feedback, 34, 186–8

ocean basin, 37
ocean ridge, 5, 7, 12, 36–7, 41, 69–74, 105–6, 134, 196
oceanic crust, 61–2
oceans, 194–201
    composition, 197–200
    general features, 194–5
    growth with time, 195
    oxygen isotopes in, 199–200
    temperature, 200
offscraping accretion, 87–8
oldest rocks, 147–50
olistostrome, 83–4
one-plate planet, 256
ophiolites, 70–4, 117, 169
    Precambrian, 74
organic evolution, 27–8
orogen, 39, 44, 52–3, 92–8, 106
orogeny, 156
Os isotopes, 122–3
outer planets, 235
oxygen:
    controls in atmosphere, 187–8
    isotopes, 199–200
    geologic indicators, 188–90
    growth in atmosphere, 191
oxyphile element, 244
ozone, 181, 191

paleoclimate indicators, 202
paleoclimates, 201–9
paleomagnetism, 15–16, 17–23
paleosol, 190
Pangea, 23, 29–32, 166–7
passive margin, 7, 12, 77–9, 106–7
passive rift, 80–2
Permian/Triassic extinctions, 224
petrotectonic assemblage, 69, 153–4
phosphates, 201
photolysis, 181, 246
photosynthesis, 187, 212
piercing point, 29
Pilbara craton, 148
planetary crusts, 255
planetary heat sources, 249–50

planetary rings, 235
planets, 229–35
plate:
    Archean, 176–7
    boundaries, 6–7
    driving forces, 17
    motions, 13–17
    velocities, 14–17, 23–4
plate tectonics, 1, 2, 6, 153–4
platform, 39, 43
Pn velocity, 40
Poisson's ratio, 55–6
polar wander, 21–3
polarity interval, 18–19
Precambrian climate, 203–6
Precambrian shield, 39, 43, 58
primitive crust, 144–7
primitive mantle, 119
prokaryote, 190, 213
P–T–t path, 51–2

rare earth elements, 58, 164–5, 166–8, 174–5
Rayleigh-Bernard convection, 134
Rayleigh number, 134, 136
recycling, 159–61
redbeds, 188
reduced heat flow, 48–9
refractory element, 244
refractory inclusions, 241
regression, 77–8, 195–7
remnant arc, 85–6
restite, 118
retroarc foreland basin, 86, 96
reworking, 64
rifts, 38, 42, 52, 79–83
RNA, 210–11
rock magnetism, 17–18
Rodinia, 29, 32–3
runaway greenhouse, 246

satellites, 235–6
sea level, 195–7
seafloor spreading, 1, 5, 14–15
seawater:
    rare earths, 167–8
    Sr isotopes, 165–6
secular changes, 163–74
seismic:
    anisotropy, 120–1
    discontinuity, 4–5
    reflection, 57, 154
    tomography, 10–11, 111–13, 126
    zone, 8
shear zones, 64–6
sheeted dykes, 70–1
shocked quartz, 221
siderophile element, 244
660-km discontinuity, 125–6, 178
SNC meteorites, 241
Solar Nebula, 241, 247–50
Solar System, 228–42, 244–5
spinel, 125
spinifex texture, 102–3
Sr isotopes, 130–1, 165–6
stress province, 13

stromatolite, 213–15
subchron, 18
subduction:
    effect, 175
    erosion, 89
    geochemical component, 72, 91
    rock assemblages, 83–6
subduction zone, 8–11, 83–6, 87, 175
    flip, 12
    high stress, 87
    low stress, 87
subduction accretion, 87
submarine plateau, 11, 38, 42, 74–6
submarine ridge, 38
sulphur isotopes, 194
superchron, 18, 208
supercontinent cycle, 29–30, 129, 157–8, 166–7, 196–7, 201
supercontinents, 22–3, 28–33, 129, 157–8, 193, 196–7, 201, 207
superplumes, 208–9
suture, 64–6, 95–6
S-wave splitting, 121

terranes, 53, 62–6, 92–3
terrestrial meteorite age, 242
thermal history, 144
thermal remanent magnetization, 17, 21–2
tillite, 202
tonalite, 104, 174–5
trace elements, 164–8, 174–5
transcurrent fault, 154
transform fault, 7, 8, 12
transgression, 77–8, 195–7
transition zone, 123–6
transitional crust, 38, 42–3
trench, 38, 83
trench–ridge interaction, 12
triple junction, 8, 9
tsunami, 224
TTG, 104, 174–5
T-Tauri solar wind, 183, 250

U–Pb isotopic ages, 64
underplating, 123
ultramafic rocks, 117
upper continental crust, 163–8
upper mantle, 110
uraninite, 189–90

Venus, 231–5
    atmosphere, 246–7
    origin, 256
    surface composition, 233
    tectonics, 233–4
volcanic front, 89–90
volcanic island, 37, 76–7
volatile element, 243–4

wadsleyite, 124
whole-mantle convection, 134, 136–7
Wilson cycle, 12–13

Xe anomaly, 184

zircon ages, 64